Hopf Algebras and Galois Module Theory

Mathematical
Surveys
and
Monographs

Volume 260

Hopf Algebras and Galois Module Theory

Lindsay N. Childs
Cornelius Greither
Kevin P. Keating
Alan Koch
Timothy Kohl
Paul J. Truman
Robert G. Underwood

AMERICAN
MATHEMATICAL
SOCIETY

Providence, Rhode Island

2020 *Mathematics Subject Classification.* Primary 16T05, 11R33;
Secondary 11S15, 12F10, 20B35, 16T25.

For additional information and updates on this book, visit
www.ams.org/bookpages/surv-260

Library of Congress Cataloging-in-Publication Data

Names: Childs, Lindsay, author.
Title: Hopf algebras and Galois module theory / Lindsay N. Childs, Cornelius Greither, Kevin P. Keating, Alan Koch, Timothy Kohl, Paul J. Truman, Robert G. Underwood.
Description: Providence, Rhode Island: American Mathematical Society, [2021] | Series: Mathematical surveys and monographs, 0076-5376; Volume 260 | Includes bibliographical references and index.
Identifiers: LCCN 2021021477 | ISBN 9781470465162 (paperback) | 9781470467371 (ebook)
Subjects: LCSH: Hopf algebras. | Galois modules (Algebra) | AMS: Associative rings and algebras – Hopf algebras, quantum groups and related topics – Hopf algebras and their applications. | Number theory – Algebraic number theory: global fields – Integral representations related to algebraic numbers; Galois module structure of rings of integers. | Field theory and polynomials – Field extensions – Separable extensions, Galois theory. | Group theory and generalizations – Permutation groups – Subgroups of symmetric groups. | Associative rings and algebras – Hopf algebras, quantum groups and related topics – Yang-Baxter equations.
Classification: LCC QA613.8 .C47 2021 | DDC 512/.55–dc23
LC record available at https://lccn.loc.gov/2021021477

10 9 8 7 6 5 4 3 2 1   26 25 24 23 22 21

Contents

CHAPTER 1

Introduction: What is this book about?

1.1. Background: Hopf-Galois theory and Galois module theory

Hopf-Galois theory was first described by Chase and Sweedler in [**CS69**] a half-century ago. To motivate the definition, we begin by recalling classical Galois theory.

Galois theory. Let K be a field, L be a field containing K and having dimension n as a K-vector space: $[L : K] = n$, and suppose L is normal and separable over K: that means, L is the splitting field of some polynomial in $K[x]$ with no repeated roots in an algebraic closure of K. Then the group G of K-algebra automorphisms of L, called the Galois group of L/K, has order n, and the fixed field of L under the action of G,

$$L^G = \{x \in L \mid g(x) = x \text{ for all } g \in G\},$$

is equal to K.

The Fundamental Theorem of Galois Theory states that the function from the set of subgroups G' of G to the set of fields E with $K \subseteq E \subseteq L$, given by $G' \mapsto L^{G'}$, is one-to-one and onto.

Galois theory dates from the 1830s, but it took over a century before E. Artin [**Art42**] described the theory in module-theoretic terms. The idea is that if L/K is a Galois extension with Galois group G, then L becomes a module over the group algebra $K[G]$ where G acts as K-algebra automorphisms of L. Then L/K is a Galois extension if and only if the map $j : L[G] \to \mathrm{End}_K(L)$ given by $j(sg)(x) = sg(x)$ for s, x in L, g in G, is an isomorphism of K-vector spaces. In other words, the elements of G can be viewed as linear transformations on L, and every K-linear transformation of L can be written as an L-linear combination of elements of G.

One example of a purely module-theoretic result in Galois theory is the Normal Basis Theorem, which goes back at least to Hilbert (1897): Let L/K be a Galois extension of fields with Galois group G. Then there is an element s in L so that as a K-vector space, $\{g(s) \mid g \in G\}$ is a K-basis of L. This translates to: L is a free $K[G]$-module of rank one.

Hopf algebras. Hopf algebras arose in topology around 1941 in work of Heinz Hopf, and began to be studied in algebra in the 1960's. To define a Hopf algebra, it is convenient to first recall the linear dual of a module.

Let R be a commutative ring, M, N projective R-modules of finite rank, and let $M^* = \mathrm{Hom}_R(M, R)$, the linear dual of M. If $f : M \to N$ is an R-module homomorphism, then f induces $f^* : N^* \to M^*$ by

$$f^*(\varphi) : M \to R \text{ is } \varphi \circ f : M \to N \to R$$

for φ in N^*.

For a commutative ring R, define an R-bialgebra H to be an R-algebra, finitely generated and projective as an R-module (hence has a multiplication $m : H \otimes_R H \to H$ and a unit map $i : R \to H$ mapping 1_R to 1_H) that is also equipped with a comultiplication $\Delta : H \to H \otimes_R H$ and a counit map $\varepsilon : H \to R$ so that the dual $\varepsilon^* : R \to H^*$ of the counit map ϵ, and the multiplication map $H^* \otimes_R H^* \to H^*$, defined as the composition of the monomorphism from $H^* \otimes_R H^*$ to $(H \otimes_R H)^*$ with the dual map Δ^*, make H^* into an R-algebra. In particular, Δ^* is associative, so

$$(\Delta \otimes 1)\Delta = (1 \otimes \Delta)\Delta : H \to H \otimes_R H \otimes_R H,$$

meaning that $\Delta : H \to H \otimes H$ is coassociative.

An example is a group ring $R[G]$ for G a finite group. The counit is defined by $\varepsilon(g) = 1$ for all g in G, and the comultiplication map Δ is defined by $\Delta(g) = g \otimes g$.

A group ring $R[G]$ also has a "coinverse" map, or "antipode" $S : R[G] \to R[G]$ by $S(g) = g^{-1}$. Recalling that $\Delta(g) = g \otimes g$ for g in G, then $gg^{-1} = 1$ for g in G is the relation: $m(1 \otimes S)\Delta(g) = 1 = i\varepsilon(g)$. Generalizing, an R-Hopf algebra H is an R-bialgebra with an antipode map S, that is, an antihomomorphism: $S : H \to H$ satisfying $m(1 \otimes S)\Delta(h) = 1 = i\varepsilon(h)$ for all h in H.

Hopf-Galois extensions. Back to L/K a Galois extension of fields. If G is the Galois group of L/K, then because G acts as automorphisms of L, $g(st) = g(s)g(t)$ and $g(1) = 1$ for g in G, s, t in L. This idea generalizes to actions by a Hopf algebra: if L is an H-module, then L is an H-module algebra if

$$h(st) = m(\Delta(h)(s \otimes t)), \quad h(1) = \varepsilon(h) \cdot 1.$$

Then L/K is an H-Hopf-Galois extension if L is an H-module algebra and the map

$$j : L \otimes H \to \mathrm{End}_K(L)$$

given by $j(s \otimes h)(x) = sh(x)$ for s, x in L, h in H, is an isomorphism of K-vector spaces.

The H-Hopf Galois extension where $H = K[G]$ will be called the classical Hopf-Galois extension; all others are non-classical.

The Galois correspondence. For L/K Hopf-Galois with Hopf algebra H, Chase and Sweedler obtained a Galois correspondence from K-sub-Hopf algebras J of H to intermediate fields by: J maps to the "fixed field"

$$L^J = \{s \in L \mid h(s) = \varepsilon(h)(s) \text{ for all } h \text{ in } J\}.$$

Then $\dim_K J = \dim_{(L^J)} L$.

The Galois correspondence is one-to one, but in contrast to classical Galois theory, for many non-classical Hopf-Galois structures the Galois correspondence is not onto. See Chapter 7.

Motivation. Chase and Sweedler's motivation for [**CS69**] was "a hope that results of this type should shed some light on inseparable extensions of fields and ramified extensions of rings". The theory does apply to purely inseparable extensions, but Chase, Shatz and others soon realized that a K-algebra of the same K-dimension as L was too small to yield good information about a purely inseparable extension L/K, and went on to study more general automorphism schemes. But Hopf-Galois theory does have interesting potential to study ramified extensions of local fields, as Byott first showed in examples in 1997-2002.

Ramification. Here we give a brief introduction to ramification and Galois module theory. The subject will be studied in much greater depth in later chapters.

A global field of characteristic zero is a finite extension K of the rational numbers $\mathbb{Q}$. The ring of integers $\mathfrak{O}_K$ of K is the set of all elements a of K that are roots of monic polynomials with coefficients in $\mathbb{Z}$. $\mathfrak{O}_K$ is a Dedekind domain: every ideal of $\mathfrak{O}_K$ factors uniquely into a product of prime ideals of $\mathfrak{O}_K$.

Let L be a finite extension of K. If $\mathfrak{p}$ is a prime ideal of $\mathfrak{O}_K$, then the ideal $\mathfrak{p}\mathfrak{O}_L$ factors uniquely into a product of prime ideals of $\mathfrak{O}_L$:

$$\mathfrak{p}\mathfrak{O}_L = \mathfrak{P}_1^{e_1} \cdots \mathfrak{P}_g^{e_g}.$$

The prime ideal $\mathfrak{p}$ of $\mathfrak{O}_K$ ramifies in L if some $e_i > 1$. Each prime ideal $\mathfrak{p}$ of K contains a unique rational prime p. Then $\mathfrak{p}$ ramifies tamely if all of the exponents e_i are coprime to p, while if p divides some e_i, then $\mathfrak{p}$ ramifies wildly. Thus an extension L/K of global fields is unramified if no prime of $\mathfrak{O}_K$ ramifies in $\mathfrak{O}_L$ (that is, all $e_i = 1$), tamely ramified if no prime of $\mathfrak{O}_K$ ramifies wildly in $\mathfrak{O}_L$, and wildly ramified otherwise.

For each prime number p, one can construct the p-adic integers $\mathbb{Z}_p$ and the p-adic rational numbers $\mathbb{Q}_p$. The ring $\mathbb{Z}_p$ has a unique prime ideal (p), and there is a p-adic valuation on elements of $\mathbb{Z}_p$: $v_p(a)$ is the smallest power of the ideal (p) that contains a.

A finite extension K of $\mathbb{Q}_p$ is a local field, and the p-adic valuation extends to K, so that $\mathfrak{O}_K$ also has a unique prime ideal $\mathfrak{p}_K$.

If L/K is a finite extension of local fields, then $\mathfrak{p}_K\mathfrak{O}_L = \mathfrak{p}_L^{e_p}$ for some exponent e_p. Then L/K is unramified if $e_p = 1$, tamely ramified if e_p and p are coprime, and wildly ramified if p divides e_p. So ideal theory for extensions of local fields is quite a bit less cluttered than for extensions of global fields.

Galois module theory. The paradigm for Galois module theory is the Normal Basis Theorem: If L/K is a Galois extension of fields with Galois group G, then L is a free $K[G]$-module of rank 1.

The first important result in local Galois module theory was by Noether [**Noe32**]: let L/K be a Galois extension of local fields with Galois group G. Then $\mathfrak{O}_L$ is a free rank one $\mathfrak{O}_K[G]$-module if and only if L/K is tamely ramified. If L/K is wildly ramified, one can still try to understand $\mathfrak{O}_L$ as an $\mathfrak{O}_K[G]$-module, but it seems unnatural, because $\mathfrak{O}_K[G]$ is too small.

Instead, define the associated order $\mathfrak{A} = \mathfrak{A}_G$ of $\mathfrak{O}_L$ in $K[G]$:

$$\mathfrak{A}_G = \{a \in K[G] \mid a\mathfrak{O}_L \subseteq \mathfrak{O}_L\},$$

then $\mathfrak{A}_G$ contains $\mathfrak{O}_K[G]$ and $\mathfrak{A}_G = \mathfrak{O}_K[G]$ if and only L/K is tamely ramified.

So for wildly ramified extensions, the idea is to try to understand $\mathfrak{O}_L$ as an $\mathfrak{A}_G$-module. More generally, given an ideal $\mathfrak{J}$ of $\mathfrak{O}_L$, we can define its associated order $\mathfrak{A}_\mathfrak{J}$ and ask if $\mathfrak{J}$ is free as an $\mathfrak{A}_\mathfrak{J}$ -module.

Some old results. The earliest result was a global result by Leopoldt [**Leo59**]: if L is an abelian Galois extension of $K = \mathbb{Q}$, then $\mathfrak{O}_L$ is a free $\mathfrak{A}_G$-module of rank 1. But that result is not valid if one omits "abelian" or "$K = \mathbb{Q}$". [**CH86**] obtained a local result: if L is a Galois extension of the local field K and the associated order $\mathfrak{A}_G$ is a Hopf order in $K[G]$, then $\mathfrak{O}_L$ is a free $\mathfrak{A}_G$-module of rank one. In [**CM94**] this was extended to L/K an H-Hopf-Galois extension of local fields, where $\mathfrak{A}_G$ is replaced by the associated order $\mathfrak{A}_H$ in H.

Byott obtained some highly interesting results that help justify the study of Hopf-Galois structures in local Galois module theory. Suppose L/K is a wildly ramified Galois extension of local fields with Galois group G, and L/K is also a Hopf-Galois extension with K-Hopf algebra H. Byott [**Byo00**], [**Byo02**], obtained examples where the associated order $\mathfrak{A}_H$ is an $\mathfrak{O}_K$-Hopf order, hence $\mathfrak{O}_L$ is $\mathfrak{A}_H$-free of rank one, but the associated order $\mathfrak{A}_G$ is not a Hopf order and $\mathfrak{O}_L$ is not $\mathfrak{A}_G$-free. His examples in [**Byo00**] are described in [**Chi00**] (and were a major inspiration for writing [**Chi00**].)

Byott's examples have motivated research in several directions in local Galois module theory. This book will describe much of this research in the 20 years since the appearance of [**Chi00**].

1.2. Hopf-Galois structures since 2000

Chapters 2-6 are devoted to the study of Hopf-Galois structures on Galois extensions of fields. Prior to [**GP87**], interest in the subject was limited, because Galois extensions of fields have, à priori, a very nice Galois theory. But [**GP87**] and then [**Byo96**] gave group-theoretic methods for studying Hopf-Galois extensions, and then came Byott's examples [**Byo00**]. So there has been a steady stream of research on the question, given a G-Galois extension L/K, what are the possible types of Hopf-Galois structures on L/K, and how many are there of each type?

Greither and Pareigis [**GP87**] showed that Hopf-Galois structures on a G-Galois extension correspond to regular subgroups of $\mathrm{Perm}(G)$ normalized by $\lambda(G)$, the image of G under the left regular representation. If N is such a subgroup then we refer to the isomorphism class of N as the type of the corresponding Hopf-Galois structure But $\mathrm{Perm}(G)$ is a large group with many subgroups, and is difficult to work in. So most of the research on finding Hopf-Galois structures has involved "Byott's translation", namely, the strategy, presented in [**Byo96**], of looking for Hopf-Galois structures of type N on a G-Galois extension by looking for regular subgroups isomorphic to G in $\mathrm{Hol}(N)$, the normalizer of the image $\lambda(N)$ of the left regular representation of N in $\mathrm{Perm}(N)$.

But then something completely unexpected happened.

In 2006, Caranti, Dalla Volta and Sala [**CDVS06**] showed that for N an elementary abelian p-group, hence isomorphic to a finite dimensional vector space over $\mathbb{F}_p$, regular subgroups of the affine group $\mathrm{Aff}_n(\mathbb{F}_p) \cong \mathrm{Hol}(N)$ correspond to radical algebra structures on the additive group $\mathbb{F}_p^n$. An $\mathbb{F}_p$-algebra A with multiplication $\cdot$ is a radical algebra if given the operation $\circ$ defined by $a \circ b = a + b + a \cdot b$, then $(A, \circ)$ is a group, called the circle group of A. The correspondence between regular subgroups of $\mathrm{Hol}(N)$ and circle groups of radical algebras with additive group N was generalized in [**FCC12**] to N an arbitrary finite additive abelian p-group, and was applied to yield a nice result on the existence of Hopf-Galois structures on G-Galois extensions of fields where G is a finite abelian p-group.

In 2007, Rump [**Rum07a**] defined the concept of left brace as a generalization of a radical algebra, and used it to characterize certain set-theoretic solutions of the Yang-Baxter equation in mathematical physics. By 2016 Bachiller recognized that the main result of [**FCC12**] could be generalized from radical algebras to left braces. In 2017 the notion of left brace was generalized to that of a skew left brace, a brace without the requirement that the "additive group" be abelian, and papers of Guarneri and Vendramin[**GV17**] and Smoktunowitz, Vendramin and Byott [**SV18**] appeared on Arxiv.math giving a precise connection between Hopf-Galois structures and skew braces: every G-Galois extension of fields with a Hopf-Galois structures of type N yields a skew left brace with additive group N and circle group G, and every skew brace with additive group N and circle group G defines Hopf-Galois structures of type N on a G-Galois extension of fields.

Thus questions about Hopf-Galois structures of type N on G-Galois extensions, and techniques used to find them, translate into questions about the existence and number of skew braces with given additive and circle groups, and vice versa.

Chapter 2 of this book describes the connection between the two theories.

Chapter 3 describes a variety of results on the existence of pairs (G, N) for which a G-Galois extension of fields has a Hopf-Galois structure of type N, equivalently, on the existence of skew braces $(B, \circ, \star)$ with additive group $B^\star = (B, \star) \cong N$ and circle, or adjoint group $B^\circ = (B, \circ) \cong G$.

Chapter 4 describes results obtained from assuming that $\lambda(G)$ is contained in the subgroup $\mathrm{InHol}(N) = \lambda(N) \cdot \mathrm{Inn}(N)$ where $\mathrm{Inn}(N)$ is the group of inner automorphisms of N, and discusses bi-skew braces, which are skew braces with either group playing the role of the additive group.

Chapter 5 is concerned with the question of how many Hopf-Galois structures are there of a given type on a G-Galois extension.

Chapter 6 reviews results of Kohl determining the number of Hopf-Galois structures on a G-Galois extension where G and N are groups of order mp where p is a prime, and m satisfies $\gcd(m, p) = 1$ and two other conditions that are always true if $p > m$. Under those conditions, one can avoid Byott's translation, by finding that if $G = \mathcal{P}\mathcal{Q}$ with $\mathcal{P}$ the p-Sylow subgroup of G, then any regular subgroup N of $\mathrm{Perm}(G)$ normalized by $\lambda(G)$ is contained in the normalizer in $\mathrm{Perm}(G)$ of $\mathcal{P}$. Thus the search for regular subgroups of $\mathrm{Perm}(G)$ yielding Hopf-Galois structures can be carried out in a small subgroup $\mathrm{Norm}_{\mathrm{Perm}(G)}(\mathcal{P})$ of $\mathrm{Perm}(G)$.

In contrast to the classical Fundamental Theorem of Galois Theory, for most Hopf-Galois structures on Galois extensions of fields, the Galois correspondence does not map onto all intermediate fields. Chapter 7 provides a proof of a special case of the Galois correspondence theorem of Chase and Sweedler [**CS69**] for Hopf-Galois structures on separable extensions of fields. Then, using skew brace ideas, a variety of results is presented that yield estimates or exact counts of the proportion of intermediate fields that are in the image of the Galois correspondence for a Hopf-Galois structure on a Galois extension of fields. This chapter closes with a brief survey of results, mostly due to Crespo, Rio and Vela, on Hopf-Galois structures on separable but not necessarily normal extensions of fields.

Chapter 8 introduces the concept of normality for a Hopf-Galois extension and provides a skew brace setting for the general Galois correspondence for a Hopf-Galois structure. This chapter includes the concept of normal Hopf-Galois extension, which until recently had been missing from the Galois correspondence for Hopf-Galois structures.

Chapter 9 is on descent theory. Galois descent is a key to Greither and Pareigis' original study of Hopf-Galois structures on separable field extensions, but it is also of use for understanding the algebra structure of the K-Hopf algebras that appear in that Hopf-Galois theory.

The final chapter in the first part of the book is on Hopf-Galois theory applied to purely inseparable extensions of fields. One of the original motivations for Hopf-Galois theory was to generalize Jacobson's Galois theory for purely inseparable extensions of fields. Chapter 10 reviews some of the attempts to develop such a theory, and provides evidence to suggest that a Greither-Pareigis approach towards Hopf-Galois extensions would be unlikely to succeed.

1.3. Galois module theory since 2000

One objective of studying Hopf-Galois structures on an extension L/K of local or global fields is to broaden the array of ways that the ring $\mathfrak{O}_L$ and its ideals may be viewed as Galois modules. The second half of this book is devoted to results related to those applications.

Let L/K be an H-Hopf-Galois extension of local or global fields with valuation rings or rings of integers $\mathfrak{O}_L, \mathfrak{O}_K$, resp. The idea of Hopf-Galois module theory is to try to understand the structure of $\mathfrak{O}_L$ (or more generally, of a (fractional) ideal of $\mathfrak{O}_L$) as a module over its associated order in H. If L/K is an extension of p-adic fields, and the associated order $\mathfrak{A}$ of $\mathfrak{O}_L$ in H is a Hopf order, then $\mathfrak{O}_L$ is a free $\mathfrak{A}$-module of rank 1. Presenting ideas around that fact was the main objective of [**Chi00**]. The most interesting cases are totally ramified p-extensions of p-adic fields.

The result in [**CH86**] that if $\mathfrak{A}$ is a Hopf order in $K[G]$, then $\mathfrak{O}_L$ is a locally free $\mathfrak{A}$-module, was applied to Kummer extensions L/K of degree p in [**Chi87**], which had the title, "Taming wild extensions with Hopf algebras". Those results were aided by the fact that $\mathfrak{O}_K$-Hopf algebras of degree p had been classified by Tate and Oort [**TO70**].

Extensions of degree p^2 were subsequently handled by Greither [**Gre92**], who classified Hopf orders over $\mathfrak{O}_K \supseteq \mathbb{Q}_p$ in $K[G]$ for $G = C_{p^2}$ where K contains a primitive p-th root of unity. Let e be the absolute ramification index of K, then $e' = e/(p-1)$ is an integer. Greither described the so-called realizable Hopf orders, those $\mathfrak{O}_K$-Hopf algebras A for which there is a G-Galois extension L/K so that $A = \mathfrak{A}$, the associated order of $\mathfrak{O}_L$ in $K[G]$, and also described the valuation ring $\mathfrak{O}_L$ of the corresponding G-Galois extension on which A acts. The Hopf orders found in [**Gre92**] are known as Greither orders. They involved a condition on parameters that was subsequently relaxed in work of Underwood [**Und94**], who constructed dual Greither orders, and then was extended to the case $G = C_p \times C_p$, involving both Greither orders and dual Greither orders, by Byott in [**Byo02**].

One of the main motivations for [**Chi00**] was to lay the groundwork to present some examples of Byott [**Byo99**] that showed that if L/K is a G-Galois extension of local fields, there can exist a H-Hopf algebra structure on L/K for H a non-classical K-Hopf algebra, so that $\mathfrak{O}_L$ is not free over its associated order in $K[G]$, but is free over its associated order in H. So one question became: Are there many examples of this phenomenon?

Extending a partial result in [**Chi96**], Byott, in [**Byo02**], produced a complete analysis for wildly ramified Galois extensions of local fields of degree p^2.

Let L/K be a wildly ramified G-Galois extension of local fields where K is a finite extension of $\mathbb{Q}_p$. Then L/K has (lower) ramification breaks $t_1 = pj - 1, t_2 = p^2 i - 1$ where i and j satisfy $0 \leq i, j \leq e' = e/(p-1)$ where e is the absolute ramification index of p in K. Hence the possible pairs (i, j) lie in an $e' \times e'$ square in the i, j-plane. Applying Byott's analysis, if $p = 3$ and $e' = 60$ for example, then of the 3600 possible pairs (i, j), there are approximately 180 pairs (i, j) for which the associated order in $K[G]$ is Hopf, and 9 pairs (i, j) where the associated order in $K[G]$ is not Hopf but the associated order in a non-classical Hopf-Galois structure is Hopf. (For more details, see Chapter 11 or [**Byo02**]).

Thus what Byott showed in [**Byo02**] is first, that examples of local fields L/K where the associated order of $\mathfrak{O}_L$ is a Hopf order are relatively rare, and second, that even when the associated order of $\mathfrak{O}_L$ in some H-Hopf-Galois structure on L/K is Hopf, cases where that occurs in a non-classical Hopf-Galois structure and not in the classical structure, as in [**Byo99**], are even more rare. Based on the p^2 case, non-classical Hopf-Galois structures apparently seldom yield better local Galois module results than the classical Hopf-Galois structure does.

Beyond the rarity of examples that motivated the Hopf algebra orientation to local Galois module theory in 2000, there are two other substantial issues.

One was that despite considerable efforts, especially by Underwood and his collaborators, constructing realizable Hopf orders in $K[C_{p^n}]$ for $n > 2$ turned out to be very difficult. A collection of realizable orders in $K[C_{p^3}]$ was found in [**Und08**] but for $n \geq 3$ the description of realizable Hopf orders remains a difficult problem. Chapter 12 describes the present state of the art on constructing Hopf orders of degree p^n, $n > 2$, mostly for K a local field of characteristic zero. Section 12.13 considers the case where K has characteristic p, and even there the algebra structure of the Hopf orders for $n = 3$ is unclear.

Thus even if one knows that $\mathfrak{A}$ is a Hopf order in $K[G]$ so that $\mathfrak{O}_L$ is $\mathfrak{A}$-free, if what $\mathfrak{A}$ looks like as an $\mathfrak{O}_K$-algebra is not understood, then knowing that $\mathfrak{O}_L$ is $\mathfrak{A}$-free is not going to be particularly satisfying.

Scaffold theory may shed some light on this issue. See below.

The other limitation, which is intrinsic, as Byott illustrated in the p^2 case in [**Byo02**] and showed in general in [**Byo97**], is that the fact that the associated order of $\mathfrak{O}_L$ in $K[G]$ is a Hopf order imposes serious constraints on the ramification breaks of L/K. If L/K is a totally ramified Galois extension of local fields of degree p^n, and $\mathfrak{A}$ is a Hopf order in $K[G]$, then the first ramification break b_1 satisfies either $b_1 = -1 + sp^n$ with $1 \leq s \leq e' = e/(p-1)$, or all the ramification breaks b_r satisfy $b_r = p^r e'$. The second case is exceptional and implies that $\mathfrak{A}$ is not a local ring. The first case, which implies that $\mathfrak{O}_L$ is in fact an $\mathfrak{A}$-Hopf-Galois extension of $\mathfrak{O}_K$, also implies that $b_i \equiv -1 \pmod{p^n}$ for all $i = 1, \ldots, n$. Moreover, when $\mathfrak{A}$ is a Hopf order, then the different $\mathfrak{D}_{L/K}$ of $\mathfrak{O}_L/\mathfrak{O}_K$ is the ideal $\mathfrak{D}_{L/K} = \pi_K^c \mathfrak{O}_L$ for some integer c. These results are all in Chapter 8 of [**Chi00**].

In [**Bon00**], Bondarko proved a converse to the first case. Suppose L/K is abelian of degree p^n, totally ramified, the ramification numbers satisfy $b_i \equiv -1 \pmod{p^n}$ and the different $\mathfrak{D}_{L/K} = \pi_K^c \mathfrak{O}_L$ for some integer c. Then $\mathfrak{A}$ is an $\mathfrak{O}_K$-Hopf order in $K[G]$ and $\mathfrak{O}_L$ is a free $\mathfrak{A}$-module of rank 1. The method of proof involves identifying L/K as a Kummer extension for a formal group defined over $\mathfrak{O}_K$.

Finally: for most Galois extensions L/K of local or global fields, the associated order $\mathfrak{A}$ of $\mathfrak{O}_L$ or ideals of L is not Hopf. Chapter 11 also describes some tools for analyzing Hopf-Galois module structure (and making comparisons between different structures) in situations in which the associated order is not necessarily a Hopf order. The chapter includes work on extensions that are tame (in the classical sense): the evidence to date suggests that for these extensions the ring of integers is (locally) free over its associated order in all of the Hopf-Galois structures admitted by the extension. It also examines some recent work of S. Taylor and Truman on Hopf-Galois module structure of extensions of global fields, including the interesting case where the extension L/K is Hopf-Galois but not normal, hence not classically Galois.

Scaffolds and semi-stable extensions. The ramification restrictions on a Galois extension L/K of p-adic fields with Galois group a finite p-group that would make the associated order $\mathfrak{A}$ of $\mathfrak{O}_L$ a Hopf order make it evident that for most such extensions, the associated order is not Hopf. So other methods have been developed.

One by Bondarko, involves the concept of stable and semi-stable extensions, developed in [**Bon02**] and [**Bon06**]. When L/K is stable, he found necessary and sufficient conditions for any ideal $\mathfrak{P}^h$ of $\mathfrak{O}_L$ to be free over its associated order in $K[G]$. Bondarko's approach is described and generalized in Chapter 14.

The other, first described by Elder and then elaborated in collaboration with Byott, involves the concept of scaffold. For a Galois scaffold, the idea is to take a totally ramified G-Galois extension L/K of p-adic fields of degree p^n, and try to find a K-basis $\{\lambda_t : t \in \mathbb{Z}/p^n\mathbb{Z}\}$, a set $\{\psi_i\}$ of elements of $K[G]$ and a set $b_1, \ldots, b_n$ of shift parameters (which in the Galois case are typically the ramification numbers of L/K) so that

(i) $v_L(\lambda_t) = t$ for all t in $\mathbb{Z}$;

(ii) $\lambda_{t_1}\lambda_{t_2}^{-1}$ is in K whenever $t_1 \equiv t_2 \pmod{p^n}$;

(iii) $\psi_i(1) = 0$ for $1 \leq i \leq n$;

(iv) $\psi_i \cdot \lambda_t = \lambda_{t+p^n b_i}$ or 0 up to a certain level of precision (more precisely, modulo $\lambda_{t+p^n b_i}\mathfrak{c}$ for some fixed ideal $\mathfrak{c}$ which defines the precision). The choice of whether $\psi_i \cdot \lambda_t = \lambda_{t+p^n b_i}$ or 0 depends on the shift parameters b_i.

For details and a slight refinement of the definition, see [**BCE18**, Definition 2.6] or Definition 15.1.

The idea behind this definition is that if one can find p^n elements of $K[G]$ that take on all valuations between 0 and $p^n - 1$, then they will form a basis of $\mathfrak{O}_L$ because L/K is totally ramified. And if the ψ_i all shift the elements λ_t to either zero or K-multiples of other elements in the set $\{\lambda_t\}$, then the elements ψ_i and their powers have a chance of becoming a basis of the associated order $\mathfrak{A}$ of $\mathfrak{O}_L$. When that is the case, then $\mathfrak{O}_L$ will be free over its associated order $\mathfrak{A}$.

The concept of scaffold is studied in more detail in Chapter 15.

Connections with realizable Hopf orders. A striking consequence of scaffold theory has been a potential breakthrough in the problem of describing realizable Hopf orders over $\mathfrak{O}_K$ in $K[G]$ for G a finite abelian p-group.

As noted above, $\mathfrak{O}_K$-Hopf algebras that are realizable, that is, can be the associated orders of valuation rings $\mathfrak{O}_L$ in $K[G]$ for G cyclic of order p^n, are completely understood for $n = 1$ and $n = 2$ but not well-understood for $n \geq 3$.

But scaffold theory as developed in [**BE18**] applies to cyclic p-extensions L/K of p-adic fields under certain conditions on the break numbers b_i.

In case all the $b_i \equiv -1 \pmod{p^n}$ and a certain combinatorial equality holds, then one obtains an explicit basis of the associated order $\mathfrak{A}$, $\mathfrak{O}_L$ is free over $\mathfrak{A}$ and the different is generated by a power of the parameter π of K.

Those are the conditions needed by Bondarko to show that the associated order $\mathfrak{A}$ is obtained from a formal group, hence is a Hopf order, and hence is a realizable Hopf order.

Thus the scaffold theory of [**BE18**] provides a set of generators for realizable Hopf orders for $n \geq 1$, and in particular, for $n \geq 3$.

This connection was used in [**EU17**] to describe realizable Hopf orders in $K[C_{p^3}]$ for K of characteristic p, and shows up in Chapter 15.

Chapters 13-15. Chapters 13-15 are devoted to p-extensions of p-adic fields (local fields) that are semistable, in the sense of Bondarko, or have a Galois scaffold, in the sense of Byott and Elder. A recent paper of Keating [**Kea20**] associates to a semi-stable Galois extension L/K of local fields a certain precision, and shows that if L/K has a scaffold with precision $\mathfrak{c}$, then L/K is semistable with precision $\mathfrak{c}$, while if L/K is semistable, then it has a scaffold of precision 1. Thus scaffold theory applies to Bondarko's semi-stable extensions.

Chapter 13 extends classical ramification theory for Galois extensions of local fields to the case of a separable extension L/K with Galois closure $\widetilde{L}/K$, where L/K is a Hopf-Galois extension.

Chapter 14 describes the concepts of stable and semistable p-extensions of p-adic fields due to Bondarko [**Bon00**], [**Bon02**], [**Bon06**] and extends the theory to the setting of Hopf-Galois extensions.

Chapter 15 introduces the scaffold theory of Elder and Byott ([**BE18**], [**BCE18**]) and, extending [**Kea20**], shows that if a Hopf-Galois extension L/K of local fields has a scaffold of a suitable precision, then it is semistable.

1.4. What's not in this book

There are three areas not well covered in this book. One, global Galois module theory, is discussed only in part of Chapter 11. For the most part, this reflects the (lack of) expertise of the authors. Another, Hopf-Galois structures on non-normal separable extensions, is discussed in the last section of Chapter 7 but, as noted there, does not fit so naturally in the first half of the book because it has not been connected to brace theory. The other is the substantial body of research on braces that has accumulated in the 13 years since [**Rum07a**]. The authors only realized that brace theory might be of interest to Hopf-Galois extensions and Galois module theory with Byott's presentation at the conference "Hopf algebras and Galois module theory" at the University of Nebraska at Omaha in 2017. Several survey papers on braces and skew braces have been written since then, notably [**GV17**], [**Ven19**] and [**SV18**], but there is a large literature on braces that up to this point has not been connected to Hopf-Galois theory.

This book is a snapshot of the state of Hopf-Galois theory and Galois module theory at a particular point in time, namely, the summer of 2020. The idea for this work was to provide a 20 year update of [**Chi00**], and at the same time, present a survey of work in the 50 years since the appearance of the original concept of Hopf-Galois extension in [**CS69**]. While we were writing this, the pace of research in the area has, if anything, accelerated. To authors of recent work that has not been included in this book, we apologize, and hope that your research will become a important part of a 10 year, or 20 year update of this book, one that perhaps you will (co)-author!

Acknowledgments

The authors of this book wish to express their profound gratitude to Griff Elder for his leadership in this area over the past decade, in particular for the organization of a series of week-long conferences/workshops at the University of Nebraska at Omaha in 2012, 2013, 2014, 2016, 2017, 2018 and 2019, and the virtual conference in 2020. The influence of these conferences on this work has been so great that for a long while our working title was "The Omaha Book"! We would like to highlight the extensive influence of Nigel Byott on much of the material presented in this book, and to thank him for organizing the 2015 conference at the University of Exeter. Our thanks also to UNO, the University of Exeter, and the London Mathematical Society for their support of the conferences mentioned above, and to Keele University for hosting the web archive of proceedings of those conferences from 2013 to the present. We are very grateful to Ina Mette of the American Mathematical Society for her patience and support during the development of this project, and we thank the anonymous referees for their insightful comments on the manuscript.

The other authors would like to express here our thanks to Paul Truman for proposing this project at Omaha in 2018 and for taking the lead as corresponding author and as coordinator and chief copy editor for the book.

Part I: Hopf-Galois Extensions

CHAPTER 2

Hopf-Galois structures on Galois extensions of fields, regular subgroups, and skew braces

2.1. Introduction

This chapter describes the connection between Hopf-Galois structures on Galois extensions of fields and skew braces.

We briefly summarize the connection here. The remainder of this chapter will discuss the connection in more detail.

In 1987 the first step was obtained by Greither and Pareigis [**GP87**]. If L/K is a G-Galois extension of fields and is a Hopf-Galois extension by some K-Hopf algebra H, then they showed that $L \otimes_K H = L[N]$ for some regular subgroup N of $\mathrm{Perm}(G)$. That subgroup is normalized by the image $\lambda(G)$ of the left regular representation map $\lambda : G \to \mathrm{Perm}(G)$, so H and its action on L can be recovered by Galois descent. Thus there is a bijection between Hopf-Galois structures on L/K and regular subgroups N of $\mathrm{Perm}(G)$ normalized by $\lambda(G)$. The group N need not be isomorphic to G. By slight abuse of language, if H corresponds to $N \subset \mathrm{Perm}(G)$, we say that the Hopf-Galois structure given by H on L/K has type N (viewed as an abstract group).

In 1996 the next step was obtained by Byott [**Byo96**], building on an observation in [**Chi89**] that if H is a Hopf-Galois structure of type N on a G-Galois extension of fields, then in $\mathrm{Hol}(N)$, the normalizer of $\lambda(N)$ in $\mathrm{Perm}(N)$, there is a subgroup T isomorphic to G that corresponds to H. Byott systematically described the relationship between the set of regular subgroups of $\mathrm{Perm}(G)$ isomorphic to N and normalized by $\lambda(G)$ and the set of regular subgroups of $\mathrm{Hol}(N)$ isomorphic to G. That analysis transformed the problem of finding or counting Hopf-Galois structures on L/K of type N on a G-Galois extension into that of finding or counting regular subgroups isomorphic to G in $\mathrm{Hol}(N)$. This was a highly significant advance, because $\mathrm{Hol}(N) \cong N \rtimes \mathrm{Aut}(N)$ is in general a far smaller and more tractable group than $\mathrm{Perm}(G) \cong \mathrm{Perm}(N)$.

Since 1996 dozens of papers have examined various strategies for constructing or counting or ruling out Hopf-Galois structures of type N on G-Galois extensions of fields. The vast majority of them have worked within the holomorph of N.

In 2006, Caranti, Dalla Volta and Sala [**CDVS06**] observed that regular subgroups of the affine group of an elementary abelian group (identified as $\mathbb{F}_p^n$ for $\mathbb{F}_p$ the field of p elements) give rise to radical algebra structures on the additive group $\mathbb{F}_p^n$. The affine group may be viewed as the holomorph of the elementary abelian group $\mathbb{F}_p^n$. The connection between regular subgroups of a holomorph and radical algebras was generalized in [**FCC12**] to the case where the additive group is any finite abelian p-group N: radical algebra structures on N correspond to regular subgroups of $\mathrm{Hol}(N)$. That paper identified the connection between Hopf-Galois structures of type N on a G-Galois extension and radical algebras with additive group N and circle (or "adjoint") group G. This connection was further generalized to arbitrary finite abelian groups and commutative radical algebras in Section 5 of [**Byo13**].

Meanwhile, in 2007, Rump [**Rum07a**] introduced the concept of a left brace, generalizing the concept of a radical ring with its adjoint group. His objective was to find solutions of the Yang-Baxter equation arising, e.g. in mathematical physics. Since the appearance of Rump's paper, dozens of papers have studied braces, often from the point of view of ring theory.

In 2016, Bachiller [**Bac16**], extending [**FCC12**], showed that for N any abelian additive group, regular subgroups of $\mathrm{Hol}(N)$ correspond to left brace structures on N, and noted the connection between a left brace $(B, \circ, +)$ and a Hopf-Galois structure of type $(B, +)$ on a Galois extension of fields with Galois group $(B, \circ)$.

In 2017, Guarneri and Vendramin [**GV17**] defined a skew left brace to be a brace with additive group not necessarily abelian and showed that skew braces give solutions to the Yang-Baxter equation.

In 2018, in the appendix to [**SV18**], Byott and Vendramin completed the description of the connection between skew braces and Hopf-Galois structures on Galois extensions of fields. The bottom line is that every Hopf-Galois structure of type N on a G-Galois extension of fields defines a skew brace with additive group isomorphic to N and circle group isomorphic to G, and every skew brace with additive group isomorphic to N and circle group isomorphic to G defines a collection of Hopf-Galois structures of type N on a G-Galois extension of fields.

In the remainder of this chapter we present the details of the connection just described between Hopf-Galois extensions on Galois extensions of fields and skew braces.

2.2. Greither-Pareigis theory

Let L/K be a G-Galois extension of fields. To classify Hopf-Galois structures on L/K, the first step is to associate to a Hopf-Galois structure a regular subgroup of $\mathrm{Perm}(G)$. This was done by Greither and Pareigis [**GP87**], using Galois descent. This is covered fairly thoroughly in [**Chi00**] and is also treated in Chapter 9, so here we will just give a summary of the results.

A subgroup N of $\mathrm{Perm}(G)$ is a regular subgroup if for every g, h in G there is a unique η in N so that $\eta[g] = h$. Thus $|N| = |G|$. To check regularity of N it suffices to check that $|N| = |G|$ and $\{\eta[e] \mid \eta \in N\} = G$ where e is the identity element of G.

Suppose L/K has a Hopf-Galois structure by a K-Hopf algebra H. Then $L \otimes_K H$ is a K-Hopf algebra that acts on $L \otimes_K L$, and since L/K is a Galois extension with Galois group G, $L \otimes_K L \cong \mathrm{Hom}_L(L[G], L)$. (This is an interpretation of the fact that if L/K is a Galois extension with Galois group G, then $L = K[x]/(p(x))$ for some polynomial with distinct roots in any algebraic closure of K, hence $L \otimes_K L = L[x]/(p(x)) \cong L \times L \times \cdots \times L$ as rings.) Then $L \otimes_K L$ is an $L \otimes_K H$-Hopf-Galois extension. But then the L-Hopf algebra $L \otimes_K H$ must be generated by grouplikes (see Section 7.1). So $L \otimes_K H \cong L[N]$, a group algebra, where $N \subset \mathrm{Perm}(G)$ is a regular subgroup: that is, $N[g] = \{\eta[g] \mid \eta \in N\} = G$ for every g in G. And because $L[N] \cong L \otimes_K H$ and N is normalized by the image $\lambda(G)$ of the left regular representation $\lambda : G \to \mathrm{Perm}(G)$ given by $\lambda(g)[x] = gx$, one can recover H

as $L[N]^G$, where G acts on L via action of the Galois group, and acts on N via conjugation by elements of $\lambda(G)$ in $\mathrm{Perm}(G)$.

To summarize:

THEOREM 2.1. *There is a bijection between Hopf-Galois structures on a G-Galois extension L/K of fields and regular subgroups of* $\mathrm{Perm}(G)$ *normalized by $\lambda(G)$, as follows: if $j : H \otimes L \to L$ is an action of a K-Hopf algebra on a G-Galois extension L/K of fields so that L/K is an H-Hopf-Galois extension, then there is a unique subgroup N of* $\mathrm{Perm}(G)$ *so that $L \otimes_K H \cong L[N]$, $1 \otimes j$ is the corresponding action of $L[N]$ on $L \otimes_K L$, and N is normalized by $\lambda(G)$ in* $\mathrm{Perm}(G)$. *Moreover, H and j may be recovered by*

$$\begin{aligned} H &= (L[N])^G \\ &= \left\{\sum s_\eta \eta \;\middle|\; \sum s_\eta \eta = \sum g(s_\eta)\lambda(g)\eta\lambda(g^{-1}) \textit{ for all } g \in G, s_\eta \in L, \eta \in N\right\}, \end{aligned}$$

acting on L/K via the restriction of $1 \otimes j$ on H.

An equivalent description of the action of H on L will be given as Proposition 4.14, based on the Byott translation theory discussed beginning in Section 2.3.

DEFINITION 2.2. If H is a K-Hopf algebra that acts on L and L/K is H-Hopf-Galois, then $L \otimes_K H = L[N]$ and the *type* of H is the isomorphism type of N.

The bijection between Hopf-Galois structures on a G-Galois extension L/K and regular subgroups N of $\mathrm{Perm}(G)$ normalized by $\lambda(G)$ turns the determination of Hopf-Galois structures on L/K into the group-theoretic problem: determine the regular subgroups of $\mathrm{Perm}(G)$ that are normalized by $\lambda(G)$.

In general, this is a hard problem. But it becomes a less hard problem by the following observation, from [**Chi89**].

PROPOSITION 2.3. *If N is a regular subgroup of* $\mathrm{Perm}(G)$ *that is normalized by $\lambda(G)$, then there is a regular subgroup isomorphic to G in* $\mathrm{Perm}(N)$ *that normalizes $\lambda(N)$.*

The normalizer of $\lambda(N)$ in $\mathrm{Perm}(N)$ is the holomorph $\mathrm{Hol}(N) = \lambda(N) \rtimes \mathrm{Aut}(N)$.

This observation, by itself, is useful for restricting the possible types of Hopf-Galois structures on a G-Galois extension L/K. Pick a group N of order n, the order of G. If there is no regular subgroup T isomorphic to G in $\mathrm{Hol}(N)$, then L/K cannot have a Hopf-Galois structure of type N.

Since $\mathrm{Hol}(N)$ is typically a much smaller group than $\mathrm{Perm}(G) \cong \mathrm{Perm}(N)$, the problem of excluding Hopf-Galois structures of type N on a G-Galois extension L/K by working in $\mathrm{Hol}(N)$ becomes feasible in many cases. Here are three notable examples:

- Kohl's result that if G is cyclic of order p^n, p odd, then L/K has no Hopf-Galois structures of non-cyclic type, follows immediately from his proof [**Koh98**, Theorem 4.5], that if N is a non-cyclic group of order p^n, p odd, then $\mathrm{Hol}(N)$ contains no element of order p^n.
- Featherstonhaugh's Theorem ([**Fea03**], [**FCC12**]) partially generalized Kohl's theorem: Let N be an abelian p-group of order p^m where p is sufficiently larger than m. Then every abelian regular subgroup of $\mathrm{Hol}(N)$ is isomorphic to N, so a G-Galois extension has no Hopf-Galois structures of abelian type N for $N \ncong G$.

• Byott's result [**Byo04a**] that if G is a non-abelian simple group, then a G-Galois extension has no Hopf-Galois extensions of type N for $N \not\cong G$, was obtained by showing that if N is a group of the same order as G but not isomorphic to G, then there is no regular embedding of G into $\mathrm{Hol}(N)$. The proof depends on consequences of the classification of finite simple groups. (See Section 3.5 for further discussion of this result.)

The observation of Proposition 2.3 concerning regular subgroups of $\mathrm{Hol}(N)$, used to deny the existence of Hopf-Galois structures of given types, has a converse, given in [**Byo96**]:

THEOREM 2.4. *Let G, N be two groups of the same order n. If* $\mathrm{Hol}(N)$ *has a regular subgroup isomorphic to G, then a G-Galois extension has a Hopf-Galois structure of type N.*

In fact, [**Byo96**] quantified that observation. We'll prove Theorem 2.4 following Theorem 2.8, below.

2.3. Byott translation theory

The next three sections are devoted to a presentation of the connection between Hopf-Galois structures of type N on a G-Galois extension L/K of fields, and regular subgroups of $\mathrm{Hol}(N)$ that are isomorphic to G. The theory, from [**Byo96**] is also described in Section 7 of [**Chi00**]. We present it here because it is essential for understanding the connection between Hopf-Galois structures and skew braces.

Let N, G be two groups of equal order with identities e_G, e_N, respectively. Let

$$\mathcal{E}^G(G,[N]) = \{\text{regular subgroups } M \cong N \text{ of } \mathrm{Perm}(G) \text{ normalized by } \lambda(G)\},$$

and let

$$e(G,[N]) = |\mathcal{E}^G(G,[N])|$$

which

$$= \#\{\text{Hopf-Galois structures of type } N \text{ on a } G\text{-Galois extension}\}$$

by Greither-Pareigis theory. Let

$$\begin{aligned}\mathcal{R}_N(N,[G]) &= \{\text{regular subgroups } T \cong G \text{ of } \mathrm{Hol}(N)\}\\ &= \{\text{regular subgroups } T \cong G \text{ of } \mathrm{Perm}(N) \text{ that normalize } \lambda(N)\ \},\end{aligned}$$

and let $r_N(N,[G]) = |\mathcal{R}_N(N,[G])|$.

To understand the relationship of $e(G,[N])$ and $r_N(N,[G])$, we set

$$\mathrm{Bij}_0(G,N) = \{ \text{ bijective maps from } G \text{ to } N \text{ that take } e_G \text{ to } e_N\}.$$

The map $a \mapsto a^{-1}$ gives a bijection between $\mathrm{Bij}_0(G,N)$ and $\mathrm{Bij}_0(N,G)$.

Given a in $\mathrm{Bij}_0(N,G)$, we get a function $\alpha_a : N \to \mathrm{Perm}(G)$ by

$$\alpha_a(m)[g] = a\lambda(m)a^{-1}[g]$$

for g in G, m in N, where $\lambda : N \to \mathrm{Perm}(N)$ is the left regular representation map defined by $\lambda(m)[n] = mn$ for n in N. Then α_a is a homomorphism:

$$\begin{aligned}\alpha_a(mn)[g] &= (a\lambda(mn)a^{-1})[g]\\ &= (a\lambda(m)\lambda(n)a^{-1})[g]\\ &= (a\lambda(m)a^{-1}a\lambda(n)a^{-1})[g]\\ &= \alpha_a(m)\alpha_a(n)[g].\end{aligned}$$

Also, α_a is injective, and regular, because

$$\alpha_a(m)[e_G] = a\lambda(m)a^{-1}[e_G] = a\lambda(m)[e_N] = a(m),$$

and $a : N \to G$ is bijective, so $\{a(m) \mid m \in N\} = G$.

So sending a to α_a gives a map from $\mathrm{Bij}_0(N, G)$ to

$$\mathrm{Reg}(N, \mathrm{Perm}(G)) = \{\alpha : N \to \mathrm{Perm}(G) : \ \alpha \text{ is a regular embedding}\}.$$

The map is 1-1. For if $\alpha_a = \alpha_{a'}$, then $\alpha_a(m)[e_G] = \alpha_{a'}(m)[e_G]$, so $a(m) = a'(m)$ for all m in N, so $a = a'$.

The map from $\mathrm{Bij}_0(N, G)$ to $\mathrm{Reg}(N, \mathrm{Perm}(G))$ is also onto. For we have

PROPOSITION 2.5. *Let $\alpha : N \to \mathrm{Perm}(G)$ be a regular embedding, and let $a : N \to G$ be defined by $a(m) = \alpha(m)[e_G]$. Then a is in $\mathrm{Bij}_0(N, G)$ and $\alpha = \alpha_a$.*

PROOF. To show that $\alpha(m)[g] = a\lambda(m)a^{-1}[g]$ for all g in G, we show that, for m and x in N,

$$a^{-1}\alpha(m)a(x) = \lambda(m)(x),$$

as follows:

$$\begin{aligned} a^{-1}\alpha(m)a(x) &= a^{-1}\alpha(m)\alpha(x)[e_G] \\ &= a^{-1}\alpha(mx)[e_G] \\ &= a^{-1}a(mx) \\ &= mx \\ &= \lambda(m)(x). \end{aligned}$$

□

The bijection between $\mathrm{Bij}_0(N, G)$ and $\mathrm{Bij}_0(G, N)$, defined by mapping $a : N \to G$ to $b = a^{-1} : G \to N$, yields a bijection

$$s : \mathrm{Reg}(N, \mathrm{Perm}(G)) \to \mathrm{Reg}(G, \mathrm{Perm}(N)).$$

If $\alpha : N \to \mathrm{Perm}(G)$ is a regular embedding with induced bijection a, then the corresponding regular embedding $s(\alpha) = \beta : G \to \mathrm{Perm}(N)$ is defined to be the regular embedding induced by the inverse bijection $b = a^{-1}$: if $\alpha(m) = a^{-1}\lambda(m)a$, then $\beta(g) = a\lambda(g)a^{-1} = b^{-1}\lambda(g)b$.

PROPOSITION 2.6. *If $\theta : N' \to N$ is an isomorphism of groups, and α_a is in $\mathrm{Reg}(N, \mathrm{Perm}(G))$, then $\alpha_{a\theta}$ is in $\mathrm{Reg}(N', \mathrm{Perm}(G))$ and $\alpha_{a\theta}(N') = \alpha_a(N)$ in $\mathrm{Perm}(G)$.*

PROOF. Let $\theta : N' \to N$ be an isomorphism and $a : N \to G$ a bijection. Then $a\theta$ is in $\mathrm{Bij}_0(N', G)$. The corresponding map $\alpha_{a\theta} : N' \to \mathrm{Perm}(G)$ is then defined for m' in N', g in G as

$$\begin{aligned} \alpha_{a\theta}(m')(g) &= a\theta\lambda(m')\theta^{-1}a^{-1}(g) \\ &= a\theta(m'(\theta^{-1}a^{-1})(g)) \\ &= a(\theta(m')a^{-1})(g) \\ &= a\lambda(\theta(m'))a^{-1}(g) \\ &= \alpha_a(\theta(m'))(g). \end{aligned}$$

So the image of $\alpha_{a\theta}$ in $\mathrm{Perm}(G)$ is the same as that of α:

$$\alpha_{a\theta}(N') = a\lambda(\theta(N'))a^{-1} = a\lambda(N)a^{-1} = \alpha_a(N).$$

□

As for the map $s(\alpha) = \beta : G \to \mathrm{Perm}(N)$ corresponding to α, we have

PROPOSITION 2.7. *Let $\alpha : N \to \mathrm{Perm}(G)$ be a regular embedding and $\theta : N' \to N$ be an isomorphism. Let $s(\alpha) = \beta : G \to \mathrm{Perm}(N)$. Let α correspond to a in $\mathrm{Bij}(N, G)$, let $b = a^{-1}$ correspond to β in $\mathrm{Bij}(G, N)$, and let*

$$\beta_\theta = s(\alpha_{a\theta}) : G \to \mathrm{Perm}(N').$$

Then in $\mathrm{Perm}(N')$,

$$\beta_\theta(\gamma) = \theta^{-1}\beta(\gamma)\theta.$$

PROOF. The map $\alpha_{a\theta}$ in $\mathrm{Reg}(N', \mathrm{Perm}(G))$ corresponds to $a\theta$ in $\mathrm{Bij}_0(N', G)$, so $s(\alpha_\theta) = \beta_\theta : G \to \mathrm{Perm}(N')$ corresponds to $\theta^{-1}a^{-1} = \theta^{-1}b$. Then for $\beta : G \to \mathrm{Perm}(N)$, defined from $b = a^{-1}$,

$$\beta(g)(m) = b\lambda(g)b^{-1}(m) = a^{-1}\lambda(g)a(m)$$

so in $\mathrm{Perm}(N')$,

$$\begin{aligned} \beta_\theta(g) &= \theta^{-1}b\lambda(g)b^{-1}\theta \\ &= \theta^{-1}\beta(g)\theta. \end{aligned}$$

□

In particular, Propositions 2.6 and 2.7 are valid when $N' = N$ and θ is in $\mathrm{Aut}(N)$.

2.4. Actions by the left regular representations

Given groups G, N of the same order, we will be interested in regular embeddings α in $\mathrm{Reg}(N, \mathrm{Perm}(G))$ whose image is either normalized by $\lambda(G)$ or that normalizes $\lambda(G)$.

Let α correspond to a in $\mathrm{Bij}_0(N, G)$ and β correspond to $b = a^{-1}$ in $\mathrm{Bij}_0(G, N)$. Suppose $\alpha(N) \subset \mathrm{Perm}(G)$ is normalized by $\lambda(G)$: for all m in N, g in G there is some m' in N so that

$$\alpha(m') = \lambda(g)\alpha(m)\lambda(g)^{-1}.$$

Then

$$a\lambda(m')a^{-1} = \lambda(g)a\lambda(m)a^{-1}\lambda(g)^{-1}.$$

So

$$\begin{aligned} \lambda(m') &= a^{-1}\lambda(g)a\lambda(m)a^{-1}\lambda(g)^{-1}a \\ &= b\lambda(g)b^{-1}\lambda(g)b\lambda(m)^{-1}b^{-1} \\ &= \beta(g)\lambda(m)\beta(g)^{-1}. \end{aligned}$$

The argument is reversible. We conclude:

PROPOSITION 2.8. *Let α in $\mathrm{Reg}(N, \mathrm{Perm}(G))$ correspond to β in $\mathrm{Reg}(G, \mathrm{Perm}(N))$. Then $\beta(G)$ normalizes $\lambda(N)$, that is, $\beta(G)$ is in $\mathrm{Hol}(N) \subset \mathrm{Perm}(N)$, if and only if $\alpha(N)$ is normalized by $\lambda(G)$.*

This result implies Theorem 2.4, that if $\mathrm{Hol}(N)$ has a regular subgroup isomorphic to G, then $\mathrm{Perm}(G)$ has a regular subgroup M that is isomorphic to N and normalized by $\lambda(G)$, hence corresponds to a Hopf-Galois structure of type N on any G-Galois extension.

2.5. Counting

By Proposition 2.8, the bijection

$$s : \mathrm{Reg}(N, \mathrm{Perm}(G)) \to \mathrm{Reg}(G, \mathrm{Perm}(N))$$

restricts to a bijection from

$$\begin{aligned}\mathrm{Reg}^G(N, \mathrm{Perm}(G)) &= \{\alpha \in \mathrm{Reg}(N, \mathrm{Perm}(G)) \mid \alpha(N) \in \mathcal{E}^G(G, [N])\}\\ &= \{\alpha \in \mathrm{Reg}(N, \mathrm{Perm}(G)) \mid \alpha(N) \text{ is normalized by } \lambda(G)\}\end{aligned}$$

to

$$\begin{aligned}\mathrm{Reg}_N(G, \mathrm{Perm}(N)) &= \{\beta \in \mathrm{Reg}(G, \mathrm{Perm}(N)) \mid \beta(G) \in \mathrm{Hol}(N)\}\\ &= \mathrm{Reg}(G, \mathrm{Hol}(N))\\ &= \{\beta \in \mathrm{Reg}(G, \mathrm{Perm}(N)) \mid \beta(G) \text{ normalizes } \lambda(N)\}.\end{aligned}$$

Now for M in $\mathcal{E}^G(G, [N])$, that is, for M a regular subgroup of $\mathrm{Perm}(G)$ isomorphic to N and normalized by $\lambda(G)$, let α in $\mathrm{Reg}^G(N, \mathrm{Perm}(G))$ have image $\alpha(N) = M$. Then for θ in $\mathrm{Aut}(N)$, $\alpha\theta(N) = M$, and if $\alpha, \alpha' : N \to M$, then $\theta = \alpha^{-1}\alpha'$ is in $\mathrm{Aut}(N)$. So the number $e(G, [N])$ of elements of $\mathcal{E}^G(G, [N])$ satisfies

PROPOSITION 2.9.

$$e(G, [N]) = \frac{|\mathrm{Reg}^G(N, \mathrm{Perm}(G))|}{|\mathrm{Aut}(N)|}.$$

Similarly, given T in $\mathcal{R}_N(N, [G])$, that is, given T a regular subgroup of $\mathrm{Perm}(N)$ that normalizes $\lambda(N)$, let β in $\mathrm{Reg}_N(G, \mathrm{Perm}(N))$ have image $\beta(G) = T$. Then, as above, for ψ in $\mathrm{Aut}(G)$, $\beta\psi(G) = T$ and if $\beta(G) = \beta'(G)$, then $\beta^{-1}\beta'$ is in $\mathrm{Aut}(G)$. So the number $r_N(N, [G])$ of elements of $\mathcal{R}_N(N, [G])$ satisfies

$$r_N(N, [G]) = \frac{|\mathrm{Reg}(G, \mathrm{Hol}(N))|}{|\mathrm{Aut}(G)|}.$$

Since

$$|\mathrm{Reg}(G, \mathrm{Hol}(N))| = |\mathrm{Reg}^G(N, \mathrm{Perm}(G))|,$$

we have Byott's counting formula:

THEOREM 2.10.

$$r_N(N, [G]) \cdot |\mathrm{Aut}(G)| = e(G, [N]) \cdot |\mathrm{Aut}(N)|.$$

This enables us to count $e(G, [N])$, the number of Hopf-Galois structures of type N on a G-Galois extension of fields, by finding $r_N(N, [G])$, the number of regular subgroups of $\mathrm{Hol}(N)$ that are isomorphic to G.

2.6. Working with regular subgroups of Hol(N)

Now we present an alternative way to count Hopf-Galois structures of type N on a G-Galois extension.

Given a G-Galois extension L/K and a group N of the same order as G, let $\alpha, \alpha' : N \to \mathrm{Perm}(G)$ be regular embeddings such that $\alpha(N)$ and $\alpha'(N)$ are normalized by $\lambda(G)$. Let $\beta, \beta' : G \to \mathrm{Hol}(N)$ be the corresponding regular embeddings. Then $\alpha(N) = \alpha'(N)$ if and only if there is some automorphism θ of N so that $\alpha' = \alpha\theta$, and so $\beta'(\gamma) = \theta^{-1}\beta(\gamma)\theta$ for all γ in G by Proposition 2.7. So define an equivalence relation on regular embeddings from G to $\mathrm{Hol}(N)$:

DEFINITION 2.11. Two regular embeddings $\beta, \beta' : G \to \mathrm{Hol}(N)$ are equivalent, $\beta \sim \beta'$, if and only if there is some θ in $\mathrm{Aut}(N)$ so that $\alpha' = \alpha\theta$, if and only if the corresponding embeddings β, β' satisfy

$$\beta'(g) = C(\theta^{-1})(\beta(g)) = \theta^{-1}\beta(g)\theta$$

for all g in G.

Thus $\beta \sim \beta'$ if and only if they yield the same Hopf-Galois structure of type N on a G-Galois extension L/K, if and only if there is some θ in $\mathrm{Aut}(N)$ so that the corresponding embeddings α and α' satisfy $\alpha' = \alpha\theta$, if and only if

$$\beta'(g) = \theta^{-1}\beta(g)\theta$$

for all g in G.

So to find Hopf-Galois structures on L/K, we can partition the set of regular subgroups of $\mathrm{Hol}(N)$ isomorphic to G into orbits, conjugacy classes under the action of $\mathrm{Aut}(N)$, pick some convenient representative T of each orbit, and determine the regular embeddings of G into T.

EXAMPLE 2.12. (From [**Chi05**, pp. 295-6]) Let G and N be elementary abelian p-groups of rank n, where we can conveniently view N as the $\mathbb{F}_p$-vector space $\mathbb{F}_p^n$. Then $\mathrm{Hol}(N) = \lambda(N) \rtimes \mathrm{Aut}(N) \cong \mathbb{F}_p^n \rtimes \mathrm{GL}_n(\mathbb{F}_p)$, which may be viewed as the affine group

$$\mathrm{Hol}(\mathbb{F}_p^n) = \mathrm{Aff}_n(\mathbb{F}_p) = \left\{ \begin{pmatrix} \mathrm{GL}_n(\mathbb{F}_p) & \mathbb{F}_p^n \\ 0 & 1 \end{pmatrix} \right\} \subset \mathrm{GL}_{n+1}(\mathbb{F}_p).$$

Now regular subgroups of $\mathrm{Hol}(\mathbb{F}_p^n)$ are p-groups, hence are conjugate under action by $\mathrm{Aut}(N) = \mathrm{GL}_n(\mathbb{F}_p)$ to regular subgroups of U_{n+1}, the p-Sylow subgroup of $\mathrm{Aff}_n(\mathbb{F}_p)$ consisting of upper triangular matrices with unit diagonal. So we can focus on regular subgroups T contained in U_{n+1}. (We'll continue this example later in this section.)

In general, let $\beta : G \to \mathrm{Hol}(N)$ with $\beta(G) = T$. Then $\beta' : G \to \mathrm{Hol}(N)$ has $\beta'(G) = T$ if and only if $\beta' = \beta\psi$ for some ψ in $\mathrm{Aut}(G)$. We ask: given T, how many Hopf-Galois structures correspond to T?

The number of Hopf-Galois structures of type N is in bijective correspondence with the number of equivalence classes of regular embeddings $\beta : G \to \mathrm{Hol}(N)$, where $\beta' \sim \beta$ if and only if $\beta' = C(\theta)\beta$ for some θ in $\mathrm{Aut}(N)$. If we restrict this equivalence relation to regular embeddings of G to $\mathrm{Hol}(N)$ with image T, then we are interested in θ in $\mathrm{Aut}(N)$ so that $C(\theta)(T) = T$.

DEFINITION 2.13. For a given regular subgroup T of $\mathrm{Hol}(N)$, the stabilizer $\mathrm{Sta}(T)$ of T is

$$\mathrm{Sta}(T) = \{\theta \in \mathrm{Aut}(N) \mid C(\theta)T = T\}.$$

PROPOSITION 2.14. *Given a Galois extension with Galois group G of order n, a group N of order n and a regular embedding $\beta_0 : G \to \mathrm{Hol}(N)$ with image T, then the set of equivalence classes of regular embeddings $\beta : G \to \mathrm{Hol}(N)$ with image T is bijective with the set of right cosets of $\beta_0^{-1}\mathrm{Sta}(T)\beta_0$ in $\mathrm{Aut}(G)$.*

PROOF. Given T and a regular embedding $\beta_0 : G \to T$, then all other regular embeddings of G onto T have the form $\beta_0\psi$ for ψ in $\mathrm{Aut}(G)$. For ψ, ψ' in $\mathrm{Aut}(G)$, we have $\beta_0\psi' \sim \beta_0\psi$ if and only if there is some θ in $\mathrm{Aut}(N)$ so that $C(\theta)\beta_0\psi = \beta_0\psi'$, if and only if $\beta_0^{-1}C(\theta)\beta_0\psi = \psi'$. So conjugating by β_0^{-1} maps Sta(T) into Aut(G), and the number of equivalence classes of embeddings $\beta : G \to \mathrm{Hol}(N)$ with $\beta(G) = T$ is equal to the number

$$\frac{|\operatorname{Aut}(G)|}{|\operatorname{Sta}(T)|}.$$

of right cosets of Sta(T) in Aut(G). □

Proposition 2.14 then yields:

THEOREM 2.15. *The number of Hopf-Galois structures of type N on a Galois extension L/K with Galois group G is equal to*

$$\sum_{T\in C} |\operatorname{Aut}(G)|/|\operatorname{Sta}(T)|$$

where C is a set of representatives of the orbits (conjugacy classes) of regular subgroups of Hol(N) *isomorphic to G under conjugation by elements of* Aut(N).

PROOF. Consider the set $\mathcal{T}$ of regular subgroups of Hol(N) isomorphic to G. Under conjugation by elements of Aut(N), $\mathcal{T}$ is partitioned into orbits. Let $T_1, \ldots, T_r$ be a set of regular subgroups that represent the orbits under conjugation. Let $\beta_i : G \to T_i$ be regular embeddings for $i = 1, \ldots, r$. Then by Proposition 2.14, for each $i = 1, \ldots, r$, the set of equivalence classes of regular embeddings $\beta : G \to T_i$ is bijective with the set of right cosets of Sta(T_i) in Aut(G). So the set of equivalence classes of regular embeddings $\beta : G \to \mathrm{Hol}(N)$ is equal to

$$\sum_{T\in C} |\operatorname{Aut}(G)|/|\operatorname{Sta}(T)|,$$

which is the number of Hopf-Galois structures of type N on a G-Galois extension. □

EXAMPLE 2.16. Continuing Example 2.12, let $G = N$ be elementary abelian groups of order p^3, $p > 3$. We look at regular subgroups of $\mathrm{Hol}(C_p^3)$, which we view as $\mathrm{Aff}_3(\mathbb{F}_p) = \begin{pmatrix} \mathrm{GL}_3(\mathbb{F}_p) & \mathbb{F}_p^3 \\ 0 & 1 \end{pmatrix} \subset \mathrm{GL}_4(\mathbb{F}_p)$. As noted in Example 2.12, we may focus on the regular subgroups of the p-Sylow subgroup U_4, the group of upper triangular matrices of $\mathrm{GL}_4(\mathbb{F}_p)$ with diagonal I. [**Chi05**] found that there are five regular subgroups $J_1, \ldots, J_5$ of U_4, generated by

$$\gamma_i = \begin{pmatrix} I + A_i & \epsilon_i \\ 0 & 1 \end{pmatrix} \text{ for } i = 1, 2, 3$$

where $\{\epsilon_1, \epsilon_2, \epsilon_3\}$ is the standard basis of $\mathbb{F}_p^3$ and

$$A_1 = 0, A_2 = \begin{pmatrix} 0 & d & a \\ 0 & 0 & 0 \\ 0 & 0 & 0 \end{pmatrix}, A_3 = \begin{pmatrix} 0 & a & b \\ 0 & 0 & c \\ 0 & 0 & 0 \end{pmatrix},$$

with $cd = 0$ and for each J_i, the values of a, b, c, d are:

$$J_1 : (a, b, c, d) = (0, 0, 0, 0);$$
$$J_2 : (a, b, c, d) = (0, 1, 0, 0);$$
$$J_3 : (a, b, c, d) = (0, 1, 0, 1);$$
$$J_4 : (a, b, c, d) = (1, 0, 1, 0);$$
$$J_5 : (a, b, c, d) = (0, 1, 0, d), d \text{ a non-square of } \mathbb{F}_p.$$

Each group has a different stabilizer, and with $\mathrm{Aut}(G) \cong \mathrm{GL}_3(\mathbb{F}_p)$ **[Chi05]** found that

i	$\lvert \mathrm{Aut}(G)\rvert / \lvert \mathrm{Sta}(J_i)\rvert$
1	1
2	$(p^3 - 1)(p - 1)$
3	$(p^3 - 1)p(p - 1)/2$
4	$(p^3 - 1)p(p + 1)/2$
5	$(p^3 - 1)(p^3 - p)$

which sum to give

$$e(C_p^3, C_p^3) = p^6 + p^5 - p^2$$

the number of Hopf-Galois structures of type C_p^3 on a Galois extension of fields with Galois group C_p^3 for $p > 3$ an odd prime.

For extensions of this result, see Sections 5.3 and 5.4.

2.7. Radical algebras

A ring A (with addition $+$ and multiplication $\cdot$ but without a multiplicative identity) is a radical ring if A is a group under the circle operation $\circ$ on A defined by $a \circ b = a + b + a \cdot b$. The group $(A, \circ)$ is called the circle group or the adjoint group of A.

For N an abelian p-group, a regular abelian subgroup T of $\mathrm{Hol}(N)$ corresponds to a radical commutative algebra structure A on the additive group N, by **[CDVS06]**, so that $T = \lambda_\circ(A)$, the left regular representation of $(A, \circ)$ in $\mathrm{Perm}(A)$.

In that setting **[CDVS06]** showed that isomorphism classes of algebras correspond bijectively to conjugacy classes of regular subgroups of $\mathrm{Hol}(N)$. Thus for a given algebra A with circle group G and a G-Galois extension L/K, the number of Hopf-Galois structures of type N on L/K is equal to $\lvert \mathrm{Aut}(G)\rvert / \lvert \mathrm{Sta}(T)\rvert$ where $\mathrm{Sta}(T)$ consists of automorphisms of $(A, +)$ that preserve the image $\lambda(A, \circ) = T$ in $\mathrm{Hol}(N)$ of the circle group of A, hence are automorphisms of the circle group, hence of the multiplicative group of A: that is, $\mathrm{Sta}(T) \cong \mathrm{Aut}_{alg}(A)$.

So the number of Hopf-Galois structures corresponding to a given radical ring A is equal to

$$\mathrm{Aut}(A, \circ)/\,\mathrm{Aut}_{alg}(A).$$

The results of **[CDVS06]** connecting regular subgroups and radical algebra structures turn out to be a special case of results on skew braces, to which we now turn.

2.8. Skew (left) braces

Here is the definition of a skew brace.

DEFINITION 2.17. A skew brace (always "left") is a finite set B with two operations, $\star$ and $\circ$, so that $(B,\star)$ is a group (the "additive group"), $(B,\circ)$ is a group (the circle group, or adjoint group), and the (left) compatibility condition

$$a\circ(b\star c)=(a\circ b)\star a^{-1}\star(a\circ c)$$

holds for all a,b,c in B. Here a^{-1} is the inverse of a in $(B,\star)$. Denote the inverse of a in $(B,\circ)$ by $\overline{a}$.

It is a fact (easily shown from the compatibility condition) that in a skew brace, the identity elements of the groups $(B,\circ)$ and $(B,\star)$ coincide.

When describing a skew brace $(B,\circ,\star)$ our general convention is that $B^\star=(B,\star)$ is the additive group and $B^\circ=(B,\circ)$ is the adjoint, or circle group. So we write $(B,\text{circle operation},\text{additive operation})$. Under the relationship between skew braces and Hopf-Galois structures on Galois extensions of fields, $G=B^\circ$ becomes the Galois group, and $N=B^\star$ the type of the Hopf-Galois structure.

A (left) brace is a skew brace with abelian additive group. When describing braces, $\star$ will typically be replaced by $+$.

A function $f:B=(B,\circ,\star)\to B'=(B',\circ',\star')$ is a homomorphism of skew braces if f is a group homomorphism from $(B,\circ)$ to $(B',\circ')$ and a group homomorphism from $(B,\star)$ to $(B',\star')$

Braces and skew braces generalize radical rings. A radical ring $(A,+,\cdot)$ with circle operation $\circ$ as in Section 2.7 is a brace $(A,\circ,+)$ with additive group $(A,+)$ and circle group $(A,\circ)$: left distributivity in the ring A translates to the (left) compatibility condition for a skew brace. One can check that a left brace that also satisfies the right compatibility condition

$$(a+b)\circ c=(a\circ c)-c+(a\circ c)$$

is a radical ring with multiplication defined by $a\cdot b=a\circ b-a-b$.

We'll consider radical rings briefly in Section 4.8 and more extensively in chapter 5.

Associated to a set B with two group operations $\circ$ and $\star$ are the two left regular representation maps:

$$\lambda_\star:B\to\mathrm{Perm}(B),\ \lambda_\star(b)(x)=b\star x,$$
$$\lambda_\circ:B\to\mathrm{Perm}(B),\ \lambda_\circ(b)(x)=b\circ x.$$

Then there is a characterization of skew braces, Proposition 1.9 of [**GV17**]:

PROPOSITION 2.18. *$(B,\circ,\star)$ is a skew brace if and only if the group homomorphism $\lambda_\circ:(B,\circ)\to\mathrm{Perm}(B)$ has image in* $\mathrm{Hol}(B,\star)=\lambda_\star(B)\mathrm{Aut}(B,\star)\subset\mathrm{Perm}(B)$.

PROOF. For all b,x,y in B, the left skew brace property:

$$b\circ(x\star y)=(b\circ x)\star b^{-1}\star(b\circ y)$$

holds if and only if

$$b^{-1}\star(b\circ(x\star y))=(b^{-1}\star(b\circ x))\star(b^{-1}\star(b\circ y)),$$

if and only if

$$\lambda_\star(b^{-1})\lambda_\circ(b)(x \star y) = (\lambda_\star(b^{-1})\lambda_\circ(b)(x)) \star (\lambda_\star(b^{-1})\lambda_\circ(b)(y)),$$

if and only if

$$\Theta_b = \lambda_\star(b^{-1})\lambda_\circ(b) \text{ is in } \mathrm{Aut}(B, \star) \text{ for all } b \text{ in } B,$$

if and only if

$$\lambda_\circ(b) = \lambda_\star(b)\Theta_b \text{ is in } \mathrm{Hol}(B, \star).$$

□

Propositions 2.8 and 2.18 imply

COROLLARY 2.19. *$(B, \circ, \star)$ is a skew brace if and only if the group homomorphism $\lambda_\star : (B, \star) \to \mathrm{Perm}(B)$ has image $\lambda_\star(B)$ that is normalized by $\lambda_\circ(B)$.*

PROOF. Given a set $B = (B, \circ, \star)$ with two group operations where $(B, \circ) = G$, $(B, \star) = N$, B is a skew brace if and only if the map $\lambda_\circ : (B, \circ) \to \mathrm{Perm}(B)$ by $\lambda_\circ(g)(x) = g \circ x$ has image in $\mathrm{Hol}(B, \star)$. The corresponding element of $\mathrm{Bij}_0(G, N)$ is the map sending g to $\lambda_\circ(g)(e) = g \circ e = g$, the identity map. Similarly, $\lambda_\star : (B, \star) \to \mathrm{Perm}(B)$, sending n to $\lambda_\star(n)(z) = n \star z$, is in $\mathrm{Reg}(N, \mathrm{Perm}(B))$, and the corresponding element of $\mathrm{Bij}_0(N, G)$ is the map sending n to $\lambda_\star(n)(e) = n \star e = n$, also the identity map. So $\lambda_\star$ corresponds to $\lambda_\circ$ under the canonical bijection $s : \mathrm{Reg}(N, \mathrm{Perm}(G)) \to \mathrm{Reg}(G, \mathrm{Perm}(N))$ given by $s(\lambda_\star) = \lambda_\circ$. The result then follows immediately from Proposition 2.8. □

2.9. Connecting skew braces with Hopf-Galois structures

The next two results show the close connection between Hopf-Galois structures on Galois extensions of fields and skew braces.

PROPOSITION 2.20. *Let $(B, \circ, \star)$ be a skew brace. If L/K is a G-Galois extension with $G \cong (B, \circ)$, then L/K has a Hopf-Galois structure of type $(B, \star)$.*

PROOF. If $(B, \circ, \star)$ is a skew brace, then by Proposition 2.18, the image of the left regular representation map $\alpha = \lambda_\star : (B, \star) \to \mathrm{Perm}(B)$ is normalized by $\lambda_\circ(B)$. So α corresponds to a Hopf-Galois structure of type $(B, \star)$ on L/K. □

Conversely, we have:

PROPOSITION 2.21. *Let L/K be a G-Galois extension. Suppose L/K is a Hopf-Galois extension of type $N = (N, \star)$. There is an operation $\circ$ on N so that $(N, \circ) \cong G$ and $(N, \circ, \star)$ is a skew brace.*

PROOF. Since L/K is a Hopf-Galois extension of type N, there is a regular subgroup M of $\mathrm{Perm}(G)$ and a regular embedding $\alpha : N \to \mathrm{Perm}(G)$ with image M.

Since $M \cong (N, \star)$ is a regular subgroup of $\mathrm{Perm}(G)$, let $\alpha : (N, \star) \to \mathrm{Perm}(G)$ be a regular embedding with image M. The map $a : N \to G$ by $a(n) = \alpha(n)[e]$ is then a bijection. Using a, define a new group operation $\circ$ on N by $m \circ n = a^{-1}(a(m) \cdot a(n))$ where $\cdot$ is the multiplication in G. Then a is an isomorphism from $(N, \circ)$ to $(G, \cdot)$. To show that $(N, \circ, \star)$ is a skew brace, we show that $\lambda_\circ(N)$ is contained in $\mathrm{Hol}(N, \star)$. Now $\alpha(m)(g) = a\lambda_\star(m)a^{-1}(g)$, and (cf. Proposition 2.8) corresponds to $\beta : G \to \mathrm{Hol}(N, \star)$, defined by

$$\beta(g)(x) = a^{-1}\lambda_\cdot(g)a(x)$$

for x in N. The isomorphism $a : N \to G$ then yields the map $\beta a : N \to \mathrm{Hol}(N)$ which for n, x in N is

$$\begin{aligned}\beta(a(n))(x) &= a^{-1}\lambda.(a(n))a(x)\\ &= a^{-1}(a(n) \cdot a(x))\\ &= n \circ x.\end{aligned}$$

So the map βa is the map $\lambda_\circ$, and so $(N, \circ, \star)$ is a skew brace. □

2.10. Isomorphic skew braces

Here is a characterization of when two skew braces with the same additive group are isomorphic.

PROPOSITION 2.22. *Two skew brace structures $(B, \circ, \star)$ and $(B, \circ', \star)$ on the group $(B, \star)$ are isomorphic if and only if $\lambda_{\circ'}(B)$ is conjugate to $\lambda_\circ(B)$ in* $\mathrm{Perm}(B)$ *by some θ in* $\mathrm{Aut}(B, \star)$.

PROOF. Suppose $\theta : (B, \circ, \star) \to (B, \circ', \star)$ is an isomorphism of skew braces. Then for all b, c in B,

$$\theta(b) \circ' \theta(c) = \theta(b \circ c)$$

so

$$\lambda_{\circ'}(\theta(b))(\theta(c)) = \theta(\lambda_\circ(b)(c)) = \theta(\lambda_\circ(b)(\theta^{-1}\theta(c))),$$

so

$$\lambda_{\circ'}(\theta(b)) = \theta\lambda_\circ(b)\theta^{-1}.$$

Hence

$$\lambda_{\circ'}(B) = \theta\lambda_\circ(B)\theta^{-1}.$$

Conversely, let θ be in $\mathrm{Aut}(B, \star)$ and suppose

$$\theta\lambda_\circ(B)\theta^{-1} = \lambda_{\circ'}(B).$$

Then for each b there exists b' so that

$$\lambda_{\circ'}(b')(x) = \theta\lambda_\circ(b)\theta^{-1}(x).$$

Setting $x = e$ shows that $b' = \theta(b)$. Then for all c in B,

$$\begin{aligned}\lambda_{\circ'}(\theta(b))(\theta(c)) &= \theta\lambda_\circ(b)\theta^{-1}(\theta(c)),\\ \theta(b) \circ' \theta(c) &= \theta(b \circ c),\end{aligned}$$

and so $\theta : (B, \circ) \to (B, \circ')$ is an isomorphism, hence a skew brace isomorphism from $(B, \circ, \star)$ to $(B, \circ', \star)$. □

This result was first proved in the case where the skew brace arises from a radical F-algebra, F any field, in **[CDVS06]**, and is also found as Proposition 4.3 of **[GV17]**.

Switching the roles of $\star$ and $\circ$ in the above proof gives:

COROLLARY 2.23. *Two skew brace structures $(B, \circ, \star)$ and $(B, \circ, \star')$ on the group $(B, \circ)$ are isomorphic if and only if $\lambda_{\star'}(B)$ is conjugate to $\lambda_\star(B)$ in* $\mathrm{Perm}(B)$ *by some θ in* $\mathrm{Aut}(B, \circ)$.

Now we can describe the relationship between the number of Hopf-Galois structures of type N on a G-Galois extension and the number of isomorphism types of skew braces $(B, \circ, \star)$ with $B^\circ \cong G$ and $B^\star \cong N$.

The following result is in **[SV18]** and **[NZ19]**.

PROPOSITION 2.24. *Let $B = (B, \circ, \star)$ be a skew brace. The number $e_B(G, N)$ of Hopf-Galois structures of type $N \cong (B, \star)$ on a G-Galois extension L/K with $G \cong (B, \circ)$ corresponding to the given skew brace B is*

$$e_B(G, N) = \frac{|\operatorname{Aut}(G)|}{|\operatorname{Aut}_{sb}(B)|}.$$

Here $\operatorname{Aut}_{sb}(B)$ denotes the group of skew brace automorphisms of the skew brace B, that is, bijective maps $B \to B$ that are group homomorphisms on the additive groups and also on the adjoint groups.

PROOF. Given $(B, \circ, \star)$, a skew brace, we have a Hopf-Galois structure uniquely defined by $\lambda_\star(B) \subset \operatorname{Perm}(B)$ normalized by $\lambda_\circ(B)$. Let $\theta \in \operatorname{Aut}(B, \circ)$. Then θ defines an operation $\star'$ on $(B, \circ)$ by

$$\theta(g \star h) = \theta(g) \star' \theta(h),$$

and $\lambda_{\star'}(B) = \theta\lambda_\star(B)\theta^{-1}$. From the definition of $\star'$, $\theta : (G, \circ, \star) \to (G, \circ, \star')$ is a skew brace isomorphism. Since $(G, \circ, \star')$ is a skew brace, $\lambda_{\star'}(B)$ is normalized by $\lambda_\circ(B)$ in $\operatorname{Perm}(B)$ by Corollary 2.19. So $\lambda_{\star'}(B)$ yields a Hopf-Galois structure on L/K.

Now $\lambda_{\star'}(B) = \lambda_\star(B)$ if and only if $\star' = \star$, if and only if θ is an automorphism of $(B, \star)$. For suppose $\lambda_{\star'}(B) = \lambda_\star(B)$. Then for all b in B there is some c in B so that $\lambda_{\star'}(b) = \lambda_\star(c)$ in $\operatorname{Perm}(B)$. Then

$$\lambda_{\star'}(b)[e] = \lambda_\star(c)[e],$$

so $b = c$ and then $b \star' x = b \star x$ for all b, x in B. Hence $\star' = \star$.

So the number $e_B((B, \circ), (B, \star))$ of Hopf-Galois structures of type $(B, \star)$ on L/K with Galois group $(B, \circ)$ arising from the given skew brace isomorphism type of $(B, \circ, \star)$ is equal to $|\operatorname{Aut}(B, \circ)|/|\operatorname{Aut}_{sb}(B, \circ, \star)|$. □

COROLLARY 2.25. *The number $e(G, [N])$ of Hopf-Galois structures of type N on a G-Galois extension of fields is given by*

$$e(G, [N]) = \sum_B e_B(G, N)$$

where B runs through all isomorphism types of skew braces B with given circle group $(B, \circ) = G$ and $(B, \star) \cong N$.

These results demonstrate the close connection between the classification of Hopf-Galois structures on a given G-Galois extension of fields and the classification of skew braces with given circle group. In particular, any result that finds pairs of groups (G, N) where $e(G, [N])$ is or is not zero is also a result on the existence or non-existence of skew braces with circle group G and additive group N. Such issues have been a persistent theme in both theories since their inception (in [**GP87**] and [**Rum07a**], respectively).

CHAPTER 3

(Non)-existence results on Hopf-Galois structures

3.1. Introduction

A consistent theme in the study of Hopf-Galois structures on Galois extensions of fields L/K since work of Byott [**Byo96**] and Kohl [**Koh98**] is to consider the question: given the Galois group G of L/K, what are the possible types of Hopf-Galois structures on L/K?

With the definition of a brace (in 2007) and a skew brace (in 2017), a similar question arose: given a skew brace $(B, \circ, \star)$ with given additive group $B^\star = (B, \star)$, what are the possible isomorphism types of the adjoint (or circle) group $B^\circ = (B, \circ)$ of a skew brace $(B, \circ, \star)$?

Every skew brace $(B, \circ, \star)$ yields Hopf-Galois structures of type $N \cong (B, \star)$ on a G-Galois extension L/K of fields where $G \cong (B, \circ)$, and conversely, any Hopf-Galois structure of type $N = (N, \star)$ on a G-Galois extension yields an operation $\circ$ on N so that $(N, \circ) \cong G$ and $(N, \circ, \star)$ is a skew brace. So both questions are interesting. Adopting a neutral point of view, let us call a pair of groups (G, N) of the same finite order n *realizable* if there exists a skew brace $(B, \circ, \star)$ so that $(B, \circ) \cong G$ and $(B, \star) \cong N$, or equivalently, if every G-Galois extension of fields has a Hopf-Galois structure of type N.

Recall from the last chapter that $e(G, [N])$ is the number of Hopf-Galois structures of type N on a G-Galois extension. So a pair (G, N) is realizable if and only if $e(G, [N]) > 0$.

This chapter focuses on realizability results. We'll consider aspects of the quantitative question about the size of $e(G, [N])$ in later chapters.

All groups and braces are finite in this chapter.

The first and most trivial realizability result is that for every group G, (G, G) is realizable. From a Hopf-Galois viewpoint this is because a G-Galois extension of fields is an H-Hopf-Galois extension where $H = K[G]$, the group ring of the Galois group G acting on L where G acts by the Galois action. This Hopf-Galois structure is called the classical Hopf-Galois structure. From a skew brace point of view the classical Hopf-Galois structure corresponds to the skew brace structure on the group $G = (G, \circ)$ where $\star = \circ$: to check the relation, we have

$$a \circ (b \star c) = (a \circ b) \star a^{-1} \star (a \circ c)$$

which follows from the associativity of $\circ$.

The earliest classification result is that of Byott [**Byo96**], who showed that a G-Galois extension of fields L/K has a unique Hopf-Galois structure, namely the classical structure just described, if and only if the order n of G is coprime to $\varphi(n)$ (Euler's totient). (Then $n = 2$ or is odd and G must be cyclic.)

The realizability question gets much more interesting when one looks for cases where $G \not\cong N$. This chapter surveys many of these results.

3.2. p-groups, p an odd prime, cyclic case

One of the earliest results was that of Kohl. In [**Koh98**], he proved

THEOREM 3.1. *Let $G \cong C_{p^n}$ be cyclic of order p^n for p an odd prime. If (C_{p^n}, N) is realizable, then $N \cong G$.*

The proof reduces to showing that if N is a non-cyclic group of order p^n, then $\mathrm{Hol}(N)$ has no element of order p^n, which implies that there is no regular embedding of G into $\mathrm{Hol}(N)$. A proof of Kohl's result may be found also in Proposition 8.6 of [**Chi00**], or in Theorem 4.1 of [**Byo07**].

Soon after creating the concept of a brace, Rump considered the reverse problem to Kohl's. [**Rum07b**] studied braces $(B, \circ, +)$ where $B^+ \cong C_{p^n}$ ("cyclic braces") for p prime and asked: given a brace B with additive group C_{p^n}, what could the adjoint group B° be? For Hopf-Galois theorists, the question translates to: for which Galois groups G of order p^n could L/K have a Hopf-Galois structure of cyclic type?

The Corollary to Theorem 2 of [**Rum07b**] is:

THEOREM 3.2. *If p is an odd prime, B is a brace of order p^n and $N = (B, +)$ is cyclic of order p^n, then the adjoint group $G = B^\circ$ is also cyclic.*

Rump called such a brace B a bicyclic brace.

This result was independently proved by Crespo and Salguero [**CS19**] in the Hopf-Galois setting: their result specializes to: if L/K is G-Galois and has a Hopf-Galois extension of type N, cyclic of odd prime power order, then $G \cong N$.

3.2.1. Non-cyclic abelian p-group case.

Let N be a finite abelian p-group. The p-rank of N is the dimension of N/pN as an $\mathbb{F}_p$-vector space, that is, the number of summands when N is written as a direct sum of cyclic groups.

Refining a result of Featherstonhaugh [**Fea03**], [**FCC12**] obtained:

THEOREM 3.3. *Let L/K be a G-Galois extension with G abelian of order p^n. Suppose L/K has an H-Hopf-Galois structure of type N, an abelian group of order p^n and p-rank m where $m + 1 < p$. Then $G \cong N$.*

Since an abelian group G of order p^n has p-rank $m \leq n$, this implies that if G is an abelian group of order p^n with $n + 1 < p$, and N is an abelian group, then (G, N) is realizable if and only if $N \cong G$.

The strategy of the proof in [**FCC12**] was to extend work of Caranti, Dalla Volta and Sala [**CDVS06**]. They showed that if $N = (N, +)$ is any abelian p-group of order p^n and G is any regular subgroup of $\mathrm{Hol}(N) \subset \mathrm{Perm}(N)$, necessarily also of order p^n, then N becomes a commutative radical ring in such a way that G is isomorphic to the adjoint group $(N, \circ) = N^\circ$ of N.

The main theorem of [**FCC12**] then says that if N has p-rank m and $m+1 < p$, then for every element g of N, the order of g in the additive group $N^+ = (N, +)$ of the radical ring is equal to the order of g in the adjoint group N°. Since N° and N^+ are abelian, this implies that $N^\circ \cong N^+$.

Every radical algebra defines a left brace. So it is natural (in retrospect) to try to extend this result to the left brace setting, where $(N, \circ, +)$ is a left brace with p^n elements. This was done by Bachiller [**Bac16**]:

THEOREM 3.4. *Let $(N, +)$ be an abelian p-group of p-rank m, and $(N, \circ, +)$ be a left brace with additive group $N^+ = (N, +)$. If $m + 1 < p$, then for every element*

g of N, the order of g in N^+ is equal to the order of g in N°. Hence if N° is abelian, then $N^\circ \cong N^+$.

The proof in [**Bac16**] is very similar to that in [**FCC12**].

REMARK 3.5. Bachiller's theorem is illustrated in his work on the classification of braces of order p^3, p odd [**Bac15**]. There are three abelian groups of order p^3, namely C_p^3 of exponent p, $C_p \times C_{p^2}$ of exponent p^2, and C_{p^3} of exponent p^3. There are two non-abelian groups of order p^3, namely the Heisenberg group H_3 of exponent p and the group $M_3(p)$ of exponent p^2. Theorem 3.2 of [**Bac15**] classifies all braces of order p^3 with $p > 2$. A brace with additive group of exponent p, resp. p^2, resp. p^3 must have a circle group with the same exponent, except when $p = 3$. There exists one brace with additive group $C_3 \times C_9$ and circle group C_3^3, and one with additive group C_p^3 and circle group $C_3 \times C_9$.

This applies to Hopf-Galois theory and extends the result of [**FCC12**] to Galois extensions with a non-abelian Galois group G. Let L/K be a G-Galois extension of degree p^n. Suppose L/K has a Hopf-Galois structure of type N where N is an abelian p-group of order p^n and p-rank m where $m + 1 < p$. Then these data yield a left brace $(N, \circ, +)$ where $N^+ = N$ and $N^\circ \cong G$. Then for every g in N, the additive order of g is equal to the order of g in $(N, \circ) \cong G$. If G is abelian, then $G \cong N$.

REMARK 3.6. [**FCC12**] and, as we just observed, [**Bac15**], have examples that show that the condition $m + 1 \leq p$ is sharp.

[**Chi07**] considers the case where $N = \mathbb{F}_p^m$ and $G = 1+M$, the group of principal units under multiplication of the finite ring $R = \mathbb{F}_p[x]/(x^{m+1})$. Then $R = \mathbb{F}_p + M$ where $M = \langle x\rangle/\langle x^{m+1}\rangle$, a radical $\mathbb{F}_p$-algebra with circle operation $a \circ b = a + b + ab$. So $(M, \circ, +)$ is a brace with $(M, +) = \mathbb{F}_p^m$ and $(M, \circ) \cong (1 + \langle x\rangle, \cdot) \subset R$. [**Chi07**] shows that

$$G \cong C_p^{d_1} \times C_{p^2}^{d_2} \times \ldots \times C_{p^e}^{d_e}$$

where $d_1 + 2d_2 + \ldots + ed_e = m$ and

$$d_k = \left\lfloor \frac{m}{p^{k-1}} \right\rfloor - 2\left\lfloor \frac{m}{p^k} \right\rfloor + \left\lfloor \frac{m}{p^{k+1}} \right\rfloor$$

for all k. Thus $G \cong C_p^m$ if and only if $m < p$. An example: for $p = 3, m = 10$,

$$G \cong C_3^5 \times C_9 \times C_{27}.$$

3.2.2. The elementary abelian case. Returning to Bachiller's theorem, the case where $(B, +)$ is elementary abelian can be proved rather efficiently.

PROPOSITION 3.7. *Let $(N, \circ, +)$ be a brace with $N^+ = \mathbb{F}_p^m$ (hence elementary abelian). Assume $m + 1 \leq p$. Then N° has exponent p.*

PROOF. The map

$$\lambda_\circ : N^\circ \to \mathrm{Hol}(N, +) \cong \mathbb{F}_p^m \rtimes \mathrm{Aut}(\mathbb{F}_p^m)$$

is a one-to-one homomorphism. Now $\mathrm{Aut}(\mathbb{F}_p^m) = \mathrm{GL}_m(\mathbb{F}_p)$, so we can identify $\mathrm{Hol}(N, +)$ as the affine group

$$\mathrm{Aff}_m(\mathbb{F}_p) = \begin{pmatrix} \mathrm{GL}_m(\mathbb{F}_p) & \mathbb{F}_p^m \\ 0 & 1 \end{pmatrix} \subset \mathrm{GL}_{m+1}(\mathbb{F}_p).$$

The group $\lambda_\circ(N)$ has order p^m, so its image in $\mathrm{GL}_{m+1}(\mathbb{F}_p)$ is a p-group. So we can conjugate it into the p-Sylow subgroup $U_{m+1}(\mathbb{F}_p)$ consisting of unipotent matrices, $(m+1)\times(m+1)$-matrices of the form $I+N$, where I is the identity matrix and N is a strictly upper triangular matrix. The matrices have entries in $\mathbb{F}_p$, so $(I+N)^p = I+N^p$. Now $N^{m+1} = 0$ and $m+1 \leq p$, so $N^p = 0$. Thus $(I+N)^p = I$. Since the map from $\lambda_\circ(N,\circ)$ to $U_{m+1}(\mathbb{F}_p)$ is a one-to one homomorphism, $\lambda_\circ(N,\circ)$ has exponent p. □

The result shows that if one wants to classify braces of order p^m for $p > m$ with additive group isomorphic to $\mathbb{F}_p^m$, it suffices to look at circle groups of order p^m and exponent p. There turn out to be many possibilities. For example, Vaughan-Lee [**VL15**] showed that for $m = 8$, the number of groups of order p^8 and exponent p is

$$p^4 + 2p^3 + 20p^2 + 147p + (3p+29)\gcd(p-1,3) + 5\gcd(p-1,4) + 1246$$

for $p \geq 11$.

One might ask: Given any group G of order p^m and exponent p with $m+1 < p$, is there a brace with additive group C_p^m and circle group isomorphic to G?

This would be the converse of Proposition 3.7. It's true for $m = 3 < p$ by Bachiller's classification [**Bac15**]: for $m = 3$, $(B,\circ)$ can be either of the two groups of exponent p, C_p^3 or the Heisenberg group H_3.

But Bachiller showed that the converse is false. He found a group of order p^{10} and exponent p which is not the circle group of a brace with additive group $\mathbb{F}_p^{10}$. See [**Bac16**]. The construction of that group involves Lie theory.

In his "Problems on skew braces" paper [**Ven19**], Vendramin asks

Problem 31: What is the minimum cardinality of a solvable group which is not the adjoint group of a left brace?

Equivalently, what is the minimum cardinality of a solvable group G so that a G-Galois extension of fields has no Hopf-Galois structure of abelian type?

3.3. 2-groups

The case $p = 2$ is different, as [**Byo07**] showed. Let $G = C_{2^n}$. For $n = 1$ the pair (G,N) of groups is realizable only for $N = C_2$, of course. But for $n = 2$, (C_{2^n}, N) is realizable for N cyclic or $= C_2 \times C_2$, and for $n > 2$ N can be C_{2^n}, the cyclic group of order 2^n, or D_{2^n}, the dihedral group of order 2^n, or Q_{2^n}, the quaternion group of order 2^n (and there are equal numbers of Hopf-Galois structures of each type).

From the brace point of view, Rump [**Rum07b**] proved that for $N = C_{2^n}$, the pair $(G,N) = (G, C_{2^n})$ is realizable if and only if G is a group of order 2^n with a cyclic subgroup of index 2. Thus for $n = 3$, G has one of the following types:

$$C_{2^n}, C_2 \times C_{2^{n-1}}, D_{2^n} \text{ (dihedral)}, Q_{2^n} \text{ (quaternion)}.$$

For $n \geq 4$ two types of semidirect products $G = C_{2^{n-1}} \rtimes C_2$ can also occur, where

$$G = \langle x, y \mid x^{2^n - 1} = y^2 = 1, yxy^{-1} = x^{-1+2^{m-2}} \rangle$$

or

$$G = \langle x, y \mid x^{2^n - 1} = y^2 = 1, yxy^{-1} = x^{1+2^{m-2}} \rangle.$$

For each of these groups G (and no others), (G, C_{2^n}) is realizable.

3.4. Groups of composite order n that decompose

Given that the case for $|N| = p^n$ is more or less known, at least for N abelian and sufficiently large primes p, it is natural to ask about the case where N is nilpotent, that is, the direct sum of its p-Sylow subgroups.

Here, the key result is a decomposition of skew braces B with nilpotent additive group, from Smoktunowicz and Vendramin [**SV18**]:

THEOREM 3.8. *Suppose B is a skew brace of order n with a nilpotent additive group $B^\star$ (which we write multiplicatively, with identity 1) and adjoint group B°. Let $B_p = (B_p)^\star$ be the p-Sylow subgroup of $B^\star$. Then in the decomposition of $B = \prod_{p|n} B_p$ as additive groups, each B_p is a skew brace, as is the product $\prod_{p\in J} B_p$ for any set J of distinct primes dividing n.*

PROOF. Let p^e be the order of the p-Sylow subgroup B_p of $B^\star$. Let a be in B_p, then $a^{p^e} = 1$ in $B^\star$. For b in B, $\lambda_\circ(b) = \lambda_\star(b)\theta_b$ where θ_b is in $\mathrm{Aut}(B^\star)$. Then $1 = \theta_b(a^{p^e})$, and so $\theta_b(a)$ is in B_p for all b in B. In particular, it is true for b in B_p. Now we show that B_p is closed under $\circ$ and under inverses in B°. For a_1, a_2 in B_p, $a_1 \circ a_2 = a_1 \star \theta_{a_1}(a_2)$, and we showed above that $\theta_{a_1}(a_2)$ is in B_p for any a_1 in B. Also for a in B_p, $0 = a \circ \overline{a} = a \cdot \theta_a(\overline{a})$, so $a^{-1} = \theta_a(\overline{a})$ is in B_p, and so $\overline{a} = \theta_a^{-1}(a^{-1})$ is in B_p. Thus B_p is a skew brace, and $B_p = B_p^\circ$ is a p-subgroup of B°. Since the direct product of skew braces is easily seen to be a skew brace, the last claim follows. □

For each prime p_j dividing n, let $n = p_j^e p_j'$ where p_j' is coprime to p_j. Then for each j, the product $B^\circ_{p_j'} = B^\circ_{p_1} \cdots B^\circ_{p_{j-1}} B^\circ_{p_{j+1}} \cdots B^\circ_{p_k}$ is a subgroup of B° of order p_j'. By Hall's Theorem, B° is solvable. [**Rob93**, Section 9.1.8] As a consequence, Corollary 1.23 of [**SV18**] is:

COROLLARY 3.9. *If B is a skew brace with nilpotent additive group $B^\star$, then B° is solvable.*

The corresponding result for Hopf-Galois structures was first proven by Byott [**Byo15**].

The decomposition of Theorem 3.8 also yields a result of Etingof, Schedler and Soloviev [**ESS99**] that predated the existence of braces. Translated into brace language and noting that a brace has an abelian, hence nilpotent, additive group, that result is:

THEOREM 3.10. *Let B be a brace. Then B° is solvable.*

The decomposition of Theorem 3.8 implies that if $G = \prod_{p|n} G_p$ and $N = \prod_{p|n} N_p$ are nilpotent groups of the same order, then (G, N) is realizable if and only if (G_p, N_p) is realizable for all p dividing n. This follows from the following quantitative result, from [**Byo13**]:

THEOREM 3.11. *Let $e(G, [N])$ be the number of Hopf-Galois structures of type N on a G-Galois extension. Suppose G is nilpotent of order n. Then for each nilpotent group N of order n, $e(G, [N]) = \prod_{p|n} e(G_p, [N_p])$.*

Vendramin lists the question of a possible reverse of Corollary 3.9 as Problem 47 in [**Ven19**]:

Let $B = (B, \circ, \star)$ be a skew brace with additive group $B^\star$. Suppose B° is a nilpotent group. Must $B^\star$ be solvable?

See Section 3.8 below.

3.4.1. Cyclic braces. Before leaving the decomposition of Theorem 3.8, we note that the same strategy was used by Rump in [**Rum19**] to characterize all cyclic braces, that is, braces with cyclic additive group. Proposition 10 of [**Rum19**] is essentially the same as Theorem 3.8 above, and use of it reduces the problem to the case where a brace has order a power of p.

Let B be a (finite) brace with additive group B^+ and adjoint group B°. Extending his work in [**Rum07b**], Rump characterized the possible adjoint groups of a cyclic brace (that is, a brace with cyclic additive group), as follows.

DEFINITION 3.12. A finite group G is 2-nilpotent if any 2-Sylow subgroup of G has a normal complement. G is Sylow-cyclic if its Sylow subgroups are cyclic. G is almost Sylow-cyclic if its Sylow subgroups of odd order are cyclic, while the 2-Sylow subgroup is either trivial or has a cyclic subgroup of index 2.

Corollary 1 of Theorem 2 of [**Rum19**] is:

THEOREM 3.13. *A finite group G is the adjoint group B° of a brace B with a cyclic additive group if and only if G is solvable, 2-nilpotent and almost Sylow-cyclic.*

To get a sense of what that implies, Example 4 of [**Rum19**] shows that there are 36 cyclic braces B of order 80. Of the 52 groups of order 80, 23 are realizable as adjoint groups of cyclic braces. All six groups of order 16 with a cyclic subgroup of index 2 occur as a 2-Sylow subgroup of some adjoint group. Those six groups are the cyclic group C_{16}, the abelian group $C_8 \times C_2$, the dihedral group D_8, the generalized quaternion group Q_4, and the two groups

$$M_8 = \langle a, b \mid a^8 = b^2 = 1, bab^{-1} = a^5\rangle$$
$$SD_8 = \langle a, b \mid a^8 = b^2 = 1, bab^{-1} = a^3\rangle.$$

Thus 23 of the 52 groups of order 80 can be the Galois group of a field extension that has a Hopf-Galois structure of cyclic type.

[**Rum19**] also considers the case where G is cyclic and asks about N:

THEOREM 3.14. *Let B be a brace of order $n = 2^r q$ with q odd, and with additive group N and adjoint group G. If G is cyclic, then B splits into a direct product $B = \prod_{p|n} B_p$ of braces where B_p° is the p-primary component of G. For all odd p, B_p has cyclic additive group, so B_p is bicyclic. For $p = 2$ and $r \neq 2$ the 2-primary component B_2 is cyclic. For $n = 2^2 q$, the additive group of B_2 can be isomorphic to either C_4 or $C_2 \times C_2$.*

Thus, since N is necessarily abelian (because B is a brace), N is the product of its p-primary components, hence is cyclic unless the 2-Sylow subgroup is isomorphic to the Klein 4-group.

If we specialize this theorem to the case $n = 2^r$, it says that if $r \neq 2$, then any brace with cyclic adjoint group must have a cyclic additive group.

As noted in Section 3.3, [**Byo07**, Corollary 5.3] says that if G is cyclic of order 2^r, $r > 2$, then L/K admits Hopf-Galois structures of type $N = C_{2^r}, D_{2^r}, Q_{2^r}$ in equal number. But since a brace has an abelian additive group, N must be abelian in [**Rum07b**], so the two results agree, as do the results for $r = 2$. Byott's result applies also to skew braces.

3.5. Cases where G must be isomorphic to N

The earliest result where (G, N) is realizable only for $N = G$ is a trivial consequence of Byott's Uniqueness Theorem in [**Byo96**], mentioned in Section 3.1.

Also belonging in this section is Kohl's result [**Koh98**] (Theorem 3.1) cited earlier for G a cyclic p-group for p odd.

An outstanding set of examples is given by:

3.5.1. Byott's simple groups theorem. Byott [**Byo04a**] showed

THEOREM 3.15. *Let G be a non-abelian simple group. Then $e(G, [N]) = 0$ for all $N \ncong G$.*

We outline the highly non-trivial proof.

Suppose G embeds as a regular subgroup of $\mathrm{Hol}(N)$ for some group N of the same order as G but not isomorphic to G. If M is a maximal characteristic subgroup of N, then, since G is simple, G embeds as a transitive subgroup of $\mathrm{Hol}(N/M)$, and N/M is a characteristically simple group, hence isomorphic to a direct sum of m copies of a simple group T.

So let $\beta : G \to \mathrm{Hol}(H) = H \rtimes \mathrm{Aut}(H)$ where $H = T^m$. Letting $\varphi : \mathrm{Hol}(H) \to \mathrm{Hol}(H)/H \cong \mathrm{Aut}(H)$ be the canonical map, then $\varphi\beta$ must be injective or trivial since G is simple. If $\varphi\beta$ is trivial, then β is a 1-1 homomorphism from G to H, so projecting H onto some direct factor T implies that there is a 1-1 homomorphism from G to T. But $|G| = |N| \geq |N/M| = |T^m| \geq |T|$, so if $G \ncong N$, this is impossible. Thus G maps into $\mathrm{Aut}(H)$.

Byott then shows that T is abelian.

Suppose T is nonabelian. Then T has trivial center.

He shows that $\mathrm{Aut}(H) = \mathrm{Aut}(T^m) \cong \mathrm{Aut}(T)^m \rtimes S_m$, by showing that if α is an automorphism of H then α is uniquely determined by the m^2 homomorphisms $\alpha_{i,j} = \pi_j \alpha \iota_i$ where ι_i is the canonical inclusion from the ith component T to H and π_i is the canonical projection from H to the ith component T.

Now suppose $\alpha_{i,j}$ is an automorphism. Then for each i, there is a unique $\sigma(i)$ so that $\alpha_{\sigma(i),i}$ is non-trivial and an automorphism of T. For $\iota_i(t_1)$ and $\iota_h(t_2)$ commute in H for all $i \neq h$. So $\alpha_{i,j}(t_1) = \pi_j \alpha \iota_i(t_1)$ and $\alpha_{h,j}(t_2) = \pi_j \alpha \iota_h(t_2)$ commute in T. Since $\alpha_{i,j}$ maps T onto T, for all t_2 in T, $\alpha_{h,j}(t_2)$ is in the center of T, so is trivial. Thus each automorphism α corresponds to a permutation σ in S_m.

Now suppose G embeds in $\mathrm{Aut}(H) = \mathrm{Aut}(T)^m \rtimes S_m$. Suppose then G maps 1-1 into S_m. Then setting $v_p(S)$ to be the power of p that divides the cardinality of the set S, then for each prime p dividing $|T|$, we have

$$v_p(G) = v_p(N) \geq v_p(N/M) = v_p(H) = m v_p(T) \geq m > v_p(S_m).$$

So G cannot embed in S_m. Thus G must map trivially to S_m since G is simple, hence G embeds in $\mathrm{Aut}(T)^m$. But then G maps to $\mathrm{Out}(T)^m$, hence to $\mathrm{Out}(T)$. Since T is simple, $\mathrm{Out}(T)$ is solvable by Schreier's Conjecture. So the image of G in $\mathrm{Out}(T)$ is trivial, and so G embeds in T^m, and hence there is a non-trivial map from G to some direct factor T of H. But that can happen only if $G \cong N = T$.

Thus T must be abelian, hence $T \cong \mathbb{F}_p$ for some prime p. So G embeds via $\varphi\beta$ to a subgroup of $\mathrm{Aut}(T^n) = \mathrm{GL}_n(p)$, and so G must have a subgroup of index p^n, namely the stabilizer of any element of T.

The remainder of the proof relies heavily on consequences of the classification of finite simple groups. A result of Guralnick shows that G must be one of four special cases $G = \mathrm{PSL}_2(11), M_{11}, M_{23}, \mathrm{PSU}_4(2)$, or the family $G = A_n$ with $n = p^m \geq 5$, or the family $G = \mathrm{PSL}_n(q)$. For all but the last family, G cannot embed into $\mathrm{GL}_m(p)$ because $|G| > |\mathrm{GL}_m(p)|$ or because $m = 1$, hence G is abelian. The argument for the last family involves application of results from the representation theory of Chevalley groups of Guralnick and Tiep, Landazuri and Seitz, and Seitz and Zalesskii.

When combined with the result of Carnahan and Childs [**CC99**] that $e(G, G) = 2$ for a non-abelian simple group G, this gives the complete story on Hopf-Galois structures on Galois extensions with Galois group a simple non-abelian group: a given Hopf-Galois structure is either the classical structure by the Galois group or the canonical non-classical structure given by H_λ, corresponding to the regular subgroup $\lambda(G) \subset \mathrm{Perm}(G)$.

Byott's description of $\mathrm{Aut}(H)$ where $H = T^n$ for T a finite non-abelian simple group was subsequently used by C. Tsang [**Tsa19a**] to count the Hopf-Galois structures on a Galois extension with Galois group H: see Section 4.3.

3.5.2. Other examples where (G, N) is realizable only for $N \cong G$. Tsang [**Tsa19c**] showed that if $G = 2A_n$ is the double cover of the alternating group A_n, then for $n \geq 5$, (G, N) is realizable only for $N = G$.

In [**Byo13**], Byott determined conditions on n so that every Hopf-Galois structure on a cyclic Galois extension of order n has cyclic type. Namely, let $n = \prod p^{\nu_p}$ and let $r(n) = \prod_{p|n} p$, the radical of n.

THEOREM 3.16. *Suppose n as above with $\nu_p \leq 2$ for all p, 4 does not divide n and for all p, q dividing n, p does not divide $q^{\nu_q} - 1$. Then every Hopf-Galois structure on a C_n-Galois extension has type C_n, and $e(C_n, C_n) = \frac{n}{r(n)}$.*

He also found that if $n \equiv 4 \pmod 8$ then $e(C_n, C_n) = \frac{n}{2r(n)}$, and if $n = 2^e n'$ where $e \geq 3$ and n' is odd, then

$$e(C_n, C_n) = e(C_n, D_{2^e} \times C_{n'}) = e(C_n, Q_{2^e} \times C_{n'}) = \frac{n}{2r(n)}.$$

But the cube-free condition on n is not necessary. In section 6 of [**BC12**] Byott found another non-trivial class of examples of cyclic groups G where (G, N) is realizable only for $N \cong G$:

PROPOSITION 3.17. *Let G be a cyclic group of odd order $n = p^3 q$ where*

(i) *q divides $p^3 - 1$*
(ii) *p and $q - 1$ are coprime*
(iii) *q and $p^2 - 1$ are coprime.*

Then $e(G, G) = p^2$ and $e(G, [N]) = 0$ for all $N \ncong G$.

There are 12 pairs (p, q) satisfying (i)-(iii) with $7 \leq p \leq 37$, including $(p, q) =$ (7, 19), (11, 7), (23, 79), (37, 7), ...

3.6. Realizability when $G = S_n$ or A_n

Byott's simple groups theorem applies of course to the case where $G = A_n$, the alternating group, for $n \geq 5$. "Most" polynomials of degree n with rational coefficients have a Galois group isomorphic to S_n or A_n, the symmetric or the alternating group. So there has been interest in determining the types of Hopf-Galois structures on Galois extensions with Galois group $G = S_n$ or A_n. Of course for $n \geq 5$, (A_n, N) is realizable only for $N \cong A_n$. Other results:

For $n = 5$ or $n \geq 7$, [**CC99**] showed that (S_n, N) is realizable for $N = S_n$ or $A_n \times C_2$.

Tsang [**Tsa19b**] proved:

THEOREM 3.18. *For $n = 5$ or $n \geq 7$, (S_n, N) is realizable only for $N = S_n$ and $A_n \times C_2$.*

The case $n = 5$ had previously been done in [**CRV18**].

For $n = 6$, [**CC99**] and [**Tsa19b**] together yield

THEOREM 3.19. *(S_6, N) is realizable for $N = S_6, A_6 \times C_2$ and M_{10} and not realizable for any other group.*

M_{10} is the Mathieu group of degree 10. It has a subgroup of index 2 isomorphic to A_6. So this result supports Byott's conjecture that if G is not solvable and (G, N) is realizable, then N is not solvable. (See Section 3.8 below.)

Crespo, Rio and Vela [**CRV18**] also look at the case $n = 4$:

THEOREM 3.20. *(A_4, N) is realizable only for $N = A_4$ and $C_3 \times V_4$; (S_4, N) is realizable only for $N = S_4$ and the direct products $A_4 \times C_2, S_3 \times V_4$ and $C_6 \times V_4$.*

3.7. Cases where given G, N can be any group with $|G| = |N|$

Non-trivial examples include

- G has order pq where p, q are primes and $p \equiv 1 \pmod q$ [**Byo04b**]
- $G = \mathrm{Hol}(C_p)$ where p is a safeprime, so $|G| = n = 2pq$ with q prime [**Chi03**]. For such n Kohl [**Koh13**] determined $e(G, [N])$ for all 36 possible choices (G, N) of order $n = 2pq$ where p and $q = (p-1)/2$ are prime. Only four choices were not realizable. See Subsection 6.4.2 below. These results were generalized by Alabdali and Byott [**AB20b**, Section 9], who let n be the product of any three distinct primes. They also found that exactly four of the 36 possible pairs were not realizable.
- G is any cyclic group of square-free order [**AB18**]. This follows from a result of Murty and Murty [**MM84**, Lemma 3.5] that every group N of square-free order n decomposes into a semidirect product. Then for $G = C_n$, a Hopf-Galois structure of type N can be constructed via the method of fixed point free pairs of homomorphisms (see Chapter 4).

Alabdali and Byott [**AB18**] went far beyond that existence result for G square-free of order n to get a precise count of how many Hopf-Galois structures of given type there are on a cyclic Galois extension of square-free order. See Theorem 4.30. For more results on Hopf-Galois structures on groups of squarefree order, see also [**AB20b**].

In [**Koh19**], Kohl determined the number of pairs (G, N) of groups of order n for which $e(G, [N]) = 0$ for each $n \leq 180$. For example, there are 45 groups of order 108, giving $45^2 = 2025$ pairs (G, N): he found that $e(G, [N]) = 0$ for 1298 of those pairs.

Kohl found all $n \leq 180$ for which $e(G, [N]) > 0$ for every pair (G, N) of groups of order n. They are the 138 numbers < 180 missing in Table 7 of [**Koh19**]. These include (of course) all n where there is only one group of order n (n prime and also n squarefree where Dickson's Theorem 4.29 [**Dic05**] applies, such as $n = 15, 33, 35, 51, \ldots$), and all $n = pq$ where p and q are primes and $p \equiv 1 \pmod{q}$ (c.f. [**Byo04b**]). The other numbers $n < 60$ where $e(G, [N]) > 0$ for every pair of groups (G, N) of order n are $n = 28, 44$.

3.8. If G is abelian or nilpotent, then N is ...?

In this section we give Tsang's proof [**TQ20**] of:

THEOREM 3.21. *Let L/K be a G-Galois extension of fields and suppose L/K is also a Hopf-Galois extension of type N.*

1) *If G is abelian, then N is metabelian.*
2) *If G is nilpotent, then N is solvable.*

As we will discuss below, 1) was previously known. But the second conclusion of Tsang's Theorem 3.21 was not. It is the reverse of Theorem 3.9, and solves Vendramin's Problem 47 in [**Ven19**].

Tsang's proof uses the following facts:

LEMMA 3.22.

(i) *If G is a product of two nilpotent groups, then G is solvable.*
(ii) *Any subgroup or quotient of a solvable group is solvable.*
(iii) *If G is a product of two abelian groups, then G is metabelian.*
(iv) *Any subgroup or quotient of a metabelian group is metabelian.*

(i) is due to Kegel [**Keg61**]. (ii) is well-known. (iii) is proved by Ito [**Itô55**]; a proof of Ito's result may be found as Lemma 6.1 of [**Byo15**]. As for (iv):

LEMMA 3.23. *Every subgroup of a metabelian group is metabelian.*

PROOF. Let W be metabelian and N be a subgroup of W. Since W is metabelian, W has an abelian normal subgroup H so that W/H is abelian. For the subgroup N of W, $N \cap H$ is abelian and is normal in N. Also,

$$N/(N \cap H) \cong NH/H \subset W/H,$$

so $N/(N \cap H)$ is abelian. Thus N is metabelian. □

Here is Tsang's proof of Theorem 3.21.

PROOF. Suppose $\beta : G \to \mathrm{Hol}(N) \cong \lambda(N) \rtimes \mathrm{Aut}(N)$ is a regular embedding. Let $T = \beta(G)$. Define $b : G \to N$ by $b(g) = \beta(g)[e]$, where e is the identity element of N. Then $b : G \to N$ is a bijection by the regularity of β. Let $\pi : \mathrm{Hol}(N) \to \mathrm{Aut}(N)$ be the canonical projection and let $\pi\beta(g) = \theta_\beta(g)$, then $\theta_\beta : G \to \mathrm{Aut}(N)$ is an homomorphism.

Now $\beta(g) = \lambda(h)\theta_\beta(g)$ for some h in G. Applying $\beta(g)$ to the identity element e of N and noting that $\theta_\beta(g)$ is an automorphism of N, we have

$$b(g) = \beta(g)[e] = \lambda(h)\theta_\beta(g)[e] = \lambda(h)[e] = h,$$

and so $\beta(g) = \lambda(b(g))\theta_\beta(g)$.

We show that $\lambda(N) \cdot \pi(T) \subseteq T \cdot \pi(T)$: for n in N, let g in G so that $b(g) = n$. Then for h in G, $\theta_\beta(h) = \pi(\beta(h))$ is in $\pi(\beta(G)) = \pi(T)$. Then

$$\begin{aligned}\lambda(b(g))\theta_\beta(h) &= \lambda(b(g))\theta_\beta(g)\theta_\beta(g^{-1})\theta_\beta(h)\\ &= \beta(g)\theta_\beta(g^{-1}h) \text{ is in } T \cdot \pi(T).\end{aligned}$$

So $\lambda(b(G)) = \lambda(N) \subseteq \beta(G)\pi(\beta(G)) = T \cdot \pi(T)$. Now if G, hence T is abelian, resp. nilpotent, then $\pi(T)$ is abelian, resp. nilpotent, and so $T \cdot \pi(T)$ is metabelian, resp. solvable, hence $\lambda(N) \cong N$ is metabelian, resp. solvable (all by Lemma 3.22). □

REMARK 3.24. The first result related to Theorem 3.21 was by Byott. In [**Byo15**] he gave two proofs of

THEOREM 3.25. *Let L/K be a G-Galois extension of fields with G abelian. Suppose L/K is also a Hopf-Galois extension of type N. Then N is solvable.*

His first proof used consequences of the classification of finite simple groups. His second proof adapted results of C. H. Li [**Li03**] related to Li's solution of the problem of W. Burnside (1901) to classify primitive finite permutation groups that contain a regular abelian subgroup.

Byott's second proof reached the conclusion that N is solvable by showing that N is a subgroup of a metabelian group. So Theorem 3.21, 1) follows immediately from Byott's second proof and Lemma 3.23.

The first explicit statement of Theorem 3.21, 1) was Corollary 4.7 of [**Nas19**].

THEOREM 3.26. *Let $(A, \circ, \star)$ be a skew brace with additive group $A^\star$. If A° is abelian, then $A^\star$ is metabelian.*

Nasybullov derived it as an application of results on two-sided skew braces. That proof is not at all similar to those of Byott or Tsang.

3.8.1. Byott's conjecture. In [**Byo15**], Byott wrote: "We do not have any examples where an extension with insoluble Galois group G admits a Hopf-Galois structure of soluble type"

In [**TQ20**], Tsang and Qin referred to that remark as

Byott's conjecture: (G, N) is not realizable for G insolvable and N solvable.

Tsang and Qin [**TQ20**] presents extensive evidence in support of Byott's conjecture.

One might also ask the reverse question: Could (G, N) be realizable for G solvable and N insolvable?

But the answer to that question is yes: Byott showed in [**Byo15**] that $(A_4 \times C_5, A_5)$ is realizable, using the method of fixed point free pairs of homomorphisms. See Proposition 4.30 of Chapter 4.

For that example, and other examples of Byott where (G, N) is realizable and the two groups have different composition factors, see Chapter 4.

3.9. Kohl's non-existence theorem

Of an entirely different character than anything else in this chapter is a result of Kohl [Ko19]:

THEOREM 3.27. *Let G and N be groups of the same order. If for some m, N has more characteristic subgroups of order m than G has subgroups of order m, then (G, N) is not realizable.*

Kohl's proof uses the Galois correspondence for Hopf-Galois extensions and the Fundamental Theorem of Galois Theory for classical Galois extensions, so is deferred to Section 7.2.

3.10. The case G metabelian and radical algebras

In Section 3.8 we showed that if G is abelian, and (G, N) is realizable, then N must be metabelian. Here we consider the question: For which groups G is there an abelian N so that (G, N) is realizable? Equivalently, if L/K is G-Galois, does L/K have an abelian Hopf-Galois structure? The question seems to be open. Note that the corresponding question for G solvable has a negative answer: there is a p-group G for which there is no abelian group N so that (G, N) is realizable, by Bachiller's counterexample [**Bac16**] (Section 3.2).

Here are some related results.

Let A be a finite radical (hence nilpotent) algebra. Then $(A, \circ, +)$ is a brace, so (A°, A^+) is realizable. So if a group G is isomorphic to the adjoint group of a radical ring, then, since A^+ is abelian, the answer to the question is affirmative.

Recall that a finite nilpotent group G has a central series of finite length of the form

$$G \supset G_1 \supset \ldots \supset G_{c-1} \supset G_c = \{e\}$$

where G_{i-1}/G_i is contained in the center of G/G_i for $i = 1, \ldots, c$. The smallest c for which a central series exists is the class of G. [**Hal59**, Chapter 10]

Suppose A is a finite nilpotent ring so that $A^{n-1} \neq 0$ while $A^n = 0$ (that is, the product of any n elements of A is 0, but there is a product of $n-1$ elements of A which is non-zero). Then A° is a finite nilpotent group of class at most $n-1$, because for each k, A^k is a normal subgroup of A° and (recalling that $a \circ b = a + b + ab$)

$$A \supset A^2 \supset \ldots \supset A^{n-1} \supset A^n = 0$$

is a central series of A°.

In particular, if A is a finite nilpotent ring with $A^3 = 0$, then A° is a nilpotent group of class 2, hence metabelian. So such an A yields a realizable pair (A°, A^+) where A^+ is abelian (of course) and A° is metabelian.

Ault and Watters [**AW73**] give a partial answer to our question.

THEOREM 3.28. *Let G be a nilpotent group of class 2. There exists a radical algebra A with $A^3 = 0$ and adjoint group $A^\circ \cong G$. Hence there is an abelian group N so that (G, N) is realizable.*

But there exist realizable pairs of p-groups (G, N) where N is abelian and G is nilpotent of class n for any n. R. Kruse [**Kru70**] gives the following example of a radical algebra A with p^{2n} elements so that A° is nilpotent of class n. Let N be the additive group of a radical algebra A on two generators over $\mathbb{Z}/p^n\mathbb{Z}$ with p^{2n} elements, generated as an algebra by a, b with $a^2 = ab = pa, b^2 = ba = pb$ and $p^n a = p^n b = 0$. Then the adjoint group $G = A^\circ$ has a shortest upper central series

$$G = A \supset pA \supset \ldots \supset p^{n-1}A \supset p^n A = 1$$

where $c = n$ is the class of the group A°.

On the other hand, Kruse [**Kru70**] has shown that there exists a nilpotent group G of class 3 that is not the circle group of a radical ring. For such a group G, could there be an abelian group N so that (G, N) is realizable?

In general, the question "if G is metabelian, does there exist N abelian so that (G, N) is realizable?" appears to be open.

3.11. Other realizability results

3.11.1. If (G, N) is realizable then (N, G) is also? The answer is often no. For example, as Kohl [**Koh13**] showed, for $n = 2pq$ where $p = 2q + 1$ and p, q are odd primes, then (G, N) is not realizable in the four (of 36) cases where G is D_{pq} or $D_q \times C_p$ and $N = C_p \rtimes C_{p-1}$ or $(C_p \rtimes C_q) \times C_2$. See Subsection 6.4.2 below.

Also, for $p = 2$, there are cases where G is cyclic of order 2^n and N is not cyclic, and (G, N) is realizable but (N, G) is not, as can be seen by comparing Byott's results in the former case with Rump's in the latter case.

Also, as noted in Section 3.9, $(A_4 \times C_5, A_5)$ is realizable, but $(A_5, A_4 \times C_5)$ is not because A_5 is simple (Theorem 3.15).

3.11.2. Bi-skew braces. But there are circumstances where if (G, N) is realizable, then so is (N, G). One setting where this occurs is where a skew brace corresponding to the pair (G, N) is a *bi-skew brace*. A bi-skew brace is a set B with two group operations $\circ$ and $\star$ so that B is a skew brace with either group acting as the additive group. Then both $(B^\circ, B^\star)$ and $(B^\star, B^\circ)$ are realizable.

There are at least three non-trivial general settings for bi-skew braces. One is where N is the semidirect product of two groups, $N = H \rtimes J$, and $G = H \times J$. Another is where N is the additive group and G is the adjoint (circle) group of a finite $\mathbb{F}_p$-algebra A with $0 = A^3 = \{a_1a_2a_3 \mid a_1, a_2, a_3 \in A\}$. A third is where G is any finite group with an abelian endomorphism (i. e. an endomorphism with an abelian image). See Sections 7 and 8 of Chapter 4 for details.

A final observation related to Section 3.10: any finite group G of square-free order n is metabelian, in fact is a semidirect product of abelian groups. So (C_n, G) is realizable by the method of fixed point free pairs of homomorphisms (see Chapter 4), and yields a bi-skew brace, hence (G, C_n) is realizable. So any G-Galois extension with G metabelian of squarefree order has an abelian Hopf-Galois structure.

CHAPTER 4

Hopf-Galois structures arising from fixed point free pairs of homomorphisms

4.1. Introduction

Let L/K be a G-Galois extension of fields. In Chapter 2 we showed that Hopf-Galois structures on L/K of type N correspond to regular subgroups of $\mathrm{Perm}(G)$ that are isomorphic to N and normalized by $\lambda(G)$. If $\alpha : N \to M \subset \mathrm{Perm}(G)$ is an isomorphism and M is normalized by $\lambda(G)$, then α corresponds to a regular embedding β from G to the holomorph $\mathrm{Hol}(N) \subset \mathrm{Perm}(N)$. The condition that two regular embeddings $\alpha, \alpha' : N \to \mathrm{Perm}(G)$ have the same image defines an equivalence relation on the corresponding regular embeddings $\beta, \beta' : G \to \mathrm{Hol}(N)$:

DEFINITION 4.1. Two regular embeddings β and β' are equivalent, $\beta \sim \beta'$, if there exists θ in $\mathrm{Aut}(N)$ so that for every g in G, $\theta\beta(g)\theta^{-1} = \beta'(g)$.

Inside $\mathrm{Aut}(N)$ is $\mathrm{Inn}(N)$, the group of inner automorphisms of N, the image in $\mathrm{Aut}(N)$ of the conjugation map $C : N \to \mathrm{Aut}(N)$ given by $C(g)(x) = gxg^{-1}$. The map C yields an exact sequence

$$1 \to Z(N) \to N \to \mathrm{Aut}(N) \to \mathrm{Out}(N) \to 1$$

where the kernel of C is $Z(N)$, the center of N, and the image of $C : N \to \mathrm{Aut}(N)$ is the group $\mathrm{Inn}(N)$ of inner automorphisms of N, a normal subgroup of $\mathrm{Aut}(N)$. The group $\mathrm{Out}(N) = \mathrm{Aut}(N)/\mathrm{Inn}(N)$ is the group of outer automorphisms of N.

Let $\mathrm{InHol}(N) = \lambda(N)\,\mathrm{Inn}(N) \subset \mathrm{Hol}(N)$.

In this chapter we primarily focus on regular embeddings of G into $\mathrm{InHol}(N)$. For abelian groups this focus is unhelpful, because $\mathrm{Inn}(N)$ is trivial and we obtain only the classical Galois structure. But for N nonabelian we obtain some interesting results, as we'll see below.

In Section 4.8 we return to radical algebras and skew braces.

We begin by describing $\mathrm{InHol}(N)$.

PROPOSITION 4.2. $\mathrm{InHol}(N) = \lambda(N)\rho(N) \subset \mathrm{Perm}(N)$.

PROOF. For m, n, x in N, in $\mathrm{Perm}(N)$ we have

$$\lambda(m)C(n)(x) = mnxn^{-1} = \lambda(mn)\rho(n)(x),$$

while

$$\lambda(m)\rho(n)(x) = mxn^{-1} = mn^{-1}nxn^{-1} = \lambda(mn^{-1})C(n)(x).$$

□

Note that $\lambda(N)$ and $\rho(N)$ centralize each other in $\mathrm{Perm}(N)$. Thus

PROPOSITION 4.3. *The map*

$$j : N \times N \to \mathrm{InHol}(N) = \lambda(N)\rho(N)$$

given by $j : (m,n) \mapsto \lambda(m)\rho(n)$ *is surjective, and is an isomorphism of groups if and only if* $Z(N) = \{e\}$.

PROOF. The last proof shows that j is surjective. Then $j(m,n)$ is the identity permutation of N if and only if $\lambda(m)\rho(n)(x) = x$ for all x in N, if and only if $mxn^{-1} = x$ for all x. Setting $x = e$ yields $m = n$, and $mxm^{-1} = x$ for all x in N if and only if m is in $Z(N)$. So the kernel of j is the set $\{(m,m) \in N \times N \mid m \in Z(N)\}$. □

4.2. Fixed point free pairs of homomorphisms

Most of this section comes from [**BC12**].

Let L/K be a G-Galois extension where G has order n, and let N be a group of order n. To find Hopf-Galois structures of type N on L/K, we can look for regular embeddings of G into $\lambda(N)\rho(N) = \mathrm{InHol}(N)$. To do so, we look at pairs (f_1, f_2) of homomorphisms from G to N with the following property.

DEFINITION 4.4. Given two homomorphisms $f_1, f_2 : G \to N$ where G and N have the same finite order, (f_1, f_2) is a fixed point free pair if $f_1(g) = f_2(g)$ only for $g = e_G$, the identity element of G.

PROPOSITION 4.5. *Let* $\beta : G \to \lambda(N)\rho(N) \subset \mathrm{Perm}(N)$ *be given by* $\beta(g) = \lambda(f_1(g))\rho(f_2(g))$ *for homomorphisms* $f_1, f_2 : G \to N$. *Then* β *is regular if and only if* (f_1, f_2) *is a fixed point free pair.*

PROOF. A homomorphism $\beta : G \to \mathrm{Perm}(N)$ is regular if and only if

$$\{\beta(g)(e_N) \mid g \in G\} = N,$$

if and only if

$$\{g \in G \mid \beta(g)(e_N) = e_N\} = \{e_G\}.$$

Letting

$$\beta(g) = \lambda(f_1(g))\rho(f_2(g)),$$

we have

$$\begin{aligned}\{g \in G \mid \beta(g)(e_N) = e_N\} &= \{g \in G \mid \lambda(f_1(g))\rho(f_2(g))(e_N) = e_N\} \\ &= \{g \in G \mid f_1(g)f_2(g)^{-1} = e_N\} \\ &= \{g \in G \mid f_1(g) = f_2(g)\}\end{aligned}$$

So β is regular if and only if the only g in G for which $f_1(g)f_2(g)^{-1} = e_N$ is $g = e_G$, if and only if (f_1, f_2) is a fixed point free pair. □

For a fixed point free pair of homomorphisms (f_1, f_2), let β_{f_1,f_2} be the corresponding regular embedding of G into $\lambda(N)\rho(N)$.

If the center $Z(N)$ of N is non-trivial, then the map from fixed point free pairs (f_1, f_2) to β_{f_1,f_2} need not be one-to-one:

PROPOSITION 4.6. *Let* (f_1, f_2) *and* (f_1', f_2') *be two fixed point free pairs from* G *to* N. *Then* $\beta_{f_1,f_2} = \beta_{f_1',f_2'}$ *if and only if there is a homomorphism* $\zeta : G \to Z(N)$ *so that* $f_1'(g) = \zeta(g)f_1(g)$ *and* $f_2'(g) = \zeta(g)f_2(g)$.

PROOF. If $\beta_{f_1,f_2}(g) = \beta_{f_1',f_2'}(g)$ for all g, then for all x in N,

$$f_1(g)xf_2(g)^{-1} = f_1'(g)xf_2'(g)^{-1},$$

so

$$f_1(g)^{-1}f_1'(g)x = xf_2(g)^{-1}f_2'(g).$$

Setting $x = e_N$ yields

$$f_1(g)^{-1}f_1'(g) = f_2(g)^{-1}f_2'(g)$$

Setting $\zeta(g) = f_1(g)^{-1}f_1'(g) = f_2(g)^{-1}f_2'(g)$, then $\zeta(g)x = x\zeta(g)$ for all x in N. Thus $\zeta : G \to Z(N)$ is a homomorphism and $f_1'(g) = f_1(g)\zeta(g)$ and $f_2'(g) = f_2(g)\zeta(g)$, as claimed. The converse is routine. □

Now we check equivalence of β_{f_1,f_2} and $\beta_{f_1',f_2'}$. Recall from Definition 4.1 that β_{f_1,f_2} and $\beta_{f_1',f_2'}$ are equivalent, $\beta_{f_1,f_2} \sim \beta_{f_1',f_2'}$, if there exists θ in $\mathrm{Aut}(N)$ so that $\theta\beta_{f_1,f_2}(g)\theta^{-1} = \beta_{f_1',f_2'}(g)$ for all g in G.

PROPOSITION 4.7. *Given two fixed point free pairs (f_1, f_2) and (f_1', f_2'), $\beta_{f_1,f_2} \sim \beta_{f_1',f_2'}$ if and only if there exists an automorphism θ of N and a homomorphism $\zeta : G \to Z(N)$ so that $f_1'(g) = \theta(f_1(g))\zeta(g)$ and $f_2'(g) = \theta(f_2(g))\zeta(g)$.*

PROOF. Suppose $\beta_{f_1,f_2} \sim \beta_{f_1',f_2'}$. Then there is some θ in $\mathrm{Aut}(N)$ so that for all g in G,

$$\lambda(f_1'(g))\rho(f_2'(g)) = \theta\lambda(f_1(g))\rho(f_2(g))\theta^{-1}.$$

Thus for x in N,

$$f_1'(g)xf_2'(g)^{-1} = \theta(f_1(g))x\theta(f_2(g)^{-1}).$$

So for $x = e_N$ we have

$$\theta(f_1(g)^{-1})f_1'(g) = \theta(f_2(g)^{-1})f_2'(g)$$

Let $\zeta(g)$ denote either side of this last equality. Then ζ is a homomorphism from G to N and $\zeta(g)$ commutes with all x in N, so $\zeta(g)$ maps to the center of N, and $f_2'(g) = \theta(f_2(g))\zeta(g)$, $f_1'(g) = \theta(f_1(g))\zeta(g)$. The converse is routine. □

This result yields the following counting formula:

COROLLARY 4.8. *Suppose L/K is a G-Galois extension, and let N be a group with trivial center with $|N| = |G|$. Let $e_f(G, N)$ be the number of Hopf-Galois structures of type N arising from fixed point free pairs of homomorphisms from G to N, and let $p_f(G, N)$ denote the number of fixed point free pairs of homomorphisms from G to N. Then*

$$p_f(G, N) = |\mathrm{Aut}(N)|e_f(G, N).$$

4.3. The case $N = G$

When $G = N$, fixed point free pairs of homomorphisms are pairs of endomorphisms of G. In this section we consider examples in that setting.

EXAMPLE 4.9. To begin with the most trivial examples, let G be any finite group, let $\iota : G \to G$ be the identity map and $\varepsilon : G \to G$ be the trivial map, $\varepsilon(g) = 1$ for all g in G. Then (ι, ε) is a fixed point free pair of homomorphisms, obviously. For the pair (ε, ι), then

$$\beta(g)(x) = \lambda(\varepsilon(g))\rho(\iota(g)) = \rho(g).$$

So in this case $\beta = \rho$, and the pair (ε, ι) yields the classical Hopf-Galois structure by the Galois group. . On the other hand, letting $f_1 = \iota$, $f_2 = \varepsilon$, then

$$\beta(g) = \lambda(\iota(g))\rho(\varepsilon(g)) = \lambda(g),$$

so the pair (ι, ε) yields the canonical non-classical Hopf-Galois structure by the K-Hopf algebra H_λ. (But see Proposition 4.15 for other possible Hopf-Galois structures by H_λ.)

For a regular embedding $\beta : G \to InHol(G)$ defined by a fixed point free pair (f_1, f_2) of endomorphisms of G,

$$\beta(g)(x) = \lambda(f_1(g))\rho(f_2(g))(x) = f_1(g)xf_2(g)^{-1},$$

we'll write $\beta \leftarrow (f_1, f_2)$. If $\beta' \leftarrow (f_1', f_2')$, we'll say that (f_1', f_2') and (f_1, f_2) are equivalent if the corresponding maps β and β' are equivalent, as described in Definition 4.1.

The following result is [**CC07**, Theorem 4.5]:

THEOREM 4.10. *Let G be a finite group with trivial center. Suppose that in every fixed point free pair (f_1, f_2) of endomorphisms of G one of f_1, f_2 is an automorphism. If f_2 is an automorphism, then every fixed point free pair (f_1, f_2) of endomorphisms is equivalent to the pair $(f_1 f_2^{-1}, \iota)$ where $\iota : G \to G$ is the identity automorphism. The number of equivalence classes of fixed point free pairs of endomorphisms is equal to twice the number of non-surjective fixed point free endomorphisms of G, plus the number of fixed point free automorphisms of G.*

PROOF. Let $f_1, f_2 : G \to G$ form a fixed point free pair of endomorphisms, and suppose θ is an automorphism of G. Then (f_1, f_2) is equivalent to $(\theta f_1, \theta f_2)$.

In particular, if $\beta \leftarrow (f_1, f_2)$ and f_2 is an automorphism, then β is equivalent to $\beta' \leftarrow (f', \iota)$ where $f' = f_2^{-1} f_1$. If f_1 is also an automorphism, then β is also equivalent to $\beta'' \leftarrow (\iota, f'')$ where $f'' = f_1^{-1} f_2$. Thus if G has a fixed-point free automorphism θ, then the pairs (θ, ι) and (ι, θ^{-1}) are equivalent, so each fixed point free automorphism of G yields a single regular embedding of G into InHol(G) up to equivalence. Otherwise, for a given non-surjective fixed point free endomorphism f, the pair (f, ι) is equivalent only to $(\theta f, \theta)$ for some automorphism θ of G, hence is never equivalent to a pair in which the right element of the pair is not an automorphism. □

EXAMPLE 4.11. Let G be a simple non-abelian group. To determine the number of regular embeddings of G to Hol(G) up to equivalence involves two consequences of the classification of finite simple groups, namely (1) the group Out(G) of outer automorphisms of G is solvable, and (2) G has no fixed-point free automorphisms. Fact (1) implies that for G simple non-abelian, every regular embedding β of G to Hol(G) followed by the canonical map from Hol(G) to Aut(G) to Out(G) = Aut(G)/ Inn(G) must be trivial. Hence the image of β must lie in InHol(G). So β arises from a fixed point free pair of homomorphisms $f_1, f_2 : G \to G$. Now since G is simple, f_1 and f_2 must either be an automorphism or be trivial. The equivalence of embeddings under the action of Aut(G) yields that we can assume one of f_1 and f_2 is the identity, which then implies that the other is trivial by Fact (2). So Theorem 4.10 implies that there are only two regular embeddings of G into Hol(G), up to equivalence, namely those given by λ and $\rho : G \to$ Perm(G), the left and right regular representations of G, as in Example 4.9. (This result is in [**CC99**]; see also [**Chi00**, Section 10].)

EXAMPLE 4.12. Cindy Tsang [**Tsa19a**] has a remarkable extension of the Carnahan-Childs result [**CC99**] that $e(G, G) = 2$ for G a finite non-abelian simple group. Her result is:

THEOREM 4.13. *Let $G = T \times T \times \cdots \times T$, a direct product of k copies of a finite non-abelian simple group. Then $e(G, G) = 2^k(k|\operatorname{Aut}(T)| + 1)^{k-1}$.*

We outline the proof.

Let T_i be the i-th factor of G for $i = 1, \ldots, k$ and let T_0 be the trivial group.

To obtain this result, the author proves that there are no regular subgroups of $\operatorname{Hol}(G)$ isomorphic to G that are not contained in $\operatorname{InHol}(G)$. That argument involves showing that if the image of G in $\operatorname{Aut}(G)$ via the map $G \to \operatorname{Hol}(G) \to \operatorname{Aut}(G)$, where the latter map is the canonical projection, is not contained in $\operatorname{Inn}(G)$, then the order of the simple group T can only be divisible by the primes 2 and 3, contradicting Burnside's theorem that a group of order divisible by at most two primes must be solvable. Thus every Hopf-Galois structure arises from a fixed point free pair endomorphisms of G.

Therefore $e(G, G)$ is equal to the number of equivalence classes of fixed point free pairs of homomorphisms from G to G.

To consider fixed point free pairs, Tsang starts with the description of the automorphisms of G described by Byott in [**Byo04a**] (see Section 3.5): let T, G as above, and let α be an automorphism of G. Let $\iota_i : T \to G$ and $\pi_j : G \to T$ be the inclusion from the ith summand of G to G, resp., the projection from G to the j-th summand of G, and let $\alpha_{i,j} = \pi_j \alpha \iota_i : T \to T$. Then for each j with $1 \leq j \leq k$ there is a unique $i = \sigma(j)$ so that $\alpha_{i,j} = \alpha_{\sigma(j),j}$ is an automorphism of T.

Tsang adapts Byott's observation to obtain all regular embeddings $\beta : G \to \operatorname{InHol}(G)$ by looking at pairs of maps (f, g) for $f, g : G \to G$ of the form

$$f(x_1, x_2, \ldots, x_k) = (f_1(x_{\theta_f(1)}), \ldots, f_k(x_{\theta_f(k)}))$$
$$g(x_1, x_2, \ldots, x_k) = (g_1(x_{\theta_g(1)}), \ldots, g_k(x_{\theta_g(k)}))$$

where $\theta_f = (\theta_f(1), \ldots, \theta_f(k))$ is some k-tuple of elements of $\mathbb{G}_{0,k} = \{0, 1, \ldots, k\}$.

Here $f_i : T_{\theta_f(i)} \to T_i$ is the trivial map if $\theta_f(i) = 0$ and f_i is an isomorphism if $\theta_f(i) > 0$. (Byott's analysis looks only at automorphisms of G. To study endomorphisms one needs to include the case that an endomorphism $f : G \to G$ can be a non-isomorphism, hence trivial, on some components T.)

Similarly for θ_g. The data (θ_f, θ_g) then yield an undirected graph $\Gamma_{(f,g)}$ on the set of vertices $\mathbb{G}_{0,k}$. The graph has k edges, between the vertex $\theta_f(i)$ and $\theta_g(i)$ for $i = 1, \ldots, k$.

Tsang proves that if T has no fixed point free automorphisms, then a pair of maps $(f, g) : G \to G$ is fixed point free if and only if the undirected graph $\Gamma_{(f,g)}$ is a tree. Thus fixed point free pairs correspond to trees. The formula given above for $e(G, G)$ is then an application of a counting formula for trees on $k + 1$ vertices of [**Cla58**].

For other related work of Tsang, see [**Tsa19c**],[**Tsa20**],[**Tsa21**].

4.4. The action of $L[N]^G$ on L

Let L/K be a Galois extension of fields with Galois group G, let N be a group with $|G| = |N|$ and suppose $\beta : G \to \text{Hol}(N) \subset \text{Perm}(N)$ is a regular embedding, with corresponding embedding $\alpha : N \to \text{Perm}(G)$. Then N gives rise to a Hopf-Galois structure on L/K by the K-Hopf algebra $H = L[\alpha(N)]^G$. We describe this action.

PROPOSITION 4.14. *Let L/K, G, N, β, H be as above. Then H acts on L as follows. For $\xi = \sum_{n\in N} s_n\alpha(n)$ in H where s_n is in L, then for s in L,*

$$\xi(s) = \sum_{n\in N} s_n b^{-1}(n^{-1})(s)$$

Here b^{-1} is the inverse of the bijection $b : G \to N$ given by $b(g) = \beta(g)(e)$, where e is the identity element of N (See Chapter 2).

PROOF. The regular embedding $\beta : G \to \text{Hol}(N)$ yields the regular embedding $\alpha : N \to \text{Perm}(G)$ via

$$\alpha(n)(g) = b^{-1}(\lambda(n)(b(g))).$$

For s in L, the image of s in $\text{Map}(G, L)$ is $\sum_g g(s)x_g$ (where $\{x_g \mid g \in G\}$ is a dual basis to N) and ξ in $H \subset L[\alpha(N)]$ acts on $L \subset \text{Map}(N, L)$ by

$$\begin{aligned}\xi\left(\sum_{g\in G} g(s)x_g\right) &= \sum_n s_n\alpha(n)\left(\sum_g g(s)x_g\right)\\ &= \sum_{n,g} s_n g(s)x_{\alpha(n)(g)}.\end{aligned}$$

Since H maps L to itself, $\xi(\sum_g g(s)x_g)$ has the form $\sum_g g(c)x_g$, the image in GL of an element c of L. Thus for s in L, $\xi(s) = c$, the coefficient of x_1 in the last sum.

To understand what c is, we note that $b : G \to N$ is a bijection that maps 1 in G to the identity element e of N, since $b(g) = \beta(g)(e)$ and β is a homomorphism. So

$$1 = \alpha(n)(g) = b^{-1}(\lambda(n)b(g))$$

iff

$$nb(g) = e,$$

iff

$$g = b^{-1}(n^{-1}).$$

So the coefficient of x_1 is

$$\sum_{(n,g),\alpha(n)(g)=1} s_n g(s) = \sum_{n\in N} s_n b^{-1}(n^{-1})(s).$$

and therefore, given $\xi = \sum_{n\in N} s_n\alpha(n)$ in H, s in L,

$$\xi(s) = \sum_{n\in N} s_n b^{-1}(n^{-1})(s) \text{ in } L.$$

□

The description of the action of H on L is particularly clear when the Hopf-Galois structure comes from a pair of abelian fixed point free endomorphisms. An endomorphism $f : G \to G$ of a group G is abelian if $f(G)$ is an abelian subgroup of G. Caranti [**Car13**] proved that any abelian fixed point free endomorphism ψ on a finite group G has a unique quasi-inverse $\overline{\psi}$ which is also an abelian fixed point free endomorphism of G. The endomorphisms ψ and $\overline{\psi}$ satisfy the relationship

$$\psi(\overline{\psi}(g)) = \overline{\psi}(g)\psi(g).$$

Suppose L/K is G-Galois and has a Hopf-Galois structure of type G defined by a fixed point free pair of endomorphisms $f_1, f_2 : G \to G$. Then $\beta : G \to \mathrm{Hol}(G)$ is defined by

$$\begin{aligned}\beta(g) &= \lambda(f_1(g))\rho(f_2(g))\\ &= \lambda(f_1(g)f_2(g^{-1}))C(f_2(g))\\ &= \rho(f_2(g)f_1(g^{-1}))C(f_1(g)),\end{aligned}$$

as is easily checked.

Then Proposition 4.14 specializes to:

PROPOSITION 4.15. *Let L/K be a G-Galois extension that has a Hopf-Galois structure of type G, corresponding to a regular embedding $\beta : G \to \mathrm{Hol}(G)$. Suppose β is defined by a fixed point free pair of endomorphisms (f_1, f_2). If $f_1 = \iota$, the identity endomorphism, then the Hopf-Galois structure on L/K corresponding to $\alpha(G)$ is given by the canonical non-classical K-Hopf algebra $H = L[\lambda(G)]^{\lambda(G)}$, which acts on L via an action uniquely determined by f_2.*

PROOF. We first identify H.

Recall that a regular embedding $\beta : G \to \mathrm{Hol}(N)$ corresponds to a regular embedding $\alpha : N \to \mathrm{Perm}(G)$ where $\alpha(N)$ is normalized by $\lambda(G)$. The action of $\lambda(G)$ on $\alpha(N)$ translates to an action of $\beta(G)$ on $\lambda(N)$, by

$$C(b)(\lambda(g)\alpha(n)\lambda(g^{-1})) = \beta(g)\lambda(n)\beta(g^{-1}).$$

The K-Hopf algebra H corresponding to $\alpha(N)$ is

$$\begin{aligned}H &= (L[\alpha(N)])^{\lambda(G)}\\ &= \left\{\sum_{n\in N} s_n\alpha(n) \;\middle|\; \sum_{n\in N} s_n\alpha(n) = \sum_{n\in N} g(s_n)\lambda(g)\alpha(n)\lambda(g^{-1})\right\}\end{aligned}$$

for all g in G. Applying $C(a^{-1})$ to the elements of $\alpha(N)$ on both sides of that defining equation gives

$$\begin{aligned}\sum_n s_n C(a^{-1})\alpha(n) &= \sum_n g(s_n)C(a^{-1})\lambda(g)\alpha(n)\lambda(g^{-1})\\ &= \sum_{n\in N} g(s_n)C(a^{-1})\lambda(g)C(a^{-1})\alpha(n)C(a^{-1})\lambda(g^{-1}),\end{aligned}$$

and since $C(a^{-1})\alpha = \lambda$ and $C(a^{-1})\lambda = \beta$, the defining equation for H becomes

$$\sum_n s_n\lambda(n) = \sum_n g(s_n)\beta(g)\lambda(n)\beta(g^{-1})$$

for all g in G. So

$$H = (L[\alpha(N)])^{\lambda(G)} \cong (L[\lambda(N)])^{\beta(G)}.$$

Now suppose $G = N$ and $\beta(g) = \lambda(f_1(g))\rho(f_2(g))$ for a fixed point free pair (f_1, f_2) of endomorphisms of N. To see how $\beta(G)$ acts on $\lambda(N)$ (and recalling that $\rho(N)$ centralizes $\lambda(N)$ in $\mathrm{Perm}(N)$), we have, for x, n and g in G,

$$\begin{aligned}\beta(g)\lambda(n)\beta(g^{-1})(x) &= \lambda(f_1(g))\rho(f_2(g))\lambda(n)\rho(f_2(g)^{-1})\lambda(f_1(g)^{-1})(x)\\ &= \lambda(f_1(g))\lambda(n)\lambda(f_1(g)^{-1})\rho(f_2(g))\rho(f_2(g)^{-1})(x)\\ &= \lambda(f_1(g))\lambda(n)\lambda(f_1(g)^{-1})(x).\end{aligned}$$

Thus if f_1 is the identity automorphism ι of G, then

$$H = \left\{\sum_{g\in G} s_g g \;\middle|\; \sum_{g\in G} s_g g = \sum_{g\in G} n(s_g)s_g\lambda(gng^{-1}) \text{ for all } n \text{ in } G\right\} = H_\lambda.$$

Now L lives inside $GL = \mathrm{Map}(G, L)$ by $s \mapsto \sum_g g(s)x_g \in GL = \{\sum_{g\in G} Lx_g\}$, and as in the proof of Proposition 4.14, we can identify an element of GL that is known to be in L by just taking the coefficient of x_1.

Suppose $\beta(g) = \lambda(g)\rho(f_2(g))$, where f_2 is an abelian fixed point free endomorphism of G. Then, by [**Car13**], f_2 has an unique quasi-inverse, $\overline{f}_2$, with the property that if $\beta(g) = \lambda(g)\rho(f_2(g))$, then $\alpha(g) = \lambda(g)\rho(\overline{f}_2(g))$. So by Proposition 4.14,

$$a(g) = \alpha(g)(e) = g\overline{f}_2(g) = b^{-1}(g)$$

so by Proposition 4.14,

$$\begin{aligned}\xi(s) &= \sum_{n\in G} s_n b^{-1}(n^{-1})(s)\\ &= \sum_{n\in G} s_n n^{-1}\overline{f}_2(n^{-1})(s).\end{aligned}$$

The action of H_λ on L is therefore determined by the abelian fixed point free endomorphism f_2.

Thus each abelian fixed point free endomorphism f_2 of G yields a distinct Hopf-Galois structure of H_λ on L. □

4.5. Fixed point free endomorphisms

In this section we look at examples of fixed point free pairs of homomorphisms from G to G.

EXAMPLE 4.16. Let $G = S_n$, the symmetric group on n symbols, where $n \geq 5$. Let f be an endomorphism on S_n. Then f is either an automorphism or $f(A_n) = 1$. Given a fixed point free pair of endomorphisms of S_n we can assume one is the identity, and as shown in [**CC99**] or [**Chi00**, (10.3)], the other, f, maps A_n to the identity, in which case f maps every odd permutation to a fixed element h in S_n of order 2. Now f is a fixed point free endomorphism of S_n iff $\{f(g)g^{-1} \mid g \text{ in } S_n\} = S_n$. For g in A_n, $f(g)g^{-1} = g^{-1}$ is even. For g an odd permutation, $f(g)g^{-1} = hg^{-1}$ is odd if and only if h is even. So for f to be fixed point free, h must be even. The converse is also easy to see.

Thus the number of fixed point free endomorphisms of S_n for $n \geq 5$ is equal to the number of even permutations of order 2 in S_n, that is the number of permutations in S_n that are products of an even number of disjoint 2-cycles in S_n. See [**CC99**] for this and related results, which yield the conclusion [**CC99**, Corollary

11] that if L/K is a G-Galois extension with $G \cong S_n$, $n \geq 5$, then L/K has at least $(n!)^{1/2}$ Hopf-Galois structures.

EXAMPLE 4.17. The following class of examples are from [**Chi11**, Section 5].

Let G be the semidirect product $A \rtimes \langle\theta\rangle$ where A is a finite abelian group of order n, written additively, θ is an element of $\mathrm{Aut}(A)$ of order d and $\gcd(n, d) = 1$. One can show that G has trivial center if and only if $\theta - 1$ is injective on A.

Define an endomorphism ψ of G by

$$\psi(f, 1) = (0, 1)$$
$$\psi(0, \theta) = (h, \theta^s)$$

for some h in A and assume $\theta^s - 1$ is injective on A. Then ψ is fixed point free if and only if ψ has a cyclic image and $\gcd(s - 1, d) = 1$. Such a ψ is then an abelian fixed point free endomorphism of G, hence has a quasi-inverse $\overline{\psi}$, defined by

$$\overline{\psi}(f, 1) = (0, 1)$$
$$\overline{\psi}(0, \theta) = (g, \theta^t)$$

where t satisfies the equation

$$(s - 1)(t - 1) \equiv 1 \pmod{d}$$

and g satisfies the relation

$$(\theta^s - 1)(g) = (\theta^t - 1)(h).$$

EXAMPLE 4.18. The above result is applied to various examples in [**CC07**], where A is cyclic of odd order. In particular, let $G = C_h \rtimes C_k = \langle x, y\rangle$ where h is odd, k divides $\varphi(h)$ and $yx = x^b y$ for b an element of $\mathbb{Z}_h^\times$ of order k. Assume $\gcd(b - 1, k) = 1$, so that G has trivial center. Then the endomorphisms ψ of G have the form

$$\psi(x) = x^m$$
$$\psi(y) = x^r y^s,$$

and ψ is fixed point free if and only if $\gcd(m - 1, h) = 1$ and $\gcd(s - 1, k) = 1$. If ψ is an automorphism, then $s = 1$, so every automorphism has a fixed point.

In case h is the power of an odd prime, then an endomorphism of G has a fixed point unless $m = 0$. Thus a fixed point free endomorphism is determined by r and s and has a cyclic image, hence is abelian. Theorem 6.5 of [**CC07**] is then

THEOREM 4.19. *Let $G = C_h \rtimes C_k$ with $h = p^e$, p odd, and $k = qp^f$ with $q > 1$ coprime to p. The number of fixed point free endomorphisms of G is*

$$\varphi(p^f)p^f + \varphi(p^f)(\varphi(q) - 1)p^e.$$

For example, if $h = 25$ and $k = 10$, then there are 20 fixed point free endomorphisms, all abelian. (We refer to [**CC07**] for proofs of the above facts.)

Elaborating on ideas arising in [**Chi13**], Caranti, in [**Car13**], considers semidirect products $G = N \rtimes A$ where N is normal in G, A is abelian and the operation is

$$n_1 a_1 n_2 a_2 = n_1(a_1 n_2 a_1^{-1}) a_1 a_2 = n_1 n_2^{a_1} a_1 a_2.$$

for n_1, n_2 in N, a_1, a_2 in A.

He obtains fixed-point-free endomorphisms of G as follows. Let α be a fixed point free automorphism of A and η be a nilpotent abelian endomorphism of N

such that the commutator subgroup $[G, G] < \ker(\eta)$ and $[\alpha(A), \eta(N)] = 1$. Define an endomorphism $\psi : G \to G$ by $\psi|_A = \alpha, \psi|_N = \eta$. Then ψ is an abelian fixed-point-free endomorphism of G.

EXAMPLE 4.20. (from [**Car13**, Section 8]). Let $N = H_3$, the Heisenberg group of order p^3, p an odd prime: $N = \langle a, b\rangle$ with $[a, b] = c, [a, c] = [b, c] = 1$. Let $A = \langle \eta \rangle$, where η in $\mathrm{End}(N)$ is defined by $\eta(a) = b, \eta(b) = 1$. Define α_s in $\mathrm{Aut}(A)$ by $\alpha_s(\eta) = \eta^s$ for any $s \not\equiv 0 \pmod p$.

Let $G = N \cdot A$. Define ψ_s in $\mathrm{End}(G)$ by $\psi_s|_A = \alpha_s, \psi_s|_N = \eta$. Then $[G, G] = \langle b, c\rangle \subset \ker(\eta)$ and $[\eta(N), \alpha_s(A)] = 1$. So ψ_s is an abelian fixed-point-free endomorphism of G. There are $p - 1$ such ψ_s, all inequivalent, which yield $p - 1$ different actions of H_λ on a G-Galois extension.

[**Car13**] shows that for any finite group G, a fixed point free endomorphism ψ of G is abelian if and only if ψ has a quasi-inverse $\overline{\psi}$.

4.6. Examples with $N \neq G$

Recall that a pair of groups (G, N) is realizable if there exists a Hopf-Galois structure of type N on a G-Galois extension, or equivalently, there exists a skew brace $(B, \circ, \star)$ with $(B, \circ) \cong G$ and $(B, \star) \cong N$. Fixed-point free pairs of homomorphisms from G to N have yielded large collections of examples of realizable pairs (G, N). We survey some examples in the remainder of this chapter.

4.6.1. N a complete group. A group N is a complete group if the center $Z(N)$ of N is trivial and every automorphism of N is inner. Then every Hopf-Galois structure of type N on a G -Galois extension arises from a regular embedding of G into $\mathrm{InHol}(N) = \lambda(N)\,\mathrm{Inn}(N)$ which, since $Z(N)$ is trivial, corresponds to a unique fixed point free pair of homomorphisms from G to N.

Well-known examples of finite complete groups are $\mathrm{Aut}(A)$ where A is simple and non-abelian, S_n for $n \geq 3, n \neq 6$, and $\mathrm{Hol}(C_m)$ where m is odd. The case $N = \mathrm{Hol}(C_p)$ for p prime and $p - 1 = 2q$ with q prime (so p is a safeprime, q a Sophie Germain prime) was studied in [**Chi03**] and in [**BC12**], where it was shown that for $N = \mathrm{Hol}(C_p)$, the numbers $e_f(G, N)$ of Hopf-Galois structures arising from fixed point free pairs of homomorphisms are as in the following table:

G	$e_f(G, \mathrm{Hol}(C_p))$
$\mathrm{Hol}(C_p)$	$2(p(q-2)+1)$
D_{pq}	0
$(C_p \rtimes C_q) \times C_2$	$2p(q-1)$
$D_p \times C_q$	$2p(q-1)$
$D_q \times C_p$	0
C_{2pq}	$2(q-1)$

By comparison, for $G = \mathrm{Hol}(C_p)$, $e(G, N) > 0$ for every N of order $2pq$. But there are no fixed point free pairs of homomorphisms from G to N for $N \not\cong G$, because of [**BC12**], Proposition 5:

PROPOSITION 4.21. *If G has a unique minimal nontrivial normal subgroup and $N \not\cong G$, then there are no fixed point free pairs of homomorphisms from G to N.*

PROOF. Suppose H is the unique minimal nontrivial normal subgroup of G. If N is not isomorphic to G, then every homomorphism from G to N must have a non-trivial kernel and therefore maps H to the identity of N. Thus no pairs of homomorphisms from G to N are fixed point free. □

This applies to $G = C_{p^e} \rtimes U$ for p odd, $e \geq 1$ and any subgroup U of $\mathrm{Aut}(C_p^e)$ because $C_p < C_{p^e}$ is the unique minimal normal subgroup of G.

4.6.2. Zappa-Szép products. A large collection of examples where fixed point free pairs of homomorphisms yield Hopf-Galois structures of type N on a Galois extension with Galois group G arise in connection with what are sometimes called Zappa-Szép products.

DEFINITION 4.22. A finite group N with identity e and subgroups N_L and N_R is an *internal Zappa-Szép product* if $N = N_L N_R$ and $N_L \cap N_R = e$. Thus every element n of N can be uniquely written as $n = n_L n_R$ for n_L in N_L, n_R in N_R.

Other terminology in the literature for a Zappa-Szép product: a finite group N has complementary subgroups if there are two subgroups N_L and N_R of N so that if $|N_L| \cdot |N_R| = |N|$ and $N_L \cap N_R = e$. Or: a group N admits an exact factorization through subgroups N_L and N_R if $N = N_L N_R$ and $N_L \cap N_R = \{e\}$. [**SV18**, Section 2.1].

PROPOSITION 4.23. *Let N be a Zappa-Szép product $N = N_L N_R$, and let $G = N_L \times N_R$. Then every Galois extension with Galois group G has a Hopf-Galois structure of type N.*

PROOF. The maps $f_1(n_L, n_R) = n_L, f_2(n_L, n_R) = n_R$ from G to N form a fixed point free pair of homomorphisms from G to N, and $\beta : G \to \mathrm{InHol}(N) = \lambda(N)\rho(N)$ given by

$$\beta(n_L, n_R) = \lambda(f_1((n_L, n_R)))\rho(f_2((n_L, n_R)))$$

is a regular embedding of G into N, for

$$\beta(N_L \times N_R)(e) = \{n_L n_R^{-1} \mid n_L \in N_L, n_R \in N_R\} = N$$

by definition of a Zappa-Szép product. □

REMARK 4.24. Connecting these examples with skew braces, since a Zappa-Szép product $N = N_L N_R$ with those maps yields a Hopf-Galois structure of type N on a G-Galois extension where $G = N_L \times N_R$, one also obtains a skew brace $(N, \circ, \cdot)$, where $(N, \cdot)$ is the given group operation on $N = N_L N_R$, and $(N, \circ)$ is the direct product operation induced by β where

$$\begin{aligned}\beta(n_L, n_R)\beta(m_L, m_R)(x) &= \beta(n_L, n_R)(m_L x m_R^{-1}) \\ &= n_L m_L x m_R^{-1} m_R^{-1} \\ &= (n_L m_L) x (n_R m_R)^{-1} \\ &= \beta(n_L m_L, n_R m_R)(x).\end{aligned}$$

There are many examples that illustrate that a G-Galois extension of fields can have Hopf-Galois structures where the isomorphism type of N can range widely.

The most obvious examples of Zappa-Szép products are semidirect products. So we have:

PROPOSITION 4.25. *Let $N = HJ$ be a semidirect product, where H is normal in N and J acts on H by conjugation:*

$$(h_1j_1)(h_2j_2) = h_1(j_1h_2j_1^{-1})j_1j_2 = h_1^{j_1}h_2j_1j_2$$

for h_1, h_2 in H, j_1, j_2 in J. Then any Galois extension with Galois group $H \times J$ has a Hopf-Galois structure of type N.

This was systematically used in [**BC12**] to prove:

THEOREM 4.26. *Let p be an odd prime and L/K a G-Galois extension with G a non-cyclic abelian p-group of order p^n, $n \geq 3$. Then L/K has a non-abelian Hopf-Galois structure.*

PROOF. View $G = \mathbb{Z}_{p^{n_1}} \times \mathbb{Z}_{p^{n_2}} \times \ldots \mathbb{Z}_{p^{n_k}}$ (where $\mathbb{Z}_m$ = the additive group of integers modulo m) where $k \geq 1$ and $1 \leq n_1 \leq n_2 \leq \ldots \leq n_k$. To prove the theorem, it suffices to construct a non-trivial semidirect product $H \rtimes J$. Let $H = \mathbb{Z}_{p^{n_2}} \times \ldots \mathbb{Z}_{p^{n_k}}$, $J = \mathbb{Z}_{p^{n_1}}$. If $H \cong \mathbb{F}_p^{k-1}$ is elementary abelian, then let A in $\mathrm{GL}_{k-1}(\mathbb{F}_p) = \mathrm{Aut}(H)$ have order p and let $\alpha : J \to \mathrm{Aut}(H)$ with $\alpha(1) = A$. Then $N = H \rtimes_\alpha K$ is a non-trivial semidirect product. If $p^{n_k} \geq p^2$, then let a be coprime to p and have order p modulo p^{n_k}. Let $\alpha : J \to \mathrm{Aut}(H)$ be such that $\alpha(1)$ is diagonal multiplication by a. Then α is an automorphism of H which is non-trivial at least on the factor $\mathbb{Z}_{p^{n_k}}$. So $N = H \rtimes_\alpha J$ is a non-trivial semidirect product. □

A similar construction yields:

THEOREM 4.27. *Let L/K be a G-Galois extension with G an abelian group of even order $2^e q$ where $e \geq 1$ and $q > 1$ is odd. Then L/K has a Hopf-Galois structure of non-abelian type.*

PROOF. Let J be a cyclic direct factor of N of order a power of 2, and H the complementary direct factor. Since H has an odd direct factor, multiplication by -1 is a non-trivial automorphism of H. So we define $\alpha : J \to \mathrm{Aut}(H)$ by sending a generator of K to multiplication by -1. Then $N = H \rtimes_\alpha J$ is a non-trivial semidirect product. As before, the result follows by Proposition 4.25. □

One might ask the question, given a G-Galois extension with G abelian, does it have a non-abelian Hopf-Galois structure? The answer, in general, is "no". Of course, Byott's Uniqueness Theorem [**Byo96**] gives some examples. More generally, there is an old theorem of Dickson:

THEOREM 4.28. [**Dic05**] *Every group of order n is abelian if and only if $n = p_1^{e_1}p_2^{e_2}\cdots p_k^{e_k}$ where all $e_i \leq 2$ and p_i does not divides $p_j^{e_j} - 1$ for all i, j.*

Clearly any $n > 4$ must be odd to satisfy the hypotheses.

More generally, Byott's Theorems 3.16 and 3.17 (from [**Byo13**] and [**BC12**]) give two collection of cyclic groups G for which $e(G, N) = 0$ for $N \not\cong G$.

4.6.3. N of square-free order. Let n be square-free, and $n = de$ with $\gcd(d, e) = 1$. Let k have order d modulo e. Then

$$N(d, e, k) = \langle g, h \mid g^e = h^d = 1, hgh^{-1} = g^k \rangle$$

is a semidirect product. Murty and Murty proved [**MM84**, Lemma 3.5] that every group N of square-free order n has the form $N(d, e, k)$. That result and Proposition 4.25 imply

THEOREM 4.29. *Let L/K be Galois with group G cyclic of squarefree order n. Then L/K has a Hopf-Galois structure of type N for every group N of order n.*

A special case of the setting where N has squarefree order was studied in [**Chi03**], [**Koh13**] and [**BC12**] in the case where $n = 2pq$ where $p - 1 = 2q$ and p, q are prime (that is, p is a safeprime, or q is a Sophie Germain prime). Let $e(G, N)$ denote the number of Hopf-Galois structures of type N on a Galois extension with Galois group G.

Childs, in [**Chi03**] showed that if $G = \mathrm{Hol}(C_p)$, then for each of the six groups N of order n, $e(G, N) > 0$. Kohl, in [**Koh13**], determined $e(G, N)$ for all 36 possible choices of the number of Hopf-Galois structures, and Section 4 of [**BC12**] determined how many of the Hopf-Galois structures counted by $e(G, N)$ came from fixed point free pairs of homomorphisms from G to N.

Alabdali and Byott extended this work to the case where $|G| = n$ is squarefree, determining $e(G, N)$ for G cyclic of order n and any N in [**AB18**]:

THEOREM 4.30. *Let L/K be G-Galois with G cyclic of squarefree order n. Let N be a group of order $n = ed$ with center of order z and commutator subgroup of order g, and $d = n/(gz)$ Then L/K admits precisely $2^{\omega(g)}\varphi(d)$ Hopf-Galois structures of type N.*

Here $\omega(g)$ is the number of prime factors of g, and $\varphi(d)$ is Euler's totient function. In [**AB20b**] this result was extended to study $e(G, N)$ for all G and N of squarefree order n. They also counted all skew braces B of squarefree order in [**AB20c**, Theorem 2.2].

4.6.4. Non-semidirect products. Fixed point free pairs apply to Zappa-Szép products that are not semidirect products. Let $N = HJ$ be a Zappa-Szép product, let $G = H \times J$. Then the two homomorphisms $f_H, f_J : G \to N$ by $f_H(h, j) = h$, $f_J(h, j) = j$ form a fixed point free pair, hence any G-Galois extension has a Hopf-Galois structure of type N.

Byott [**Byo15**, Section 7] obtained some very nice results on the existence of Hopf-Galois structures by using the method of fixed point free pairs on some Zappa-Szép examples.

PROPOSITION 4.31. *There exists a G-Galois extension with a Hopf-Galois structure of type N where the composition factors of G are not the same as those for N.*

EXAMPLE 4.32. Let $N = S_n$, viewed as symmetries of a set of $n \geq 3$ points, let $H \cong S_{n-1}$ be the stabilizer of a point, and let $J = C_n$ be the cyclic subgroup generated by a fixed n-cycle. Then $N = HJ$ is a Zappa-Szép product (but not a semidirect product). So an $S_{n-1} \times C_n$-Galois extension has a Hopf-Galois structure of type S_n. In particular, for $n \geq 5$, S_n has a composition series with the simple group A_n as a composition factor, but A_n does not appear as a composition factor for $S_{n-1} \times C_n$.

PROPOSITION 4.33. *There exists a G-Galois extension with a Hopf-Galois structure of type N where G is solvable and N is a non-abelian simple group.*

EXAMPLE 4.34. Let $N = A_n$ with $n \geq 5$ and odd. Let A_{n-1} be the stabilizer of a point. Since n is odd, there is an n-cycle in A_n, and so $N = A_n = A_{n-1}C_n$ is a Zappa-Szép product. Thus a G-Galois extension with $G = A_{n-1} \times C_n$ has a Hopf-Galois structure of type A_n. For $n = 5$, G is solvable, while $N = A_5$ is simple.

4.7. Bi-skew braces and semidirect products

The examples of Hopf-Galois structures of type $H \rtimes J$ on $H \times J$-Galois extensions arising from fixed point free pairs of homomorphisms, described in previous sections, yield an interesting class of skew braces.

DEFINITION 4.35. A bi-skew brace is a set B with two operations, $\star$ and $\circ$ so that $(B, \star)$ is a group, $(B, \circ)$ is a group, and $(B, \circ, \star)$ is a skew brace with either group acting as the additive group. Thus both compatibility conditions

$$a \circ (b \star c) = (a \circ b) \star a^{-1} \star (a \circ c)$$

and

$$a \star (b \circ c) = (a \star b) \circ \overline{a} \circ (a \star c)$$

must hold for all a, b, c in B.

Given a bi-skew brace with $(B, \circ) = G, (B, \star) = N$, every G-Galois extension has a Hopf-Galois structure of type N, and every N-Galois extension has a Hopf-Galois structure of type G.

Recall from Chapter 2 that a pair (G, N) of groups of order n is realizable if every G-Galois extension of fields has a Hopf-Galois structure of type N.

If $B = (B, \circ, \star)$ is a skew brace with additive group $(B, \star)$, then $((B, \circ), (B, \star))$ is realizable.

If B is a bi-skew brace, then also $((B, \star), (B, \circ))$ is realizable.

In **[Chi19b]** two large classes of bi-skew braces were identified.

One class arises from certain finite radical rings, and will be discussed in Section 4.8. The other class involves semidirect products of finite groups.

THEOREM 4.36. *Let $N = N_L \rtimes N_R$ be a semidirect product of two finite groups N_L and N_R, where N_L is normal in N. Then N is a bi-skew brace.*

PROOF. Denote the group operation in N by $\cdot$, which we will often omit. (So for x, y in N, $xy = x \cdot y$.) In N an element x_R of N_R acts on y_L in N_L by conjugation: writing $x = x_L x_R^{-1}, y = y_L y_R^{-1}$, then

$$\begin{aligned} xy &= x_L x_R^{-1} y_L y_R^{-1} \\ &= x_L (x_R^{-1} y_L x_R) x_R^{-1} y_R^{-1}. \end{aligned}$$

Define a direct product operation $\circ$ on N by

$$\begin{aligned} x \circ y &= x_L x_R^{-1} \circ y_L y_R^{-1} \\ &= x_L y_L y_R^{-1} x_R^{-1} \\ &= x_L y x_R^{-1}. \end{aligned}$$

Then $(N, \circ, \cdot)$ is a bi-skew brace. We observed in Remark 4.24 that $(N, \circ, \cdot)$ is a skew brace with additive group $(N, \cdot)$ because $(N, \cdot)$ is a Zappa-Szép product. So to show that $(N, \circ, \cdot)$ is a bi-skew brace, we need only check the skew brace relation with $(N, \circ)$ as the additive group:

$$(*) \qquad x \cdot (y \circ z) = (x \cdot y) \circ \overline{x} \circ (x \cdot z)$$

where if $x = x_L x_R^{-1}$, then $\overline{x} = x_L^{-1} x_R$ is the $\circ$-inverse of x.

The left side of $(*)$ is

$$x \cdot (y \circ z) = x y_L z y_R^{-1},$$

while the right side of $(*)$ is

$$\begin{aligned}(xy)\circ\overline{x}\circ(xz) &= xy\circ x_L^{-1}x_R\circ xz\\ &= xy\circ x_L^{-1}xzx_R\\ &= xy\circ x_R^{-1}zx_R\\ &= (xy)_Lx_R^{-1}zx_R(xy)_R^{-1}\\ &= (x_L(x_R^{-1}y_Lx_R))x_R^{-1}zx_R(x_R^{-1}y_R^{-1})\\ &= xy_Lzy_R^{-1}.\end{aligned}$$

Hence formula $(*)$ holds and $(N,\circ,\cdot)$ is a bi-skew brace. $\square$

In [**CRV16a**], Crespo, Rio and Vela constructed so-called *induced* Hopf-Galois structures (see Section 8.3), by proving:

THEOREM 4.37. *Let $N = H \rtimes J$ be a semidirect product and let $G = H \times J$. Then every N-Galois extension of fields has a Hopf-Galois structure of type G.*

Proposition 4.36 and the results in Section 2.10 imply that a semi-direct product $N = H \rtimes J$ yields a bi-skew brace B with $(B,\circ) \cong G = H \times J$ and $(B,\star) \cong N$, which implies that every G-Galois extension has a Hopf-Galois structure of type N, and also (Theorem 4.37) that every N-Galois extension has a Hopf-Galois structure of type G. (Our proof could not be more different from the proof in [**CRV16a**]. But see Chapter 11 for discussion on induced structures as applied to Galois module theory.)

EXAMPLE 4.38. Let n be squarefree. Then every group N of order n is a semidirect product, $N = H \rtimes J$, so (C_n, N) is realizable, as noted in Theorem 4.29. It then follows from Theorem 4.36 that (N, C_n) is also realizable, that is,

COROLLARY 4.39. *Every Galois extension of squarefree order n has a Hopf-Galois structure of cyclic type.*

EXAMPLE 4.40. Moving away from the abelian case, an obvious class of examples is to take $N = \mathrm{Hol}(M) = M \rtimes \mathrm{Aut}(M)$ for any finite group M. Letting $G = M \times \mathrm{Aut}(M)$, we have that both (G,N) and (N,G) are realizable. If N is non-abelian, then both G and N are non-abelian. Thus there are non-trivial bi-skew braces in which neither group is abelian.

For a small example, let $N = H_{(p)} \rtimes \mathrm{Aut}(H_{(p)})$, and let $G = H_{(p)} \times \mathrm{Aut}(H_{(p)})$. Here $H_{(p)}$ is the Heisenberg group over $\mathbb{F}_p$, which can be identified as the subgroup of $\mathrm{GL}_3(\mathbb{F}_p)$ consisting of upper triangular 3×3 matrices with diagonal entries $= 1$, and $\mathrm{Aut}(H_{(p)})$ is isomorphic to $C_p^2 \rtimes \mathrm{GL}_2(\mathbb{F}_p)$ [**NZ19**]. This example illustrates that there exist non-trivial bi-skew braces with both additive and circle groups non-abelian.

REMARK 4.41. As noted in [**Chi19b**], the fact that semidirect products yield bi-skew braces does not extend to general Zappa-Szép products $N = HJ$ where neither H nor J is normal in N. A small example (found in [**Byo15**]) is to let $N = S_4 = S_3 \cdot C_4$ where $S_3 = \mathrm{Perm}(\{1,2,3\})$ and C_4 is generated by the cycle $(1,2,3,4)$. Letting $\star$ be the usual operation in S_4 and $\circ$ be the operation in S_4 induced from the componentwise operation on $S_3 \times C_4$, $(S_4,\circ,\star)$ is a skew brace. But the skew brace defining relation for $(S_4,\star,\circ)$ can fail to hold with $(S_4,\circ)$ as the additive group.

4.8. Bi-skew braces and nilpotent rings

Bi-skew braces can arise outside of the setting of fixed point free pairs of homomorphisms, as we will show in this section.

Given a finite ring A without unit, define the operation $\circ$ on A by $a \circ b = a + b + ab$ (where as usual we omit $\cdot$ for multiplication). Then the operation $\circ$ on A is associative, and the identity element of $(A, \circ)$ is the additive identity 0. Then A is a radical ring if and only if A with the operation $\circ$ is a group, often denoted by A° and called the adjoint group, or circle group of A. The concept of brace generalizes the concept of radical ring: in fact, a radical ring with the circle operation defined as above is the same as a two-sided brace, that is, a left brace also satisfying $(a + b) \cdot c = a \cdot c - c + b \cdot c$ for all a, b, c in A.

Define $A^n = \{a_1 \cdot a_2 \cdot \ldots \cdot a_n \mid a_1, \ldots a_n \text{ in } A\}$ Then

PROPOSITION 4.42. *A finite radical ring $A = (A, +, \cdot)$ with the operation $\circ$ defined by $a \circ b = a + b + ab$ is a bi-skew brace if and only if $A^3 = 0$.*

PROOF. One can verify that for all a, b, c in A,

$$a \circ b \circ c = (a + b + c) + (ab + ac + bc) + abc.$$

We need to show the skew brace condition where $(A, \circ)$ is the additive group:

$$a + (b \circ c) = (a + b) \circ \overline{a} \circ (a + c).$$

If it holds, then it holds modulo A^4, and modulo A^4, the circle inverse $\overline{a}$ of a is $-a + a^2 - a^3$ (where $a^n = a \cdot a \cdot \cdots \cdot a$ (n factors)).

Now the left side of the skew brace defining relation is

$$a + b + c + b \cdot c.$$

It is a routine computation to show that this is equal to the right side if and only if $b \cdot a \cdot c = 0$. □

We know that if $(A, +, \cdot)$ is a radical algebra with additive group $A^+ = (A, +)$ and adjoint group $A^\circ = (A, \circ)$, then (A°, A^+) is realizable. If also $A^3 = 0$, then also (A^+, A°) is realizable and A° is nilpotent of class 2, hence metabelian (see Section 3.11). But A° need not be a semidirect product. Here is an example, from [**Chi19b**, Remark 7.9]:

EXAMPLE 4.43. Let A be the nilpotent $\mathbb{F}_p$-algebra of dimension 6, p an odd prime, with an $\mathbb{F}_p$-basis consisting of x, y, z, a, b, c and relations $xy = a, yz = b, zx = c$ and all other products of the basis elements are $= 0$. Then $A^2 = \langle a, b, c \rangle$, a subalgebra of A on which the circle operation coincides with addition. It is easy to see that $(A^2, \circ)$ is the center of $(A, \circ)$, and since modulo A^2, multiplication is trivial on A, the circle operation on $(A, \circ)/(A^2, \circ)$ is addition. So $(A, \circ)$ is metabelian. But if A° were isomorphic to a semidirect product of two abelian subgroups, then the set A° would be the direct product of two abelian groups. But any subgroup of A° that contains two elements $g = rx + sy + tz + h, g' = r'x + s'y + t'z + h'$ for h, h' in the center of A, where g, g' are linearly independent modulo the subspace generated by a, b, c, cannot be abelian. So A° cannot be a semidirect product of two abelian subgroups.

REMARK 4.44. Caranti [**Car20**] has obtained examples of bi-skew braces that go beyond the examples in [**Chi19b**]. Also, Koch [**Koc21a**] has shown that if $G =$

$(C, \star)$ is a finite group (with $z\star z^{-1} = 1$) and $\psi : G \to G$ is an abelian endomorphism (that is, $\psi(G)$ is an abelian subgroup of G), then defining an operation $\circ$ on G by

$$w \circ z = w \star \psi(w^{-1}) \star z \star \psi(w),$$

then $(G, \circ)$ is a group and $(G, \circ, \star)$ is a bi-skew brace.

EXAMPLE 4.45. Let $(G, \star) = C_{17} \rtimes_b C_{16} = \langle x\rangle \rtimes_b \langle y\rangle$ with $x^{17} = 1 = y^{16}$ and $yx = x^b y$, where b has order 16 modulo 17. Let $\psi : G \to G$ by $\psi(x) = 1, \psi(y) = y^{-1}$. Then $x \circ z = x \star z$ for all z in G, $y \circ y = y \star x$ and

$$\begin{aligned} y \circ x &= y\psi(y^{-1})x\psi(y) \\ &= y \star y \star x \star y^{-1} \\ &= y^2 \star x \star y^{-2} \star y \\ &= x^{b^2} y = x^{b^2} \circ y. \end{aligned}$$

So $(G, \circ) \cong C_{17} \rtimes_{b^2} C_{16}$. Setting $\circ = \circ_1$ and repeating the same construction successively to $(G, \circ_1)$ with the same abelian endomorphism ψ yields $(G, \circ_2) \cong C_{17} \rtimes_{b^4} C_{16}$, $(G, \circ_3) \cong C_{17} \rtimes_{b^8} C_{16}$, and $(G, \circ_4) \cong C_{17} \rtimes_{b^{16}} C_{16} = C_{17} \times C_{16}$, the direct product. Setting $\circ_0 = \star$, it turns out that $(G, \circ_i, \circ_j)$ is a bi-skew brace for all pairs i, j with $0 \le i, j \le 4$. See [**Koc21a**] and [**Koc21b**] for the theory behind examples like this and additional examples.

For more discussion about Hopf-Galois structures arising from nilpotent rings, see Chapter 5.

CHAPTER 5

Quantitative results

5.1. Introduction

In this chapter we look at some results describing the number of Hopf-Galois structures of a given type on a Galois extension.

Varying notation slightly from Chapter 3:

DEFINITION 5.1. Given two groups G and N with $|G| = |N|$, let $e(G, N)$, resp. $e_b(G, N)$ be the number of Hopf-Galois structures of type N on a G-Galois extension, resp. the number of isomorphism types of skew braces $(B, \circ, \star)$ with $(B, \circ) \cong G$ and $(B, \star) \cong N$.

In Chapter 3 we called the pair (G, N) of finite groups of the same order realizable if and only if $e(G, N) > 0$, or equivalently, $e_b(G, N) > 0$. In this chapter we focus more on the size of $e(G, N)$, and in particular, $e(G, G)$.

Every G-Galois extension L/K has at least one Hopf-Galois structure, the classical Hopf-Galois structure given by the K-Hopf algebra $K[G]$. Under the Greither-Pareigis correspondence, the regular subgroup of $\mathrm{Perm}(G)$ corresponding to the classical structure is the image $\rho(G)$ of the right regular representation $\rho : G \to \mathrm{Perm}(G)$ given by $\rho(g)(x) = xg^{-1}$. Since $\lambda(G)$ centralizes $\rho(G)$, the Hopf algebra $H = L[\rho(G)]^G = K[\rho(G)]$ is just the group algebra of the Galois group with the usual action on L. So for every group G, $e(G, G) > 0$.

Chapter 3 described some groups G for which the classical Galois structure is the only Galois structure, and some non-abelian groups G where $e(G, G) = 2$. But examples like those are relatively uncommon. As we'll see in this chapter, for many Galois groups G, the number of Hopf-Galois structures on a Galois extension of fields can be quite large.

But first we'll review some counts from Chapter 3.

5.1.1. G cyclic of order p^n. In [**Koh98**], Kohl proved that for G cyclic of odd order p^n, then $e(G, G) = p^{n-1}$ and $e(G, N) = 0$ for $N \not\cong G$. A proof may be found in [**Chi00**, Theorem 9.1]

Byott [**Byo07**] completed Kohl's results by determining the number of Hopf-Galois structures on a G-Galois extension with $G \cong C_{2^n}$. In contrast to the case for p odd, he showed [**Byo07**, Corollary 5.3] that for $n \geq 3$, $e(C_{2^n}, N) = 2^{n-2}$ when N is the cyclic group C_{2^n}, the dihedral group D_{2^n}, and the quaternion group Q_{2^n}. For $n = 1$ there is only the classical (cyclic) Hopf-Galois structure. For $n = 2$ there is the classical cyclic Hopf-Galois structure and also a unique Hopf-Galois structure of elementary abelian type. [**Byo07**] includes a proof of Kohl's results.

5.1.2. Galois field extensions of degree pq. Byott [**Byo04b**] determined the Hopf-Galois structures of type N on a G-Galois field extension of degree pq, where p and q are primes and $p \equiv 1 \pmod q$. There are two groups of order pq up to isomorphism, the cyclic group $C = C_{pq}$ and the non-abelian metacyclic group $M = C_p \rtimes C_q$. To sum up his results:

PROPOSITION 5.2. *For M, C as above, $e(C,C) = 1, e(C,M) = 2(q-1)$, $e(M,C) = p$ and $e(M,M) = 2 + 2p(q-2)$.*

The method for determining $e(G,N)$ was to search for regular subgroups of $\mathrm{Hol}(N)$ isomorphic to G. The search was organized by looking at regular subgroups T isomorphic to G in $\mathrm{Hol}(N) = \lambda(N) \rtimes \mathrm{Aut}(N)$ whose image $\theta(T)$ under the canonical projection θ from $\mathrm{Hol}(N)$ to $\mathrm{Aut}(N)$ had a fixed order m in $\mathrm{Aut}(N)$, for $m = 1, p, q$ and pq.

This organizational approach was later used by Nejabati Zenouz in [**Nej18**] to determine left skew braces and Hopf-Galois structures of degree p^3.

5.2. Regular subgroups and nilpotent algebras

In the next four sections we will study the number of Hopf-Galois structures on an abelian Galois extension of fields whose type is an elementary abelian p-group C_p^n.

Recall from Chapter 2 the connection between $e(G,N)$ and $e_b(G,N)$. Given a skew brace $(B, \circ, \star)$, the number of Hopf-Galois structures associated to that skew brace is, by Proposition 2.24,

$$|\mathrm{Aut}(B,\circ)|/|\mathrm{Aut}_{sb}(B,\circ,\star)|.$$

We proved (Proposition 2.22) that two skew brace structures $(B, \circ, \star)$ and $(B, \circ', \star)$ on the group $(B, \star)$ are isomorphic if and only if $\lambda_{\circ'}(B)$ is conjugate to $\lambda_\circ(B)$ in $\mathrm{Perm}(B)$ by some θ in $\mathrm{Aut}(B, \star)$.

As a consequence (Corollary 2.25), the number $e(G,N)$ of Hopf-Galois structures of type N on a G-Galois extension is

$$e(G,N) = \sum_{T \in \mathcal{C}} |\mathrm{Aut}(G)|/|\mathrm{Sta}(T)|$$

where $\mathcal{C}$ runs through representatives T of the conjugacy classes of regular subgroups of $\mathrm{Hol}(N)$ isomorphic to G under the action of $\mathrm{Aut}(N)$, and $\mathrm{Sta}(T)$ is the set of automorphisms of N that conjugate T to itself.

Let $N = C_p^n$, the elementary abelian p-group of rank n, which we will view as an n-dimensional $\mathbb{F}_p$-vector space $(\mathbb{F}_p^n, +)$. Then

$$\mathrm{Hol}(N) \cong \begin{pmatrix} GL_n(\mathbb{F}_p) & \mathbb{F}_p^n \\ 0 & 1 \end{pmatrix} = \mathrm{Aff}_n(\mathbb{F}_p).$$

Since a regular subgroup T of $\mathrm{Hol}(N)$ is a p-group, T may be conjugated by an element of $\mathrm{Aut}(N) = \mathrm{GL}_{n+1}(\mathbb{F}_p)$ into any suitable p-Sylow subgroup of $\mathrm{GL}_{n+1}(\mathbb{F}_p)$, in particular into U_{n+1}, the group of upper triangular $n+1 \times n+1$ matrices with diagonal I. So we may, and will, assume that all regular subgroups of $\mathrm{Hol}(N)$ are contained in

$$\begin{pmatrix} U_n & \mathbb{F}_p^n \\ 0 & 1 \end{pmatrix}.$$

Caranti, et. al. associated to a regular subgroup T of $\mathrm{Hol}(N)$ a nilpotent ring structure on N.

For $N = (\mathbb{F}_p^n, +)$ (or more generally for N any finite abelian p-group), let T be a regular abelian subgroup of $\mathrm{Hol}(N)$. Then T yields a function (not a homomorphism)

$$\tau : N \to \mathrm{Hol}(N)$$

where for x in N, $\tau(x)$ is the unique element

$$\begin{pmatrix} I + A_x & x \\ 0 & 1 \end{pmatrix}$$

of T whose last column is $(x, 1)^t$.

This defines a multiplication on $(N, +)$ by $x \cdot y = A_x y$, which makes $(N, +, \cdot)$ into a commutative nilpotent ring. Moreover, the circle multiplication on N defined by

$$x \circ y = x + y + A_x y$$

for all x, y in N defines a group structure on the set $\mathbb{F}_p^n$ so that $(\mathbb{F}_p^n, \circ) \cong T$ by the map

$$x \mapsto \begin{pmatrix} I + A_x & x \\ 0 & 1 \end{pmatrix}.$$

Conversely, every commutative nilpotent ring structure on $\mathbb{F}_p^n$ arises from an abelian regular subgroup of $\mathrm{Aff}_n(\mathbb{F}_p)$ [**CDVS06**]. Thus there is a bijection between abelian regular subgroups of $\mathrm{Aff}_n(\mathbb{F}_p)$ and commutative nilpotent $\mathbb{F}_p$-algebras of $\mathbb{F}_p$-dimension n.

PROPOSITION 5.3. *Two commutative nilpotent $\mathbb{F}_p$-algebras are isomorphic if and only if the corresponding regular subgroups of* $\mathrm{Hol}(N) = \mathrm{Aff}_n(\mathbb{F}_p)$ *are in the same orbit under conjugation by elements of* $\mathrm{Aut}(N) = \mathrm{GL}_n(\mathbb{F}_p)$.

The equivalence is in [**CDVS06**].

5.3. Elementary abelian p-groups

Let $N = C_p^n$, an elementary abelian p-group, which we view as the vector space $\mathbb{F}_p^n$, so that $\mathrm{Hol}(N)$ can be identified as the affine group $\mathrm{Aff}_n(\mathbb{F}_p)$ of $\mathbb{F}_p^n$ as in Section 5.2, namely, as the subgroup of $\mathrm{GL}_{n+1}(\mathbb{F}_p)$ of matrices of the form $\begin{pmatrix} A & v \\ 0 & 1 \end{pmatrix}$ where A is in $\mathrm{GL}_n(\mathbb{F}_p)$, and v is an element of N viewed as a column vector in $\mathbb{F}_p^n$.

In [**Chi05**] the regular subgroups of $\mathrm{Hol}(N)$ isomorphic to N were identified for $n = 1, 2, 3$. The first step for any n is to choose a suitable set of regular subgroups of $\mathrm{Hol}(N)$ that represent the orbits of regular subgroups under conjugation by elements of $\mathrm{Aut}(N) = \mathrm{GL}_n(\mathbb{F}_p)$. So, as noted in Section 5.2, we choose a basis of N so that elements of $\mathrm{Hol}(N)$ are upper triangular $n + 1 \times n + 1$ matrices with diagonal entries all $= 1$.

For $n = 2$, up to equivalence there turn out to be two regular subgroups of $\mathrm{Hol}(N)$ isomorphic to N:

$$T_0 = \left\langle \begin{pmatrix} 1 & 0 & 1 \\ 0 & 1 & 0 \\ 0 & 0 & 1 \end{pmatrix}, \begin{pmatrix} 1 & 0 & 0 \\ 0 & 1 & 1 \\ 0 & 0 & 1 \end{pmatrix} \right\rangle$$

and

$$T_1 = \left\langle \begin{pmatrix} 1 & 1 & 0 \\ 0 & 1 & 1 \\ 0 & 0 & 1 \end{pmatrix}, \begin{pmatrix} 1 & 0 & 1 \\ 0 & 1 & 0 \\ 0 & 0 & 1 \end{pmatrix} \right\rangle:$$

T_1 is the group generated by the matrices $I+J$ and $I+J^2$ where J is the 3×3 Jordan block $\begin{pmatrix} 0 & 1 & 0 \\ 0 & 0 & 1 \\ 0 & 0 & 0 \end{pmatrix}$.

The stabilizer $\mathrm{Sta}(T_0) \cong GL_2(\mathbb{F}_p)$, so there is up to equivalence a single regular embedding of $N = C_p^2$ with image T_0. But $\mathrm{Sta}(T_1) = \left\{\begin{pmatrix} a^2 & b \\ 0 & a \end{pmatrix}\right\}$, so has order $p(p-1)$, and so the number of regular embeddings of N with image T_1 is $|\mathrm{GL}_2(\mathbb{F}_p)|/p(p-1) = p^2 - 1$.

The results for $n = 3$ were described in Example 2.16. There are five equivalence classes of regular subgroups of $\mathrm{Hol}(N)$ that are isomorphic to N. The orders of the stabilizers are respectively, $|\mathrm{GL}_3(\mathbb{F}_p)|, p^3(p-1)^2, 2p^2(p^2-1), 2p^2(p-1)^2$ and $(p-1)p^2$. Dividing $|GL_3(\mathbb{F}_p)| = (p^3-1)(p^3-p)(p^3-p^2)$ by these stabilizer orders and summing the resulting numbers of embeddings yields Theorem 7.2 of [**Chi05**]:

PROPOSITION 5.4. *For $p > 3$ there are $p^6+p^5-p^2$ equivalence classes of regular embeddings of C_p^3 into $\mathrm{Hol}(C_p^3)$, hence $p^6+p^5-p^2$ Hopf-Galois structures of type C_p^3 on a G-Galois extension with $G \cong C_p^3$.*

The $n = 3$ results have been generalized in two ways.

The $n = 3$ case with stabilizer of order $(p-1)p^2$ is the case where the regular subgroup T of $\mathrm{Hol}(\mathbb{F}_p^3)$ is the subgroup generated by the 4×4 Jordan matrix $I+J$. In turn, this corresponds under the translation between regular subgroups and radical algebras to the radical $\mathbb{F}_p$-algebra A with a single generator x with $x^4 = 0$. The circle group of A is isomorphic to the group of principal units of the ring (with identity) $\mathbb{F}_p[x]/(x^4)$.

All of the other regular subgroups of dimension p^3 correspond to radical algebras A with two generators and with $A^2 = \{xy \mid x, y \in A\}$ of $\mathbb{F}_p$ dimension 1 or 0. (The case $\dim A^2 = 0$ is the algebra with trivial multiplication.)

The case where G has order p^3 and is isomorphic to the regular subgroup T of $\mathrm{Hol}(C_p^3)$ consisting of the principal units of $\mathbb{F}_p[x]/(x^4)$ was generalized in [**Chi07**] by replacing $\mathbb{F}_p[x]/(x^4)$ by $\mathbb{F}_p[x]/(x^{n+1})$ for any n. Doing so gives a lower bound on the number of Hopf-Galois structures of type C_p^n on a G-Galois extension. This is Theorem 7 of [**Chi07**]:

THEOREM 5.5. *Let T be the group of principal units of the $\mathbb{F}_p$-algebra $\mathbb{F}_p[x]/(x^{n+1})$. The number of Hopf-Galois structures of elementary abelian type N of order p^n on a G-Galois extension is $> p^s$ where $s \geq \frac{(n-1)^3}{3} - n$ if $p \geq 5$ or $n \geq 25$.*

For example, for $p = 3, n = 10$, the number of Hopf-Galois structures corresponding to T exceeds 368×10^{18}.

[**Chi07**] includes a description of the group G of principal units of $\mathbb{F}_p[x]/(x^{n+1})$ as a direct sum of cyclic p-groups. That group has exponent p if and only if $n < p$. See Remark 3.6.

5.4. Almost trivial algebras

The prime $p > 2$ in this section.

The four equivalence classes of regular subgroups of $\mathrm{Hol}(C_p^3)$ that do not correspond to a group of principal units, above, correspond to commutative nilpotent $\mathbb{F}_p$-algebras $A = \mathbb{F}_p[x_1, x_2, x_3]$ that are almost trivial, in the sense of [**Chi19a**]:

DEFINITION 5.6. An $\mathbb{F}_p$-dimension n commutative nilpotent $\mathbb{F}_p$-algebra A is *almost trivial* if $A^3 = 0$ and $\dim(A^2) = 1$.

In this section we show how to obtain a precise count of the number of Hopf-Galois structures of elementary abelian type C_p^n for all $n \geq 2$ corresponding to an almost trivial commutative $\mathbb{F}_p$-algebra A. Since $A^3 = 0$, necessarily the adjoint group of A is also elementary abelian (by Caranti's Lemma, Proposition 5.8 below), so each almost trivial $\mathbb{F}_p$-algebra yields a collection of Hopf-Galois structures of type C_p^n on a G-Galois extension with $G \cong C_p^n$.

An almost trivial $\mathbb{F}_p$-algebra A has an $\mathbb{F}_p$-basis $(x_1, \ldots, x_{n-1}, x_n)$ with $x_i x_j = \varphi_{ij} x_n$, so A is determined by that basis and the coefficients φ_{ij}, which we can collect into a single $n \times n$ matrix $\Phi = (\varphi_{ij})$ that we call the structure matrix of A. Since $A^3 = 0$, $\varphi_{ni} = \varphi_{in} = 0$ for all i. For such an algebra A we can obtain a precise count of the number $e_A(G, G)$ of Hopf-Galois structures corresponding to A by determining the regular subgroups of $\mathrm{Hol}(A, +) = \mathrm{Hol}(\mathbb{F}_p^n)$.

Since A is commutative, the structure matrix Φ is symmetric. So by a change of basis, we can assume that the structure matrix $\Phi = \mathrm{Diag}(d_1, \ldots, d_{n-1}, 0)$, that is, there is a basis $(x_1, \ldots, x_n)$ so that $x_i x_j = 0$ for $i \neq j$ and $x_i^2 = d_i x_n$. Moreover, we may assume that there is some $k < n$ so that $d_1 = \ldots = d_{k-1} = 1, d_k = s, d_{k+1} = \ldots = d_n = 0$ where s is either 1 or some non-square in $\mathbb{F}_p$.

For $n = 3$ the almost trivial algebras are the algebras with diagonal structure matrices $\mathrm{Diag}(0,0,0), \mathrm{Diag}(1,0,0), \mathrm{Diag}(1,1,0)$ and $\mathrm{Diag}(1,s,0)$, where s is a non-square in $\mathbb{F}_p$.

With this basis we can then identify

$$\mathrm{Hol}(C_p^n) = \lambda(C_p^n) \rtimes \mathrm{Aut}(C_p^n) \subset \mathrm{Perm}(C_p^n)$$

(where $\lambda : G \to \mathrm{Perm}(G)$ is the left regular representation, as usual) as $\mathbb{F}_p^n \rtimes \mathrm{GL}_n(\mathbb{F}_p) = \mathrm{Aff}_n(\mathbb{F}_p) \subset \mathrm{GL}_{n+1}(\mathbb{F}_p)$, where

$$\mathrm{Aff}_n(\mathbb{F}_p) = \begin{pmatrix} \mathrm{GL}_n(\mathbb{F}_p) & \mathbb{F}_p^n \\ 0 & 1 \end{pmatrix} \subset \mathrm{GL}_{n+1}(\mathbb{F}_p).$$

Recalling that for $\overline{r} = \sum r_i x_i$ and $\overline{s} = \sum s_i x_i$ in A, we have

$$\lambda_\circ(\overline{r})(\overline{s}) = \overline{r} + \overline{s} + \overline{r} \cdot \overline{s},$$

we have

$$\begin{aligned} \lambda_\circ(\overline{r})(x_j) &= \lambda_\circ(\textstyle\sum r_i x_i)(x_j) \\ &= \sum r_i x_i + x_j + \sum r_i x_i x_j \\ &= \sum r_i x_i + x_j + d_j r_j x_n. \end{aligned}$$

So $\overline{r} = \sum r_i x_i$ in $\lambda(A) \subset \mathrm{Aff}_n(\mathbb{F}_p)$ corresponds to the matrix

$$\lambda_\circ(\overline{r}) = \begin{pmatrix} 1 & 0 & \cdots & 0 & 0 & r_1 \\ 0 & 1 & \cdots & 0 & 0 & r_2 \\ & & \vdots & & & \\ 0 & 0 & \cdots & 1 & 0 & r_{n-1} \\ r_1 d_1 & r_2 d_2 & \cdots & r_{n-1} d_{n-1} & 1 & r_n \\ 0 & 0 & \cdots & 0 & 0 & 1 \end{pmatrix},$$

in $GL_{n+1}(\mathbb{F}_p)$, acting on $\mathbb{F}_p^n$ where $\mathbb{F}_p^n$ is identified as the set of vectors $\begin{pmatrix}\overline{v}\\1\end{pmatrix}$ in $\mathbb{F}_p^{n+1}$.

The number of Hopf-Galois structures corresponding to the isomorphism type of the radical algebra A is then equal to

$$\frac{|\mathrm{GL}_n(\mathbb{F}_p)|}{|\mathrm{Sta}(\lambda_\circ(A))|}$$

where

$$\mathrm{Sta}(\lambda_\circ(A)) = \left\{P \in \mathrm{GL}_n(\mathbb{F}_p) \mid \begin{pmatrix}P & 0\\0 & 1\end{pmatrix}\lambda_\circ(A) = \lambda_\circ(A)\begin{pmatrix}P & 0\\0 & 1\end{pmatrix}\right\}.$$

This follows by **[CDVS06]**, which showed that two nilpotent $\mathbb{F}_p$-algebras on $(\mathbb{F}_p^n, +)$ are isomorphic if and only if the corresponding regular subgroups of $\mathrm{Aff}_n(\mathbb{F}_p)$ are conjugate by an element of $\mathrm{Aut}(N) = \begin{pmatrix}\mathrm{GL}_n(\mathbb{F}_p) & 0\\0 & 1\end{pmatrix}$ in $\mathrm{Aff}_n(\mathbb{F}_p)$. (See also Chapter 2, section 6.)

So let

$$\begin{aligned}\overline{v}_s &= (r_1, r_2, \ldots, r_{k-1}, sr_k)^t,\\ \overline{v} &= (r_1, r_2, \ldots, r_{k-1}, r_k)^t,\\ \overline{w} &= (r_{k+1}, \ldots, r_{n-1})^t\end{aligned}$$

(column vectors of elements of $\mathbb{F}_p$: t denotes transpose). It is convenient to write elements of $T = \lambda_\circ(A)$ as block matrices of the form

$$T = \left\{\lambda_\circ(r) = \begin{pmatrix}I & 0 & 0 & \overline{v}\\0 & I & 0 & \overline{w}\\\overline{v}_s^t & 0 & 1 & r_n\\0 & 0 & 0 & 1\end{pmatrix} \middle| \; r \in \mathbb{F}_p^n\right\}$$

where the diagonal entries are identity matrices of size $k\times k$, $(n-1-k)\times(n-1-k)$, 1×1 and 1×1, respectively.

The stabilizer $\mathrm{Sta}(\lambda_\circ(A))$ of T is the set of $(n+1)\times(n+1)$ matrices

$$Q = \begin{pmatrix}P & 0\\0 & 1\end{pmatrix}$$

where P in $\mathrm{GL}_n(\mathbb{F}_p)$ has the form

$$P = \begin{pmatrix}P_{11} & P_{12} & P_{13}\\P_{21} & P_{22} & P_{23}\\P_{31} & P_{32} & P_{33}\end{pmatrix}$$

so that Q has blocks of the same size as in the matrix $\lambda_\circ(r)$. A study of the equation $QT = TQ$ yields that P_{12}, P_{13} and $P_{23} = 0$, and

$$Q = \begin{pmatrix}P_{11} & 0 & 0 & 0\\P_{21} & P_{22} & 0 & 0\\P_{31} & P_{32} & P_{33} & 0\\0 & 0 & 0 & 1\end{pmatrix}$$

is in the stabilizer of $\lambda_\circ(A)$ if and only if

$P_{33} = q$ is in $\mathrm{GL}_1(\mathbb{F}_p)$;
P_{32} is $1\times(n-1-k)$ and arbitrary;
P_{31} is $1\times k$ and arbitrary;

P_{22} is in $\mathrm{GL}_{n-1-k}(\mathbb{F}_p)$;
P_{21} is $(n-1-k) \times k$ and is arbitrary; and
P_{11} is in $\mathrm{GL}_k(\mathbb{F}_p)$ and satisfies $P_{11}^t D_s P_{11} = qD_s$, where $D_s = \mathrm{Diag}(1, 1, \ldots, s)$.

To determine the possible P_{11} there are three cases: k is odd; k is even and $s = 1$; k is even and s is a non-square. Each case yields P_{11} as a fixed matrix times an element of an orthogonal group which varies with the case: $\mathrm{GO}_{2m+1}(\mathbb{F}_p)$, $\mathrm{GO}_{2m}^+(\mathbb{F}_p)$ or $\mathrm{GO}_{2m}^-(\mathbb{F}_p)$, respectively, of known size (c.f [**Wil09**]). The bottom line is that we can get a precise value of the ratio $|\mathrm{GL}_n(\mathbb{F}_p)|/|\mathrm{Sta}(\lambda_\circ(A))|$ in all three cases.

To sum up, we have, from [**Chi19a**]:

THEOREM 5.7. *Let A be a commutative nilpotent $\mathbb{F}_p$-algebra of dimension n with $A^3 = 0$, $\dim(A^2) = 1$ and structure matrix of rank k. Let L/K be an N-Galois extension with $N \cong (A, +)$. Then the number of Hopf-Galois structures on L/K of type $(A, \circ)$ is a polynomial function of p of degree*

$$\frac{(2n-k)(k+1)}{2} - 1.$$

The number of Hopf-Galois structures increases with k and is maximal when $k = n-1$, that is, when A is as far as possible from having $A^2 = 0$.

In the cases $n = 2, 3$, the stabilizer of the regular subgroup corresponding to the group of principal units of $\mathbb{F}_p[x]/(x^{n+1})$ has order that is a polynomial in p of the same or lower degree than the stabilizers of the other regular subgroups. That suggests the conjecture that the number of Hopf-Galois structures corresponding to any given nilpotent $\mathbb{F}_p$-algebra A of dimension n is $\leq$ the number of Hopf-Galois structures corresponding to the algebra $A = \mathbb{F}_p[x]/(x^{n+1})$ (as described in [**Chi07**]). (Of course as $n > 3$ grows, so does the number of $\mathbb{F}_p$-algebras for which neither Theorem 5.5 nor Theorem 5.7 applies.)

5.5. Asymptotic results on $e(C_p^n, C_p^n)$ for large n

Following [**Chi15**], we consider the size of $e(G, N)$ where $N \cong G \cong C_p^n$ for large n.

Section 2.6 gives the number of Hopf-Galois structures of type C_p^n on a G-Galois extension L/K with $G \cong C_p^n$, namely

$$\sum_{T \in \mathcal{C}} |\operatorname{Aut}(G)|/|\mathrm{Sta}(T)|,$$

where the number of summands $|\mathcal{C}|$ is the number of isomorphism types of nilpotent $\mathbb{F}_p$-algebras A whose circle group is isomorphic to C_p^n, and $T \cong \lambda_\circ(A)$ for a representative A in $\mathcal{C}$.

That collection of algebras includes the isomorphism types of rank n $\mathbb{F}_p$-algebras A so that $A^3 = 0$, for we have Caranti's Lemma:

PROPOSITION 5.8. *Let $N = (\mathbb{F}_p^n, +)$. Let A is a commutative nilpotent $\mathbb{F}_p$-algebra of dimension n with $A^e = 0$. If $e \leq p$ then the regular subgroup T of $Hol(N)$ defined by the circle operation on A is isomorphic to N.*

PROOF. Let A be is a commutative nilpotent $\mathbb{F}_p$-algebra of dimension n with $A^e = 0$. Then the circle operation on A defined by $a \circ b = a + b + a \cdot b$ makes $(\mathbb{F}_p^n, \circ)$ into a group isomorphic to T. Defining

$$m_\circ a = a \circ a \circ \ldots \circ a \ (m \text{ factors}),$$

we obtain the formula: for a in $(\mathbb{F}_p^n, +)$,

$$p_\circ a = pa + \sum_{i=2}^{p-1} \binom{p}{i} a^i + a^p$$

in A.

Now $e \leq p$, so $a^p = 0$, and so $p_\circ a = 0$ for all a in $(\mathbb{F}_p^n, \circ)$, hence $(\mathbb{F}_p^n, \circ) \cong T$ is isomorphic to N. $\square$

This result with $e = 3$ and $p \geq 3$ yields a lower bound on $|\mathcal{C}|$ for large n as $n \to \infty$. For the number (of isomorphism types) of $\mathbb{F}_p$-algebras whose circle group is C_p^n is $\geq$ the number of $\mathbb{F}_p$-algebras A with $A^3 = 0$, and that number is given by Kruse and Price [**KP69**] as

$$p^{\frac{2}{27}n^3 - \frac{4}{9}n^2 + O(n)}$$

On the other hand, Proposition 10.7 of [**Poo08b**] says that for large n, the number of rank n commutative local $\mathbb{F}_p$-algebras is equal to

$$p^{\frac{2}{27}n^3 + O(n^{8/3})}$$

as $n \to \infty$. The number of rank n nilpotent $\mathbb{F}_p$-algebras is the same as the number of rank $n+1$ local $\mathbb{F}_p$-algebras. Not all have circle group isomorphic to C_p^n, so Poonen's bound is an upper bound on $|\mathcal{C}|$.

So $\log_p(|\mathcal{C}|)$ is asymptotic to $\frac{2}{27}n^3$ as $n \to \infty$.

For each isomorphism type A of $\mathbb{F}_p$-algebra, the corresponding number of Hopf-Galois structures is equal to $|\mathrm{GL}_n(\mathbb{F}_p)|/|\mathrm{Sta}(T)|$ where $T \cong \lambda_\circ(A) \subset \mathrm{Hol}(\mathbb{F}_p)$. Since $\mathrm{Sta}(T) \subset \mathrm{Aut}(N)$,

$$1 \leq |\mathrm{Sta}(T)| \leq |\mathrm{GL}_n(\mathbb{F}_p)| < p^{n^2},$$

hence each summand lies between 1 and p^{n^2}. The size of the summands does not asymptotically affect the size of the sum as $n \to \infty$. Therefore:

THEOREM 5.9. *The number $e_n = e(C_p^n, C_p^n)$ of Hopf-Galois structures of type C_p^n on a G-Galois extension with $G \cong C_p^n$ satisfies $\log_p(e_n) \sim p^{\frac{2}{27}n^3}$ as $n \to \infty$.*

5.6. Other counting results

In this section we note some other results on $e(G, N)$.

We noted in Section 4.6 the work of Alabdali and Byott [**AB18**], [**AB20b**] on determining $e(G, N)$ where $|G|$ is squarefree.

5.6.1. Results for other n. [**Byo96**], see also Theorem 2.5 of [**Byo02**], showed that if L/K has degree p^2, p an odd prime, then

$$e(C_{p^2}, C_{p^2}) = p$$
$$e(C_p^2, C_p^2) = p^2.$$

[**Byo02**] studied the Galois module theory for this set of Hopf-Galois structures: see Subsection 11.2.2.

The appendix by Byott and Vendramin in [**SV18**] contains tables of $e(G, N)$ for $|G| = |N| = 8$ and 12, and the number of Hopf-Galois structures on Galois extensions of degree n for composite n, $6 \leq n \leq 46$ (omitting $n = 32$).

Alabdali and Byott [**AB20b**, Theorem 9.6] determine $e(G, N)$ for all groups G and N where $|G| = |N|$ is the product of three distinct primes.

5.6.2. Related classifications for small n. In view of the known connection between Hopf-Galois extensions, and skew braces (see Chapter 2), classifications of radical algebras, braces and skew braces are of evident interest.

We note the following work:

• D. Bachiller [**Bac15**] has a classification of left braces of order p^n, $n \leq 3$. Thus he is interested in skew braces with commutative additive group.

For $n = 1$ there is, up to isomorphism, only one group of order p^n, namely the cyclic group C_p, and $e_{sb}(C_p, C_p) = 1$.

For $n = 2$, N can be C_{p^2} or $C_p \times C_p$, and $e_{sb}(C_{p^2}, C_{p^2}) = 2, e_{sb}(C_p^2, C_p^2) = 2$ for $p > 2$, while $e_{sb}(G, N) = 0$ if $G \not\cong N$.

For $n = 3$ and $p \geq 3$ there are five groups of order p^3: the three abelian groups $C_{p^3}, C_{p^2} \times C_p$ and C_p^3, the Heisenberg group $M(p)$ of exponent p which can be realized as the group of upper triangular 3×3 matrices with unit diagonal in $\mathrm{GL}_3(\mathbb{F}_p)$, and the group $M_3(p) \cong C_{p^2} \rtimes C_p$. For $p \geq 3$, the main theorems of [**Bac15**] yield the following tables: For $p > 3$:

$G\backslash N$	C_{p^3}	$C_{p^2} \times C_p$	$(C_p)^3$
C_{p^3}	3	0	0
$C_{p^2} \times C_p$	0	8	0
$(C_p)^3$	0	0	5
$M(p)$	0	$4p+2$	0
$M_3(p)$	0	0	$2p$

For $p = 3$:

$G\backslash N$	C_{p^3}	$C_{p^2} \times C_p$	$(C_p)^3$
C_{p^3}	3	0	0
$C_{p^2} \times C_p$	0	7	1
$(C_p)^3$	0	1	5
$M(p)$	0	13	0
$M_3(p)$	0	1	6

This work was generalized to skew braces of order p^3 by Nejabati Zenouz [**Nej18**], who determined the automorphism groups of the skew braces and determined the number of Hopf-Galois structures corresponding to each of the skew braces. In [**NZ19**] he found that for $p > 3$,

$$\begin{aligned} e_b(M(p), M(p)) &= 2p^2 - p + 3 \\ e_b(C_p^3, M(p)) &= 2p + 1 \\ e(M(p), M(p)) &= (2p^3 - 3p + 1)p^2 \\ e(C_p^3, M(p)) &= (p^3 - 1)(p^2 + p - 1)p^2 \end{aligned}$$

• W. DeGraaf [**DG18**] has a classification of nilpotent F-algebras (up to isomorphism) of dimension n for F any field and $n \leq 4$. There is one algebra of dimension 1 and two algebras of dimension 2 (the trivial algebra and the algebra $A = \langle a, b \mid a^2 = b \rangle$.) For $n = 3$ and $F = \mathbb{F}_q$, q odd, there are $q + 6$ isomorphism types, described by generators and relations in Section 5 of [**DG18**], and for $n = 4$ there are $5q + 20$ types lying in 26 families and listed in Section 3 of [**DG18**].

• B. Poonen [**Poo08a**] has a classification of commutative local algebras with identity of rank ≤ 6 over an algebraically closed field k, described as quotients of a polynomial ring modulo an ideal whose generators are given. For characteristic $p > 3$, there are 1, 1, 2, 4, 9, 25 algebras of rank 1, 2, 3, 4, 5, 6, respectively. For rank > 6 there are infinitely many isomorphism classes of commutative local algebras. Each nilpotent algebra A (without identity) of rank $n-1$ corresponds to a local algebra (with identity) of rank n by $A \to k \oplus A$.

• D. A. Suprunenko and R. I. Tyskevic [**ST68**] constructed a class of commutative nilpotent F-algebras A with $\dim(A) = 6, \dim(A^2) = 4$ and $A^3 = 0$. These algebras were studied over $\mathbb{F}_p$ in [**Chi15**], where it was shown (Theorem 9.3) that for $p = 6k+1$ or $6k+5$ then those algebras form k isomorphism types. Hence, in contrast to the cases where $\dim(A) \leq 4$ and the number of isomorphism types of commutative nilpotent $\mathbb{F}_p$-algebras is bounded independent of the prime p, for $n \geq 6$ the number of isomorphism types goes to infinity as (at least) a linear function of p.

It appears to be an open question whether the number of isomorphism types of commutative nilpotent algebras A of dimension 5 over the field $\mathbb{F}_p$ is bounded by a constant independent of p.

CHAPTER 6

Enumeration of Hopf-Galois structures on Galois extensions of degree mp

6.1. Introduction

For G, M two groups of the same order, let $R(G,[M])$ be the set of regular subgroups of $B = \mathrm{Perm}(G)$ isomorphic to M and normalized by $\lambda(G)$, the image of the left regular representation $\lambda : G \to B$. By Greither-Pareigis theory, a group N in $R(G,[M])$ corresponds to a unique K-Hopf algebra structure of type M on a G-Galois extension L/K of fields. Thus to find Hopf Galois structures on L/K we can try to find the elements of $R(G,[M])$.

But finding regular subgroups of type M of $\mathrm{Perm}(G)$ normalized by $\lambda(G)$ appears to be a difficult problem, because $\mathrm{Perm}(G)$ is in general a large group with many subgroups. Thus nearly all work on finding or counting Hopf Galois extensions has utilized the idea that if N is a regular subgroup of $\mathrm{Perm}(G)$ normalized by $\lambda(G)$ of type M, then $\lambda(G)$ is isomorphic to a regular subgroup of $\mathrm{Hol}(M)$, the normalizer in $\mathrm{Perm}(M)$ of $\lambda(M)$. The condition that N be normalized by $\lambda(G)$ therefore translates the problem into one of finding regular subgroups isomorphic to G in a relatively small, well-understood subgroup $\mathrm{Hol}(M) = \lambda(M)\,\mathrm{Aut}(M)$ of $\mathrm{Perm}(M) \cong \mathrm{Perm}(G)$. This idea is often called Byott's translation, because of [**Byo96**], and is presented in Chapter 2 and elsewhere in this book.

Most results on Hopf Galois structures have been obtained by Byott's translation.

This chapter describes an exception. Kohl examined groups of order mp, where p is a prime, m is coprime to p, with unique Sylow p-subgroup. Then, by the Schur-Zassenhaus Lemma, any such group is a semidirect product $P \rtimes_\alpha Q$ where $|P| = p, |Q| = m$. So for G the Galois group, $\lambda(G) = \mathcal{P}\mathcal{Q}$ where $\mathcal{P}$ and $\mathcal{Q}$ are subgroups of $\mathrm{Perm}(G)$, and any subgroup N in $R(G,[M])$ has the same form: $N = P(N)Q(N) \cong P(N) \rtimes Q(N)$ for $|P(N)| = p$ and $|Q(N)| = m$.

One further restriction is needed: p does not divide $|\mathrm{Aut}(Q)|$ for any group Q of order m. Then the key result about regular subgroups N normalized by $\lambda(G)$ in $\mathrm{Perm}(G)$ is that any such group N is contained in the normalizer in $\mathrm{Perm}(G)$ of the Sylow p-subgroup $\mathcal{P}$.

Under these conditions, the problem of looking for regular subgroups N isomorphic to M in $\mathrm{Perm}(G)$ is replaced by the problem of looking for regular subgroups N isomorphic to M in the subgroup $\mathrm{Norm}_B(\mathcal{P})$ of $B = \mathrm{Perm}(G)$. As we'll see, this normalizer group has the form $\mathcal{V}\mathcal{U}\mathcal{S}$ where $\mathcal{S} \cong S_m$, $\mathcal{V} \cong C_p^m$ and $\mathcal{U} \cong \mathbb{F}_p^\times$. Thus $\mathcal{V}$ and $\mathcal{U}$ are abelian.

This group can in turn be identified as a subgroup of the affine group $\mathrm{Aff}_m(\mathbb{F}_p) < \mathrm{GL}_{m+1}(\mathbb{F}_p)$. Thus $\mathrm{Norm}_{\mathrm{Perm}(G)}(\mathcal{P})$ is a group in which one can explicitly obtain regular subgroups isomorphic to M.

In this chapter we describe Kohl's approach to finding Hopf Galois structures. The results come from three articles with increasing levels of generality [**Koh07**] for $m = 4$, [**Koh13**] for $p > m$, and [**Koh16**] for more general choices of p and m, but with sufficient constraints to insure the applicability of the methods developed earlier.

6.2. Preliminaries

Throughout this discussion we assume that L/K is a G-Galois extension which is also Hopf-Galois where $[L : K] = mp$ with p prime and $\gcd(m,p) = 1$. As such, for any K-Hopf algebra acting on L/K one has that $H = L[N]^G$ where N is a regular subgroup of $B = \mathrm{Perm}(G)$, normalized by $\lambda(G)$, where now *both* $\lambda(G)$ and N are semidirect products, namely $\lambda(G) = \mathcal{P}\mathcal{Q}$ and $N = P(N)Q(N)$. The key to the method for enumerating the Hopf-Galois structures on these extensions is the relationship between $\mathcal{P}$ and $P(N)$ for any $N \in R(G, [M])$.

For subgroups (in particular Sylow p-subgroups) of these regular permutation groups, their cycle structure is crucially important.

DEFINITION 6.1. For a set Z, a subgroup $K \leq \mathrm{Perm}(Z)$ is *semiregular* if, for each $k \in K$, the length of the orbit of every $z \in Z$ under the action of k is the same. That is, if k is written as a product of disjoint cycles, all cycles have the same length.

So, in particular, any subgroup of a regular permutation group is semiregular. Note also, that if $K \leq \mathrm{Perm}(Z)$ is semiregular then $|K| \leq |Z|$ since semiregular subgroups act fixed point freely. As such, if $|K| = |Z|$ then K is regular.

We assume that p and m are such that any group of order mp has a unique Sylow p-subgroup. Additionally, we need to assume that p does not divide the order of the automorphism group of any group of order m, which also is automatic if $p > m$, but still holds in other cases too. The import of this is in how it affects the Sylow p-subgroup of the holomorph of these groups of order mp. To be slightly formal, we define the following subsets of $\mathbb{N} \times \mathbb{N}$

$$F_Q = \{(p, m) \mid p \text{ prime}, \ \gcd(p, m) = 1, p \nmid |\mathrm{Aut}(Q)| \text{ for all groups } Q \text{ of order } m\}$$
$$F_S = \{(p, m) \mid p \text{ prime}, \ \gcd(p, m) = 1, n_p = 1 \text{ for all groups of order } mp\}$$

where n_p denotes the number of Sylow p-subgroups of a group. The program in [**Koh07**],[**Koh13**], and [**Koh16**] for enumerating Hopf-Galois structures on Galois extensions of order mp may be used for those $(p, m) \in F_Q \cap F_S$. If $p > m$ (as was the setting of [**Koh07**] and [**Koh13**]) then $(p, m) \in F_Q \cap F_S$ automatically. However $F_Q \cap F_S$ is non-empty for many p, m where $p < m$, for example if $p = 5$ and $m = 8$.

We begin with the following basic fact.

LEMMA 6.2. [**Koh16**, Lemma 1.4] *If* $|G| = mp$ *and* $(p, m) \in F_Q \cap F_S$ *where* $G \cong P \rtimes_\alpha Q$ *for* P, Q *as above, and* $\alpha : Q \to \mathrm{Aut}(P)$, *then*

(a) If α *is trivial then* $p \nmid |\mathrm{Aut}(G)|$.

(b) If α *is non-trivial, then* $p \mid |\mathrm{Aut}(G)|$ *but* $p^2 \nmid |\mathrm{Aut}(G)|$.

In particular, if the Sylow p-subgroup of G is not a direct factor then for $P = \langle \pi \rangle$ one has that $c_\pi \in \mathrm{Aut}(G)$ (conjugation by π) is the Sylow p-subgroup of $\mathrm{Aut}(G)$, which acts trivially on P. As such, if the Sylow p-subgroup of G is not a direct factor then the Sylow p-subgroup of $\mathrm{Hol}(G)$ is $P\langle c_\pi\rangle \cong C_p \times C_p$. If $G = P \times Q$ then the Sylow p-subgroup of $\mathrm{Hol}(G)$ is P itself.

The key to enumerating all $N \in R(G,[M])$ is to relate the Sylow p-subgroup $\mathcal{P}$ of $\lambda(G)$ and the Sylow p-subgroup $P(N)$ of N. We assume that $(p,m) \in F_Q \cap F_S$ and the principal result is the following.

THEOREM 6.3. [**Koh16**, Theorem 1.3] *If $(p,m) \in F_Q \cap F_S$ and $G = \mathrm{Gal}(L/K)$, where $|G| = [L : K] = mp$ and $\mathcal{P}$ is the Sylow p-subgroup of $\lambda(G)$, then for $N \in R(G,[M])$ we have $N \leq \mathrm{Norm}_B(\mathcal{P})$.*

6.3. Twisted wreath products

The utility of this result is that the N which give rise to Hopf-Galois structures reside inside a relatively small subgroup of the larger ambient symmetric group B and this subgroup has a well understood structure which lends itself to computing those regular subgroups contained therein. We outline this development here.

As G has order mp then in the left regular representation $\lambda(G) \leq B$ one has the Sylow p-subgroup $\mathcal{P} = \langle \pi \rangle = \langle \pi_1\pi_2\cdots\pi_m\rangle$ where the π_i are disjoint p-cycles. By regularity, for each i from 1 to m one may find $\gamma_i \in G$ such that $\pi_i = (\gamma_i, \pi(\gamma_i), \pi^2(\gamma_i), \ldots, \pi^{p-1}(\gamma_i))$. We can define $\mathcal{V} = \langle \pi_1, \ldots, \pi_m\rangle$ which is elementary abelian of order p^m, and for each $i,j \in \{1,\ldots,m\}$ we can define $\sigma_{ij} = (\gamma_i,\gamma_j)\cdot(\pi(\gamma_i),\pi(\gamma_j))\cdots(\pi^{p-1}(\gamma_i),\pi^{p-1}(\gamma_j))$ where $\mathcal{S} = \langle \sigma_{ij}\rangle \cong S_m$. This copy of S_m acts on the supports of the p-cycles π_i which are blocks whose union is all of G. Also, since each π_i is order p then there exists a product of m disjoint $p-1$ cycles $u_1\cdots u_m$ where $u_i\pi_i u_i^{-1} = \pi_i^u$ for $\langle u\rangle = U_p \cong \mathrm{Aut}(\mathbb{Z}_p)$. If we let $\mathcal{U} = \langle u_1 \cdots u_m\rangle$ then $\mathcal{U} \cong U_p$ which also acts in a uniform fashion on the π_i. What we have then is the following descriptions of $\mathrm{Cent}_B(\mathcal{P})$ and $\mathrm{Norm}_B(\mathcal{P})$, namely

$$\mathrm{Cent}_B(\mathcal{P}) = \mathcal{VS} \cong C_p^m \rtimes S_m \cong C_p \wr S_m$$
$$\mathrm{Norm}_B(\mathcal{P}) = \mathcal{VUS} \cong C_p^m \rtimes (U_p \times S_m)$$

where the centralizer is an ordinary wreath product and the normalizer is a so-called "twisted" wreath product where the twist arises from the additional action by $\mathcal{U}$ acting diagonally on the p-cycles in $\mathcal{V}$.

In order to enumerate $N \in R(G,[M])$, as a subgroup of $\mathrm{Norm}_B(\mathcal{P})$, it is necessary to have a more computationally convenient representation of this twisted wreath product.

6.3.1. Affine representation. Another way to view $\mathrm{Norm}_B(\mathcal{P})$ is as the subgroup $\mathbb{F}_p^m \rtimes (U_p \times S_m)$ of

$$\mathrm{Aff}(\mathbb{F}_p^m) \cong \mathbb{F}_p^m \rtimes \mathrm{GL}_m(\mathbb{F}_p) \cong \mathrm{Hol}(\mathbb{F}_p^m)$$

namely as the set of 3-tuples $(\hat{a}, (v, \alpha)) = (\hat{a}, v, \alpha)$ where $\hat{a} \in \mathbb{F}_p^m$ and $v \in U_p$ represents scalar multiplication, and $\alpha \in S_m$ represents coordinate shift on the vectors in $\mathbb{F}_p^m$, so that

$$(\hat{a}, v, \alpha)(\hat{b}, w, \beta) = (\hat{a} + v\alpha(\hat{b}), vw, \alpha\beta)$$

is the product. Additionally, the elements of $\mathrm{Cent}_B(\mathcal{P})$ are naturally those tuples of the form $(\hat{a}, 1, \alpha)$, since the centralizer does not include the "twist" by the action of scalar multiplication by elements of U_p. To see this directly, observe that the elements of $\mathbb{F}_p^m \rtimes (U_p \times S_m)$ corresponding to $\mathcal{P}$ are of the form $(c[1,1,\ldots,1],1,I)$

for some $c \in \mathbb{F}_p$. As such, for any $\beta \in S_m$ we have $\beta(c[1,1,\ldots,1]) = c[1,1,\ldots,1]$. For $p > m$ any order p subgroup must be of the form $\langle(\hat{a},1,I)\rangle$ for $\hat{a} \in \mathbb{F}_p^m \setminus \{\hat{0}\}$. However, for $p < m$ there do exist order p elements coming from the S_m component, but as is shown in [**Koh16**], these never generate the Sylow p-subgroup of a regular subgroup normalized by $\lambda(G)$.

As to those $N \in R(G,[M])$, each is embedded as a regular subgroup of $\mathcal{VUS}$. Therefore, each is representable as a subgroup $P(N)Q(N)$ of $\mathbb{F}_p^m \rtimes (U_p \times S_m)$, where $P(N)$ has order p and $Q(N)$ has order m where $P(N)$ is normalized by $Q(N)$. More fundamentally, $P(N)$ must be normalized by $\mathcal{Q}$. (We note that $\mathcal{P}$ centralizes $P(N)$ so therefore certainly normalizes it by the above observation about the elements of order p in $\mathrm{Norm}_B(\mathcal{P})$). As such, one of the necessities for using this method is to first enumerate the possibilities for $P(N)$, in particular those that are normalized by $\mathcal{Q}$.

We note a basic fact about $\mathcal{Q}$ viewed as a subgroup of $\mathbb{F}_p^m \rtimes (U_p \times S_m)$, namely that if $t : \mathbb{F}_p^m \rtimes (U_p \times S_m) \to S_m$ is the natural projection map, that is, $t(\hat{a},v,\alpha) = \alpha$, then $t(\mathcal{Q})$ is a regular subgroup of S_m, since the elements of S_m act on the coordinates of the "vectors" in the $\mathbb{F}_p^m$ component in the semidirect product. We also need to determine all linear characters of $\mathcal{Q}$ in $\mathbb{F}_p$, namely those homomorphisms $\chi : \mathcal{Q} \to \mathbb{F}_p^\times = U_p$. If we define elementary basis vectors $\hat{v}_i = [0,\ldots,1,\ldots,0] \in \mathbb{F}_p^m$ corresponding to $\pi_i \in \mathcal{V}$ then we have the following

THEOREM 6.4. [**Koh13**, Theorem 2.1] *Any semiregular subgroup of $\mathcal{V}$ of order p, that is normalized by $\mathcal{Q}$ is generated by*

$$\hat{p}_\chi = \sum_{q \in \mathcal{Q}} \chi(q)\hat{v}_{q(1)}$$

where $\chi : \mathcal{Q} \to U_p$ is a linear character of $\mathcal{Q}$.

We note that the subscript "$q(1)$" makes sense in light of the comment about $t(\mathcal{Q})$ being a regular subgroup of S_m, so that, in particular, the orbit of 1 under $t(Q)$ is all of $\{1,\ldots,m\}$. We also should point out that the precursor to this result is that any $P(N)$ is, in fact, a subgroup of $\mathcal{V}$. In [**Koh07**] and [**Koh13**] the assumption that $p > m$ guaranteed this since in $\mathbb{F}_p^m \rtimes (U_p \times S_m)$ no element of the form $(\hat{a},v,\alpha)$ can have order p unless $v = 1$ and $\alpha = I$, which is *not* true if $p < m$. That is, there are semiregular, elementary abelian subgroups of $\mathbb{F}_p^m \rtimes (U_p \times S_m)$ of order p^m that lie outside of $\mathbb{F}_p^m$. However, [**Koh16**, Theorem 1.6] shows that regardless of whether $p > m$ or not, the fact that $\mathcal{Q}$ normalizes $\mathcal{P}$ and $P(N)$ forces $P(N) \leq \mathcal{V}$.

The last component necessary to show that any $N \in R(G,[M])$ must lie in $\mathrm{Norm}_B(\mathcal{P})$ is the condition that $p \nmid |\mathrm{Aut}(Q)|$. As seen above, this implies that the Sylow p-subgroup of the holomorph of a group of order mp (for those $(p,m) \in F_Q \cap F_S$) is either cyclic or elementary abelian of order p^2. For any $N \in R(G,[M])$ one has that either $N \leq \mathrm{Cent}_B(\mathcal{P})$ or $N^{opp} \leq \mathrm{Cent}_B(\mathcal{P})$, where $N^{opp} = \mathrm{Cent}_B(N)$. This is based on the fact, [**Koh07**, Proposition 4.3], that for any N regular of order mp, the Sylow p-subgroup of $\mathrm{Norm}_B(N)$ is $P(N) = \mathcal{P}$, if $N = P(N) \times Q(N)$, or of the form $P(N)P(N^{opp}) \cong C_p \times C_p$ for $N \not\cong P(N) \times Q(N)$. In either case $P(N) = \mathcal{P}$ or $P(N^{opp}) = \mathcal{P}$. From this condition on $P(N)$ we find that either $N \leq \mathrm{Cent}_B(\mathcal{P})$ or $N^{opp} \leq \mathrm{Cent}_B(\mathcal{P})$ and since N and N^{opp} mutually centralize each other we conclude that indeed $N \leq \mathrm{Norm}_B(\mathcal{P})$.

6.3.2. Centralizer vs. normalizer. One other simplification that can be made, is based on the parity between N and N^{opp} for $N \in R(G,[M])$, namely that $N \in R(G,[M])$ if and only if $N^{opp} \in R(G,[M])$. As we've seen, N is a split extension where $N = P(N)Q(N)$. For a given N, either $P(N) = \mathcal{P}$ or $P(N^{opp}) = \mathcal{P}$. If $N \cong P(N) \times Q(N)$ then $P(N) = P(N^{opp}) = \mathcal{P}$, which automatically implies that $N \leq \mathrm{Cent}_B(\mathcal{P})$. And in general, if $P(N^{opp}) = \mathcal{P}$ then $N \leq \mathrm{Cent}_B(\mathcal{P})$ which is a smaller group than $\mathrm{Norm}_B(\mathcal{P})$. So one may search for just those N contained in the centralizer, and for each such N simply compute N^{opp} which will either equal N (if N is abelian) or will be yet another member of $R(G,[M])$. That is, if $R(G,[M];\mathrm{Cent}_B(\mathcal{P}))$ is the set of those $N \in R(G,[M])$ contained in the centralizer then

$$R(G,[M]) = R(G,[M];\mathrm{Cent}_B(\mathcal{P})) \cup R(G,[M];\mathrm{Cent}_B(\mathcal{P}))^{opp}$$

where

$$R(G,[M];\mathrm{Cent}_B(\mathcal{P}))^{opp}$$

is the set of opposites of each N in $R(G,[M];\mathrm{Cent}_B(\mathcal{P}))$. Again, these sets are identical if $P(N)$ is a direct factor, which is a function of the particular isomorphism class $[M]$. The other consequence of this division, is that calculating those $N \leq \mathrm{Cent}_B(\mathcal{P})$ requires one to compute all possible $P(N)$, namely the totality of all $\hat{p}_\chi$ for all characters $\chi : \mathcal{Q} \to U_p$. In contrast, if one focuses on those N such that $P(N) = \mathcal{P}$ then one avoids the computation of the linear characters, but at the expense of needing to find generators of $Q(N)$ within the larger ambient group, $\mathrm{Norm}_B(\mathcal{P})$.

6.4. Enumeration within S_{mp}

The general strategy then for actually enumerating $R(G,[M])$ is to utilize the fact that regularity is not tied to left regular representations. This is due to the fact, quoted in [**MBD61**] and elsewhere, that any two regular subgroups of a given ambient symmetric group are isomorphic if and only if they are conjugates. Indeed, if we let $\beta \in B$ then $\beta\lambda(G)\beta^{-1}$ is also a regular subgroup of B, and if $\lambda(G) \leq \mathrm{Norm}_B(N)$ then $\beta\lambda(G)\beta^{-1} \leq \beta\,\mathrm{Norm}_B(N)\beta^{-1} = \mathrm{Norm}_B(\beta N\beta^{-1})$, and similarly, *every* such N may be conjugated into the collection of those regular subgroups of B normalized by $\beta\lambda(G)\beta^{-1}$, where both collections must have the exact same size. Put simply, if Γ_1 and Γ_1 are isomorphic regular subgroups of B, and if $R(\Gamma_1)$ and $R(\Gamma_2)$ are the associated collections of regular subgroups normalized by Γ_1 and Γ_2 then there exists $\beta \in B$ such that $\beta\Gamma_1\beta^{-1} = \Gamma_2$ which gives rise to a bijection $R(\Gamma_1) \to R(\Gamma_2)$ given by $N \mapsto \beta N\beta^{-1}$. To further simplify matters, one may replace $\lambda(G)$ and $B = \mathrm{Perm}(G)$ by an arbitrary regular subgroup $G \leq S_{mp}$ and enumerate those $N \leq S_{mp}$ normalized by this G. We outline this strategy in detail below.

Since groups of order mp subject to (p,m) being in $F_Q \cap F_S$ are semidirect products, we can simplify the enumeration of $R(G,[M])$ a bit. If we let G and $[M]$ vary over all pairings of groups of the same order, then each G (which represents the $\mathrm{Gal}(L/K)$) gives rise to a Sylow p-subgroup $P(G)$ (i.e. the analog of $\mathcal{P}$ for $\lambda(G)$).

We start by choosing a *fixed* semiregular $\mathcal{P}$ embedded in S_{mp}, namely

$$\begin{aligned}&\langle \pi_1\pi_2\cdots\pi_m\rangle \text{ where}\\ &\pi_i = (1+(i-1)p, 2+(i-1)p, \ldots, ip)\end{aligned}$$

and then, knowing the structure of the particular isomorphism class of group G, find/construct an appropriately chosen $\mathcal{Q} = Q(G)$ such that $G \cong \mathcal{P}\mathcal{Q}$. We then determine the linear characters of each $\mathcal{Q}$ and use these to determine $\hat{p}_\chi$ which are the generators of potential $P(N)$ for $N \in R(G, [M])$. Beyond this, the enumeration of $N \in R(G, [M])$ amounts to taking each candidate $P(N)$, and determining $Q(N)$ (both as subgroups of $\mathrm{Norm}_B(\mathcal{P})$) which normalize $P(N)$, and then determine which $P(N)Q(N)$ are regular.

6.4.1. Groups of order $4p$. We review the relatively simple case where $m = 4$ and $p > 3$ as in [**Koh07**]. The π_i for $i = 1, \ldots, 4$ are as follows

$$\begin{aligned}\pi_1 &= (1, 2, \ldots, p)\\ \pi_2 &= (p+1, p+2, \ldots, 2p)\\ \pi_3 &= (2p+1, 2p+2, \ldots, 3p)\\ \pi_4 &= (3p+1, 3p+2, \ldots, 4p)\end{aligned}$$

For the possible groups of order 4 which, when joined with $\mathcal{P} = \langle \pi_1\pi_2\pi_3\pi_4 \rangle$, yield the different groups of order $4p$, we must be attentive to the fact that the order 4 subgroup will either centralize the order p subgroup, or the order 4 subgroup will act non-trivially. Here are the groups of order $4p$:

$$\begin{aligned}C_{4p} &\cong \langle \pi, \sigma \mid \pi^p, \sigma^4, [\pi, \sigma] \rangle\\ C_p \times V &\cong \langle \pi, \delta, \tau \mid \pi^p, \delta^2, \tau^2, [\pi, \delta], [\pi, \tau] \rangle\\ E_p &\cong \langle \pi, \sigma \mid \pi^p, \sigma^4, \sigma\pi\sigma^{-1}\pi^{-\zeta} \rangle\\ D_{2p} &\cong \langle \pi, \delta, \tau \mid \pi^p, \delta^2, \tau^2, [\pi, \delta], \pi\tau\pi\tau^{-1} \rangle\\ Q_p &\cong \langle \pi, \sigma \mid \pi^p, \sigma^4, \sigma\pi\sigma^{-1}\pi \rangle\end{aligned}$$

where E_p is defined only if $p \equiv 1 \pmod 4$ and $\zeta \in U_p$ has order 4. Based on the π_i given above, we may define

$$\begin{aligned}\tilde{\sigma} &= (1, p+1, 2p+1, 3p+1) \cdots (p, 2p, 3p, 4p)\\ \tilde{\delta} &= (1, p+1) \cdots (p, 2p) \cdot (2p+1, 3p+1) \cdots (3p, 4p)\\ \tilde{\tau} &= (1, 2p+1) \cdots (p, 3p) \cdot (p+1, 3p+1) \cdots (2p, 4p)\end{aligned}$$

in $\mathcal{S} \leq \mathrm{Norm}_B(\mathcal{P}) = \mathcal{V}\mathcal{U}\mathcal{S}$ which (when viewed as acting on the blocks arising from the π_i) correspond to the following elements of $S_4 \leq \mathbb{F}_p^4 \rtimes (U_p \times S_4)$

$$\begin{aligned}\sigma &= (1, 2, 3, 4)\\ \delta &= (1, 2)(3, 4)\\ \tau &= (1, 3)(2, 4)\end{aligned}$$

and therefore we can define the following groups G:

$$\begin{aligned}C_{4p} &\cong \langle ([1, 1, 1, 1], 1, I), (\hat{0}, 1, \sigma) \rangle\\ C_p \times V &\cong \langle ([1, 1, 1, 1], 1, I), (\hat{0}, 1, \delta), (\hat{0}, 1, \delta) \rangle\\ E_p &\cong \langle ([1, 1, 1, 1], 1, I), (\hat{0}, \zeta, \sigma) \rangle\\ D_{2p} &\cong \langle ([1, 1, 1, 1], 1, I), (\hat{0}, 1, \delta), (\hat{0}, -1, \delta) \rangle\\ Q_p &\cong \langle ([1, 1, 1, 1], 1, I), (\hat{0}, -1, \sigma) \rangle\end{aligned}$$

from each isomorphism class, as subgroups of $\mathbb{F}_p^4 \rtimes (U_p \times S_4)$. For the groups of order 4 we have the following character tables.

	1	δ	τ	$\delta\tau$
χ_1	1	1	1	1
χ_2	1	1	-1	-1
χ_3	1	-1	1	-1
χ_4	1	-1	-1	1

	1	σ	σ^2	σ^3
ψ_1	1	1	1	1
ψ_2	1	-1	1	-1

If $p \equiv 1 \pmod 4$ then one has an element $\zeta \in U_p$ of order 4 and thus two additional characters

	1	σ	σ^2	σ^3
ψ_3	1	ζ	-1	ζ^{-1}
ψ_4	1	ζ^{-1}	-1	ζ

where $\zeta^{-1} = -\zeta$. Therefore, the potential order p semiregular subgroups are

$$
\begin{aligned}
P_1 &= \langle([1,1,1,1],1,I)\rangle \\
P_2 &= \langle([1,1,-1,-1],1,I)\rangle \\
P_3 &= \langle([1,-1,1,-1],1,I)\rangle \\
P_4 &= \langle([1,-1,-1,1],1,I)\rangle \\
P_5 &= \langle([1,-1,\zeta,\zeta^{-1}],1,I)\rangle \\
P_6 &= \langle([1,-1,\zeta^{-1},\zeta],1,I)\rangle
\end{aligned}
$$

which are the possible Sylow p-subgroups $P(N)$ for $N \in R(G,[M])$. The enumeration of $R(G,[M])$ is a matter of finding, for each possible $P(N) = P_i$ a complementary subgroup $Q(N)$ and determining if $N = P(N)Q(N)$ is regular, and, most importantly, if it is normalized by G, all embedded in $\mathrm{Norm}_B(\mathcal{P})$. As seen in [**Koh07**],[**Koh13**], and [**Koh16**], it becomes a mix of techniques in linear algebra, combinatorics, and group theory to determine $R(G,[M])$. For the $4p$ case, we have the following, which includes the distribution of the $P(N)$ amongst $P_1,\dots,P_6$, where we observe that, when $Q(N) \not\triangleleft N$, the number of N where $P(N) = \mathcal{P} = P_1$ is equal to the total number of those N where $P(N) \neq \mathcal{P}$. For the result below, $R(G)$ is the union of the $R(G,[M])$, as $[M]$ varies over the all the isomorphism classes of groups of order $|G|$.

PROPOSITION 6.5. [**Koh07**, Proposition 6.1] *The distribution of the $R(G,[M])$ are as follows, subject to the constraints imposed by the class of p mod 4, specifically that for $p \equiv 3$, $R(E_p) = R(G,[E_p]) = \emptyset$ for all G.*

- $|R(C_{4p})| = 10$ *if* $p \equiv 1 \pmod 4)$ *or* 6 *if* $p \equiv 3 \pmod 4$

	P_1	P_2	P_3	P_4	P_5	P_6
$R(C_{4p},[C_{4p}])$	1	0	0	0	0	0
$R(C_{4p},[C_p \times V])$	1	0	0	0	0	0
$R(C_{4p},[E_p])$	2	0	0	0	1	1
$R(C_{4p},[D_{2p}])$	1	1	0	0	0	0
$R(C_{4p},[Q_p])$	1	1	0	0	0	0

- $|R(C_p \times V)| = 16$

	P_1	P_2	P_3	P_4	P_5	P_6
$R(C_p \times V, [C_{4p}])$	*3*	*0*	*0*	*0*	*0*	*0*
$R(C_p \times V, [C_p \times V])$	*1*	*0*	*0*	*0*	*0*	*0*
$R(C_p \times V, [E_p])$	*0*	*0*	*0*	*0*	*0*	*0*
$R(C_p \times V, [D_{2p}])$	*3*	*1*	*1*	*1*	*0*	*0*
$R(C_p \times V, [Q_p])$	*3*	*1*	*1*	*1*	*0*	*0*

- $|R(E_p)| = 8p + 2$ *if* $p \equiv 1 \pmod 4$

	P_1	P_2	P_3	P_4	P_5	P_6
$R(E_p, [C_{4p}])$	p	*0*	*0*	*0*	*0*	*0*
$R(E_p, [C_p \times V])$	p	*0*	*0*	*0*	*0*	*0*
$R(E_p, [E_p])$	p+*1*	*0*	*0*	*0*	p	*1*
$R(E_p, [D_{2p}])$	p	p	*0*	*0*	*0*	*0*
$R(E_p, [Q_p])$	p	p	*0*	*0*	*0*	*0*

- $|R(D_{2p})| = 12p + 4$

	P_1	P_2	P_3	P_4	P_5	P_6
$R(D_{2p}, [C_{4p}])$	*3*p	*0*	*0*	*0*	*0*	*0*
$R(D_{2p}, [C_p \times V])$	p	*0*	*0*	*0*	*0*	*0*
$R(D_{2p}, [E_p])$	*0*	*0*	*0*	*0*	*0*	*0*
$R(D_{2p}, [D_{2p}])$	*2*p+*1*	*1*	p	p	*0*	*0*
$R(D_{2p}, [Q_p])$	*2*p+*1*	*1*	p	p	*0*	*0*

- $|R(Q_p)| = 6p + 4$ *if* $p \equiv 1 \pmod 4$ *or* $2p + 4$ *if* $p \equiv 3 \pmod 4$

	P_1	P_2	P_3	P_4	P_5	P_6
$R(Q_p, [C_{4p}])$	p	*0*	*0*	*0*	*0*	*0*
$R(Q_p, [C_p \times V])$	p	*0*	*0*	*0*	*0*	*0*
$R(Q_p, [E_p])$	*2*p	*0*	*0*	*0*	p	p
$R(Q_p, [D_{2p}])$	*1*	*1*	*0*	*0*	*0*	*0*
$R(Q_p, [Q_p])$	*1*	*1*	*0*	*0*	*0*	*0*

6.4.2. Groups of order $(2q+1)2q$**.** In [**Koh13**] a similar analysis is done for the groups of order mp with $p > m$. In particular, two classes of cases are analyzed. In the first, a recapitulation of the enumeration done by Byott in [**Byo04c**] for $m = q$, prime where either $p \equiv 1 \pmod 4$, or $p \not\equiv 1$, is done as a test of these techniques, as they fit for these m and p, and the results are tabulated here for the $2 \cdot 2$ different pairings $R(G, [M])$, where the column corresponds to G and the row to $[M]$:

$G \setminus M$	C_{pq}	$C_p \rtimes C_q$
C_{pq}	1	$2(q-2)$
$C_p \rtimes C_q$	p	$2(p(q-2)+1)$

In addition, the enumeration of $R(G, [M])$ was performed for the case where $p = 2q + 1$, for q a prime, making p a "safe prime" and $m = \varphi(p) = p - 1$, whence $n = mp = p(p-1) = 2q(2q+1)$. In this case, there are six different classes of groups, C_{mp}, $C_p \times D_q$, $F \times C_2$, $C_q \times D_p$, D_{pq} and $\mathrm{Hol}(C_p)$ where $F \cong C_p \rtimes C_q$, and the enumeration of all $R(G, [M])$ is summarized here:

$G \setminus M$	C_{mp}	$C_p \times D_q$	$F \times C_2$	$C_q \times D_p$	D_{pq}	$\mathrm{Hol}(C_p)$
C_{mp}	1	2	$2(q-1)$	2	4	$2(q-1)$
$C_p \times D_q$	q	2	0	$2q$	4	0
$F \times C_2$	p	$2p$	$2(p(q-2)+1)$	$2p$	$4p$	$2p(q-1)$
$C_q \times D_p$	p	$2p$	$2p(q-1)$	2	4	$2p(q-1)$
D_{pq}	qp	$2p$	0	$2q$	4	0
$\mathrm{Hol}(C_p)$	p	$2p$	$2p(q-1)$	$2p$	$4p$	$2(p(q-2)+1)$ (*)

where we note that the case marked (*) was also considered by Childs in [**Chi03**]. We note again, $R(G,[M])$ is odd only for groups where the order p subgroup is a direct factor, since in that case $P(N) = P(N^{opp})$, for example $R(G,[C_{mp}])$. For $M = F \times C_2, C_q \times D_p, D_{pq}$ and $\mathrm{Hol}(C_p)$, the counts are even, accounting for the distribution of those N for which $P(N) = \mathcal{P}$, which equals the distribution of those where $P(N) \neq \mathcal{P}$.

6.4.3. Metacyclic groups. We note that the groups of order $4p$ and $2q(2q+1)$ above are all metacyclic, and for the safeprime examples $m = p-1 = 2q$. A closely related class of examples we can therefore consider are the metacyclic groups $C_p \rtimes C_m$ where $m|p-1$. The enumeration can be done using similar analyses as were used to enumerate the safeprime cases.
If we let $\mathbb{F}_p^\times = \langle b \rangle$ and define h and u by the conditions $mh = p-1$ and $b^h = u$ then for e a proper divisor of m let

$$\begin{aligned} F_e &= C_p \rtimes_e C_m \\ &= \langle x, y | x^p = y^m = 1, yx = x^{u^e} y \rangle \end{aligned}$$

and for $e = 1, m$ we have $F_1 = \mathrm{Hol}(C_p)$ and $F_m = C_p \times C_m \cong C_{mp}$.
The values of $|R(G,[M])| = |R(F_e,[F_d])|$ as d, e vary over the divisors of m are as follows.

THEOREM 6.6.

$$|R(F_e,[F_d])| = \begin{cases} 2p\varphi(m/d) & e \neq d, d \neq m \\ p & e < m, d \neq m \\ p & e < m, d = m \\ 2(p(\varphi(m/d)-1)+1) & d = e < m \\ 1 & d = e = m \end{cases}$$

We should point out that two of the safe-prime examples are not of the form $C_p \rtimes C_m$, namely $D_{pq} \cong C_{pq} \rtimes C_2$, and $C_p \times D_q$. However, the other four are, as these correspond to the four divisors of $m = 2q$, namely $1, 2, q, 2q$. For these four, the counts above are in sync, for example for $d = e = 1$ we have

$$2(p(\varphi(2q/1)-1)+1) = 2(p(q-2)+1)$$

which agrees with the above table for $R(\mathrm{Hol}(C_p), [\mathrm{Hol}(C_p)])$.

6.4.4. Groups of order $\mathbf{p_1p_2p_3}$. In [**Koh16**], $n = |G|$ is the squarefree number $p_1p_2p_3$ where $p_1 < p_2 < p_3$ are primes with "p" being p_3 and $m = p_1p_2$. It is a standard result that $(p,m) \in F_S$ when $p > m$, but it is also possible when $p < m$. Moreover, the automorphism groups of any group of order p_1p_2 all have order relatively prime to p_3. The groups of order $p_1p_2p_3$ are iterated semidirect products, the number of which depend on the congruence classes of p_3 mod p_1 and p_2 and the class of p_2 mod p_1, generalizing the enumeration of groups of order pq. In [**Koh16**] the author focused on the case where $p_3 \equiv 1 \pmod{p_1}$, $p_2 \equiv 1 \pmod{p_1}$, but $p_3 \not\equiv 1 \pmod{p_2}$ as this forces $p < m$ which contrasts with the previously studied collections of groups of order mp. What one finds is that the enumeration of $R(G,[M])$ follows some similar "motifs" as seen in the safeprime case above. However, for $R(G,[M])$, where G and M are semidirect products $C_{p_3p_2} \rtimes C_{p_1}$, the order p_1 generator acts non-trivially on the generators of order p_3 *and* p_2. In total there are p_1+2 cases overall, p_1-1 of which correspond to elements of order p_1 in $U_{p_2p_3}$ which arise when $p_3 \equiv 1 \bmod p_2$ and p_3. So for these, not only does one have the larger centralizer $\mathrm{Norm}_B(\mathcal{P})$, where $\mathcal{P} \cong C_{p_3}$, but the complementary group of order $m = p_1p_2$, itself has a Sylow subgroup. This Sylow subgroup (of order p_2) is normalized by the complementary group and so this complement is contained within a smaller twisted wreath product. So for any $N \in R(G,[M])$, this leads to a separate, but similar, enumeration problem for the possible $Q(N)$ that may arise.

THEOREM 6.7 ([**Koh16**]). *If we define*

$$f(a,b) = 2(a(b-2)+1)$$
$$g(a,b) = 2a(b-1)$$

then the values of each $|R(G,[M])|$ *are as given here:*

$Gm \setminus M$	$C_{p_3p_2p_1}$	$C_{p_3} \times (C_{p_2} \rtimes C_{p_1})$	$C_{p_2} \times (C_{p_3} \rtimes C_{p_1})$	$C_{p_3p_2} \rtimes_i C_{p_1}$
$C_{p_3p_2p_1}$	1	$g(1,p_1)$	$g(1,p_1)$	$2g(1,p_1)$
$C_{p_3} \times (C_{p_2} \rtimes C_{p_1})$	p_2	$f(p_2,p_1)$	$g(p_2,p_1)$	$2f(p_2,p_1)$
$C_{p_2} \times (C_{p_3} \rtimes C_{p_1})$	p_3	$g(p_3,p_1)$	$f(p_3,p_1)$	$2f(p_3,p_1)$
$C_{p_3p_2} \rtimes_j C_{p_1}$	p_3p_2	$p_3f(p_2,p_1)$	$p_2f(p_3,p_1)$	-

i,j	$\lvert R(C_{p_3p_2} \rtimes_j C_{p_1}, [C_{p_3p_2} \rtimes_i C_{p_1}])\rvert$
$j = i, -i$	$2(p_3+p_2+(2p_1-5)p_2p_3+1)$
$j \neq i, -i$	$2(2p_3+2p_2+(2p_1-6)p_2p_3)$

6.4.5. Square free groups of order $\mathbf{p_1p_2\cdots p_n}$ in general. For groups of square-free order in general, namely $n = p_1p_2\cdots p_r$ for $p_1 < \cdots < p_r$ we can let $p = p_r$ and $m = p_1\cdots p_{r-1}$, the question is whether $(p,m) \in F_Q \cap F_S$. That $(p,m) \in F_S$ is automatic (as in the $r=3$ case above), since p_r is the largest prime divisor, but the question to consider is whether $(p,m) \in F_Q$. To address this, we use the following small but very useful fact due to Birkhoff and Hall in [**BH36**].

THEOREM 6.8. *If* $|G| = p_1^{n_1}p_2^{n_2}\cdots p_r^{n_r}$ *and* $\theta(p^n) = \prod_{i=0}^{n-1}(p^n - p^i)$ *then*
(a) $|\mathrm{Aut}(G)|$ *divides* $\theta(p_1^{n_1})\cdots\theta(p_r^{n_r})|G|^{r-1}$,
(b) if G *is solvable,* $|\mathrm{Aut}(G)|$ *divides* $\theta(p_1^{n_1})\cdots\theta(p_r^{n_r})|G|$,
(c) if G *is nilpotent,* $|\mathrm{Aut}(G)|$ *divides* $\theta(p_1^{n_1})\cdots\theta(p_r^{n_r})$.

For our purposes, we note that $\theta(p) = p-1$, and that all squarefree groups are solvable. So if $|G| = p_1p_2\cdots p_r$ where $p_1 < \cdots < p_r$ then the Sylow p_r-subgroup is unique and $p = p_r \nmid |\mathrm{Aut}(Q)|$ where $|Q| = p_1\cdots p_{r-1} = m$, since by the above, $|\mathrm{Aut}(Q)|$ divides $(p_1-1)\cdots(p_{r-1}-1)p_1\cdots p_{r-1}$. Thus this program may be applied to *all* groups of square-free order.

6.5. Block systems and Hopf-Galois structures

In the above discussion, we used the term *block* as synonymous with the supports of the cycle structure of the Sylow p-subgroup of a regular permutation group of order mp. Moreover, we utilized the fact that the groups that contain these subgroups, act on these blocks, and indeed the structure of $\mathrm{Cent}_B(\mathcal{P})$ and $\mathrm{Norm}_B(\mathcal{P})$

can be framed in terms of how they act on these blocks, being maximal with respect to the property of centralizing or normalizing the semiregular subgroups which give rise to these blocks.

However, the combinatorial nature of block structure can be utilized in and of itself to aid in the enumeration of the regular permutation groups that give rise to Hopf-Galois structures. This was applied recently to the case of enumeration of $R(D_n, [D_n])$ for D_n the n-th dihedral group, of order $2n$. The key feature is how normal, and, in particular, characteristic subgroups of regular permutation groups corresponds to the block structure of these groups and their normalizers.

If $N \leq \mathrm{Perm}(Z)$ is a regular permutation group, then the regularity implies that any nontrivial $K \leq N$ is semiregular. By regularity, for any $z \in Z$, one has $\mathrm{Orb}_N(z) = Z$ of course, and for $K \leq N$, one has $|\mathrm{Orb}_K(z)| = |K|$ and more generally that Z is partitioned into $[N : K]$ orbits of size $|K|$. If $K \triangleleft N$ then the normality of K in N implies an action of N on this set of orbits. For a group G which acts on N by automorphisms, this implies an action of G on these orbits as well. One may infer information about the action of G on N by means of studying the accompanying action on these orbits. If we start with a given group $N \leq \mathrm{Perm}(Z)$ then any subgroups of it give rise to orbits, and if G acts on N as well, then one may correlate the orbits with respect to subgroups of G as compared to those of equal sized subgroups of N. And this correlation may therefore, in turn, dictate if and how G acts on N by conjugation.

As mentioned above, we somewhat casually used the term block as synonymous with the supports of the cycles that make up the generator(s) of the Sylow p-subgroup of a regular permutation group of order mp, but for this more general situation, we remind the reader of the formal definition.

DEFINITION 6.9. If $N \leq \mathrm{Perm}(Z)$ is a transitive subgroup, then a block is a subset $X \subseteq Z$ such that for all $\eta \in N$ one has $\eta(X) = X$ or $\eta(X) \cap X = \emptyset$.

In the setting of regularity, blocks arise very naturally.

PROPOSITION 6.10. *If $N \leq \mathrm{Perm}(Z)$ is a regular subgroup, and if $K \triangleleft N$, then, for any $z \in Z$, if $X = \mathrm{Orb}_K(z)$ then X is a block for N.*

PROOF. If $X = \mathrm{Orb}_K(z) = \{k(z) \mid k \in K\}$ then for $\eta \in N$ we have $\eta(X) = \{\eta k(z) \mid k \in K\}$. If $\eta k_1(z) = k_2(z)$ for some $k_1, k_2 \in K$ then by normality we have $\eta k_1 = k_3 \eta$ so $\eta(z) = k_3^{-1} k_2(z)$. But by regularity this implies that $\eta = k_3^{-1} k_2 \in K$. □

More generally, we consider the case where $N \leq \mathrm{Perm}(Z)$ where Z is the disjoint union of blocks of N.

DEFINITION 6.11. For $N \leq \mathrm{Perm}(Z)$, a *block system* for the action of N is a partition of Z that is N-invariant.

When $N \leq \mathrm{Perm}(Z)$ is regular then normal subgroups give rise to block systems that are termed "normal block systems", and we have the following which is basically [**DM96**, Theorem 1.6A (i)].

PROPOSITION 6.12. *If $N \leq \mathrm{Perm}(Z)$ is a regular subgroup and $K \triangleleft N$ where $\eta_1 K, \eta_2 K, \ldots, \eta_m K$ are the distinct cosets, then if $X_i^K = \eta_i K(z)$ then $\{X_i^K\}$ is a normal block system for N, where each $|X_i^K| = |K|$.*

The main takeaway from this is that, for $\eta \in N$, and $\eta_i K$ a coset, one has $\eta\eta_i K = \eta_j K$ for some η_j, so that $\eta(X_i^K) = X_j^K$ and $\eta(X_i^K) = X_i^K$ if and only if $\eta \in K$, so that N acts transitively on the blocks $\{X_i^K\}$.

The reason for this is that, assuming $K \neq N$, it is semiregular, namely fixed point free, but not transitive since its size is less than $|N| = |Z|$. As such, for any given $z \in Z$, one may restrict K to $K|_{\mathrm{Orb}_K(z)}$ which will be a regular subgroup of $\mathrm{Perm}(\mathrm{Orb}_K(z))$ and therefore K embeds diagonally in $\mathrm{Perm}(X_1^K) \times \mathrm{Perm}(X_2^K) \times \cdots \times \mathrm{Perm}(X_m^K) \leq \mathrm{Perm}(Z)$ since Z is the union of the $\{X_i^K\}$. Moreover, as $K \triangleleft N$ then we can view N/K as acting on these X_i^K in the obvious way which may be viewed as a *regular* permutation group on the $\{X_i^K\}$.

Given a normal block system $\{X_i^K\}$ for Z, one may ask for the maximal subgroup $W(\{X_i^K\})$ of $\mathrm{Perm}(Z)$ which respects the $\{X_i^K\}$. For a normal block system, we have that all X_i^K have the same size, so if $|X_i^K| = n$ then $|Z| = mn$. Of interest to the dihedral groups is the case where $m = 2$ of course, but we can say a bit more, to highlight the utility of this analysis to cases beyond the dihedral groups.

A permutation in $\omega \in W(\{X_i^K\})$ must act on any given block so that either $\omega(X_i^K) = X_i^K$ or $\omega(X_i^K) \cap X_i^K = \emptyset$, but since this is true for the entire block system $\{X_i^K\}$ then W is generated by permutations which either map X_i^K to itself, or bijectively to some other X_j^K. If $X_i^K = \{x_{i1}, \ldots, x_{in}\}$ then for $\tau_{ij} = (x_{i1}, x_{j1}) \cdots (x_{in}, x_{jn})$ we have that $\Sigma_m = \langle\{\tau_{ij} \mid i \neq j\}\rangle \cong S_m$. And it's clear that the elements of Σ_m preserve the blocks $\{X_i^K\}$ but act trivially on the elements of any given X_i^K. (i.e. They act as bijections $X_i^K \to X_j^K$.) If we embed $\mathrm{Perm}(X_i^K)$ into $\mathrm{Perm}(Z)$ as the set of those permutations with support equal to X_i^K then, by a slight abuse of notation, we have the following.

PROPOSITION 6.13. *The maximal subgroup of* $\mathrm{Perm}(Z)$ *which preserves the normal block system* $\{X_i^K\}$ *is:*

$$\begin{aligned} W(\{X_i^K\}) &= \mathrm{Perm}(X_1^K)\,\mathrm{Perm}(X_2^K) \cdots \mathrm{Perm}(X_m^K) \cdot \Sigma_m \\ &\cong \mathrm{Perm}(X_1^K) \times \mathrm{Perm}(X_2^K) \times \cdots \times \mathrm{Perm}(X_m^K) \rtimes \Sigma_m \\ &\cong S_n^m \rtimes S_m \ (i.e.\ S_n \wr S_m) \end{aligned}$$

where for any $\sigma_i \in Perm(X_i^K)$ and $\tau \in \Sigma_m$, the action of $\omega = \sigma_1 \cdots \sigma_m \tau$ on $z \in Z$ is given by first determining $\tau(z)$. Since $z \in X_i$ for some i, therefore $\tau(z) \in X_j$, and so $\omega(z) = \sigma_j(z)$.

It is interesting to note the relationship of this wreath product to that seen in the mp analysis, in particular the centralizer $\mathrm{Cent}_B(\mathcal{P})$ whose structure was known by virtue of the following general fact discovered by Burnside.

PROPOSITION 6.14. [**Bur11**, Section 171] *If $\sigma = \sigma_1 \cdots \sigma_m \in S_{mn}$ where the σ_i are disjoint n-cycles then* $\mathrm{Cent}_{S_{mn}}(\langle\sigma\rangle) \cong C_n \wr S_m$.

The relation of Proposition 6.14 to the wreath product in Proposition 6.13 is that in this centralizer, the blocks are the supports of the n-cycles, but we are restricting to those permutations which preserve the ordering of the elements in each block. For the wreath product in Proposition 6.13 we are permitting the elements within a given block (of size n) to be shuffled arbitrarily. As such, the "S_n" factor is replaced by the "C_n" factor.

One may question whether this definition is dependent on how the X_i are ordered, especially given the definition of Σ_m. However, the order two generators

τ_{ij} given above for Σ_m represent particular bijections $X_i^K \to X_j^K$. But if we post-compose τ_{ij} with an element of $\text{Perm}(X_j^K)$ then we get a *different* bijection from X_i^K to X_j^K but this too is an element of $W(\{X_i\})$. As such, a different initial ordering merely gives rise to an isomorphic Σ'_m which when combined with the $\text{Perm}(X_i^K)$ yields the same group W.

So, if $K \triangleleft N$ gives rise to a normal block system then we have the following.

PROPOSITION 6.15. *If $K \triangleleft N$ where $N \leq \text{Perm}(Z)$ is regular, then $N \leq W(\{X_i^K\})$ for the block system $\{X_i^K\}$ for Z arising from K.*

Going beyond normal subgroups, if $K \triangleleft N$ is, in fact, characteristic in N then this implies that $K \triangleleft \text{Norm}_B(N)$, and so we have this obvious, yet important, fact.

PROPOSITION 6.16. *If $K \triangleleft N$ gives rise to the block system $\{X_i^K\}$ and if K is characteristic in N then $\text{Norm}_B(N) \leq W(\{X_i^K\})$.*

For Hopf-Galois structures on G-Galois extensions, we may apply the above result to a given $N \in R(G, [N])$ since if $K \triangleleft N$ is characteristic then $\text{Norm}_B(N) \leq W(\{X_i^K\})$ and $\lambda(G) \leq \text{Norm}_B(N)$. Therefore $\lambda(G) \leq W(\{X_i^K\})$ as well, which implies that $\lambda(G)$ preserves the block structure established by K. Of course, G determines $N \in R(G, [N])$ but what we see is that any such N must have block structure preserved by $\lambda(G)$.

For the case of $N \in R(G, [G])$, both $\lambda(G)$ and N have identical subgroup structures. So if $K_1(G), \ldots, K_m(G)$ are normal subgroups of G of the same size, and if amongst these, only $K_1(G)$ is characteristic, then the same is true for any $N \in R(G, [G])$. Each $\lambda(K_t(G))$ gives rise to a block system $\{X_i^{K_t(G)}\}$. For a given $N \in R(G, [G])$ there are similar normal subgroups $K_t(N)$ which give rise to block systems $\{X_i^{K_t(N)}\}$. And if we assume $K_1(N)$ is characteristic then $\text{Norm}_B(N) \leq W(\{X_i^{K_1(N)}\})$, and is the unique block structure for $\text{Norm}_B(N)$.

The application of this to Hopf-Galois structures is that, as $\lambda(G) \leq \text{Norm}_B(N) \leq W(\{X_i^{K_1(N)}\})$ then, as a collection, $\{X_i^{K_1(N)}\} = \{X_i^{K_t(G)}\}$ for some $t \in \{1, \ldots, m\}$. As such, $R(G, [G])$ may be subdivided as $R(G, [G]; \{X_i^{K_t(G)}\})$ corresponding to those N where $K_1(N)$ (the unique characteristic subgroup) gives rise to the block structure $\{X_i^{K_t(G)}\}$.

This was done for the case of $G = D_n$, the nth dihedral group of order $2n$ where D_n either has one (n odd) or three (n even) subgroups of index 2, but only one of which is characteristic, namely the only cyclic subgroup of order n. As such, $\lambda(D_n) \leq W(\{X_i^{K_t(D_n)}\})$ for either one $K_t(D_n)$ if n is odd, or three if n is even. The utility of this method is that the action of $\lambda(D_n)$ on these subgroups of index 2 is reflected in how it acts on the associated blocks, and for any N which has similar block structure. This action allows us to determine how $\lambda(G)$ acts on the block structure arising from looking at the unique characteristic subgroup of $K(N) \leq N$, and how many such characteristic subgroups exist.

When n is odd, for example, the block structure corresponding to the characteristic subgroup of $G = D_n$ forces the block structure of any $N \in R(D_n, [D_n])$ to be the same. For n even, the block structure of the characteristic subgroup $K(N) \leq N$ for any $N \in R(D_n, [D_n])$ may be any one of the three block structures induced by the three normal subgroups of $G = D_n$ of index 2. One of the other facts which is utilized is that for the case of the dihedral groups, this characteristic subgroup is the characteristic subgroup of *exactly one* N. And so the enumeration of $R(D_n, [D_n])$ is directly correlated to the enumeration of these characteristic subgroups, which are, as indicated, correlated to the block structures established by the normal index two subgroups of $G = D_n$. The end result is the following count.

THEOREM 6.17. [**Koh20**] *For the dihedral group D_n of order $2n$*

$$R(D_n, [D_n]) = \begin{cases} (\frac{n}{2}+2)|\Upsilon_n| & \text{if } v_2(n) \geq 3 \\ (\frac{n}{2}+1)|\Upsilon_n| & \text{if } v_2(n) = 2 \\ (n+1)|\Upsilon_n| & \text{if } v_2(n) = 1 \\ |\Upsilon_n| & \text{if } v_2(n) = 0 \end{cases}$$

where $\Upsilon_n = \{x \in U_n \mid x^2 = 1\}$ (the group of units mod n of exponent 2) and $v_2(n)$ is the 2-adic valuation of n.

The presence of the factor $|\Upsilon_n|$, in particular for the case where n is odd, arises from the fact that it is the size of $\mathcal{H}(D_n)$ which is the set of those regular subgroups $N \leq \mathrm{Hol}(D_n)$ which are isomorphic to D_n (i.e. $\lambda(D_n)$) and have the same normalizer as $\lambda(D_n)$. It is easy to deduce that $\mathcal{H}(D_n) \subseteq R(D_n, [D_n])$, but for n odd, they are, in fact, equal. And indeed, the enumeration of the $N \in \mathcal{H}(D_n)$ was also determined by the unique characteristic subgroups they contain, as detailed in [**Koh15**].

CHAPTER 7

On the Galois correspondence for Hopf-Galois structures

7.1. Introduction

In [**CS69**], Chase and Sweedler proved a weak form of the Fundamental Theorem of Galois Theory in a quite general setting, namely, given an R-Hopf algebra H and an H-Hopf-Galois extension of commutative rings S/R satisfying suitable conditions, there is an injective Galois correspondence from R-Hopf subalgebras to rings T with $R \subseteq T \subseteq S$, given by mapping an R-Hopf subalgebra J of H to the ring $T^J = \{t \in T \mid h(t) = \varepsilon(h)t \text{ for all } h \in J\}$. But the so-called strong form of the FTGT, that the Galois correspondence maps onto all intermediate rings between R and S, is in general not true for Hopf-Galois extensions, even when S/R is a field extension. The first explicit examples were given in [**GP87**]. See Remark 7.4 below.

So in most of this chapter we let L/K be a G-Galois extension of fields with an H-Hopf-Galois structure of type N, and want to determine, or estimate, the ratio $GC(L/K, H) =$

$$\frac{|\{\text{intermediate fields } E \text{ in the image of the Galois correspondence for } H\}|}{|\{\text{intermediate fields } E \text{ with } K \subseteq E \subseteq L\}|}.$$

We'll call $GC(L/K, H)$ the Galois correspondence ratio for the H-Hopf-Galois structure on L/K.

Of course the Galois correspondence problem is also of interest for Hopf-Galois structures on separable, non-normal extensions. Work on the non-normal problem was initiated by Greither and Pareigis [**GP87**], who defined a class of Hopf-Galois extensions, called *almost classical extensions*, for which the Galois correspondence maps onto all intermediate fields. Crespo, Rio and Vela have obtained numerous results on the existence or non-existence of Hopf-Galois structures on separable extensions L/K of small degree, and on the surjectivity of the Galois correspondence for those Hopf-Galois structures. See the last section of this chapter for more discussion of their results.

The main part of this chapter will only consider the Galois correspondence question for Galois extensions of fields, because, as we'll see, our approach is closely tied to skew braces. At this writing, there appears to be no generalization of skew braces that would relate to Hopf-Galois structures on non-normal separable field extensions.

So until otherwise stated, we assume that L/K is a G-Galois extension and study $GC(L/K, H)$ in that setting.

For L/K a G-Galois extension of fields which is also an H-Hopf-Galois extension, the denominator of the Galois correspondence ratio is equal to the number of subgroups of the Galois group G, by the Fundamental Theorem of Galois Theory. So the issue is to determine the numerator, the number of intermediate subfields E with $K \subseteq E \subseteq L$ which are the fixed fields $E = K^J$ for K-Hopf subalgebras J of H. We do that in two steps.

By Greither-Pareigis theory, $L \otimes_K H \cong L[N]$ for N a regular subgroup of $\mathrm{Perm}(G)$ normalized by $\lambda(G)$. In Section 7.2 we show that under base change from K to L, each K-Hopf subalgebra J of H becomes isomorphic to $L[N']$ for some subgroup N' of N where N' is normalized by $\lambda(G)$, and conversely. So the number of K-Hopf subalgebras of H is equal to the number of subgroups N' of N that are normalized by $\lambda(G)$.

For L/K a finite separable extension of fields which has an H-Hopf-Galois structure, Section 7.3 uses the result of Section 7.2 to give a proof of the injectivity of the Galois correspondence from K-Hopf subalgebras J of H to intermediate fields $E = L^J$ with $K \subseteq E \subseteq L$. This is the Chase-Sweedler result in the special case of a Hopf-Galois structure on a separable field extension. Working over fields permits a somewhat more direct proof than the general argument in [**CS69**].

The injectivity enables us to identify the numerator of the Galois correspondence ratio $GC(L/K, H)$ for a G-Galois extension with an H-Hopf-Galois structure as the number of $\lambda(G)$-invariant subgroups of N. Since the denominator of $GC(L/K, H)$ is equal to the number of subgroups of G, determining the Galois correspondence ratio becomes a matter of counting subgroups of G and $\lambda(G)$-invariant subgroups of N. As a consequence, the Galois correspondence ratio only depends on the skew brace associated to the type N Hopf-Galois structure on the G-Galois extension, as we'll see in Section 7.5.

Section 7.4 uses the injectivity of the Galois correspondence for Hopf-Galois structures to describe a method of Kohl ([**Koh19**]) to find examples of pairs of groups (G, N) where a G-Galois extension L/K of fields cannot have a Hopf-Galois structure of type N: in the terminology of Chapter 3, the pair (G, N) is not realizable.

Except for Section 7.10, the remainder of the chapter studies the Galois correspondence ratio in its skew brace setting, and computes the ratio for a variety of examples: Hopf-Galois structures arising from fixed point free pairs of homomorphisms (from Chapter 4), semidirect products of groups, radical rings, and commutative radical $\mathbb{F}_p$-algebras.

For later reference we recall some definitions, from [**Chi00**], Section 2.

DEFINITION 7.1. Let S be a commutative R-algebra and a finitely generated projective R-module. Suppose H is a cocommutative R-Hopf algebra and a finitely generated projective R-module.

Then S is an H-module algebra if S is an H-module and for all h in H, s, t in S, $h(st) = \sum_h h_{(1)}(s)h_{(2)}(t)$ where the comultiplication

$$\Delta : H \to H \otimes_R H$$

on H is described using Sweedler notation:

$$\Delta(h) = \sum_{(h)} h_{(1)} \otimes h_{(2)}$$

For S an H-module algebra, the smash product $S\#_R H$ is the R-algebra structure defined on the R-module $S \otimes_R H$, with multiplication

$$(s\#h)(t\#h') = sh_{(1)}(t)\#h_{(2)}h'$$

for s, t in S, h, h' in H. Then S/R is an H-Hopf-Galois extension if and only if the map

$$j : S\#H \to \mathrm{End}_R(S)$$

given by $j(s \otimes h)(t) = sh(t)$ is an R-algebra isomorphism.

It is routine that the property of being a Hopf-Galois extension is preserved under base change from R to some commutative R-algebra T which is a flat R-module. (In particular it is true if T/R is a field extension.)

7.2. K-Hopf subalgebras

Let L/K be a G-Galois extension and an H-Hopf-Galois extension. Our first task related to understanding the Galois correspondence ratio for L/K is to understand the Hopf subalgebras of H. To do so we prove a result that was first explicitly stated and proved as [**CRV16b**, Theorem 2.3].

THEOREM 7.2. *Let L/K be a G-Galois extension and an H-Hopf-Galois extension, corresponding to the regular subgroup N of* $\mathrm{Perm}(G)$ *normalized by $\lambda(G)$. Then the K-Hopf subalgebras of H correspond to the $\lambda(G)$-invariant subgroups of N, as follows: if H' is a K-Hopf subalgebra of H, then $L \otimes_K H' = L[N']$ for some subgroup N' of N that is normalized by $\lambda(G)$; conversely, if N' is a subgroup of N that is normalized by $\lambda(G)$, then $H' = (L[N'])^G$, the set of elements of $L[N']$ fixed under the action of G, is a K-Hopf subalgebra of H, where G acts on L via action of the Galois group G and on N' by conjugation by elements of $\lambda(G)$ in* $\mathrm{Perm}(G)$.

PROOF. Given N', a subgroup of $\mathrm{Perm}(G)$ normalized by $\lambda(G)$, $(L[N'])^G$ is a K-Hopf algebra by general Galois descent theory. The only issue is the converse. So suppose H' is a K-Hopf subalgebra of H. Then $L \otimes_K H'$ is an L-Hopf subalgebra of $L \otimes_K H = L[N]$ for some subgroup N of $\mathrm{Perm}(G)$. We need to show that $L \otimes_K H'$ is a group ring $L[N']$ for some $\lambda(G)$ invariant subgroup of N.

To show that $L \otimes_K H'$ is a group ring $L[N']$ for some subgroup N' of N, there is a general result:

LEMMA 7.3. *If H' is a Hopf subalgebra of a group ring $H = K[N]$, then $H' = K[N']$ for some subgroup N' of N.*

PROOF. Let $\langle\ ,\ \rangle : H \times H^* \to K$ be the duality map: $\langle h, f\rangle = f(h)$ for h in H, f in $H^* = \mathrm{Hom}_K(H, K)$. Let $N = \{g_1, \ldots, g_n\}$ and let $\{f_1, \ldots f_n\}$ be the dual basis for the basis $\{g_1, \ldots, g_n\}$ of N where $\langle g_i, f_j\rangle = \delta_{i,j}$, then $K[N]^* = \sum_{i=1}^n Kf_i$. The formula that defines the multiplication on the dual of a K-Hopf algebra:

$$\langle \Delta(g_i), f_j \otimes f_k\rangle = \langle g_i, f_j f_k\rangle$$

where $\Delta(g_i) = g_i \otimes g_i$, implies that the f_i are pairwise orthogonal idempotents of H^*, and conversely, if the f_i are pairwise orthogonal idempotents of H^* and $\{g_1, \ldots, g_n\} \subset H$ is the dual basis for $\{f_1, \ldots, f_n\} \subset H^*$ then the g_i are grouplike: that is, $\Delta(g_i) = g_i \otimes g_i$.

Now the inclusion map $\iota : H' \to H$ is a Hopf algebra homomorphism and yields a surjective Hopf algebra homomorphism $\iota^* : H^* \to H'^*$. Since H^* is spanned by pairwise orthogonal idempotents $f_1, \ldots, f_n$ and ι is an algebra homomorphism, the set $\{\iota^*(f_i) \mid \iota(f_i) \neq 0\}$ form a basis of pairwise orthogonal idempotents of H'^*. Hence H' has a basis of grouplike elements. Grouplike elements of a Hopf algebra over a field form a group ([**Chi00**], (1.6)), and so H' is a group ring $K[N']$. Since $H' \subset H$, the grouplike elements of N' are grouplike in H.

It is a standard fact that grouplike elements of a Hopf algebra are linearly independent ([**Und15**], Proposition 1.2.18). Since the elements $g_1, \ldots, g_n$ of N are grouplike and form a basis of H, there can be no grouplike elements of H other than the elements of N. Every g' in N' is a grouplike element of H', hence of H, so every g' in N' is in N. The multiplication in H' is the restriction to H' of the multiplication in H, and the set $N' \subset N$ of grouplike elements of H' form a group under that multiplication. Therefore N' is a subgroup of N. □

The rest of the proof of Proposition 7.2 follows from general Galois descent theory, for example as described in ([**Chi00**], (2.12)), as follows:

By Morita theory, there is an equivalence between the category of left K-modules and the category of left $\mathrm{End}_K(L)$-modules, given by the map $M \mapsto L \otimes_K M$. For L/K a G-Galois extension, $\mathrm{End}_K(L)$ is isomorphic to the smash product $D(L,G) = L\#_K K[G]$ by the map $j : L\#_K K[G] \to \mathrm{End}_K(L)$ given by $j(s\#g)(t) = sg(t)$. So an $\mathrm{End}_K(L)$ module is an L-module with a compatible action by G. If M is a $L\#_K K[G]$-module, then M^G, the elements of M fixed by the action of G, is a K-module. So the map from left $\mathrm{End}_K(L)$-modules and K modules is given by $M \mapsto M^G$, hence every left $\mathrm{End}_K(L)$-module has the form $M = L \otimes_K M^G$.

Now if H is a K-Hopf algebra and L/K is a G-Galois extension, then $L \otimes_K H$ is a $D(L,G)$-module. As we've seen, $L \otimes_K H = L[N]$ for some regular subgroup N of $\mathrm{Perm}(G)$ that is normalized by $\lambda(G)$, and every K-Hopf subalgebra H' of H has the property that $L \otimes_K H' = L[N']$ for N' some subgroup of N. Since the category of K-modules is equivalent to the category of $D(L,G)$-modules, $L[N']$ is a $D(L,G)$-submodule of $L[N]$. Hence $L[N']$ is closed under the action of $D(L,G)$ on $L[N]$. Since $D(L,G)$ acts on elements of N by conjugation by elements of $\lambda(G)$, N' must be a $\lambda(G)$-invariant subgroup of N, hence is also closed under conjugation by elements of $\lambda(G)$. Thus $H' = L[N']^G$.

To sum up, there is a bijection between K-Hopf subalgebras H' of H and $\lambda(G)$-invariant subgroups N' of N, given by

$$H' \mapsto L \otimes H' = L[N'], \text{ and } L[N'] \mapsto L[N']^G.$$

□

REMARK 7.4. It should be noted that this result was already used in [**GP87**]: for L/K a G-Galois extension, L/K has a Hopf-Galois structure by the K-Hopf algebra H_λ corresponding to the image of the left regular representation $\lambda : G \to \mathrm{Perm}(G)$. [**GP87**] showed that the K-Hopf subalgebras of H_λ correspond to the subgroups of $\lambda(G)$ that are normalized by $\lambda(G)$, that is, to the normal subgroups of $\lambda(G)$. Thus almost every Galois extension with a non-abelian group G has a Hopf-Galois structure of type G in which the Galois correspondence for a Hopf-Galois structure is not surjective. (The exceptions are where G is a non-abelian group in which every subgroup is normal, namely the Hamiltonian groups: groups of the form $Q \times U \times V$ where Q is the quaternion group of order 8, U is abelian of odd order and V is abelian of exponent 2 [**Hal59**, Theorem 12.5.4].)

7.3. On the Galois correspondence for Hopf-Galois structures

Now we give a proof of the injectivity of the Galois correspondence in our setting.

THEOREM 7.5. *Let E/K be a separable extension of fields with normal closure $\widetilde{L}$, let G be the Galois group of $\widetilde{L}/K$, and G' the Galois group of $\widetilde{L}/E$. Suppose E/K is an H-Hopf-Galois extension. Then the map from K-Hopf subalgebras J of H to intermediate subfields F, $K \subseteq F \subseteq E$ that sends J to E^J is injective.*

This is a special case of a general result for Hopf-Galois structures of Chase and Sweedler [**CS69**].

PROOF. In this proof, unadorned $\otimes$ means $\otimes_K$.

The idea of the proof is to base change from K to $\widetilde{L}$ and prove the result in that setting.

By Galois theory, $\widetilde{L} \otimes_K E = \mathrm{Map}(X, \widetilde{L})$ where X is the set G/G' of left cosets $gG' = \overline{g}$ of G' in G.

By Greither-Pareigis theory, $\widetilde{L} \otimes H = \widetilde{L}[N]$, where N is a regular subgroup of $\mathrm{Perm}(X)$ normalized by $\lambda(G)$. (See [**Chi00**], (6.8)).

Let J be a K-Hopf subalgebra of H and let

$$F = E^J = \{x \in E \mid h(x) = \varepsilon(h)x \text{ for all } h \text{ in } J\}.$$

It is routine to verify that $\widetilde{L} \otimes E$ is an $\widetilde{L} \otimes H$-Hopf-Galois extension of $\widetilde{L} \otimes K = \widetilde{L}$, where $\widetilde{L} \otimes J$ acts on $\widetilde{L} \otimes E$ by

$$(s \otimes h)(x \otimes y) = (s \otimes 1)(x \otimes h(y)).$$

We now show that $(\widetilde{L} \otimes E)^{\widetilde{L} \otimes J} = \widetilde{L} \otimes F$. Let $(x_1, \ldots, x_m)$ be a K-basis of $\widetilde{L}$, hence an E-basis of $\widetilde{L} \otimes E$:

$$\widetilde{L} \otimes_K E = \{\sum x_i \otimes y_i \mid y_i \in E\}.$$

Then since $s \otimes 1$ is invertible in $\widetilde{L} \otimes E$ for every $s \neq 0$ in $\widetilde{L}$,

$$(\widetilde{L} \otimes E)^{\widetilde{L} \otimes J}$$
$$= \left\{ \sum_i x_i \otimes y_i \,\middle|\, (s \otimes 1)(x_i \otimes h(y_i)) = (s \otimes 1)(x_i \otimes \varepsilon(h) y_i) \text{ for all } h \text{ in } J,\ s \text{ in } \widetilde{L} \right\},$$
$$= \left\{ \sum_i x_i \otimes y_i \,\middle|\, x_i \otimes (h(y_i) - \varepsilon(h) y_i) = 0 \text{ for all } h \text{ in } J \right\}.$$

Since the x_i are an E-basis of $\widetilde{L} \otimes E$ and $h(y) - \varepsilon(h)y = 0$ if and only if y is in E^J, this means that

$$(\widetilde{L} \otimes E)^{\widetilde{L} \otimes J} = \widetilde{L} \otimes E^J = \widetilde{L} \otimes F.$$

Recall that $\widetilde{L} \otimes H = \widetilde{L}[N]$ for some regular subgroup N of $\mathrm{Perm}(G)$, and $\widetilde{L} \otimes E \cong \mathrm{Map}(X, \widetilde{L}) = \sum_{\overline{g} \in X} \widetilde{L} e_{\overline{g}}$ where $\{e_{\overline{g}}\}$ is a dual basis for X.

Now N is a regular subgroup of $\mathrm{Perm}(X) = \mathrm{Perm}(G/G')$, so if we pick any h in X, the set Ne_h is the dual basis for X.

As we showed in Lemma 7.3, Hopf subalgebras of group rings are group rings of subgroups. So given the Hopf subalgebra J of H, $\widetilde{L} \otimes J = \widetilde{L}[N']$ for some subgroup N' of N.

Suppose we are given two K-Hopf subalgebras H_1 and H_2 of H. We want to show that $H_1 \subset H_2$ if and only if $\widetilde{L}^{H_2} \subset \widetilde{L}^{H_1}$. Base changing from K to $\widetilde{L}$, we want to show that $N_1 < N_2$ if and only if $\mathrm{Map}(X, \widetilde{L})^{\widetilde{L}[N_2]} \subset \mathrm{Map}(X, \widetilde{L})^{\widetilde{L}[N_1]}$.

Now $\mathrm{Map}(X, \widetilde{L})^{\widetilde{L}[N_1]}$ has an $\widetilde{L}$-basis consisting of the sums

$$c_{nN_1} = \sum_{m \in N_1} e_{nm}$$

for $T = \{n\}$, a transversal of N_1 in N, and similarly for $\mathrm{Map}(X, \widetilde{L})^{\widetilde{L}[N_2]}$. Call the support of c_{nN_1} the set of elements in the coset nN_1. Then two elements c_{nN_1} and $c_{n'N_1}$ have disjoint support unless $nN_1 = n'N_1$.

Now if $\mathrm{Map}(X, \widetilde{L})^{\widetilde{L}[N_2]} \subseteq \mathrm{Map}(X, \widetilde{L})^{\widetilde{L}[N_1]}$, then each c_{nN_2} in the basis for $\mathrm{Map}(X, \widetilde{L})^{\widetilde{L}[N_2]}$ is an $\widetilde{L}$-linear combination of elements c_{mN_1} for some elements m of N. Specializing to $c_{N_2} = \sum_{m \in N_2} e_m$, we have $c_{N_2} = \sum c_{nN_1}$ where the sum runs through some n in the transversal T. But then N_2, the support of c_{N_2}, is the union of the supports of c_{nN_1}. Since 1 is in N_2, c_{N_1} must be a summand of c_{N_2}. Therefore $N_1 \subset N_2$.

Since the opposite implication, if $N_1 \subset N_2$ then

$$\mathrm{Map}(X, \widetilde{L})^{E[N_2]} \subset \mathrm{Map}(X, \widetilde{L})^{E[N_1]}$$

is clear, the injectivity of the Galois correspondence from Hopf subalgebras of H to subrings of the H-Hopf-Galois extension is true after base change to $\widetilde{L}$.

To finish the proof, we observe that $\widetilde{L}$ is a free K-module, so given any homomorphism $f : M \to N$ of K-modules, f is injective if and only if $1 \otimes f : \widetilde{L} \otimes_K M \to \widetilde{L} \otimes_K N$ is injective. So the injectivity of the Galois correspondence is true before base change to $\widetilde{L}$. □

COROLLARY 7.6. *Let E/K be a separable extension which is an H-Hopf-Galois extension. If H_1 is a K-Hopf subalgebra of H, then $[E : E^{H_1}] = \dim_K H_1$.*

PROOF. As in the proof of Theorem 7.5, we base change from K to $\widetilde{L}$ where $\widetilde{L}$ is the normal closure of E/K, $\mathrm{Gal}(\widetilde{L}/K) = G, \mathrm{Gal}(\widetilde{L}/E) = G'$ and $X = G/G'$. Then $\widetilde{L} \otimes H = \widetilde{L}[N]$ for a regular subgroup N of $\mathrm{Perm}(X)$. Let H_1 be a Hopf subalgebra of H, then $\widetilde{L} \otimes H_1 = \widetilde{L}[N_1]$, and, as above, $\widetilde{L} \otimes_K E^{H_1} = \mathrm{Map}(X, \widetilde{L})^{\widetilde{L}[N_1]}$ has an $\widetilde{L}$ basis $\{c_{nN_1} = \sum_{m \in N_1} e_{nm}\}$ where n runs through a transversal of N_1 in N. So

$$\dim_K E^{H_1} = \dim_{\widetilde{L}} \widetilde{L} \otimes \widetilde{L}^{H_1} = [N : N_1] = \dim_K H / \dim_K H_1$$

and

$$\dim_K H = |N| = \dim_{\widetilde{L}}(\widetilde{L} \otimes_K E) = \dim_K E.$$

So $\dim_{E^{H_1}} E = |N_1| = \dim_K H_1$. □

With this result, we know that if L/K is a G-Galois extension and an H-Hopf-Galois extension where $L \otimes_K H = L[N]$ for N a regular subgroup of $\mathrm{Perm}(G)$ that is normalized by $\lambda(G)$, then by the last result, there is a one-to-one correspondence between subfields fixed under the action of K-Hopf subalgebras of H and the K-Hopf subalgebras of H; and the K-Hopf subalgebras of H correspond to $\lambda(G)$-invariant subgroups of N. Thus in the Galois correspondence ratio

$$GC(L/K, H) = \frac{|\{\text{fields } E \text{ in the image of the Galois correspondence for } H\}|}{|\{\text{fields } E \text{ with } K \subseteq E \subseteq L\}|},$$

the denominator is equal to the number of subgroups of the Galois group G, and the numerator is equal to the number of $\lambda(G)$-invariant subgroups of N:

$$GC(L/K,H) = GC(G,N) = \frac{|\lambda(G)\text{-invariant subgroups of } N|}{|\text{subgroups of } G|},$$

and hence depends only on G and N and how they interact. Thus the dependence of the Galois correspondence ratio on the skew brace corresponding to the Hopf-Galois structure on the Galois extension is evident, and will be made explicit beginning in Section 7.5.

EXAMPLE 7.7. Here are some initial examples where, as in [**GP87**], $N = \lambda(G)$ and so the numerator counts the normal subgroups of G.

Let G be a simple non-abelian group and consider the Hopf-Galois structure by H_λ, corresponding to the regular subgroup $\lambda(G) \subset \mathrm{Perm}(G)$. Then

$$GC(G,\lambda(G)) = \frac{2}{\text{number of subgroups of } G}.$$

Thus for the alternating group A_5 ,

$$GC(A_5,\lambda(A_5)) = \frac{2}{59}.$$

Similarly, the symmetric group S_5 has three normal subgroups, so

$$GC(S_5,\lambda(S_5)) = \frac{3}{156}$$

and

$$GC(S_8,\lambda(S_8)) = \frac{3}{151221}.$$

Evidently the Galois correspondence ratio can be very close to 0.

7.4. Kohl's application of Corollary 7.6

Before returning to the study of the Galois correspondence ratio, we present a nice result on the non-realizability of certain pairs (G,N) as the Galois group, resp. type, of a G-Galois extension L/K with a Hopf-Galois structure of type N. The proof uses the injectivity of the Galois correspondence for Hopf-Galois structures and Corollary 7.6.

Recall from Chapter 3 that $e(G,N)$ is the number of Hopf-Galois structures of type N on a G-Galois extension L/K of fields.

From [**Koh19**] is:

THEOREM 7.8. *Let G and N be two groups of order n. Suppose that for some divisor m of n, G has no subgroup of order m and N has a characteristic subgroup of order m. Then $e(G,N) = 0$. More generally, if N has more characteristic subgroups of order m than G has subgroups of order m, then $e(G,N) = 0$.*

PROOF. Suppose L/K has a Hopf-Galois structure corresponding to a subgroup M of $\mathrm{Perm}(G)$ of type N. Then the FTGT for the Hopf-Galois structure gives a one-to-one correspondence from subgroups M_1 of M that are normalized by $\lambda(G)$ in $\mathrm{Perm}(G)$ to intermediate fields E with $K \subseteq E \subseteq L$, by sending M_1 to the corresponding K-Hopf subalgebra $H_1 = L[M_1]^G$, then to the corresponding fixed field $E = L^{H_1}$. Furthermore, $|M_1| = [L:E]$.

If N has a characteristic subgroup of order m, then $M \cong N$ has a characteristic subgroup M_1 of order m, and so any automorphism of M restricts to an automorphism of M_1. In particular, M_1 is normalized by $\lambda(G)$. So by the FTGT for Hopf-Galois structures, there is an intermediate subfield E with $[L : E] = m$. But if L/K is Galois with Galois group G, then by the classical FTGT, every intermediate subfield E of L is the fixed field of a subgroup M of G, and $[L : E] = |M|$. So G has a subgroup of order m, contrary to assumption. So $e(G, N) = 0$.

More generally, the number r of intermediate fields E of L/K with $[L : E] = m$ is equal to the number of subgroups of G of order m. If N has more than r characteristic subgroups of order m, then an H-Hopf-Galois structure of type N on L/K would have more than r Hopf subalgebras H_i with $\dim_K(H_i) = m$, hence by the Galois correspondence from Hopf subalgebras of H to intermediate fields would yield more than r intermediate fields E with $[L : E] = m$, contradicting the classical Fundamental Theorem of Galois Theory. □

[**Koh19**] provides an extensive collection of examples.

For $|G|$ of order 12: suppose the Galois group G is isomorphic to the alternating group A_4. There is no subgroup of A_4 of order 6, while the quaternion group Q_3, the cyclic group C_{12} and the dihedral group D_6 have characteristic subgroups of order 6. So $e(G, N) = 0$ for $G = A_4$ and $N = Q_3, C_{12}$ or D_6. (The only other pairs (G, N) with G of order 12 where $e(G, N) = 0$ are where $(G, N) = (D_6, A_4)$ and (C_{12}, A_4).)

For $|N|$ of order 24, among the 15 groups of order 24, ten have one characteristic subgroup of order 12 and two have three characteristic subgroups, namely $C_4 \times S_3$ and $(C_6 \times C_2) \rtimes C_2$. Since $SL_2(\mathbb{F}_3)$ has no subgroups of order 12, and $C_{24}, S_4, C_3 \rtimes C_8$ and $C_2 \times A_4$ have just one subgroup of order 12, it follows that there are 20 pairs (G, N) with $e(G, N) = 0$ because of the criterion of Theorem 7.8.

[**Koh19**] includes a complete set of computations of $e(G, N)$ for $n = 24$ that show that $e(G, N) = 0$ for 76 pairs (G, N) of groups of order 24. (The largest values of $e(G, N)$ are $e(G, N) = 456$ for $(G, N) = (C_2 \times C_2 \times S_3, (C_6 \times C_2) \rtimes C_2)$ and $(C_2 \times C_2 \times S_3, (C_4 \rtimes S_3)$.) That paper also provides a table of the number of cases where $e(G, N) = 0$ and the number of those cases where Theorem 7.8 applies for selected composite numbers $n \leq 180$.

7.5. A skew brace setting

Returning to the Galois correspondence ratio, it is convenient to translate that ratio to the setting of skew braces.

For a G-Galois extension L/K with the Hopf-Galois structure corresponding to a regular subgroup N of $\mathrm{Perm}(G)$ normalized by $\lambda(G)$, let B be the corresponding skew brace (described in Chapter 2). Thus $B = (B, \circ, \cdot)$ where $(B, \circ) \cong G$ and $(B, \cdot) = N$. We can study the image of the Galois correspondence for the H-Hopf-Galois structure on a G-Galois extension L/K of fields by looking at certain sub-braces of the brace B associated to G and N, called in [**Chi17**] the $\circ$-stable ("circle-stable") subgroups of B.

DEFINITION 7.9. Given a skew brace $(B, \circ, \cdot)$ with additive group $(B, \cdot)$, a $\circ$-stable subgroup of $(B, \cdot)$ is a subgroup J of $(B, \cdot)$ so that for all x in B, y in J, $(x \circ y) \cdot x^{-1}$ is in J (where x^{-1} is the $\cdot$-inverse of x).

A $\circ$-stable subgroup J of $(B,\cdot)$ is also a subgroup of $(B,\circ)$. For if x,y are in J, then $(x\circ y)\cdot x^{-1}$ is in J, and since J is a subgroup of $(B,\cdot)$,

$$(x\circ y)\cdot x^{-1}\cdot x = x\circ y$$

is in J.

The main point in ([**Chi18**], Theorem 4.3) is:

PROPOSITION 7.10. *Let $B=(B,\circ,\cdot)$ be a skew brace, and let L/K be a $(B,\circ)$-Galois extension with a Hopf-Galois structure corresponding to the subgroup $\lambda_{\cdot}(B)$ of* $\mathrm{Perm}(B)$ *normalized by $\lambda_{\circ}(B)$. Then a subgroup J of $\lambda_{\cdot}(B)$ is normalized by $\lambda_{\circ}(B)$ if and only if J is $\circ$-stable.*

PROOF. Suppose J is normalized by $\lambda_{\circ}(B)$. Then for all h in J, g in B, there exists h' in J so that for all x in B,

$$\lambda_{\circ}(g)\lambda_{\cdot}(h)(x)=\lambda_{\cdot}(h')\lambda_{\circ}(g)(x),$$

or

$$g\circ(h\cdot x)=h'\cdot(g\circ x).$$

The left brace property for B applied to the left side of this equation yields

$$(g\circ h)\cdot g^{-1}\cdot(g\circ x)=h'\cdot(g\circ x).$$

Thus $(g\circ h)\cdot g^{-1}=h'$ in J, hence J is $\circ$-stable. The argument is reversible. □

COROLLARY 7.11. *Given a skew brace $B=(B,\circ,\cdot)$ and a $G=(B,\circ)$-Galois extension L/K of type $N=(B,\cdot)$ corresponding to the skew brace B, the Galois correspondence ratio for L/K is*

$$GC(G,N)=\frac{|\{\circ\text{-stable subgroups of }(B,\cdot)\}|}{|\{\text{subgroups of }(B,\circ)\}|},$$

and hence depends only on the skew brace B corresponding to the H-Hopf-Galois structure on the G-Galois extension L/K.

From Proposition 2.24 we know that given a skew brace $(B,\circ,\cdot)$ and a G-Galois extension L/K with $G\cong(B,\circ)$, there are $|\operatorname{Aut}(G)|/|\operatorname{Aut}_{sb}(B)|$ Hopf-Galois structures on L/K corresponding to the brace B. Corollary 7.11 shows that the Galois correspondence ratio is the same for all of those Hopf-Galois structures.

So we could call $GC(L/K,H)=GC(B)$.

Note that $GC(B)\le 1$, because a $\circ$-stable subgroup of $(B,\cdot)$ is a subgroup of $(B,\circ)$.

Koch and Truman [**KT20**] have adapted this method for studying the image of the Galois correspondence for the Hopf-Galois structure by introducing the opposite skew brace $B^{opp}=(B,\circ,\cdot')$ of the brace $B=(B,\circ,\cdot)$ corresponding to the Hopf-Galois structure. The operation $\cdot'$ on B^{opp} is defined by $a\cdot' b=b\cdot a$. If B^{opp} is the opposite brace to B, then the $\circ$-stable subgroups of $(B,\cdot)$ in B correspond to the *quasi-ideals* of B^{opp}: a quasi-ideal I of B^{opp} is a sub-brace of B^{opp} such that $x^{-1}\cdot'(x\circ y)$ is in I for all x in B^{opp}, y in I, hence satisfies $(x\circ y)\cdot x^{-1}$ in B, hence is a $\circ$-stable subgroup of $(B,\cdot)$.

In either setting, the determination of the image of the Galois correspondence becomes an internal study of the structure of the skew left brace B or, equivalently, its opposite B^{opp}. For more details see Section 8.2.

In this chapter we examine the Galois correspondence ratio in a variety of examples. Of course, in many cases determining the precise ratio is a difficult problem. Just finding the denominator, the number of subgroups of a finite group, is a known difficult problem, even for a finite abelian group that is a product of three or more cyclic groups (see [**Rem32**] for the case of three cyclic groups.) There is a formula of Hölder for the number of subgroups of a group of square-free order — see [**AB18**].

We describe results and examples (from [**Chi17**], [**Chi18**], [**CG18**] and [**Chi21**]) related to $GC(L/K, H)$ in several settings, involving Zappa-Szép groups (including semidirect products), radical algebras, and cyclic or elementary abelian p-groups. We will see that $GC(L/K, H)$ is rarely equal to 1, and we'll find classes of examples beyond those in Example 7.7, including abelian examples, where the ratio can be made arbitrarily close to 0.

7.6. Fixed point free pairs

In this section we focus on G-Galois extensions L/K where the Hopf-Galois extension arises from a fixed point free pair of homomorphisms from G to N, where $|G| = |N|$.

Recall from Chapter 4 that two homomorphisms $f, g : G \to N$ form a fixed point free pair (f, g) if and only if for all γ in G, $f(\gamma) = g(\gamma)$ only for $\gamma = e$, the identity of G.

If (f, g) is a fixed point free pair, then $\{f(\gamma)g(\gamma^{-1}) | \gamma \in G\} = N$.

Hopf-Galois structures arising from fixed point free pairs are studied in Chapter 4.

7.6.1. Zappa-Szép groups.

Recall from Chapter 4 that a group $G = (G, \cdot)$ is a Zappa-Szép group if $G = G_L G_R$ and every element of G factors uniquely into $g = hk$ for h in G_L, k in G_R. Examples include semidirect products of groups.

In the group $(G, \cdot)$ we often omit the group operation $\cdot$: so $hk = h \cdot k$, and g^{-1} is the inverse of g in $(G, \cdot)$.

Given $G = G_L G_R$, let D be the the direct product of G_L and G_R: $D = G_L \times G_R = \{(h, k)\}$ with h in G_L, k in G_R. Define $f_1, f_2 : D \to G$ by

$$f_1(h, k) = h,$$
$$f_2(h, k) = k.$$

Then (f_1, f_2) is a fixed point free pair, hence defines a Hopf-Galois structure of type $G = G_L G_R$ on a D-Galois extension of fields.

We recall how the fixed point free pair (f, g) yields a skew brace structure on $(G, \cdot)$. For $g = (h, k)$ in D, define $\beta : D \to \mathrm{Perm}(G)$ by

$$\beta(g) = \lambda(f_1(g))\rho(f_2(g)) = \lambda(h)\rho(k).$$

Then β maps into $\mathrm{Hol}(G)$. For x in G,

$$\beta(g)(x) = \beta(h, k)(x) = hxk^{-1}.$$

Then the bijection $b : D \to G$, defined by

$$b(h, k) = \beta(h, k)(e) = hk^{-1}$$

(where e is the identity element of G) has inverse $a : G \to D$: for $g = hk$ in G,

$a(g) = a(hk) = (h, k^{-1})$. Then a defines the circle operation on G as follows: for $g_1 = h_1k_1, g_2 = h_2k_2$ in G,

$$\begin{aligned} g_1 \circ g_2 &= b(a(g_1) \cdot a(g_2)) \\ &= b((h_1, k_1^{-1}) \cdot (h_2, k_2^{-1})) \\ &= b((h_1h_2, k_1^{-1}k_2^{-1})) \\ &= (h_1h_2)(k_1^{-1}k_2^{-1})^{-1} \\ &= h_1h_2k_2k_1 = h_1g_2k_1. \end{aligned}$$

Then a defines an isomorphism $(G, \circ) \cong D$. To show that $(G, \circ, \cdot)$ is a skew brace, we can check the relation: for all x, y, z in G,

$$x \circ (y \cdot z) = x_L yz x_R^{-1} = (x \circ y) \cdot x^{-1} \cdot (x \circ z).$$

So $(G, \circ, \cdot)$ is a skew brace with additive group $G = (G, \cdot)$ and $a : (G, \circ) \to D$ an isomorphism of groups.

Thus if L/K is D-Galois with $D = G_L \times G_R$ and has a Hopf-Galois structure of type $G = G_LG_R$, then

$$GC(L/K, H) = GC(B) = \frac{|\{\circ\text{-stable subgroups of } (G, \cdot)\}|}{|\{\text{subgroups of } D\}|}.$$

To determine this ratio, we have

PROPOSITION 7.12. *Let $G = (G, \cdot)$ be a group with complementary subgroups G_L and G_R. Let $D \cong (G, \circ) = G_L \times G_R$, and for $g = (h, k)$ in G define $\beta : D \to$ $\mathrm{Hol}(G)$ by*

$$\beta(g) = \lambda(h)\rho(k)$$

in $\mathrm{Hol}(G)$. Then the $\circ$-stable subgroups of $(G, \cdot)$ are the subgroups of $(G, \cdot)$ that are closed under conjugation by elements of G_L.

PROOF. Let G' be a subgroup of $G = G_LG_R$. Then G' is $\circ$-stable if for all x in G' and all $g = hk$ in G, there exists y in G' so that

$$g \circ x = yg.$$

This is true if and only if $hxk = yhk$, if and only if $y = hxh^{-1} = C(h)x$. □

Here is a class of examples, from **[Chi18]**.

PROPOSITION 7.13. *Let $G = (G, \cdot)$ be a semidirect product $G = A \rtimes \Delta$ where A is a finite simple group and Δ is some finite group, and let $G = A \times \Delta$, with operation $\circ$ as defined above. Let G' be a $\circ$-stable subgroup of G and suppose there exists some x in G' and a in A so that $a^{-1}xa \neq x$. Then $A \subseteq G'$.*

PROOF. Suppose G' is a $\circ$-stable subgroup of G. Then G' is normalized by A by Proposition 7.12. Suppose x is in G' and a is in A with $a^{-1}xa \neq x$. Then $a^{-1}xa$ is in G' and $(a^{-1}xa)x^{-1} = a^{-1}(xax^{-1})$ is a non-trivial element of A because A is normal in G. So $a^{-1}xax^{-1}$ is in $G' \cap A$, hence $G' \cap A$ is a non-trivial subgroup of A and is normalized by A. Since A is simple, $G' \cap A = A$. □

EXAMPLE 7.14. Let $G = C_p \rtimes \Delta$ where C_p is written multiplicatively with identity 1 and Δ is a non-trivial subgroup of $\mathrm{Aut}(C_p)$. Then the $\circ$-stable subgroups of G are $\{1\}$ and the subgroups of G containing C_p.

To see this, let $C_p = \langle a \rangle$ and let $G = \langle a, \delta \rangle$ with δ in Δ, where $a^p = \delta^k = 1, \delta a = a^b \delta$, and the order of δ in Δ is $k > 1$. If G' is a $\circ$-stable subgroup and $g = a^r \delta^s$ is in G' with $\delta^s \neq 1$ (so $1 \leq s < k$), then one sees easily that g is not fixed under conjugation by a^{-1}. Thus by Proposition 7.13, G' contains C_p.

Conversely, if G' is a subgroup of G containing C_p, then clearly G' is closed under conjugation by elements of C_p, so is $\circ$-stable.

EXAMPLE 7.15. Here is an example from [**Byo15**].

Let $G = S_n = A_n \rtimes C_2$ and $G = A_n \times C_2$, where A_n, S_n is the alternating, resp. symmetric group and $n \geq 5$. Then the only $\circ$-stable subgroups of G are $(1), A_n$ and S_n. So

$$GC(A_n \times C_2, S_n) = \frac{3}{\# \text{ of subgroups of } A_n \times C_2}.$$

More examples are found in [**Chi18**].

7.6.2. Semidirect products and bi-skew braces.

Given a semidirect product $G = G_L G_R = G_L \rtimes G_R$, we have a skew brace $(G, \circ, \cdot)$, where $(G, \cdot)$ is the given group operation on $G = G_L G_R$, and $(G, \circ)$ is the direct product operation. Proposition 7.12 characterizes the $\circ$-stable subgroups of $(G, \cdot)$: a subgroup G' of $(G, \cdot)$ is $\circ$-stable if and only if G' is normalized by G_L.

Recall from Section 4.7:

DEFINITION 7.16. A bi-skew brace is a set $(G, \circ, \cdot)$ that is a skew brace with either group $(G, \circ)$ or $(G, \cdot)$ acting as the additive group.

From Section 4.7 we have:

PROPOSITION 7.17. *Let $(G, \circ, \cdot)$ be as above, so that $(G, \cdot)$ is a semidirect product and $(G, \circ)$ is the corresponding direct product. Then $(G, \circ, \cdot)$ is a bi-skew brace.*

Since $(G, \circ, \cdot)$ is a bi-skew brace, a $(G, \circ)$-Galois extension L/K has an H-Hopf-Galois structure of type $(G, \cdot)$, and a $(G, \cdot)$-Galois extension L'/K' has an H'-Hopf-Galois structure of type $(G, \circ)$. So we can consider the Galois correspondence ratio for L'/K',

$$GC((G, \cdot), (G, \circ)) = \frac{|\{\cdot\text{-stable subgroups of } (G, \circ)\}|}{|\{\text{subgroups of } (G, \cdot)\}|}.$$

The $\cdot$-stable subgroups of $(G, \circ)$ are the subgroups G' of $(G, \circ)$ with the property that for all x in G', g in G, the element $(g \cdot x) \circ \overline{g}$ is in G'.

From [**Chi21**], Proposition 7.2, we have

PROPOSITION 7.18. *Let $(G, \circ, \cdot)$ be the bi-skew brace where $(G, \cdot) \cong G_L \rtimes G_R$, a semidirect product where G_R acts on G_L by conjugation, and let $\circ$ be the direct product operation on G given by $g \circ x = g_L x g_R^{-1}$ for $g = g_L g_R^{-1}$ in G. Then a subgroup G' of $(G, \circ)$ is $\cdot$-stable if and only if for every $x = x_L x_R^{-1}$ in G' and all g in G, $g x_L g^{-1} x_R^{-1}$ is in G'.*

A subgroup G' of $(G, \circ)$ is $\cdot$-stable iff for all x in G', g in G, the element $(g \cdot x) \circ \overline{g}$ is in G'. It is a routine computation to show that $(g \cdot x) \circ \overline{g} = g x_L g^{-1} x_R^{-1}$.

7.6.3. Dihedral groups of squarefree order. We illustrate these results by looking at the Galois correspondence ratios involving dihedral groups of squarefree order, from [**Chi21**].

Let m be odd and squarefree, and let $G = (G, \cdot) = \langle a, b \rangle$ be the dihedral group D_{2m} of order $2m$. Then $G \cong C_m \rtimes C_2 = \langle a, b \rangle$ with $a^n = 1, b^2 = 1$ and $bab^{-1} = a^{-1}$. Define the operation $\circ$ on G by $a_1b_1 \circ a_2b_2 = a_1a_2b_1b_2$, then $(G, \circ)$ is isomorphic to the cyclic group C_{2m}. Then $(G, \circ, \cdot)$ is a bi-skew brace, so yields a Hopf-Galois structure of type $(G, \circ)$ on a $(G, \cdot)$-Galois extension and also a Hopf-Galois structure of type $(G, \cdot)$ on a $(G, \circ)$-Galois extension. To determine the Galois correspondence ratios for the two Hopf-Galois extensions, we want to find

$$GC((G, \circ), (G, \cdot)) = \frac{|\{ \circ\text{-stable subgroups of } (G, \cdot)\}|}{|\{ \text{subgroups of } (G, \circ)\}|}$$

and

$$GC((G, \cdot), (G, \circ)) = \frac{|\{ \cdot\text{-stable subgroups of } (G, \circ)\}|}{|\{ \text{subgroups of } (G, \cdot)\}}.$$

To summarize what these ratios are: Let $m = p_1p_2 \cdots p_g$ be the product of g distinct odd primes. Then $(G, \circ)$ is the direct product of $g + 1$ cyclic groups of coprime orders, so the number of subgroups of $(G, \cdot)$ is 2^{g+1}, and the number of subgroups of the dihedral group of order $2m$ is $2^g + \sigma(m)$ where $\sigma(m)$ is the sum of the divisors of m. One sees easily that $\sigma(m) = \prod_{i=1}^{g}(1 + p_i) \geq 3^g$, so we know the denominators of the ratios.

As for the numerators, we find that every subgroup of $(G, \circ)$ is a $\cdot$-stable subgroup of $(G, \circ)$, so we have

$$\begin{aligned} GC((G, \circ), (G, \cdot)) &= \frac{|\{\cdot\text{-stable subgroups of } (G, \cdot)\}|}{|\{ \text{subgroups of } (G, \circ)\}|} \\ &= \frac{2^{g+1}}{\prod_{i=1}^{g}(1 + p_i)} \\ &\leq \frac{2^{g+1}}{3^g} = 2(\frac{2}{3})^g. \end{aligned}$$

For the other ratio, the $\cdot$-stable subgroups of $(G, \circ) = D_{2m}$ are subgroups of the cyclic group $(G, \circ)$, so have the form $H = \{(r^i, s^i)$ for r in C_m, s in C_2. Then H is $\cdot$-stable for $r = 0$ and any s, or for $s = 0$ and $r = d$ for d any divisor of m. So the number of $\cdot$-stable subgroups of $(G, \circ)$ is equal to the number of non-trivial divisors of m plus 2, or $2^g + 1$. So the other Galois correspondence ratio is

$$\begin{aligned} GC((G, \cdot), (G, \circ)) &= \frac{|\{ \cdot\text{-stable subgroups of } (G, \circ)\}|}{|\{ \text{subgroups of } (G, \cdot)\}|} \\ &= \frac{2^g + 1}{2^{g+1}} > \frac{1}{2}. \end{aligned}$$

The two Galois correspondence ratios are close to 0 and 1/2, respectively, for large g.

For more details and generalizations, see [**Chi21**].

7.7. Radical algebras and the Galois correspondence

We review radical rings from Section 4.8.

A ring $A = (A, +, \cdot)$ (without unit) is a radical ring if the operation

$$a \circ b = a + b + a \cdot b$$

for all a, b in A makes $(A, \circ)$ into a group with identity 0 (the circle group of A). If A is a nilpotent ring (for every a in A, there is some $n > 0$ so that $a^n = 0$), then A is a radical ring. Associativity of $\circ$ is clear, the identity is 0, and the inverse of a in $(A, \circ)$, denoted by $\overline{a}$, is

$$\overline{a} = -a + a^2 - a^3 + \dots,$$

a finite sum because A is nilpotent.

(Conversely, a finite radical ring is nilpotent: c.f. Section 1.2 of [**Her61**].)

Let $(A, +, \cdot)$ be a finite radical ring, and let $a \circ b = a + b + a \cdot b$ (or $a + b + ab$ if we omit the $\cdot$ as usual). Then $(A, \circ, +)$ is a skew brace. From [**Chi18**] we have

PROPOSITION 7.19. *If $(A, \circ, +)$ is the skew brace arising from a radical ring A, then the $\circ$-stable subgroups of $(A, +)$ are the left ideals of the ring A.*

PROOF. Suppose $(G', +)$ is a $\circ$-stable subgroup of $(A, +)$. Then for all g in A, g' in G', there is some h' in G' so that $g \circ g' = h' + g$, or equivalently:

$$\begin{aligned} g + g' + gg' &= h' + g \\ gg' &= h' - g' \text{ in } G'. \end{aligned}$$

Thus G' is a left ideal of A.

Conversely, if G' is a left ideal of A, then for all g' in G', g in A, gg' is in G', so $gg' + g' = h'$ is in G'. So

$$g \circ g' = g + g' + gg' = h' + g,$$

the condition that $(G', +)$ is a $\circ$-stable subgroup of $(A, +)$. □

Thus if A is a finite radical ring and L/K is a Galois extension with Galois group $(A, \circ)$ with a Hopf-Galois structure with H of type $(A, +)$, then the Galois correspondence ratio for H acting on L/K is

$$GC((A, \circ), (A, +)) = \frac{|\{\text{left ideals of } A\}|}{|\{\text{subgroups of } (A, \circ)|}.$$

EXAMPLE 7.20. Let A be the $\mathbb{F}_p$-algebra with an $\mathbb{F}_p$-basis consisting of elements a, b, c, d with multiplication given by $a^2 = c, ab = d$ and all other products of basis elements equal to 0 — the algebra denoted $A^0_{4,21}$ in [**DG18**]. A subset J of A is a left ideal of A if and only if $aJ \subset J$. It's not hard to see that there are $p^2 + 3p + 5$ left ideals of A. Since $(A, \circ)$ is a non-abelian group of order p^4, it is somewhat more difficult to determine that there are $2p^3 + 4p^2 + 3p + 5$ subgroups of $(A, \circ)$. Thus

$$GC((A, \circ), (A, \cdot)) = \frac{p^2 + 3p + 5}{2p^3 + 4p^2 + 3p + 5}.$$

See [**Chi21**, Example 4.2] for details.

7.8. Commutative examples

Let L/K be a G-Galois extension of fields with G an elementary abelian p-group of order p^n. If L/K is also H-Hopf-Galois for H of abelian type, then the corresponding skew brace is a commutative radical ring. In this section we survey some results on the Galois correspondence ratio in this setting.

Perhaps the first example where both G and H are commutative and the Hopf-Galois correspondence is not surjective is in [**CRV15a**], where $G = C_2 \times C_2$ and H has type C_4. Then G has five subgroups and C_4 has three subgroups, hence H can have at most three K-Hopf subalgebras.

Examples of this type are a $p=2$ phenomenon, because of Crespo and Salguero's result [**CS19**] that if L/K is Galois with Galois group G and has a Hopf-Galois structure of type C_{p^n} when p is an odd prime, then G must be cyclic.

In fact, we have Proposition 4.3 of [**Chi17**]:

PROPOSITION 7.21. *Let L/K be a G-Galois extension of fields with G cyclic of order p^n, p odd. Let the K-Hopf algebra H give a Hopf-Galois structure on L/K. Then H has type G and the Galois correspondence for that Hopf-Galois structure holds in its strong form.*

EXAMPLE 7.22. Let A be the radical of the ring $\mathbb{Z}/p^2\mathbb{Z}[x]/(x^2 - px)$. Letting a be the coset $x + (x^2 - px)$, then $A = \{a, 2a, \ldots, p^2a = 0\}$. Addition is $ra + sa = (r + s)a$, multiplication is by $ra \cdot sa = rspa$. Thus the circle operation on A is defined by $ra \circ sa = (r + s + rsp)a$. Then $(A, +)$ has three subgroups, $\langle a \rangle = A$, $\langle pa \rangle$ and $\{0\}$, and each is an ideal of A. Also, $(A, \circ)$ is cyclic of order p^2, so has three subgroups, $\langle a \rangle = A$, $\langle pa \rangle$ and $\{0\}$. If L/K is Galois with Galois group $G \cong (A, \circ)$, then L/K has a Hopf-Galois structure by a Hopf algebra of type $G = (A, +)$ where the corresponding nilpotent ring $(A, +, \cdot)$ has non-trivial multiplication, hence the corresponding brace is non-trivial (that is, $+$ and $\circ$ are different), but the Galois correspondence for the Hopf-Galois structure on L/K holds in its strong form:

$$GC(L/K, H) = \frac{|\text{ ideals of } A|}{|\text{subgroups of } (A, \circ)|} = \frac{3}{3} = 1.$$

For comparison, let L/K be a G-Galois extension of fields with a Hopf-Galois structure of type N where $G \cong N$ is an elementary abelian p-group of order p^n.

Let A be the corresponding skew brace where $(A, \circ) \cong G$ and $(A, +) \cong G$. Then by [**CDVS06**] A is a commutative nilpotent $\mathbb{F}_p$-algebra, where for a, b in A, $a \circ b = a + b + ab$. Thus the Galois correspondence ratio for the H-Hopf-Galois structure on L/K is

$$GC(L/K, H) = GC((A, \circ), (A, +)) = \frac{\#\text{ of ideals of } A}{\#\text{ of subgroups of } \mathbb{F}_p^n}.$$

We have ([**Chi17**], Theorem 4.1):

THEOREM 7.23. *With L/K as above, the Galois correspondence for the H-Hopf-Galois structure on L/K maps onto all intermediate fields if and only if $ab = 0$ for all a, b in A (that is, if and only if $A^2 = 0$), hence H is the classical Galois structure on L/K.*

PROOF. If $A^2 = 0$, then every subspace of A is closed under scalar multiplication by elements of A, hence is an ideal of A, so the Galois correspondence is surjective. On the other hand, suppose $A^2 \neq 0$. Then for some a, b in A, $ab \neq 0$.

Then the subspace $\mathbb{F}_p a$ does not contain ab. For if $ab = ra$ for $r \neq 0$ in $\mathbb{F}_p$, then $a = sba$ for $s \neq 0$ in $\mathbb{F}_p$. Then, since A is nilpotent, $(sb)^m = 0$ for some $m > 0$, so

$$a = (sb)a = (sb)^2 a = \ldots = (sb)^m a = 0.$$

Thus the non-zero subspace $\mathbb{F}_p a$ is not an ideal of A. So the Galois correspondence is not surjective. □

EXAMPLE 7.24. Theorem 4.2 of [**Chi17**] gives a set of examples where the Galois correspondence ratio is near 0. Let $s(n)$ be the number of subspaces of $\mathbb{F}_p^n$: then it is well-known (for example, see [**CG18**]) that $s(n) \geq p^{\delta(n)}$ where

$$\delta(n) = p^{\lfloor \frac{n^2}{4} \rfloor}.$$

Let A be the primitive n-dimensional nilpotent $\mathbb{F}_p$-algebra generated by z with $z^{n+1} = 0$. Then $(A, +) \cong (\mathbb{F}_p^n, +)$ and so the multiplication on A yields a nilpotent $\mathbb{F}_p$-algebra structure on $(G, +) = (\mathbb{F}_p^n, +)$. Let $G = (\mathbb{F}_p^n, \circ)$ where the operation $\circ$ is defined using the multiplication on A by $a \circ b = a + b + a \cdot b$.

THEOREM 7.25. *Let G be an elementary abelian p-group of order p^n. Suppose $p > n$. Let A be a primitive $\mathbb{F}_p$-algebra structure on G, and let $(G, \circ)$ be the corresponding group structure on $\mathbb{F}_p^n$. Suppose L/K is a Galois extension of fields with Galois group $G \cong (G, \circ)$. Then the primitive nilpotent $\mathbb{F}_p$-algebra A corresponds to an H-Hopf-Galois structure on L/K for some K-Hopf algebra H, and the K-Hopf subalgebras of H form a descending chain*

$$H = H_1 \supset H_2 \supset \ldots \supset H_n \supset K.$$

Hence the Galois correspondence $\mathcal{F}$ for H maps onto exactly $n + 1$ fields F with $K \subseteq F \subseteq L$, and the Galois correspondence ratio $GC(L/K, H) = (n + 1)/s(n)$

PROOF. One can show that the non-zero ideals of A are $J_i = \langle z^i \rangle$ for $i = 1, \ldots, n$. The idea is to take an ideal J, pick out of J a polynomial in z of minimal degree k, hence $\langle z^k \rangle \subseteq J$, and then show that $J = \langle z^k \rangle$.

The condition that $p > n$ is needed to insure that $(G, \circ)$ is an elementary abelian p-group of order p^n, hence isomorphic to the $\mathbb{F}_p$-vector space $\mathbb{F}_p^n$, hence the number of subgroups of $(G, \circ)$ is $s(n)$. □

For A a primitive nilpotent $\mathbb{F}_p$-algebra with $A^{n+1} = 0$, the corresponding group $(G, \circ)$ is isomorphic (by $a \mapsto 1 + a$) to the group of principal units of the truncated polynomial ring $\mathbb{F}_p[x]/(x^{n+1})$. The structure of that group is described in Remark 3.6 above (c.f. [**Chi07**]). In particular $(G, \circ)$ is an elementary abelian p-group if and only if $p > n$.

More information on finite commutative nilpotent $\mathbb{F}_p$-algebras may be found in [**Chi15**] and the references listed there, notably [**Poo08b**].

7.9. Elementary abelian p-groups

Let A be a finite commutative nilpotent $\mathbb{F}_p$-algebra. Then, as noted above, if $(A, \circ)$ is the circle group of A, then a G-Galois extension L/K with $G \cong (A, \circ)$ has an H-Hopf-Galois structure of type $N = (A, +) = \mathbb{F}_p^n$.

Let $A^r = \{a_1 a_2 \cdots a_r \mid a_1, \ldots, a_r \in A\}$. Since A is a finite ring, there is some $e > 0$ so that $A^e \neq \{0\}, A^{e+1} = \{0\}$. Denote that e by $e(A)$. If $e < p$, then $(A, \circ) \cong C_p^n \cong (\mathbb{F}_p^n, +)$ by Proposition 5.8.

Let L/K be a G-Galois extension and has a Hopf-Galois structure of type $N \cong (\mathbb{F}_p^n, +)$. Then the left brace associated to $(L/K, G, N)$ is a radical $\mathbb{F}_p$-algebra $(A, +, \cdot)$ where $(A, +) = (\mathbb{F}_p^n, +)$ and $(A, \circ) \cong G$. Let

$$s(A) = \#\text{ of subgroups of } (\mathbb{F}_p^n, +) = \#\text{ of subspaces of } \mathbb{F}_p^n$$
$$i(A) = \#\text{ of ideals of } A.$$

Then the Galois correspondence ratio for L/K is

$$GC(G, N) = GC((A, \circ), (A, +)) = \frac{i(A)}{s(A)}.$$

To estimate this ratio for a given isomorphism type of A requires an estimate of $i(A)$, the number of ideals of A. That was the subject of [**CG18**], in which $i(A)$, the number of ideals of A, was bounded above by a quantity that depends on $e = e(A)$. Assuming that $e < p$, the main result of [**CG18**] is:

THEOREM 7.26. *With A, e, $\delta(e)$ as above,*

$$i(A)/s(A) \leq \frac{2e-1}{p^{\delta(e)}}.$$

To get a sense of what this implies, slight improvements in [**CG18**] to that inequality for $e = 2, 3$ yield:

COROLLARY 7.27. *Suppose L/K is a N-Galois extension with N an elementary abelian p-group, and is an H-Hopf-Galois extension where H arises from a commutative nilpotent $\mathbb{F}_p$-algebra structure A on the additive group N. Suppose $A^e \neq 0$, $A^{e+1} = 0$ and $e < n < p$. Then $i(A)/s(A)$, the proportion of intermediate fields that are in the image of the Galois correspondence from Hopf subalgebras of H, satisfies $i(A)/s(A) < 0.01$ for*
- *$e = 2, p \geq 200$;*
- *$e = 3, p \geq 17$;*
- *$e = 4, p \geq 7$;*
- *all e, p with $5 \leq e < p$.*

As shown in the last section of [**CG18**], for many examples the inequality of the theorem substantially overestimates the number $i(A)$, hence the Galois correspondence ratio.

We refer to [**CG18**] for details.

7.10. Non-normal Hopf-Galois structures

In this section we look briefly at the Galois correspondence for separable, non-normal Hopf-Galois extensions of fields. Most of these results are due to Crespo, Rio and Vela.

We begin by reviewing Greither-Pareigis theory in the non-normal setting.

Let L/K be a separable extension of fields with normal closure $\widetilde{L}$, let $G = \mathrm{Gal}(\widetilde{L}/K)$, $G' = \mathrm{Gal}(\widetilde{L}/L)$ and $X = G/G'$, the set of left cosets of G' in G. Then there is a bijection between Hopf-Galois structures on L/K and regular subgroups N of $\mathrm{Perm}(X)$ normalized by $\lambda(G)$. This is Theorem 6.8 of [**Chi00**].

Under that bijection, if H corresponds to N, then K-Hopf subalgebras of H correspond bijectively to subgroups of N that are normalized by $\lambda(G)$.

In particular, if G' is trivial and $L = \widetilde{L}$, then the classical Hopf-Galois structure on L/K corresponds to $\rho(G)$, the centralizer of $\lambda(G)$ in $\mathrm{Perm}(G)$.

The first set of examples of non-classical Hopf-Galois structures was constructed in [**GP87**]: given a separable extension L/K with normal closure $\widetilde{L}$, let $G = \mathrm{Gal}(\widetilde{L}/K), G' = \mathrm{Gal}(\widetilde{L}/L)$. Suppose there is a Galois extension E/K so that $L \otimes_K E \cong \widetilde{L}$. Let $N = \mathrm{Gal}(\widetilde{L}/E)$. Then N is a normal subgroup of G, and $G' \rtimes N \cong G$. So $\lambda(N)$ is a regular subgroup of $\mathrm{Perm}(X)$ where X is the set of left cosets of G' in G, and $\lambda(N)$ is normalized by $\lambda(G)$, hence yields a Hopf-Galois structure on L/K.

DEFINITION 7.28. An H-Hopf-Galois extension L/K with normal closure $\widetilde{L}$ and $G = \mathrm{Gal}(\widetilde{L}/K), G' = \mathrm{Gal}(\widetilde{L}/L)$ is *almost classical* if the corresponding subgroup N of $\mathrm{Perm}(X)$ is a normal subgroup of G.

Theorem 5.2 of [**GP87**] is:

THEOREM 7.29. *Suppose L/K is an almost classical H-Hopf-Galois extension corresponding to the normal subgroup N of $G = Gal(\widetilde{L}/K)$. Then the Galois correspondence for H maps onto all intermediate fields $K \subset L' \subset L$.*

PROOF. The intermediate subfields L' correspond to subgroups of G containing G' by classical Galois theory. There is a bijection between subgroups of G containing G' and subgroups of G that are $\lambda(G)$-stable by mapping V with $G' < V < G$ to $W = V \cap N$, with inverse mapping a $\lambda(G)$-stable subgroup U of N to $UG' \cong U \rtimes G'$. So every intermediate field L' between K and L corresponds to a K-Hopf subalgebra of H. □

The smallest example found in [**GP87**] of a Hopf-Galois extension that is not almost classical is described in Corollary 4.4 of [**GP87**], where $K = \mathbb{Q}[\sqrt{-2}]$ and $L = K[\sqrt[16]{2}]$, then $[L : K] = 16, [K : \mathbb{Q}] = 2$.

[**GP87**] also looked at separable extensions L/K for $[L : K] = 2, 3, 4, 5$ and asked if they were all (Hopf) Galois. They found that the answer is "yes" for $n = 2, 3, 4$, but not for $n = 5$. The case $n = 5$ was generalized by Theorem 2 of [**Chi89**]: for $[L : K] = p$, prime, L/K is Hopf-Galois if and only if the Galois group $\mathrm{Gal}(\widetilde{L}/K)$ is solvable.

In [**CRV16b**, Example 2.1], Crespo, Rio and Vela systematically studied extensions L/K where $[L : K] = 4, 5$ and 6 with all possibilities for the Galois group $\mathrm{Gal}(\widetilde{L}/K)$ of the normal closure. Let $|\mathrm{Gal}(\widetilde{L}/K)| = n^*$. They found that for $n = 4$, every extension L/K is either Galois or almost classically Galois; for $n = 5$, L/K is Galois for $n^* = 5$ (of course), almost classically Galois for $n^* = 10$ or 20, and not Hopf-Galois for $n^* = 60$ or 120. For $n = 6$, there are sixteen possible Galois groups. For two cases L/K is Galois, for three cases L/K is almost classically Galois and in 11 cases L/K is not Hopf-Galois.

In the prime cases $n = 7$ and 11, where by [**Chi89**] all Hopf-Galois extensions have a solvable normal closure, they showed that every Hopf-Galois structure is classical or almost classical.

However, in Example 2.1 [**CRV16b**] found an example of an extension $L/\mathbb{Q}$ of degree 8 and splitting field $\widetilde{L}$ where $\mathrm{Gal}(\widetilde{L}/\mathbb{Q}) \cong S_4$ (so $n^* = 24$). The extension $L/\mathbb{Q})$ is Hopf-Galois but not almost classically Galois. By the results just described, this is an example of the smallest possible degree.

In [**CRV16b**], the authors looked at a separable extension L/K of degree pd where $p \geq 5$ is prime and d is a non-trivial divisor of $p-1$, whose normal closure $\widetilde{L}$ has as Galois group the Frobenius group of order $p(p-1)$. If $\gcd(d,(p-1)/d) > 1$ then L/K has a Hopf-Galois structure that is not almost classical, but for which the Galois correspondence is surjective. Thus the set of Hopf-Galois structures for which the Galois correspondence is surjective is strictly larger than the set of almost classical Galois structures. The smallest examples of this type are for $p = 5, d = 2$ where L/K is not almost classically Galois but has two Hopf-Galois structures, of types C_{10} and D_5, both of which have a surjective Galois correspondence.

[**CRV16b**] also has two other interesting examples.

One is a classical Galois extension L/K with Galois group D_p, the dihedral group of order $2p$, p prime. There are $2+p$ Hopf-Galois structures on L/K by [**Byo04b**], and the images of the Galois correspondence for the $2+p$ structures are all different.

The other is a separable extension L/K of degree 12 whose normal closure $\widetilde{L}$ has $\mathrm{Gal}(\widetilde{L}/K) \cong S_3 \times S_3$ of order 36. Then L/K is a Hopf-Galois extension of type D_6, not almost classical (because $S_3 \times S_3$ has no normal subgroup of order 12), and there is no Hopf-Galois structure on L/K for which the Galois correspondence is surjective.

The paper [**CRV15a**] gives a nice overview of much of this work.

CHAPTER 8

Normality in Hopf-Galois extensions

We have seen in Chapter 7 that if L/K is an H-Galois extension of fields then each Hopf subalgebra H' of H has a corresponding fixed field

$$F = L^{H'} = \{x \in L \mid h \cdot x = \varepsilon(h)x \text{ for all } h \in H'\},$$

and that this construction yields an injective (but rarely surjective) lattice inverting map from the set of Hopf subalgebras of H to the set of intermediate fields of the extension L/K. In this chapter we investigate the existence of Hopf-Galois structures on the extensions L/F and F/K, in analogy with the remaining parts of the classical Fundamental Theorem of Galois Theory.

A result of Chase and Sweedler ([**CS69**, Theorem 7.6], summarized below) implies that the F-Hopf algebra $F \otimes_K H'$ gives a Hopf-Galois structure on the extension L/F. In Section 8.1 we present results from [**KKTU19a**], which uses Greither-Pareigis theory to study the existence of a "quotient" Hopf-Galois structure on F/K in the case that L/K is Galois. A consequence of these results is that when L/K is Galois we can study the existence of such "quotient" Hopf-Galois structures via the skew brace corresponding to the Hopf-Galois structure given by H: this is the subject of Section 8.2.

It is natural to study the reverse of this situation: given a Galois extension L/K, an intermediate field F, and Hopf-Galois structures on L/F and F/K, it is possible to combine these Hopf-Galois structures to give a Hopf-Galois structure on L/K? In Section 8.3 we survey work by Crespo, Rio and Vela on *induced* Hopf-Galois structures, which provides the only known construction of this type.

We begin with the result of Chase and Sweedler mentioned above. The proof in loc. cit. is expressed in terms of *coactions* (see Chapter 10) and applies to Hopf-Galois extensions of commutative rings; here we reformulate it in terms of module actions and specialize to separable Hopf-Galois extensions of fields. In this case by Corollary 7.6 we have

$$[L : L^{H'}] = \dim_K(H'), \tag{8.1}$$

which is the Hopf-Galois analogue of the familiar fact from Galois theory that if L/K is a G-Galois extension and J is a subgroup of G then $[L : L^J] = |J|$.

PROPOSITION 8.1. *Let L/K be a separable H-Galois extension, let H' be a Hopf subalgebra of H, and let $F = L^{H'}$. Then L/F is $(F \otimes_K H')$-Galois.*

PROOF. The action of H on L yields a natural action of $F \otimes_K H'$ on L by

$$(x \otimes h) \cdot t = x(h \cdot t) \text{ for all } x \in F,\ h \in H,\ t \in L,$$

and the fact that L is an H-module algebra implies that L is an $F \otimes_K H'$-module algebra.

To show that the natural map

$$L \otimes_F (F \otimes_K H') \to \mathrm{End}_F(L)$$

is an isomorphism, we begin with the usual identification

$$L \otimes_F (F \otimes_K H') \cong L \otimes_K H'.$$

Now $L \otimes_K H'$ injects into $L \otimes_K H$, and since H gives a Hopf-Galois structure on L/K the map

$$j : L \otimes_K H \to \mathrm{End}_K(L)$$

defined by

$$j(s \otimes h)[t] = s(h \cdot t)$$

is an isomorphism. We claim that $j(L \otimes_K H') = \mathrm{End}_F(L)$. For all $h \in H'$, $x \in F$, and $s, t \in L$ we have

$$\begin{aligned} j(s \otimes h)[xt] &= s(h \cdot (xt)) \\ &= s \sum_{(h)} (h_{(1)} \cdot x)(h_{(2)} \cdot t) \\ &= s \sum_{(h)} (\varepsilon(h_{(1)})x)(h_{(2)} \cdot t) \text{ since } h_{(1)} \in H' \text{ and } x \in F \\ &= sx \sum_{(h)} (h_{(1)} \cdot 1)(h_{(2)} \cdot t) \text{ since } h_{(1)} \cdot 1 = \varepsilon(h_{(1)}) \\ &= sx(h \cdot t) \\ &= xj(s \otimes h)[t], \end{aligned}$$

and so $j(L \otimes_K H') \subseteq \mathrm{End}_F(L)$. Now by (8.1) we have $\dim_K(H') = [L : F]$, so

$$\dim_K(L \otimes_K H') = [L : K][L : F] = [F : K][L : F]^2 = \dim_K(\mathrm{End}_F(L)).$$

Since j is injective we obtain $j(L \otimes_K H') = \mathrm{End}_F(L)$, and so L/F is $(F \otimes_K H')$-Galois. ☐

8.1. Normality for Hopf-Galois structures on Galois extensions

We now specialize to the case in which L/K is a G-Galois extension. As above, we assume that L/K is H-Galois and that H' is a Hopf subalgebra of H. By Greither-Pareigis theory, we have $H = H_N = L[N]^G$ for some regular subgroup $N \leq \mathrm{Perm}(G)$ that is normalized by $\lambda(G)$, and by Lemma 7.2 we have $H' = H_P = L[P]^G$ for some subgroup $P \leq N$ that is itself normalized by $\lambda(G)$. We let $F = L^{H_P}$, and $J = \mathrm{Gal}(L/F)$; by (8.1) we have

$$|J| = \dim_K(L[P]^G) = |P|. \tag{8.2}$$

Our aim is to prove the following result [**KKTU19a**, Theorem 2.9]:

THEOREM 8.2. *If P is normal in N then, for a suitable action of G on N/P, the extension F/K is $L[N/P]^G$-Galois.*

Interpreting parts of the Fundamental Theorem of Galois Theory via group algebras, we see that if $K[G]$ gives the classical Hopf-Galois structure on L/K then L/F is $F[J]$-Galois, and if in addition J is normal in G then F/K is $K[G/J]$-Galois. Thus Theorem 8.2 is a natural analogue of these parts of the Fundamental Theorem of Galois Theory. We note, however, that in general the assumption that P is normal in N does not imply that J is normal in G, and so F/K is not necessarily a Galois extension. For this reason, the notation G/J employed in this section denotes the left coset space of J in G, not a quotient group. We shall abuse the symbol λ by using it to refer to a number of different left translation maps: the precise interpretation will depend upon the context.

A little care is needed to reconcile the results of Proposition 8.1 and Theorem 8.2 with Greither-Pareigis theory. We know that a Hopf algebra giving a Hopf-Galois structure on L/F must have the form $L[M]^J$ for some regular subgroup M of $\mathrm{Perm}(J)$ normalized by $\lambda(J)$, and that a Hopf algebra giving a Hopf-Galois structure on F/K must have the form $E[M]^{\overline{G}}$ for some regular subgroup M of $\mathrm{Perm}(\overline{G}/\overline{J})$ normalized by $\lambda(\overline{G})$, where E is the Galois closure of F/K and $\overline{G}, \overline{J}$ denote quotient groups modulo $\mathrm{Gal}(L/E)$. It is not immediately clear that the Hopf algebras produced by Proposition 8.1 and Theorem 8.2 have the required forms, but we shall see in the course of this section that each can be rewritten in an appropriate way. The key to the former is the following:

PROPOSITION 8.3. *The subgroup P of N identifies naturally with a regular subgroup of* $\mathrm{Perm}(J)$ *normalized by* $\lambda(J)$.

PROOF. By Greither-Pareigis theory the Hopf-Galois structure given by $F \otimes_K L[P]^G$ on L/F corresponds to a regular subgroup of $\mathrm{Perm}(J)$ normalized by $\lambda(J)$. To obtain this regular subgroup explicitly, we follow the proof of loc. cit. Extending scalars from F to L we have

$$L \otimes_F (F \otimes_K L[P]^G) \cong L \otimes_K L[P]^G \cong L[P]$$

and

$$L \otimes_F L \cong \mathrm{Map}(J, L),$$

and the fact that L/F is $(F \otimes_K L[P]^G)$-Galois implies that the extension of rings $\mathrm{Map}(J, L)/L$ is $L[P]$-Galois. The L-algebra $\mathrm{Map}(J, L)$ has a basis of mutually orthogonal idempotents $\{u_j \mid j \in J\}$, which are permuted regularly by the action of the grouplike elements of $L[P]$ [**Chi00**, Theorem 6.3]; hence P may be identified with a regular subgroup of $\mathrm{Perm}(J)$. However, the action of $L[P]$ on $\mathrm{Map}(J, L)$ is determined by the action of $L[P]^G$ on L; this in turn is determined by the action of $L[N]^G$ on L, which is determined by N. Therefore the identification of P with a regular subgroup of $\mathrm{Perm}(J)$ is simply the action of P (viewed as a subgroup of N) on the elements of J (viewed as a subgroup of G). Hence P acts regularly on J, as claimed. □

Proposition 8.3 implies that the extension L/F is $L[P]^J$-Galois, whereas Proposition 8.1 implies that it is $(F \otimes_K L[P]^G)$-Galois. Since $J = \mathrm{Gal}(L/F)$ there is a natural map

$$F \otimes_K L[P]^G \to L[P]^J,$$

which is bijective by Galois descent. Since in addition the action of $L[P]^J$ on L is determined by the action of $L[P]^G$ on L, we see that these are two interpretations of the same Hopf-Galois structure.

We now continue towards the proof of Theorem 8.2. Since P is a subgroup of N, it acts on G; we denote the orbit of $g \in G$ under P by $\mathrm{Orb}_P(g)$.

PROPOSITION 8.4. *For $g \in G$, we have*

$$\mathrm{Orb}_P(g) = gJ.$$

PROOF. Proposition 8.3 implies immediately that $\mathrm{Orb}_P(1_G) = J$. Now let $g \in G$. Since P is normalized by $\lambda(G)$ we have $\lambda(g)^{-1}P\lambda(g) = P$, and so

$$\begin{aligned} J &= \mathrm{Orb}_P(1_G) \\ &= \{\pi[1_G] \mid \pi \in P\} \\ &= \{\lambda(g)^{-1}\pi\lambda(g)[1_G] \mid \pi \in P\} \\ &= \{g^{-1}\pi[g] \mid \pi \in P\}. \end{aligned}$$

Therefore

$$gJ = \{\pi[g] \mid \pi \in P\} = \mathrm{Orb}_P(g).$$

□

We now impose the assumption that P is normal in N. We reiterate that this does not in general imply that J is normal in G and so, although N/P denotes a quotient group, G/J denotes merely a left coset space.

PROPOSITION 8.5. *The quotient group N/P identifies naturally with a regular subgroup of* $\mathrm{Perm}(G/J)$.

PROOF. For $\eta \in N$ and $g \in G$ we have

$$\begin{aligned} \eta[gJ] &= \{\eta[\pi[g]] \mid \pi \in P\} \text{ by Proposition 8.4} \\ &= \{\pi[\eta[g]] \mid \pi \in P\} \text{ since } P \text{ is normal in } N \\ &= \eta[g]J \text{ by Proposition 8.4 again.} \end{aligned}$$

Hence N acts transitively on G/J by $\eta[gJ] = \eta[g]J$. By Proposition 8.4, P acts trivially on G/J, and so N/P acts transitively on G/J by $(\eta P)[gJ] = \eta[g]J$. Since $|N/P| = |G/J|$, this action is in fact regular. □

PROPOSITION 8.6. *Viewed as a subgroup of* $\mathrm{Perm}(G/J)$, *the quotient group N/P is normalized by $\lambda(G)$.*

PROOF. Let $\eta P \in N/P$, $gJ \in G/J$, and $h \in G$. Then, using the action of N/P on G/J given in Proposition 8.5 we have

$$\begin{aligned} \lambda(h)\eta P\lambda(h)^{-1}[gJ] &= \lambda(h)\eta P[h^{-1}gJ] \\ &= \lambda(h)\eta[h^{-1}g]J \\ &= h\eta[h^{-1}g]J \\ &= \lambda(h)\eta\lambda(h)^{-1}P[gJ], \end{aligned}$$

and $\lambda(h)\eta\lambda(h)^{-1}P \in N/P$ since N is normalized by $\lambda(G)$ inside $\mathrm{Perm}(G)$. □

Combining Proposition 8.5 and Proposition 8.6 allows us to prove Theorem 8.2.

PROOF OF THEOREM 8.2. Let $E \subseteq L$ denote the Galois closure of F/K, and let $\overline{G}, \overline{J}$ denote quotient groups modulo $\mathrm{Gal}(L/E)$, so that $\overline{G} = \mathrm{Gal}(E/K)$ and $\overline{J} = \mathrm{Gal}(E/F)$. There is a natural isomorphism

$$\mathrm{Perm}(G/J) \cong \mathrm{Perm}(\overline{G}/\overline{J}),$$

which transports $\lambda(G)$ to $\lambda(\overline{G})$. Thus N/P, viewed as a regular subgroup of $\mathrm{Perm}(G/J)$ normalized by $\lambda(G)$, is transported to a regular subgroup of $\mathrm{Perm}(\overline{G}/\overline{J})$

normalized by $\lambda(\overline{G})$. The Hopf algebra giving the corresponding Hopf-Galois structure is $E[N/P]^{\overline{G}}$.

In order to reconcile this with the description given in the statement of Theorem 8.2 we begin with the formulation given there, and write

$$L[N/P]^G = \left(L[N/P]^{\mathrm{Gal}(L/E)}\right)^{\overline{G}}, \tag{8.3}$$

with the quotient group $\overline{G}$ acting in the natural way. Now we note that, since $\mathrm{Gal}(L/E)$ is the largest normal subgroup of G contained in J, for $h \in \mathrm{Gal}(L/E)$ and $g \in G$ we have

$$\begin{aligned} \lambda(h)gJ &= hgJ \\ &= gh'J \text{ for some } h' \in \mathrm{Gal}(L/E) \\ &= gJ \text{ since } \mathrm{Gal}(L/E) \subseteq J, \end{aligned}$$

and so the subgroup $\lambda(\mathrm{Gal}(L/E)) \leq \mathrm{Perm}(G/J)$ is trivial. Therefore (8.3) becomes

$$\begin{aligned} L[N/P]^G &= \left(L[N/P]^{\mathrm{Gal}(L/E)}\right)^{\overline{G}} \\ &= \left(L^{\mathrm{Gal}(L/E)}[N/P]\right)^{\overline{G}} \\ &= E[N/P]^{\overline{G}}, \end{aligned}$$

as claimed. □

Before presenting examples of this theory in action, we describe the action of the Hopf algebra $L[N/P]^G$ on F. By Greither-Pareigis theory, the Hopf algebra $L[N]^G$ acts on L by

$$\left(\sum_{\eta \in N} c_\eta \eta\right) \cdot x = \sum_{\eta \in N} c_\eta \eta^{-1}[1_G](x) \text{ for all } x \in L.$$

(See Section 4.4.) It follows that $L[N/P]^G$ acts on F by

$$\left(\sum_{\eta P \in N/P} c_{(\eta P)} \eta P\right) \cdot x = \sum_{\eta P \in N/P} c_{(\eta P)} \left(\eta^{-1}[1_G]J\right)(x) \text{ for all } x \in F. \tag{8.4}$$

EXAMPLE 8.7. Let p, q be prime numbers with $p \equiv 1 \pmod{q}$, and let L/K be a G-Galois extension with G isomorphic to the metacyclic group of order pq:

$$G = \langle s, t \mid s^p = t^q = 1,\ tst^{-1} = s^d \rangle,$$

where d is an integer having multiplicative order q modulo p. The extension L/K admits p Hopf-Galois structures of cyclic type [**Byo04b**, Lemma 4.1]. The corresponding subgroups of $\mathrm{Perm}(G)$ are the the cyclic subgroups

$$N_c = \langle \lambda(s)\rho(s^c t) \rangle \text{ for } c = 0, \ldots, p-1.$$

For each $c = 0, \ldots, p-1$, the group N_c has a unique subgroup of order p, namely $P = \langle \lambda(s) \rangle$, and a unique subgroup of order q, namely $Q = \langle \rho(s^c t) \rangle$. Each of these subgroups is normal and, by uniqueness, each is normalized by $\lambda(G)$; hence we obtain Hopf subalgebras $H_P = L[P]^G$ and $H_Q = L[Q]^G$ of $H_N = L[N]^G$ and, in turn, intermediate fields L^{H_P} and L^{H_Q}.

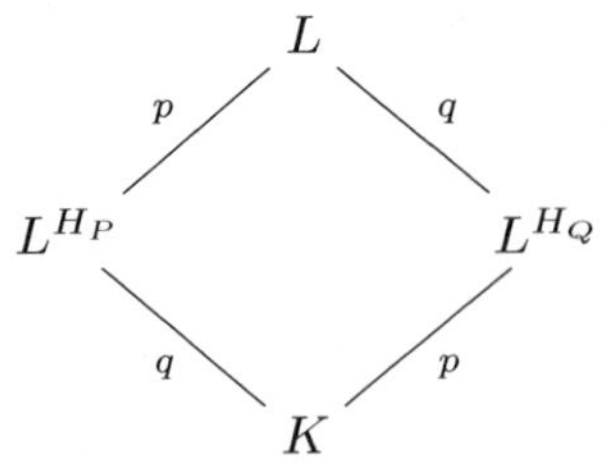

By classical Galois theory, we see that $L^{H_P} = L^{\langle s \rangle}$, which a Galois extension of K, and that $L^{H_Q} = L^{\langle s^c t \rangle}$, which is not. However, since N_c is abelian, the subgroups P, Q are both normal in N_c, and so by Theorem 8.2 the extension L^{H_P}/K is $L[N_c/P]^G$-Galois and the extension L^{H_Q}/K is $L[N_c/Q]^G$-Galois.

Since the Galois extension L^{H_P}/K has prime degree, it admits only the classical Hopf-Galois structure, which must coincide with that given by $L[N_c/P]^G$. Similarly, the non-normal extension L^{H_Q}/K has prime degree and is contained in a Galois extension with solvable Galois group, so by [**Chi89**, Theorem 2] it admits a unique Hopf-Galois structure, which must coincide with that given by $L[N_c/Q]^G$.

EXAMPLE 8.8. Let L/K be a G-Galois extension with G isomorphic to the quaternion group of order 8. Then L/K admits 6 Hopf-Galois structures of cyclic type (see [**SV18**, Table A.1]). Let N be a cyclic subgroup of $\mathrm{Perm}(G)$ corresponding to such a structure, let P be the unique subgroup of N of order 2, and let Q be the unique subgroup of N of order 4. Each of these subgroups is normal and, by uniqueness, each is normalized by $\lambda(G)$; hence we obtain Hopf subalgebras $H_P = L[P]^G$ and $H_Q = L[Q]^G$ of $H_N = L[N]^G$ and, in turn, intermediate fields L^{H_P} and L^{H_Q}.

By classical Galois theory, we see that L^{H_P}/K is a biquadratic subextension of L/K, and that L^{H_Q}/K is a quadratic subextension of L/K (the precise quadratic subfield of L that we obtain depends upon the choice of N). Since P and Q are both normal in N, by Theorem 8.2 the extension L^{H_P}/K is $L[N/P]^G$-Galois, and the extension L^{H_Q}/K is $L[N/Q]^G$-Galois.

A biquadratic extension admits precisely 4 Hopf-Galois structures [**Byo02**, Theorem 2.3]: the classical structure (of type $C_2 \times C_2$) and 3 nonclassical structures (of cyclic type). Since the quotient group N/P is cyclic, the Hopf algebra $L[N/P]^G$ gives ones of the nonclassical Hopf-Galois structures on L^{H_P}/K. The quadratic extension L^{H_Q}/K admits only the classical Hopf-Galois structure, which must coincide with that given by $L[N/Q]^G$.

Theorem 8.2 is applied in [**CRV18**] to study questions of realizability of Hopf-Galois structures of certain types on certain Galois extensions: these results appear in Chapter 3. It is also applied in [**Tay20**] to study integral module structure in Galois extensions of local or global fields: we survey this work in Chapter 11.

8.2. Skew braces and normality in Hopf-Galois extensions

We continue to suppose that L/K is a G-Galois extension. Recall from Section 2.9 that each Hopf-Galois structure on L/K of type N yields a skew brace $B_N = (B, \circ, \star)$ with $(B, \circ) \cong G$ and $(B, \star) \cong N$, and from Section 7.5 that it is possible to study the image of the Galois correspondence for H by studying $\circ$-stable subgroups of B_N. In this section we develop this idea further by giving a brace-theoretic interpretation of the results of Section 8.1. These results first appeared in [**KT20**], and make use of the notions of *opposite Hopf-Galois structure* and *opposite skew brace*.

It has been known since [**GP87**, Theorem 2.5] that if N is a regular subgroup of $\mathrm{Perm}(G)$ normalized by $\lambda(G)$ then so too is

$$N^{opp} = \mathrm{Cent}_{\mathrm{Perm}(G)}(N) = \{\eta' \in \mathrm{Perm}(G) \mid \eta'\eta = \eta\eta' \text{ for all } \eta \in N\}.$$

DEFINITION 8.9. Let N be a regular subgroup of $\mathrm{Perm}(G)$ normalized by $\lambda(G)$, and let $H = L[N]^G$. The subgroup N^{opp} is called the *opposite* subgroup of N, the Hopf-Galois structure on L/K corresponding to N^{opp} is called the *opposite* of the structure corresponding to N, and we write $H^{opp} = L[N^{opp}]^G$.

We remark that this definition extends naturally to the case in which L/K is merely separable. Opposite subgroups, and the corresponding Hopf-Galois structures, are studied in [**Koh07**], [**Tru16a**], and [**Tru18b**]. In [**KT20**] the skew braces corresponding to opposite Hopf-Galois structures on Galois extensions are studied: in [**KT20**, Proposition 3.4] it is shown that if $B_N = (B, \circ, \star)$ is the skew brace corresponding to N then the skew brace corresponding to N^{opp} is $B_{N^{opp}} = (B, \circ, \star')$, where $x \star' y = y \star x$ for all $x, y \in B$. Motivated by this, we define the *opposite* of any skew brace $B = (B, \circ, \star)$ to be the skew brace $B^{opp} = (B, \circ, \star')$, with the binary operation $\star'$ defined as above.

When viewed in the opposite skew brace, $\circ$-stable subgroups resemble the more familiar skew brace structure of an ideal. In fact, we generalize the notion of an ideal slightly:

DEFINITION 8.10. Let $B = (B, \circ, \star)$ be a skew brace.

(1) A *quasi-ideal* (QI) of B is a subgroup $I \leq (B, \star)$ such that

$$x^{-1} \star (x \circ y) \in I, \; x \in B, y \in I.$$

(2) A $\star$-*quasi-ideal* ($\star$-QI) of B is a quasi-ideal which is normal in $(B, \star)$.

(3) A $\circ$-*quasi-ideal* ($\circ$-QI) of B is a quasi-ideal which is normal in $(B, \circ)$.

(4) An *ideal* of B is a subgroup of $(B, \star)$ which is both a $\star$-QI and a $\circ$-QI.

We note immediately:

PROPOSITION 8.11. *Let I be a quasi-ideal of a skew brace $B = (B, \circ, \star)$. Then I is also a subgroup of $(B, \circ)$, hence is a sub-brace of B.*

PROOF. Clearly B is nonempty. For all $x, y, z \in B$ we have

$$x^{-1} \star (x \circ y \circ z) = x^{-1} \star (x \circ y) \star (x \circ y)^{-1} \star (x \circ y \circ z).$$

If $y, z \in I$ then for all $x \in B$ we have $x^{-1} \star (x \circ y) \in I$ and $(x \circ y)^{-1} \star (x \circ y \circ z) \in I$, and so their product under $\star$ is in I. Choosing $x = e$ we see that $y \circ z \in I$; hence I is closed under $\circ$. Since B is finite, this implies that I is a subgroup of $(B, \circ)$. □

Recall that a regular subgroup N of $\mathrm{Perm}(G)$ that is normalized by $\lambda(G)$ yields a skew brace $B = B_N$ and a Hopf algebra $H = L[N]^G$ giving a Hopf-Galois structure on L/K, and that its opposite subgroup N^{opp} yields a skew brace $B^{opp} = B_{N^{opp}}$ and a Hopf algebra $H^{opp} = L[N^{opp}]^G$ giving a Hopf-Galois structure on L/K. Each of the substructures defined above gives us insight into the intermediate fields of L/K that are realizable with respect to these Hopf-Galois structures.

LEMMA 8.12. *There is a bijection between quasi-ideals I of B and intermediate fields $K \leq L_I \leq L$ realizable with respect to H^{opp}.*

PROOF. Let I be a quasi-ideal of B. Since the underlying sets of B and B^{opp} are the same and $x^{-1} \star (x \circ y) = (x \circ y) \star' x^{-1}$ for all $x, y \in B$ we see that I is a $\circ$-stable subgroup of B^{opp}. Since $B^{opp} \cong B_{N^{opp}}$ we obtain a $\circ$-stable subgroup in $B_{N^{opp}}$, say I'. Then, as noted above, I' corresponds to a sub-Hopf algebra of H^{opp}, hence an intermediate field, say L_I, in L/K which is realizable with respect to H^{opp}. Conversely, if F is a field which is realizable with respect to H^{opp}, there is a corresponding $\circ$-stable subgroup of $B_{N^{opp}}$, hence of B^{opp}, giving us a quasi-ideal of B. □

The extension L/L_I is Galois, and by Proposition 8.1 it is Hopf-Galois with respect to the Hopf algebra $L_I \otimes_K L[I']^G$.

LEMMA 8.13. *There is a bijection between $\star$-quasi-ideals I of B and intermediate fields $K \leq L_I \leq L$ realizable with respect to H^{opp} such that L_I/K is Hopf-Galois via the K-Hopf algebra $L[N^{opp}/I']^G$, where I' is the image of I under the map $I \mapsto I'$ above.*

PROOF. A $\star$-quasi-ideal I of B yields a $\circ$-stable subgroup I' of $B_{N^{opp}}$ which is a normal subgroup of $(N^{opp}, \star')$. In other words, I' is a normal subgroup of N^{opp} that is normalized by $\lambda(G)$. The result now follows from Theorem 8.2. □

LEMMA 8.14. *There is a bijection between $\circ$-quasi-ideals I of B and intermediate fields $K \leq L_I \leq L$ realizable with respect to $H_{N^{opp}}$ such that L_I/K is Galois.*

PROOF. A $\circ$-quasi-ideal I of B yields a $\circ$-stable subgroup I' of $B_{N^{opp}}$ which is a normal subgroup of $(N^{opp}, \circ)$, and so corresponds to a normal subgroup of G. □

We summarize the results of this section (see [**KT20**, Theorem 5.6]):

THEOREM 8.15. *Let $B = B_N$, let $I \subseteq B$ be a quasi-ideal, and let I' be the corresponding $\circ$-stable subgroup of $B_{N^{opp}}$. Then there exists a field $K \leq L_I \leq L$ such that L/L_I is Hopf-Galois via the L_I-Hopf algebra $L_I \otimes_K L[I']^G$. Furthermore:*

(1) *If I is a $\star$-QI then L_I/K is also Hopf-Galois with respect to the Hopf algebra $L[N^{opp}/I']^G$.*
(2) *If I is a $\circ$-QI then L_I/K is (classically) Galois.*
(3) *If I is an ideal, then L_I/K is both Galois and Hopf-Galois in the sense mentioned above.*

Furthermore, any realizable intermediate field F is of the form L_I for some quasi-ideal I; and if F satisfies the properties (1), (2), or (3) above, then I is a $\star$-QI, $\circ$-QI, or an ideal respectively.

EXAMPLE 8.16. Let p, q be prime numbers with $p \equiv 1 \pmod q$, and let L/K be a G-Galois extension with G isomorphic to the metacyclic group of order pq, as in Example 8.7. We stated there that L/K admits p Hopf-Galois structures of cyclic type, each corresponding to a cyclic regular subgroup of $\mathrm{Perm}(G)$ that is normalized by $\lambda(G)$. By [**AB20a**] or [**CCDC20**] there is, up to isomorphism, a unique skew brace $B = (B, \circ, \star)$ of order pq with $(B, \circ)$ metacyclic and $(B, \star)$ cyclic; hence, all of these Hopf-Galois structures yield this brace. Explicitly, it is given by

$$B = \{\eta^i \pi^j \mid 0 \leq i \leq p-1,\ 0 \leq j \leq q-1\},$$

with

$$\begin{aligned}\eta^i\pi^j \circ \eta^k\pi^\ell &= \eta^{i+kd^j}\pi^{j+\ell}\\ \eta^i\pi^j \star \eta^k\pi^\ell &= \eta^{i+k}\pi^{j+\ell}\end{aligned}$$

(where d is an integer having multiplicative order q modulo p).

The group $(B, \star)$ has a unique subgroup $\langle\eta\rangle$ of order p and a unique subgroup $\langle\pi\rangle$ of order q. It is routine to verify that these are both $\star$-quasi-ideals of B. Since $\langle\eta\rangle$ is also a normal subgroup of $(B, \circ)$, it is also a $\circ$-quasi-ideal of B, and hence an ideal of B. However, $\langle\pi\rangle$ is not a normal subgroup of $(B, \circ)$, and so is not a $\circ$-quasi-ideal of B. Thus the realizable intermediate fields for a Hopf-Galois structure of cyclic type on L/K comprise (in addition to L and K) a field of degree q over K, which is a Galois and Hopf-Galois extension of K, and a field of degree p over K, which is a Hopf-Galois extension of K but not a Galois extension of K. The former is common to all Hopf-Galois structures of cyclic type on L/K; the latter depends upon the particular structure concerned. This confirms the results of Example 8.7.

EXAMPLE 8.17. Let L/K be a G-Galois extension with G isomorphic to the quaternion group of order 8, as in Example 8.8. We stated there that L/K admits 6 Hopf-Galois structures of cyclic type, each corresponding to a cyclic regular subgroup of $\mathrm{Perm}(G)$ that is normalized by $\lambda(G)$. By [**Nej18**, Theorem 1.2.6] there is, up to isomorphism, a unique skew brace $B = (B, \circ, \star)$ of order 8 with $(B, \circ) \cong Q_8$ and $(B, \star) \cong C_8$; hence, all of these Hopf-Galois structures yield this brace. Explicitly, it is given by:

$$B = \{\eta^i \mid 0 \le i \le 7\},$$

with

$$\begin{aligned}\eta^i \circ \eta^j &= \eta^{i+(-1)^i j+4ij}\\ \eta^i \star \eta^j &= \eta^{i+j}.\end{aligned}$$

The group $(B, \star)$ has a unique subgroup $\langle\eta^4\rangle$ of order 2 and a unique subgroup $\langle\eta^2\rangle$ of order 4. It is routine to verify that these are both ideals of B. Thus the realizable intermediate fields for a Hopf-Galois structure of cyclic type on L/K comprise (in addition to L and K) a field of degree 4 over K and a field of degree 2 over K, which are both Galois and Hopf-Galois extensions of K. The former is common to all Hopf-Galois structures of cyclic type on L/K; the latter depends upon the particular structure concerned. This confirms the results of Example 8.8.

8.3. Induced Hopf-Galois structures

In light of Proposition 8.1 and Theorem 8.2 the following question arises naturally: given a G-Galois extension L/K, an intermediate field F, and Hopf-Galois structures on L/F and F/K, can we classify the Hopf-Galois structures on L/K that yield the given Hopf-Galois structure on L/F (via Proposition 8.1) and the given Hopf-Galois structure on F/K (via Theorem 8.2)? That is: can we obtain a theory of extensions of Hopf-Galois structures on L/K?

The central result of this section is a theorem of Crespo, Rio, and Vela [**CRV16b**], which is the only construction of this type currently known. We note that in fact this result predates [**KKTU19a**], in which Theorem 8.2 first appeared.

THEOREM 8.18. *Let J be a subgroup of G and suppose that J has a normal complement H in G. Suppose that $A \le \mathrm{Perm}(G/J)$ corresponds to a Hopf-Galois structure on L^J/K, and that $B \le \mathrm{Perm}(J)$ corresponds to a Hopf-Galois structure* *on L/L^J. Then there is a Hopf-Galois structure on L/K whose corresponding subgroup $N \leq \mathrm{Perm}(G)$ is isomorphic to $A \times B$.*

PROOF. Let $X = G/J$ and let $\lambda_X : G \to \mathrm{Perm}(X)$ be the left translation map. Since $A \leq \mathrm{Perm}(X)$ corresponds to a Hopf-Galois structure on L^J/K, given $\alpha \in A$ and $g \in G$ there exists $\alpha' \in A$ such that $\alpha\lambda_X(g) = \lambda_X(g)\alpha'$. Similarly, let $\lambda_J : J \to \mathrm{Perm}(J)$ be the left regular representation. Since $B \leq \mathrm{Perm}(J)$ corresponds to a Hopf-Galois structure on L/L^J, given $\beta \in B$ and $j \in J$ there exists $\beta' \in B$ such that $\beta\lambda_J(j) = \lambda_J(j)\beta'$.

Since H is a complement to J in G, every element of G may be uniquely expressed in the form hj with $h \in H$ and $j \in J$, and so we may identify $X = G/J$ and H. This identifies A with a regular subgroup of $\mathrm{Perm}(H)$, and λ_X with the homomorphism $\lambda_H : G \to \mathrm{Perm}(H)$ defined by

$$\lambda_H(hj)[h'] = hjh'j^{-1} \text{ for all } h, h' \in H \text{ and } j \in J.$$

(Note that λ_H is the composition of the action of J on H by conjugation with the left regular representation of H.) The normalization property of A becomes: given $\alpha \in A$ and $g \in G$ there exists $\alpha' \in A$ such that $\alpha\lambda_H(g) = \lambda_H(g)\alpha'$. We can describe the multiplication in G in terms of the maps λ_H and λ_J: for $h, h' \in H$ and $j, j' \in J$ we have

$$\begin{aligned} hjh'j' &= hjh'j^{-1}jj' \\ &= \lambda_H(hj)[h']\lambda_J(j)[j']. \end{aligned}$$

Now we define a map $\eta : A \times B \to \mathrm{Perm}(G)$ by

$$\eta(\alpha, \beta)[hj] = \alpha[h]\beta[j], \tag{8.5}$$

and note that η is a homomorphism whose image $N = \eta(A \times B)$ is a regular subgroup of $\mathrm{Perm}(G)$. We need to show that N is normalized by the image of G under the left regular representation $\lambda_G : G \to \mathrm{Perm}(G)$. For $\alpha \in A$, $\beta \in B$, $h, h' \in H$, and $j, j' \in J$ we have:

$$\begin{aligned} \eta(\alpha,\beta)[\lambda_G(hj)[h'j']] &= \eta(\alpha,\beta)[h(jh'j^{-1})jj'] \\ &= \alpha[h(jh'j^{-1})]\beta[jj'] \\ &= \alpha[\lambda_H(hj)h']\beta[\lambda_J(j)[j']] \\ &= \lambda_H(hj)\alpha'[h']\lambda_J(j)\beta'[j'] \text{ for some } \alpha' \in A \text{ and } \beta' \in B \\ &= \lambda_G(hj)\eta(\alpha',\beta')[h'j']. \end{aligned}$$

Therefore N is indeed normalized by $\lambda_G(G)$, and hence corresponds to a Hopf-Galois structure on L/K. □

The following corollary is a restatement of Theorem 4.37, which was proved in Chapter 4 by brace-theoretic methods:

COROLLARY 8.19. *Let L/K be a G-Galois extension with G a semidirect product $G = H \rtimes J$. Then L/K admits a Hopf-Galois structure of type $H \times J$.*

PROOF. The extension L/L^J is a J-Galois extension, and so admits at least the classical Hopf-Galois structure, which has type J. Let $X = G/J$ and $\lambda_X : G \to \mathrm{Perm}(X)$ be the left translation map, and consider the subgroup $\lambda_X(H)$ of $\lambda_X(G)$. The fact that the elements of H form a system of coset representatives for J in G implies that this is a regular subgroup of $\mathrm{Perm}(X)$, and it is normalized by $\lambda(G)$

because H is normal in G. It therefore corresponds to a Hopf-Galois structure on L^J/K of type H. Applying Theorem 8.18, we see that L/K admits a Hopf-Galois structure of type $H \times J$. □

A Hopf-Galois structure on a Galois extension L/K that is obtained via the construction in Theorem 8.18 is said to be *induced*. By construction, the regular subgroup N of $\operatorname{Perm}(G)$ that corresponds to an induced Hopf-Galois structure is the direct product of two normal subgroups $\widehat{A} = \eta(A, 1_B)$ and $\widehat{B} = \eta(1_A, B)$. Examining the proof of Theorem 8.18 we see that, in addition, each of $\widehat{A}$ and $\widehat{B}$ is normalized by $\lambda(G)$, and thus we obtain Hopf subalgebras $L[\widehat{A}]^G$ and $L[\widehat{B}]^G$ of $L[N]^G$. By Proposition 8.1 the $L^H \otimes_K L[\widehat{A}]^G$ gives a Hopf-Galois structure on L/L^H, and $L^J \otimes_K L[\widehat{B}]^G$ gives a Hopf-Galois structure on L/L^J, which coincides with the Hopf-Galois structure corresponding to the regular subgroup B of $\operatorname{Perm}(J)$. Applying Theorem 8.2, we see that $L[N/\widehat{A}]^G$ gives a Hopf-Galois structure on L^H/K, and that $L[N/\widehat{B}]^G$ gives a Hopf-Galois structure on L^J/K, which coincides with the Hopf-Galois structure corresponding to the regular subgroup A of $\operatorname{Perm}(G/J)$.

The following example appears in slightly different language in [**CRV16b**, Subsection 4.3]:

EXAMPLE 8.20. Let p, q be prime numbers with $p \equiv 1 \pmod{q}$, and let L/K be a G-Galois extension with G isomorphic to the metacyclic group of order pq, as in Example 8.7. Fix $c \in \{0, \ldots, p-1\}$, and let $J = J_c = \langle s^c t\rangle$; then J is a subgroup of G of order q, and $H = \langle s\rangle$ is a normal complement of J in G. The extension L^J/K is a separable extension of prime degree whose Galois closure L/K has solvable Galois group; by [**Chi89**, Theorem 2] it admits a unique Hopf-Galois structure, corresponding to the subgroup $A = \langle \alpha\rangle \leq \operatorname{Perm}(G/J)$, where

$$\alpha[s^i J] = s^{i-1} J \text{ for } i = 0, \ldots, p-1.$$

The extension L/L^J is a J-Galois extension and has prime degree; by Byott's Uniqueness Theorem [**Byo96**, Theorem 1] it admits a unique Hopf-Galois structure, corresponding to the subgroup $B = \rho(J) \leq \operatorname{Perm}(J)$. Writing $\beta = \rho(s^c t)$, we have $B = \langle \beta\rangle$. By Theorem 8.18 we can induce a Hopf-Galois structure on L/K: the corresponding subgroup of $\operatorname{Perm}(G)$ is

$$N = N_c = \{\eta(\alpha^i, \beta^j) \mid i = 0, \ldots p-1,\ j = 0, \ldots q-1\},$$

where, expressing G as the semidirect product of H and J,

$$\eta(\alpha^i, \beta^j)[s^u (s^c t)^v] = s^{u-i}(s^c t)^{v-j}.$$

We note that each $c = 0, \ldots, p-1$ yields a different subgroup N_c: the intersection of N_c with $\rho(G)$ is precisely $\rho(J_c)$. Thus we obtain p distinct Hopf-Galois structures on L/K of cyclic type. By [**Byo04b**, Theorem 6.2], these are all of the commutative Hopf-Galois structures of cyclic type on L/K.

Other examples considered in [**CRV16b**] include Galois extensions of degree $4p$, for $p \geq 5$ prime (c.f. [**Koh07**]), those of degree $p(p-1)$, with p a safeprime (c.f. [**BC12**]), and examples arising from the Schur-Zassenhaus Theorem. In [**CRV18**] induced Hopf-Galois structures are used to study symmetric or alternating Galois extensions.

Induced Hopf-Galois structures can also yield information about non-normal extensions. A separable extension L/K with Galois closure $\widetilde{L}$ is called *almost classically Galois* if $G_L = \mathrm{Gal}(\widetilde{L}/L)$ has a normal complement in $G = \mathrm{Gal}(\widetilde{L}/K)$ (see [**GP87**, Proposition 4.1]). By combining Theorem 8.18 and Theorem 8.2, we can show that every Hopf-Galois structure on an almost classically Galois extension L/K arises as a "quotient" of some Hopf-Galois structure on $\widetilde{L}/K$:

PROPOSITION 8.21. *Let L/K be an almost classically Galois extension and suppose that L/K is $\widetilde{L}[M]^G$-Galois for some regular subgroup M of* $\mathrm{Perm}(G/G_L)$ *normalized by $\lambda(G)$. Then there exists a regular subgroup N of* $\mathrm{Perm}(G)$ *and a normal subgroup P of N, both normalized by $\lambda(G)$, such that the Hopf-Galois structure given by $\widetilde{L}[N/P]^G$ coincides with that given by $\widetilde{L}[M]^G$.*

PROOF. The extension $\widetilde{L}/L$ is a G_L-Galois extension, so certainly it admits a classical Hopf-Galois structure, corresponding to the regular subgroup $\rho(G_L) \leq \mathrm{Perm}(G_L)$. Since G_L has a normal complement in G, by Theorem 8.18 we can induce a Hopf-Galois structure on $\widetilde{L}/K$ from the classical Hopf-Galois structure on $\widetilde{L}/L$ and the Hopf-Galois structure on L/K corresponding to M. Referring to (8.5), the corresponding regular subgroup of $\mathrm{Perm}(G)$ is $N = \eta(M \times \rho(G_L))$. Let $P = \eta(1_M, \rho(G_L))$; then P is a normal subgroup of N and is normalized by $\lambda(G)$, so by Theorem 8.2 the extension L/K is $\widetilde{L}[N/P]^G$-Galois. By the discussion following Corollary 8.19, we see that this Hopf-Galois structure on L/K coincides with the one given by $\widetilde{L}[M]^G$. □

Finally, we return to the case in which L/K is Galois and consider a situation in which the structure of its Galois group is governed by arithmetic information:

PROPOSITION 8.22. *Let L/K be a G-Galois extension of local fields, and suppose that L/K is at most tamely ramified. Then every Hopf-Galois structure of abelian type on L/K is induced.*

PROOF. Suppose that L/K is $L[N]^G$-Galois for some abelian regular subgroup N of $\mathrm{Perm}(G)$ normalized by $\lambda(G)$. Let p be the residue characteristic of K, write $|N| = p^r m$ with $(p, m) = 1$, and write $N = A \times B$ with $|A| = m$ and $|B| = p^r$. Since A, B are the unique subgroups of N of their respective orders, we see that each is normalized by $\lambda(G)$. In particular, $L[B]^G$ is a Hopf subalgebra of $L[N]^G$ and $[L : L^{L[B]^G}] = p^r$; let $J = \mathrm{Gal}(L/L^{L[B]^G})$. By the results of this section, B corresponds to a Hopf-Galois structure on L/L^J and A corresponds to a Hopf-Galois structure on L^J/K.

Next we consider the structure of G. Let G_0 denote the inertia subgroup of G; then L^{G_0}/K is the maximal unramified subextension of L/K, so the quotient group $\overline{G} = G/G_0$ is cyclic. By tameness, $|G_0|$ is coprime to p, so p^r divides $|\overline{G}|$, and so $\overline{G}$ has a unique subgroup of index p^r, which is normal in $\overline{G}$. Therefore G has a unique subgroup H of index p^r, which is normal in G. Since $|J| = p^r$ we have $J \cap H = \{1\}$, so H is a normal complement to J in G.

Now we may induce a Hopf-Galois structure on L/K from the Hopf-Galois structures corresponding to A on L^J/K and to B on L/L^J. Referring to (8.5) once again, we see that the corresponding regular subgroup $\eta(A \times B)$ of $\mathrm{Perm}(G)$ coincides with N, and so the original Hopf-Galois structure on L/K is induced. □

Proposition 8.22 is applied in [**Tru18a**] to study the integral Hopf-Galois module structure of tamely ramified extensions of local fields. We will prove the main results of loc. cit. by different methods in Chapter 11.

CHAPTER 9

Descent theory, and the structure of Hopf algebras acting on separable field extensions

The intention of this chapter is twofold. While the final goal is an extensive discussion of the algebra structure of Hopf algebras that arise in Hopf-Galois theory or more generally via Galois descent of Hopf algebras, we prefer to start out by putting this in a wider perspective: the first section (= roughly a third) of this chapter will be devoted to a sketch of basic descent theory in a more general setting. In this sketch we decided to forsake extreme generality and restrict ourselves from the beginning to the affine case. Of course one loses interesting geometric applications by doing so, but since our final purpose is getting a better understanding of finite-dimensional Hopf algebras over fields, the affine case is enough for us, and of course the presentation simplifies when one can work with rings and algebras only, bypassing the language of varieties or schemes.

As we said, we will begin by treating descent in general. But there is a close and immediate link to Hopf-Galois theory, which we will mention right now, even if we have to use terminology that is only explained later. Whenever L/K is H-Galois for a K-Hopf algebra H, the algebra L is an L/K-form of the dual H^*, in the category of algebras. More explicitly: there is an isomorphism of K-algebras $L \otimes_K L \cong L \otimes_K H^*$. See Example 10.6 and Corollary 10.9.

In the ring-theoretic context, the most general type of descent theory works along so-called faithfully flat extensions of commutative rings. This general theory gives rise to several more specialized incarnations: faithfully projective descent, Galois descent, and descent along purely inseparable extensions of height one. The last-named variant will not be used explicitly in this book. (Somehow it will be in the background in Chapter 10, where inseparable field extensions and their Hopf-Galois structures are treated, simply by the universal connection mentioned in the previous paragraph.) The "faithfully projective" variant is just an intermediate stop on the road from "faithfully flat" to "Galois" and will only be mentioned in passing. On the other hand, we try to describe clearly how Galois descent arises from faithfully flat descent, and how Galois descent works; we support this (and also the more general theory) by many examples. All this is done more or less simultaneously for modules, algebras, and Hopf algebras, taking up the first section of the chapter. In that section there are no new results; an excellent reference, also for the subjects we omit, is [**KO74**]. Perhaps some of the examples cannot be found elsewhere, but we have not actively searched the literature to check this claim.

In the second section we focus on Galois descent for Hopf algebras. Again, this is in principle anything but new; it underlies the major part of the results in [**GP87**] for example. But one main impetus for revisiting this topic is the following possibly surprising fact. Hopf forms are usually (that is, only with obvious exceptions) nontrivial, i.e. not isomorphic to the Hopf algebra that gets deformed, but it was recently discovered that fairly often they are isomorphic to the original object *in the category of algebras.* In fact, the main business of this second section is getting to grips with this phenomenon, both by way of general theorems and by series of examples. In the first subsection we explain the mechanism of Galois descent for Hopf algebras in the particular case that the original object is a group ring; this simplifies things considerably. Then (§2.2) we state and prove a general sufficient criterion, Theorem 9.19, for the triviality of the algebra structure. The remaining three subsections are concerned with a variety of situations where this result does not apply. Since the general criterion gives a sufficient condition whose proof does not show the slightest sign of being reversible, one may expect that in situations where it does not apply, sometimes the algebra structure is trivial and sometimes it is not. This is indeed what happens; however, we have not found any reasonable criterion in the strict sense. In other words, we are unable to offer any quick and elegant condition that is *equivalent* to the triviality of the algebra structure. On the positive side, we hope that our series of examples will give an idea of what typically happens, and that we have succeeded in going slightly beyond a mere enumeration of cases for which the algebra structure is non-trivial. There seem to be a few rules among all the chaos, and perhaps a clearer picture will emerge in the future.

We are always interested in forms of $K[G]$. From some point in Section 9.2 onward we will impose the condition $\mathrm{char}(K) = 0$, in order to assure that $K[G]$ is semisimple. We will take up the following situations in turn: G is nonabelian of order 8 (§2.3); G is non-abelian of order $p^3 \neq 8$ (§2.4), G is cyclic of p-power order or dihedral of order $2p$, for any odd prime p (§2.5). We provide full proofs for results that are not yet in the literature. Proofs that are already available in the literature, see mainly [**KKTU19b**], [**KKTU19c**] and [**TT19**], will not be repeated as a rule, but sometimes we try to sketch the idea of proof and in some other cases we offer a different argument.

In a short third section we take up a question which was originally raised by Pareigis: Which rational group rings $\mathbb{Q}[G]$ admit absolutely semisimple forms as a Hopf algebra? We only resolve this for some typical examples; no general answer seems to be available.

9.1. General descent theory

9.1.1. The basic idea of a descent datum. All rings are assumed commutative. Let $R \subset S$ be an extension of rings (this setup will often be abridged by writing S/R); $\otimes$ stands for $\otimes_R$, and ${}_SM$ is short for $S \otimes M$. The question is: When an S-module N is given, how can we decide whether N has the form ${}_SM$ for some R-module M, and can we describe the candidates M? If $N = {}_SM$, then there is an isomorphism $g : S \otimes {}_SM \to {}_SM \otimes S$, sending $s \otimes t \otimes m$ to $s \otimes m \otimes t$, in other words: just shifting the M-entry from the right to the middle. Then the map g has two important properties. First it fits into the following diagram, in which all triangles involving the vertex M commute:

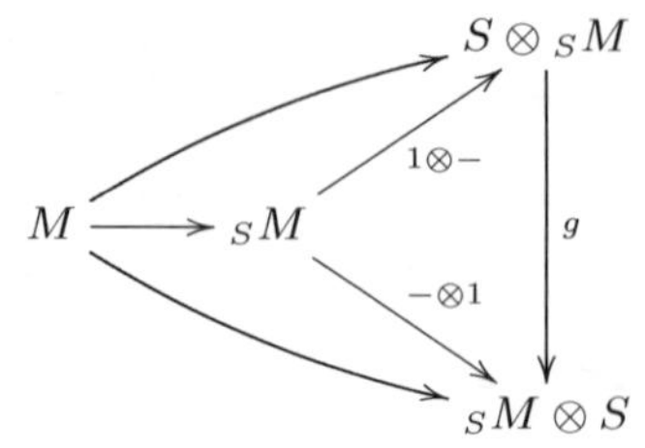

Note that we do *not* claim that the right-hand triangle (involving S_M and g) commutes. For the second property we need shorthand notation. The map g_1 : $S \otimes S \otimes {}_S M \to S \otimes {}_S M \otimes S$ arises from g via tensoring with S on the left; similarly g_3 by tensoring on the right, and g_2 by tensoring in the middle. All these maps are simple to imagine, but g_2 is the most awkward to write down: if $h : S \otimes {}_S M \to {}_S M \otimes S$ is any map and $h(s \otimes n) = \sum_i n_i \otimes s_i$, then $h_2(s \otimes t \otimes n) = \sum_i n_i \otimes t \otimes s_i$. The second property of g is then

$$g_2 = g_3 \circ g_1.$$

The effect of either side in this equation is simply shifting M from right to left; in one fell swoop if we use g_2, and in two moves, via the middle, if we use $g_3 \circ g_1$. The idea is now to recover M as the difference kernel of the pair of maps

$${}_S M \underset{-\otimes 1}{\overset{g\circ(1\otimes -)}{\rightrightarrows}} {}_S M \otimes S.$$

A minimal requirement for this to work is that the functor $M \mapsto {}_S M$ is exact, in other words, S/R should be flat. Moreover it should work in the simplest of all cases, that is, R should identify with the kernel of the map $S \to S \otimes S$, $s \mapsto 1 \otimes s - s \otimes 1$. This excludes a lot of otherwise nice extensions S/R. For instance when R is a domain and S any nontrivial localization of R seen as a subring of $\operatorname{Quot}(R)$, then $S \otimes S$ identifies with S and the map $s \mapsto 1 \otimes s - s \otimes 1$ is the zero map, so the kernel is S, and not R as one would like.

The correct condition for the most general version of descent is that S/R should be *faithfully* flat. Just for once, let us also say what this means in geometry: the corresponding map of varieties/schemes should be surjective. A convenient definition is as follows.

DEFINITION 9.1. If S/R is an extension of commutative rings, then S is called *faithfully flat* over R if S is flat over R, and for all maximal ideals $\mathfrak{m}$ of R, $\mathfrak{m}S$ is a proper ideal of S.

One has the following very useful characterization of faithful flatness: S/R is faithfully flat if and only if the functor $M \mapsto {}_S M$ preserves exactness, and detects exactness. We are now ready for the basic concept.

DEFINITION 9.2. A *descent datum* is given by an S-module N and an $S \otimes S$-linear isomorphism $g : S \otimes N \to N \otimes S$ such that $g_2 = g_3 g_1$. Descent data form a category, when we declare: A morphism of descent data (N, g) and (N', g') is an S-linear map $f : N \to N'$ making the obvious square commute (that is, $g' f_1 = f_2 g$).

We have described above a functor J from Mod-R into the category of descent data: M is sent to $N = {}_S M$, and $g : S \otimes N \to N \otimes S$ is defined explicitly as above. The basic result, which is not very hard to prove (see [**KO74**]), then goes as follows.

THEOREM 9.3. *Whenever S/R is faithfully flat, the functor J is an equivalence of categories. The inverse functor, on the level of modules, sends (N, g) to the difference kernel of the pair of maps $g \circ (1 \otimes -)$ and $- \otimes 1$ from N to $N \otimes S$.*

We have seen above that if (N, g) comes from an R-module M, then this difference kernel construction retrieves M.

Descent theory works fine if we impose extra structure on modules. That is, the preceding theorem has many variants, each applying to some category which is richer than just the category of R-modules. Then all occurring morphisms, like g

and f in the above definition, are of course required to be in the relevant category. The cases we are most interested in are the categories of R-algebras and R-Hopf algebras respectively.

Let us mention some important classes of ring extensions S/R which are faithfully flat. The first situation that comes to mind is perhaps the case where S is finitely generated free (of rank $n > 0$) over R. Slightly more generally, by obvious localization arguments, one sees: If S is finitely generated and projective (= locally free) over R, and the rank, as a locally constant function on the spectrum of R, is nowhere 0 (it need not be constant), then S/R is faithfully flat; in fact it is called faithfully projective. If R is connected (that is, R has no idempotents beside 0 and 1), then a ring extension S/R which is projective as an R-module is automatically faithfully projective, because it is not zero.

These examples will go a long way in the Hopf-algebraic setting to be studied soon, but we should add here that in geometry and commutative algebra, a different class of faithfully flat extensions plays a much greater role: extensions of the form $S = R_{f_1} \times \ldots R_{f_t}$. Here a subscript f denotes localization at the multiplicative set $\{1, f, f^2, \ldots\}$, and one assumes that the elements $f_1, \ldots, f_t$ generate the unit ideal of R. The geometric counterpart of the inclusion $R \to S$, that is, the corresponding map from the spectrum of S to the spectrum of R, is called a Zariski covering. Its surjectivity is equivalent to the property $(f_1, \ldots, f_t)_R = (1)$.

9.1.2. Twisted forms and Amitsur cohomology. In this subsection, we will omit the proofs of some very basic results, referring to [**KO74**] instead.

We take a new point of view now. Let us start with $N = {}_SM$, where the R-module M is given, and look for other modules M' obtained by descent data of the form (N, g). In other words, we try to solve the "equation"

$$S \otimes_R M' \cong S \otimes_R M$$

for M'.

DEFINITION 9.4. If the preceding "equation" holds, then M' is called an *S/R-form* (or simply an S-form) of M.

We now identify $S \otimes {}_SM$ and ${}_SM \otimes S$ with ${}_{S\otimes S}M$, the module M base-changed from R to $S \otimes S$. (One identification is obvious, the other just shifts elements of M.) This allows us to simply consider the map g in a descent datum for $N = {}_SM$ as an automorphism of ${}_{S\otimes S}M$. We need to restate the descent condition. For $i = 1, 2, 3$, let $a_i : S \otimes S \to S \otimes S \otimes S$ be the map that inserts 1 at position i. Then g_i identifies with the map $a_{i*}g$ (the basechanged map along the inclusion of rings given by a_i). The descent condition is still

$$g_2 = g_3 \circ g_1 \in \mathrm{Aut}_{S\otimes S\otimes S}\,({}_{S\otimes S\otimes S}M).$$

Every isomorphism g satisfying this condition defines a form M_g, and one can show that $M_g \cong M_h$ iff there is $f \in \mathrm{Aut}_S({}_SM)$ such that

$$h = f_2 \circ g \circ f_1^{-1}.$$

For each n, we have $n+1$ injections $a_1, a_2, \ldots, a_{n+1}$ going from $S^{\otimes n}$ to $S^{\otimes(n+1)}$; a_i inserts a 1 at position i. (So the index n is suppressed.) This can be visualized as follows:

$$S \rightrightarrows S \otimes S \overset{\longrightarrow}{\rightrightarrows} S \otimes S \otimes S \ldots$$

If F is any functor from the category of (commutative) R-algebras to the category of abelian groups, this leads to a complex

$$F(S) \to F(S \otimes S) \to F(S \otimes S \otimes S) \to \dots,$$

where each term $F(S^{\otimes n})$ is placed in degree $n-1$, and the differential from degree $n-1$ to degree n is the alternating sum of the maps $F(a_i)$ (there are $n+1$ of them). So from degree zero to one we have $F(a_1) - F(a_2)$, and from degree one to two we have $F(a_1) - F(a_2) + F(a_3)$. The complex property (concatenation of two differentials always gives zero) is easy to check. But one can say more:

PROPOSITION 9.5. *Let $F = \mathbb{G}_a$ be the underlying abelian group functor (so the beginning of the complex reads $S \to S \otimes S$, and the differential is $s \mapsto 1 \otimes s - s \otimes 1$). Then the complex is exact in each positive degree.*

In general, the cohomology groups of this complex are called Amitsur cohomology, written $\mathrm{H}^i_{S/R}(F)$. For instance, if $F = \mathbb{G}_a$, then $\mathrm{H}^0_{S/R}(F) = \ker(S \to S \otimes S) = R$.

It is an important fact that one can define H^0 and H^1 also for functors F with values in the category of all groups (not necessary abelian). One puts

$$H^0_{S/R}(F) = \{x \in F(S) \mid F(a_1)(x) = F(a_2)(x)\};$$

$$H^1_{S/R}(F) = \{y \in F(S \otimes S) \mid F(a_2)(y) = F(a_3)(y)\, F(a_1)(y)\}.$$

Note that H^1 will no longer be a group in general, only a pointed set, whose distinguished point is given by the neutral element of $F(S \otimes S)$.

In particular, if we define $\underline{\mathrm{Aut}}\, M(T) = \mathrm{Aut}_T(T \otimes_R M)$ for all R-algebras T, then by our explanations above, the first Amitsur cohomology $H^1_{S/R}(\underline{\mathrm{Aut}}\, M)$ classifies the S/R-form of a given R-module M. The same holds, mutatis mutandis, if we impose extra structure.

There is no obvious structure whose $\underline{\mathrm{Aut}}$ is $\mathbb{G}_a$. But for the multiplicative group functor $\mathbb{G}_m$ there is a very natural structure having the multiplicative group $\mathbb{G}_m$ as its functor of automorphisms. For every R-algebra T, the T-automorphism group of the free rank one module T over itself is clearly $T^\times$, so $\underline{\mathrm{Aut}}\, R = \mathbb{G}_m$. On the other hand one knows the S/R-forms of R; they are exactly the projective modules of rank one trivialized by S. The set of these modules up to R-isomorphism, with the multiplication induced by $\otimes_R$, forms the abelian group $\mathrm{Pic}(S/R)$. So we have:

PROPOSITION 9.6. $H^1_{S/R}(\mathbb{G}_m) = \mathrm{Pic}(S/R) = \ker(\mathrm{Pic}(R) \to \mathrm{Pic}(S))$.

If S/R is a field extension L/K, all these objects are trivial. This is true more generally. Let us look at L/K-forms of the free K-module (vector space) of rank n. Clearly, the automorphism functor $\underline{\mathrm{Aut}}(K^n)$ identifies with the functor $\underline{\mathrm{GL}}_n$. From the theory of vector spaces it is obvious that all L/K-forms of K^n are trivial, just looking at the dimension. This gives:

PROPOSITION 9.7. *For any field extension L/K, the Amitsur cohomology set $H^1_{L/K}(\underline{\mathrm{GL}}_n)$ is trivial.*

To add a little geometric context, let us briefly consider Zariski coverings as described at the end of the preceding subsection. There $S = S(f)$ was a direct product of finitely many localizations S_{f_i}, with $f = (f_1, \dots, f_t)$ a vector of elements of R that generate the unit ideal. Here the Amitsur cohomology of $\mathbb{G}_m$ is by no means trivial; on the contrary it is highly interesting. It can be shown that every element of the Picard group of R is trivialized by a suitable ring extension $S(f)/R$.

One may then take the inductive limit (in an appropriate sense) of the cohomology groups $H^1_{S(f)/R}(\mathbb{G}_m)$. (We do have groups, not just pointed sets, since $\mathbb{G}_m$ is abelian.) This limit group, called the first Zariski cohomology group, gives the whole of $\mathrm{Pic}(R)$ in a rather geometric way; we don't have the space here to explain in detail how "glueing of line bundles on open subsets of the spectrum" works.

EXAMPLE 9.8. This is taken from algebraic number theory. Let $R = \mathbb{Z}[\sqrt{-5}]$; this is a Dedekind ring. The ideal $2R$ is the square of the ideal $P = (2, 1 + \sqrt{-5})$. Then this ideal P is the only prime ideal dividing 2, or differently put, the only prime ideal of norm 2. It is well known and easy to check that P is not principal, that is, as an R-module P is not isomorphic to R. But P is projective as an R-module. Let us define $S = R_2 \times R_3$ (recall $R_f = R[1/f]$). We claim that $S \otimes_R P$ is S-isomorphic to S, or in other words, principal. This is equivalent to saying that R_2P and R_3P are principal. For the former this is clear since $R_2P = R_2$. For the latter, look at the principal ideal Rx, $x = 1 + \sqrt{-5}$. Since x has norm 6, Rx factors as P times an ideal dividing 3; the latter disappears after inverting 3. Therefore R_3P is principal with generator x. Thus, P is a nontrivial S-form of the free module R.

It already transpires that finding *nontrivial* forms can be harder than showing that all forms are trivial. We will postpone further examples until after the discussion of frameworks where the general description of Amitsur cohomology simplifies. The most important of these frameworks for us will be Galois descent, and this will be used all the time in the case of Hopf algebras.

9.1.3. Galois descent for general structures. If we know more about the extension S/R than just that it is faithfully flat, then we can expect more precise and amenable results. Our goal is Galois descent; this works when S/R is Galois in the sense of Chase, Harrison and Rosenberg. Note that this implies S/R projective and faithful. In the book [**KO74**] of Knus-Ojanguren, so-called "faithfully projective" descent is treated as an intermediary between "faithfully flat" and "Galois". We will short-circuit this, for two reasons. One of them is saving space. The other is that over a noetherian ring, every finitely generated flat module is already projective, so the former of the two transitions does not go a very long way anyway. It should be pointed out that the typical flat modules which are not finitely generated are localizations. If one wants faithfully flat extensions, one has to take direct sums of sufficiently many localizations; this type of extensions has already been mentioned.

So let us assume S/R is Galois in the sense of Chase, Harrison and Rosenberg [**CHR65**], with group G. Any reader who is not quite comfortable with Galois extensions of rings may retreat to the case where $L = S$ and $K = R$ are fields; the concept of Chase et al. does reduce to the classical notion of a G-Galois extension of fields. In the general ring-theoretic case as well, the map $S \otimes S \to \mathrm{Map}(G, S)$ given by $s \otimes t \mapsto (\gamma \mapsto s \cdot \gamma(t)) \in \mathrm{Map}(G, S)$ is an isomorphism of S-modules. If we identify $\mathrm{Map}(G, S)$ with the G-fold direct product S^G of S with itself, this reads: $s \otimes t$ maps to $\sum_{\gamma \in G} s \cdot \gamma(t) e_\gamma$, where $e_\gamma \in S^G$ is the basic idempotent with index γ.

This isomorphism leads to an iterated version: The map

$$S \otimes S \otimes S \to S^{G \times G}, \quad s \otimes t \otimes u \mapsto \big(s \cdot \gamma(t) \cdot \gamma\delta(u)\big)_{\gamma, \delta \in G}$$

is likewise an isomorphism over S, which operates on $S \otimes S \otimes S$ via the leftmost factor, and on $G \times G$-tuples with entries in S in the obvious way.

Now one has two maps $d_1, d_2 : S \to S^G$ given by $d_1(s)_\gamma = \gamma(s)$, and $d_2(s)_\gamma = s$. In the same vein, there are three maps $S^G \to S^{G\times G}$ given by

$$\begin{aligned} \partial_1(f)_{\gamma,\delta} &= \gamma(f_\delta); \\ \partial_2(f)_{\gamma,\delta} &= f_{\gamma\delta}; \\ \partial_3(f)_{\gamma,\delta} &= f_\gamma \end{aligned}$$

for each G-tuple $f \in S^G$.

The main point is now that the following diagram commutes, and this is not too hard to check.

$$\begin{array}{ccccc} S & \overset{a_1}{\underset{a_2}{\rightrightarrows}} & S \otimes S & \overset{a_1}{\underset{a_3}{\overset{a_2}{\Rrightarrow}}} & S \otimes S \otimes S \\ \downarrow & & \downarrow & & \downarrow \\ S & \overset{d_1}{\underset{d_2}{\rightrightarrows}} & S^G & \overset{\partial_1}{\underset{\partial_3}{\overset{\partial_2}{\Rrightarrow}}} & S^{G\times G}. \end{array}$$

The maps d_i and ∂_i are exactly those that appear in the definition of (non-abelian) group cohomology. Therefore we have:

PROPOSITION 9.9. *(a) For every group-valued functor F and every G-Galois extension S/R of rings there is an identification of pointed sets*

$$H^1_{S/R}(F) = H^1(G, F(S)).$$

(b) If in particular $F = \underline{\mathrm{Aut}}\, A$ for an R-algebra A, then $H^1(G, \mathrm{Aut}({}_SA))$ classifies the S/R-forms of A. The G-action on $\mathrm{Aut}({}_SA)$ is by conjugation involving the G-action on the tensor factor S. More explicitly, if A is finitely generated free over R, and an automorphism of ${}_SA$ is given as a square matrix with coefficients in S, then G acts on these coefficients.

Let us explain how one calculates the form attached to (the class of) a 1-cocycle. This works in the category of R-modules, R-algebras, and R-Hopf algebras by the same token; to be safe, let us assume that our R-Hopf algebras are all finitely generated and projective over R, to make the tensor product well-behaved. The following is taken from [**KO74**, Theorem II 5.3 plus the end of the proof of II 9.1].

We start with an R-structure A ("structure" standing for any of the three options just mentioned), and a cocycle $\theta : G \to \mathrm{Aut}({}_SA)$. This then does not yet give an "action" of G on ${}_SA$. We first define

$$\beta_\sigma = \theta_\sigma \circ (\sigma \otimes 1_A). \quad (\sigma \in G)$$

Then these β_σ are R-automorphisms of $S \otimes_R A$; they are no longer S-linear. But on the other hand the cocycle condition translates into the simpler relation

$$\beta_{\sigma\tau} = \beta_\sigma \beta_\tau, \quad (\sigma, \tau \in G)$$

so that G now acts on $S \otimes_R A$ by R-automorphisms. The form $B = B(\theta)$ that corresponds to $\theta = (\theta_\sigma)_\sigma$ is then simply

$$B = (S \otimes_R A)^G,$$

the fixed "structure" under this action (which depends on θ of course) of the group G on $S \otimes_R A = {}_SA$.

REMARK 9.10. There is a slightly different approach to Galois descent, via so-called Morita theory. Since this is summarized in Section 7.2 and has no immediate relevance for this chapter, we keep our more direct approach here. Both approaches have their merit.

Here is an example of descent in the category of Hopf algebras. We take $R = \mathbb{R}$, the real numbers. The extension $\mathbb{C}/\mathbb{R}$ is Galois with $G = C_2$. Let us take $A = \mathbb{R}[t, t^{-1}]$ (from a geometric view this is the ring of functions on $\mathbb{G}_m$). Then $\mathrm{Aut}({}_{\mathbb{C}}A)$ is a semidirect product of $\mathbb{C}^\times$ and C_2. A nonzero scalar λ in $\mathbb{C}$ corresponds to $t \mapsto \lambda t$. The nontrivial element of C_2 corresponds to $t \mapsto 1/t$. We find

$$H^1(G, \mathrm{Aut}({}_{\mathbb{C}}A)) = H^1(G, C_2) = \mathrm{Hom}(G, C_2)$$

has two elements. (In the first equality one uses the triviality of $H^1(G, \mathbb{C}^\times)$, see Proposition 9.7 and note that $\underline{\mathrm{GL}}_1 = \mathbb{G}_m$.) The nontrivial cocycle sends the complex conjugation τ to the nontrivial element of C_2, that is, to the automorphism $t \mapsto 1/t$. The corresponding Hopf form H is the fixed ring under the operation that conjugates all coefficients and inverts t. This fixed ring is generated by $t + t^{-1}$ and $i(t - t^{-1})$. Note the similarity with the definition of cos and sin from the exponential. This nontrivial form is well known and may be seen as the ring of functions on the unit circle over $\mathbb{R}$ (see for example [**HP86**]). The $\mathbb{R}$-variety with affine algebra H is an algebraic group of dimension 1, which is known as a non-split one-dimensional torus.

9.1.4. More examples. We begin with three examples that use the general notion of a faithfully flat extension and of Amitsur cohomology. In the third example, we will already see how things simplify when using Galois cohomology. The subsequent examples will be formulated purely within the framework of Galois descent.

EXAMPLE 9.11. Let p a prime, $R = K$ a field of characteristic p, $S = L$ an extension field of K. We are considering L/K-forms in the category of Hopf algebras. Let $A = K[x, y]/(x^p, y^p)$; this becomes a K-Hopf algebra by letting x and y be primitive. We need to determine $\underline{\mathrm{Aut}}\, A$. Every automorphism of Hopf algebras respects primitive elements. Hence every Hopf automorphism of A over K restricts to an automorphism of the K-space $Kx + Ky$. The same property, with T instead of K, also holds for all base-changed Hopf algebras ${}_TA = T[x, y]/(x^p, y^p)$. Hence the functor $\underline{\mathrm{Aut}}\, A$ identifies with the functor $\underline{\mathrm{GL}}_2$. By Proposition 9.7, the latter has trivial Amitsur cohomology for any field extension L/K, and as a consequence all L/K-forms of the Hopf algebra A are trivial.

EXAMPLE 9.12. We look at the same object A again over a field K of characteristic p, but this time just as an algebra. Let $M = xA + yA$ be the radical of A. Then every element z of M satisfies $z^p = 0$. So for any elements $z, w \in M$ one has a well-defined algebra endomorphism sending x to z and y to w, and this will be an automorphism iff z and w are independent modulo M^2. If we now let $G_i = \{\varphi \in \mathrm{Aut}(A) | \varphi \equiv id \pmod{M^i}\}$, then G_0/G_1 identifies with $GL_2(K)$, and all higher quotients G_i/G_{i+1} are isomorphic to products of copies of $(K, +)$. A similar discussion works with obvious changes when K is replaced by any K-algebra T. There exists an appropriate version of the long exact sequence in Amitsur cohomology, attached to any short exact sequence of group-valued functors. Using this and the triviality of $H^1_{L/K}(\mathbb{G}_m)$ and $H^1_{L/K}(\mathbb{G}_a)$, we obtain again that A considered as a K-algebra has no nontrivial L/K-forms, for any field extension L/K.

EXAMPLE 9.13. Now we put $p = 2$ and look at the same $A = K[x, y]/(x^2, y^2)$, but we take K of characteristic different from 2. To be on the safe side, we assume

L/K Galois (separability would be enough). The picture is now quite different. The point is that $x + y$ is no longer nilpotent of exponent 2. Stretching this observation a bit, one can show, for any automorphism φ of A (again one would have to write this down the functorial way, with K replaced by any finite extension, but let us skip that): Modulo Kxy we either have $\varphi(x) \in K^\times x$ and $\varphi(y) \in K^\times y$, or vice versa: $\varphi(x) \in K^\times y$ and $\varphi(y) \in K^\times x$. The automorphisms of the first kind form a subgroup functor $G' \subset \underline{\mathrm{Aut}}\, A$, and G' can be filtered again by subgroups such that all quotients are $\mathbb{G}_m$ or $\mathbb{G}_a$, so its first Amitsur cohomology is trivial. The entire group $\underline{\mathrm{Aut}}\, A$ is a semidirect product of G' and the two-element group C_2 generated by the automorphism that switches x and y. And now indeed $H^1_{L/K}(C_2)$ may well be non-trivial. For L/K Galois with group G, we may replace general descent theory by Galois descent. Hence we may work with the Galois cohomology group $H^1(G, C_2)$. As the action of G on C_2 is trivial, we actually have $H^1(G, C_2) = \mathrm{Hom}(G, C_2)$. If now L/K is a separable quadratic extension, we see that the latter Hom set has exactly two elements, trivial and non-trivial. Thus, there exists exactly one nontrivial L/K-form of A. It is possible to go through the details and calculate the nontrivial form. We will see such calculations in the context of Hopf algebras later. For now let us just say that for $L/K = \mathbb{C}/\mathbb{R}$, the nontrivial form of the algebra A is

$$A' = \mathbb{R}[u, v]/(uv, u^2 - v^2).$$

For the rest of the examples, we consider Galois extensions L/K with group G. The motivation for the following example is the fact that every finite étale algebra over a separably closed field $F = F^{sep}$ is a (finite) product of copies of F.

EXAMPLE 9.14. Let $A = K^n = K \times \cdots \times K$ considered as a K-algebra. We are interested in L-forms of A. The primitive idempotents of A are exactly the "unit base vectors" $e_i = (0, ..., 0, 1, 0, ..., 0)$ with 1 sitting in the i-th position. Any automorphism of the K-algebra A has to permute them. From this it is clear that $\mathrm{Aut}_K(A)$ identifies with the permutation group S_n. Of course $L \otimes_K A$ identifies with L^n. To give a 1-cocycle $G \to \mathrm{Aut}_L(L^n)$ is therefore the same as giving a group homomorphism $G \to \mathrm{Perm}\,\{1, \ldots, n\}$, or giving an action of the group G on the set $\{1, \ldots, n\}$. It is not really more general but more convenient to replace the latter set by any finite set Σ and correspondingly A be the algebra of "functions" $\mathrm{Map}(\Sigma, K) =: A_\Sigma$. One checks that two actions of G on Σ result in the same cocycle iff there is an isomorphism of G-sets between the two G-sets afforded by the two actions (i.e. a bijection from Σ to itself that transforms the one action into the other). Hence we obtain an equivalence of categories: L/K-forms of algebras A_Σ on the one hand, and the category of finite G-sets on the other hand.

From a slightly different viewpoint, commutative ring theory and a bit of field theory will tell us that L/K-forms of algebras of the form $A \times \cdots \times A$ are the same as the finite products of fields $\prod_i K_i$, where each K_i is an extension of K that is embeddable in L.

By a passage to the limit, which we will not describe in detail, one obtains another equivalence: The étale finite K-algebras on the one hand, and continuous finite Γ-sets on the other hand, where $\Gamma = \mathrm{Gal}(K^{sep}/K)$. One can moreover show that the separable field extensions of K on the one side correspond exactly to the Γ-sets with only one orbit on the other side. (The product of K-algebras corresponds to the disjoint union of Γ-sets.) If the field K is perfect, étale finite K-algebras are the same thing as finite (commutative) K-algebras without nilpotent elements, or as finite products of finite field extensions of K.

EXAMPLE 9.15. We elaborate on the preceding example. It is well known that all possible Hopf structures on the K-algebra $\mathrm{Map}(\Sigma, K)$ are induced by group structures on the set Σ. Thus we can rename Σ to N (a finite group) to begin with; then A_N is the dual Hopf algebra of the group ring $K[N]$. By an obvious modification of the preceding arguments, the Hopf automorphisms of A_N all come from group automorphisms of N. This leads to another equivalence of categories: K-Hopf algebras which are étale over K on one side, and finite groups with an Γ-action on the other side.

As a corollary we also obtain, by dualization, that all Hopf automorphisms of the group ring $K[N]$ come from group automorphisms of N; indeed this gives a natural identification $\mathrm{Aut}_K(K[N]) = \mathrm{Aut}_{gps}(N)$.

9.2. Galois descent for Hopf algebras

9.2.1. Generalities; description of the forms. We use the general technique of Galois descent along a G-Galois extension L/K of fields, now restricting ourselves to Hopf algebras. We intend to determine the forms of the very common K-Hopf algebra $H = K[N]$, the group ring of some arbitrary finite group N. These forms are classified by the first Galois cohomology of G with values in the functor $\underline{\mathrm{Aut}}_{Hopf}(K[N])$.

But here we are in a favorable situation. For any extension field E, the E-Hopf automorphisms of $E[N]$ identify with the group automorphisms of N. (This well-known fact was explained at the end of the preceding section.) In particular for $E = L$, the action of G on $\mathrm{Aut}_L(L[N])$ is the trivial one. (The automorphisms are all defined over K; in other words, if one considers them as matrices via the obvious choice of basis, the entries of the matrices are only zeros and ones.) Therefore cocycles $G \to \mathrm{Aut}_L(L[N])$ are no more and no less than group homomorphisms, and we may replace $\mathrm{Aut}_{L-Hopf}(L[N])$ by $\mathrm{Aut}(N)$.

The following fact will play a role later on. Two homomorphisms $\theta, \theta' : G \to \mathrm{Aut}(N)$ are cohomologous as 1-cocycles if and only if there exists $\nu \in \mathrm{Aut}(N)$ such that

$$\theta'_g = \nu\theta_g\nu^{-1} \quad \text{for all } g \in G.$$

In particular, the trivial cocycle is only cohomologous to itself.

Likewise, the explicit description of the form attached to a cocycle given in §1.3 simplifies: The form $H(\theta)$ associated to a 1-cocycle θ is the fixed set under the semilinear action β of G on $L[N]$, where β_g acts as g on L and as θ_g on the group N. We have already given an example (the "nonsplit torus").

Let us mention briefly one class of examples which has gained some popularity in Hopf Galois theory. Start with any G-Galois extension L/K, and put N equal to G. Then one has a distinguished homomorphism c (conjugation) from G to $\mathrm{Aut}(N) = \mathrm{Aut}(G)$, sending $g \in G$ to c_g, the conjugation with g. This cocycle is trivial iff G is abelian. So (by a remark made a few lines ago) if G is nonabelian, this cocycle does not give the trivial class in cohomology, and the corresponding form (called H_λ for historical reasons, see [**GP87**]) is not isomorphic to $H = K[G]$ as a Hopf algebra.

Now we are going to take a more general point of view. First, let us deal with the question: Under what precise condition are two L/K-forms of the group ring $K[N]$ isomorphic as K-Hopf algebras? Consider two 1-cocycles θ and θ', defined on G and taking values in $\mathrm{Aut}(N)$. As explained before, we then are looking at the abstract group N, made into a "G-group" (that is, a group with an action of G via automorphisms) in two different ways; let us call these objects (N, θ) and (N, θ') respectively.

By general descent theory, $H(\theta)$ and $H(\theta')$ are isomorphic as K-Hopf algebras if and only if θ and θ' are cohomologous. The following fact has already been stated and is well known, but let us repeat the statement and give a proof, because of its importance. For a slightly (but not totally) different presentation, see [**KKTU19b**, Theorem 2.2].

PROPOSITION 9.16. *The following conditions are equivalent:*

i) *$H(\theta)$ and $H(\theta')$ are isomorphic as K-Hopf algebras;*
ii) *The cocycles θ and θ' are cohomologous;*
iii) *The two objects (N, θ) and (N, θ') are isomorphic in the category of G-groups.*

PROOF. The equivalence of i) and ii) comes from general descent theory.

It suffices therefore to show that ii) and iii) are equivalent. Suppose iii) holds. Let $\nu : N \to N$ be an isomorphism of groups that affords an isomorphism $(N, \theta) \to (N, \theta')$ in the category of G-groups. We spell out what this means for any $g \in G$. Let $*$ (and $*'$) denote the G-action given by θ (and θ') respectively. Then we need to have

$$\nu(g * x) = g *' \nu(x). \quad (x \in N)$$

By definition this is equivalent to

$$\nu(\theta_g(x)) = \theta'_g(\nu(x)). \quad (x \in N)$$

In other words, the maps $\nu \circ \theta(g)$ and $\theta'(g) \circ \nu$ have to agree. This simply means $\theta'_g = \nu\theta_g\nu^{-1}$ for all g (the same ν all the time), and this is exactly the definition of two cohomologous 1-cocycles. The argument can be reversed. □

Now that we have a good handle on the isomorphism question in the category of *Hopf* algebras, it is tempting to ask what can be said about H_λ and all other Hopf forms in the category of *algebras*. A surprising thing happens. Rather frequently, nontrivial Hopf forms $H(\theta)$ are nevertheless "trivial", that is, isomorphic to the group ring, as K-algebras. We owe the first findings in this direction to Underwood [**Und16**]. He showed H_λ to be isomorphic to the group ring H for the case $K = \mathbb{Q}$ and $G = S_3$, by a hands-on computation. We do not want to give the show away here, but let us give a few hints. By preparatory arguments one sees that the "interesting part" both of H and H_λ is 4-dimensional and central simple over $\mathbb{Q}$. In the case of H one has a 2-by-2 matrix ring. So it suffices to find just one nonzero nilpotent element in the corresponding part of H_λ. We will give some details at the end of the fourth subsection. Of course this direct attack will not generalize too far.

9.2.2. A sufficient condition for the algebra structure to be trivial. We keep the notation of the preceding section, and we introduce a little more terminology, before stating and proving a fairly general result.

For any group N, every element $\sigma \in N$ induces an automorphism $C(\sigma)$ which is simply conjugation by N. All such automorphisms are called "inner"; they form a subgroup $\mathrm{Inn}(N)$ of $\mathrm{Aut}(N)$, and the map $C = C_N : N \to \mathrm{Inn}(N)$ sending σ to $C(\sigma) : \nu \mapsto \sigma\nu\sigma^{-1}$ is a surjective group homomorphism, whose kernel is the center $Z(N)$ of N. Later we will also be interested in automorphisms that are not inner (and hence called "outer"); in fact $\mathrm{Inn}(N)$ is normal in $\mathrm{Aut}(N)$, and one may define $\mathrm{Out}(N) = \mathrm{Aut}(N)/\mathrm{Inn}(N)$.

We continue to assume that L/K is G-Galois; we will also assume N is a finite group. We are interested in L/K-forms of the Hopf algebra $K[N]$. As said before, these are classified by 1-cocycles (equivalently: homomorphisms)

$$\theta : G \to \mathrm{Aut}_L(L[N]) = \mathrm{Aut}(N).$$

We now make some natural definitions.

DEFINITION 9.17. (1) The cocycle θ is *inner,* if its image is contained in $\mathrm{Inn}(N)$. Any form of $K[N]$ defined by an inner cocycle is called an *inner form.* If θ is not inner, we will occasionally call it an *outer* cocycle, and likewise the attached form an *outer form.*

(2) An inner cocycle θ is called *liftable,* if there exists a homomorphism $\Theta : G \to N$ such that $\theta = C_N \circ \Theta$. Any form of $K[N]$ defined by a liftable inner cocycle is called a *liftable inner form.*

REMARK 9.18. (a) Recall that two cocycles θ and θ' are cohomologous, i.e., represent the same class in $H^1(G, \mathrm{Aut}(N))$, if and only if there exists $\nu \in \mathrm{Aut}(N)$ such that $\theta'_\sigma = \nu\theta_\sigma\nu^{-1}$ for all $\sigma \in G$. From this it readily follows that the question whether θ is inner, or liftable inner, only depends on the cohomology class of θ. This justifies talking about inner forms, and liftable inner forms.

(b) The form H_λ of $K[G]$ (recall that one takes $N = G$ here) is certainly a liftable inner form. Indeed, the cocycle that defines it (recall that it sends σ to $C(\sigma)$) is in a way the *universal liftable inner cocycle.*

We are now in a position to state and prove a sufficient criterion for triviality of the algebra structure, making use of the notation just introduced. Starting from now, and for the whole rest of the chapter, we will assume that K is of characteristic zero, so that all group rings over K will be semisimple.

THEOREM 9.19. *Every liftable inner form of the Hopf algebra $K[N]$ is isomorphic to $K[N]$ as a K-algebra. In particular, $H_\lambda \cong K[G]$ as K-algebras.*

PROOF. Apart from a non-commutative generalization of Hilbert's Theorem 90, which we were unable to locate in the literature, this is in fact a fairly easy argument. We will formulate and use this Hilbert 90-type result when we need it, and postpone its proof, so as not to interrupt the main argument.

Let $\theta : G \to \mathrm{Inn}(N) \subset \mathrm{Aut}(N)$ be an inner cocycle defining the form H, and let $\Theta : G \to N$ be a lift: $\theta = C_N \circ \Theta$. Thus, for every $\sigma \in G$, θ_σ is conjugation with the group element $\Theta_\sigma \in N$.

Let $K[N] = A_1 \times A_2 \times \cdots \times A_t$ be the Wedderburn decomposition of $K[N]$, so $A_i = \mathrm{Mat}_{n_i}(E_i)$ for certain skew fields E_i and for some positive integers n_i. Let F_i denote the center of E_i (hence of $\mathrm{Mat}_{n_i}(E_i)$); every F_i is an extension field of K of finite degree. As a result, $L[N]$ decomposes as the direct product of the algebras $A_{i,L} = L \otimes_K A_i$.

Step 1: This decomposition is respected by every automorphism θ_σ. Indeed, since all these automorphisms come from inner automorphisms of N, they preserve all the central idempotents, and each component A_i has the form $e_i K[N]$ for some central idempotent e_i. So everything decomposes in terms of the components $A_{i,L}$.

Step 2: For each $i = 1, \ldots, t$, let $\theta^{(i)} : G \to \mathrm{Aut}_L(A_{i,L})$ be the 1-cocycle (=homomorphism) induced by θ, and let $\pi_i : L[G] \to A_{i,L}$ be the i-th projection. (Note: The tensored-up components $A_{i,L}$ may well be decomposable as algebras, but this does not matter.) It will then suffice to prove that the form B_i defined by $\theta^{(i)}$ is isomorphic to A_i as a K-algebra; in other words we must show that the class of $\theta^{(i)}$

in $H^1(G, \mathrm{Aut}_L(A_{i,L}))$ is trivial. Note that we are just working with algebras now; the Hopf structure has disappeared from the picture.
Step 3: For each i, there is a natural homomorphism

$$\nu_i : A_{i,L}^{\times} \to \mathrm{Aut}_L(A_{i,L}),$$

sending x to conjugation with x. Define $\Theta^{(i)}(\sigma) = \pi_i(\Theta_\sigma)$. Then $\Theta^{(i)} : N \to A_{i,L}^{\times}$ is a homomorphism, and $\theta_\sigma^{(i)}$ is simply conjugation by $\Theta^{(i)}(\sigma)$; in other words, we have

$$\nu_i \circ \Theta^{(i)} = \theta^{(i)}.$$

We now consider $\Theta^{(i)}$ as a 1-cocycle in its own right. Since A_i is a simple K-algebra, Lemma 9.20 below shows that

$$[\Theta^{(i)}] \in H^1(G, A_{i,L}^{\times})$$

is the trivial element. Since ν_i induces a homomorphism of pointed sets in cohomology, this of course implies that the class of the 1-cocycle $\theta^{(i)}$ is trivial as well. Hence the algebra form B_i of A_i is trivial, as claimed. □

It remains to state and prove our auxiliary result. We hasten to say that many cases of this are known, or even very well known (A commutative, or A a full matrix ring over K, or A a skew-field such that A_L is still a skew-field). In fact we believe that this result must be known in its entirety. But as already mentioned, we did not find any reference that satisfactorily covers all we need.

LEMMA 9.20. *Let A be a finite-dimensional simple K-algebra, and L/K be a G-Galois extension. Then $H^1(G, A_L^{\times})$ vanishes.*

PROOF. We have $A = \mathrm{Mat}_n(E)$ for a skew field E finite-dimensional over K. The idea of the proof is to appeal to Galois descent theory (which is the standard way to proceed when $E = K$, where the result is well known). We want to interpret $A^{\times}$ as the automorphism group of something. This is easy: Let $Mod - A^{op}$ denote the category of right A^{op}-modules (where as usual A^{op} is A with the order of multiplication reversed). Then $A^{\times} \cong \mathrm{Aut}_{Mod-A^{op}} A$, with $x \in A$ corresponding to the automorphism $A \ni y \mapsto xy \in A$, and where the opposite algebra A^{op} acts in the obvious way by right multiplication. Therefore $H^1(G, A_L^{\times})$ classifies the A_L^{op}-forms of the obvious A^{op}-module A. But now the module theory of A^{op} is very easily described: there is only one simple module M, and every module is a direct sum of copies of M. In other words: once you know the K-dimension of a module, you know the module. (It is not necessary to know what this unique simple module M is, but let us describe it anyway. Indeed, M is the row space $E_n = E \times \cdots \times E$ with A acting by matrix multiplication on the right: a row times a matrix gives a row.) Hence the A^{op}-module A has no nontrivial L/K-forms, and we are done. □

9.2.3. Nontrivial algebra forms in dimension 8. In this subsection we intend to demonstrate that there exist many nontrivial forms (seen as K-algebras) among the forms that are not covered by the theorem of the preceding subsection. This includes both outer forms (by this we mean of course forms given by a cocycle which is not inner), and inner forms that belong to non-liftable cocycles. We will see that forms defined by either type of cocycles may contain skew-field components in cases where the original group ring only involves matrix rings, and give one example where the reverse move from a skew field to a matrix ring occurs.

Since we are mainly interested in non-triviality phenomena caused by skew field components in our form, we will ignore the case N abelian in this subsection. The abelian case seems to be rather different; there will be abelian examples in the last subsection.

For emphasis we repeat that we assume $\mathrm{char}(K) = 0$ for the rest of the chapter.

We will first discuss *inner forms coming from non-liftable cocycles.* We cannot take $N = S_3$ here, since the homomorphism $S_3 \to \mathrm{Inn}(S_3)$ is an isomorphism, so every inner cocycle $\theta : S_3 \to \mathrm{Inn}(S_3)$ is liftable. (In fact every automorphism of S_3 is inner; this will become relevant a little later.) The same holds for all centerless groups N. But there are interesting examples involving the dihedral group of order 8. Our findings reproduce results obtained by Taylor and Truman [**TT19**] by different methods, involving explicit calculations in the natural basis of the group ring.

We have completely neglected the Hopf-Galois aspect so far (on which extensions do the Hopf algebras act?). Other authors do keep this aspect in mind or even put it in the center of the stage, for instance in the paper [**TT19**]. The following discussion is borrowed from that paper. We are interested in L-forms H of the Hopf algebra $K[D_4]$, where L/K is a Galois extension with group $G = Q_8$ (the quaternion group of order 8). Similarly as in the preceding subsection, all such H will decompose as $H_{ab} \times H_n$, where H_{ab} is a form of the abelianized Hopf algebra $K[D_4^{ab}]$ and H_n is a form of the Wedderburn component of $K[D_4]$ which belongs to the irreducible degree 2 representation. We will only be interested in H_n, which will be called "the non-abelian part of the form".

Hopf Galois theory affords various actions of the quaternion group Q_8, which plays the role of G, on the dihedral group D_4, which plays the role of N. If one is interested in the number of different Hopf Galois structures, the complete count of these actions comes to 6, see [**TT19**, Lemma 2.5]. For our purposes two of these actions are enough. The first (we will denote it by θ) arises from the choices $s = \sigma$, and $D_4 = D_{s,\lambda}$ (not $D_{s,\rho}$). The second (denoted θ') arises from the choices $s = \sigma$, and $D_4 = D_{s,\rho}$. (We guess that the four other possible choices would not lead to essentially different phenomena, but we did not check this.)

We need to describe the actions θ and θ'. Let us recall (or introduce) some notation first. We use the presentation $\langle \sigma, \tau \mid \sigma^4 = 1, \sigma^2 = \tau^2, \sigma^\tau = \sigma^3 \rangle$ for Q_8. For the dihedral group $N = D_4$, we use $\langle s, t \mid s^4 = 1 = t^2, s^t = s^3 \rangle$. Then $\mathrm{Inn}(D_4)$ is elementary abelian of order 4 (generated by the images of s and t, each of order 2), and the natural epimorphism $D_4 \to \mathrm{Inn}(D_4)$ does not split. The center of D_4 has order 2 and is generated by s^2.

Then one finds by direct calculation: The first cocycle (= homomorphism) $\theta : Q_8 \to \mathrm{Inn}(D_4)$ is given by

$$\theta_\sigma = C(s), \quad \theta_\tau = C(t),$$

and the second cocycle $\theta' : Q_8 \to \mathrm{Inn}(D_4)$ is given by

$$\theta'_\sigma = id_{D_4}, \quad \theta'_\tau = C(s).$$

Our goal is to describe the non-abelian part of the forms H and H' attached to θ and θ' respectively. Theorem 9.19 does not apply, since θ and θ' are not liftable. Indeed θ cannot factor through $D_4 \to \mathrm{Inn}(D_4)$, because this would result in an isomorphism $Q_8 \to D_4$. The argument for θ' is similar. In order to describe our results, we need certain crossed products. We only describe them in the situation we need; there they are also known under the name "generalized quaternion algebras". For a general introduction to these algebras we recommend [**Lam05**, Chapter II].

Let $K(\sqrt{y})$ be a quadratic extension of K, and let $x \in K^\times$. The crossed product algebra $(x, y)_K$ is constructed as follows. Take the K-vector space A with basis $1, z, w, zw$, and define the multiplication by the following rules:

$$z^2 = y; \; w^2 = x; \quad zw = -wz.$$

Note that for $K = \mathbb{R}$, $x = y = -1$, we recover the classical algebra of quaternions. The algebra A contains a copy of $K(\sqrt{y})$ (generated by z), and A is a $K(\sqrt{y})/K$-form of $\mathrm{Mat}_2(K)$. It is known that A is "trivial" (isomorphic to $\mathrm{Mat}_2(K)$) if and only if $x \in K$ is a norm from $K(\sqrt{y})$. Our results concerning the forms H and H' are now as follows. They have also been obtained in [**TT19**] by different methods.

THEOREM 9.21. *(a) The non-abelian part of the form H attached to θ is isomorphic to $(-a, b)_K$. (The quantities a and b will be explained in the course of the proof; they are determined by the quadratic subfields of L/K.)*

(b) The non-abelian part of the form H attached to θ' is isomorphic to $(-1, a)_K$.

It was observed by [**TT19**] that the two generalized quaternion algebras in this theorem are actually isomorphic (a non-obvious fact). Indeed θ and θ' are related to each other in a certain way (details in loc.cit.), and it is conjectured that whenever two Hopf Galois actions are related in this way, the associated Hopf forms should be isomorphic as algebras. If true, this would generalize the transition from the untwisted group ring $K[G]$ (here $G = N$) to H_λ, where we do have an isomorphism by Theorem 9.19.

We now proceed to prove Theorem 9.21. The case of θ' is quite a bit easier, and therefore we will treat it first.

The main simplifying circumstance is that θ'_σ is trivial. Therefore we may regard the form defined by θ' as a K'/K-form of $K[D_4]$, where K' is the fixed field of σ. This is a quadratic extension of K, with Galois groups $\overline{G} = G/\langle\sigma\rangle = \{1, \tau\}$. As announced earlier, we will only be working with the Wedderburn component A of $K[D_4]$ belonging to the nonabelian irreducible character; in fact $A = \mathrm{Mat}_2(K)$. We will continue to write θ' for the cocycle induced by θ' on $A_{K'}$. (Recall this is shorthand for $K' \otimes_K A$.) Then this cocycle is inner, having its values in $\mathrm{Inn}(A_{K'}) = \mathrm{PGL}(2, K')$, but not liftable. The main expedient is to consider a part of the exact sequence in cohomology stemming from the short exact sequence

$$1 \to K'^{\times} \to \mathrm{GL}(2, K') \to \mathrm{PGL}(2, K') \to 1.$$

Since $\mathrm{GL}(2, K')$ and $\mathrm{PGL}(2.K')$ are not abelian, this long sequence will only be exact in the category of pointed sets; its terms do not carry a group structure. We only need three terms of that long sequence:

$$H^1(\overline{G}, \mathrm{GL}(2, K')) \to H^1(\overline{G}, \mathrm{PGL}(2, K')) \to H^2(\overline{G}, K'^{\times}).$$

Here the second map is the connecting map, which will be written ∂. (We repeat that ∂ is only a morphism in the category of pointed sets.) By Hilbert 90, the first of the three cohomology sets in this sequence is trivial. Our objective is to calculate $\partial([\theta'])$ as an element of $H^2(\overline{G}, K'^{\times})$. This term is known to classify those central simple algebras of dimension 4 over K that become trivial over K', that is, become isomorphic to $\mathrm{Mat}_2(K')$ upon base change from K to K'. If $K' = K(\sqrt{a})$, then these generalized quaternion algebras are given, as explained above, as $(x, a)_K$; we recall that this algebra is trivial over K iff $x \in K^{\times}$ is a norm from K'. We recall moreover that $H^2(\overline{G}, K'^{\times})$ is in fact a group, and identifies with the quotient $K^{\times}/N_{K'/K}K'^{\times}$; this identification does take the algebra $(x, a)_K$ to the class of x.

To make sense of the rest of our calculation, two points have to be clarified first. We will need (1) another compatibility property, and (2) the injectivity of ∂.

(1) If φ is any 1-cocycle $\overline{G} \to \mathrm{PGL}(2, K')$, then the form of $\mathrm{Mat}_2(K)$ defined by this cocycle (remember that the automorphism group of the algebra $\mathrm{Mat}_2(E)$ over E identifies with $\mathrm{PGL}(2, E)$ for any E/K) is the same, as a central simple algebra over K, as the generalized quaternion algebra defined by the element $\partial(\varphi) \in H^2(\overline{G}, K'^\times)$. This compatibility property is clearly stated on [**NSW08**, page 355] (see also the description of crossed products on page 350), referring to [**Ker90**, chap.V, §30] for the proof.

(2) From the diagram on [**NSW08**, page 355], one sees directly that the connecting map ∂ is injective. Note the following subtlety: The exactness of the 3-term sequence of pointed sets only tells us that the preimage of the trivial element of $H^2(\overline{G}, K'^\times)$ under ∂ equals the image of the left-hand map, and hence consists of the trivial element only. This is *much* weaker than saying that ∂ is injective. But this injectivity is important for us: it ensures that we do know the form associated to θ', once we know $\partial(\theta')$.

After these clarifications, let us now calculate the image of θ' under the connecting map. First we have to make that map explicit. This goes as follows: Pick a preimage Y of θ'_τ under the map $\mathrm{GL}(2, K') \to \mathrm{PGL}(2, K')$. One cannot expect to have $Y^2 = I_2$ (otherwise indeed θ' would lift as a cocycle), but Y^2 must be a scalar matrix yI_2 because θ'_τ squares to the identity. Then $\partial([\theta'])$ is the class of the scalar $y \in K$ modulo $N_{K'/K}K'^\times$. Let us now make this concrete. Since θ'_τ is conjugation by s, it suffices to know to which matrix s corresponds in the non-abelian Wedderburn component of $K[D_4]$; this is the matrix

$$S = \begin{pmatrix} 0 & 1 \\ -1 & 0 \end{pmatrix}.$$

If we square S, we find -1 times the identity matrix, and hence $\partial([\theta'])$ is the quaternion algebra $(-1, a)_K$. This proves the second part of Theorem 9.21.

Now we take up the first cocycle θ and prove the first part of the theorem. This is harder, since the action does not factor through any quadratic subfield of the quaternionic extension L/K. We need both quadratic subfields: K' the fixed field of $\langle\sigma\rangle$, which we wrote $K(\sqrt{a})$ above, and the other one: K'' is the subfield fixed by $U = \langle\tau\rangle$, which we will represent as $K(\sqrt{b})$. Recall $\theta_\sigma = C(s)$ and $\theta_\tau = C(t)$. Now the group element t corresponds to the matrix $T = \begin{pmatrix} 0 & 1 \\ 1 & 0 \end{pmatrix}$, which has order 2. Hence θ_τ lifts to the element T of order 2 in $A = K^{2,2}$, and this tells us (Hilbert 90 again) that the restriction of the cocycle θ to the subgroup U of G has to represent the trivial class in $H^1(U, \mathrm{PGL}(2, L))$. The idea is to replace θ by a cohomologous cocycle θ^*, whose restriction to U is not only cohomologous to 1, but equal to 1. This will allow us to use the inflation-restriction sequence, and put ourselves in a situation where cohomology is again taken over a cyclic group of order 2.

The details of this program are a little delicate. We write $\alpha = \sqrt{a}$ and $\beta = \sqrt{b}$. We seek to write T in the form $Y^{-1}\tau(Y)$ (all equalities now in PGL); if we have this, then the cocycle

$$G \ni \gamma \mapsto \theta^*_\gamma = Y^{-1}\theta_\gamma\gamma(Y^{-1})$$

will be cohomologous to θ, and its restriction to U will be constant $= 1$.

For Y we take the matrix $\begin{pmatrix} \alpha & -\alpha \\ 1 & 1 \end{pmatrix}$. Then applying τ to Y swaps the columns, and one easily checks that $Y^{-1}\tau(Y) = T$.

Now one has to recall the inf-res-sequence. More precisely: we consider inflation from G/U to G and restriction from G to U. We want to write $[\theta]$ as $\inf_{G/U}^{G}([\overline{\theta}])$ for some cocycle $\overline{G} \to \mathrm{PGL}(2, K'')$, where $\overline{G} = G/U$. We know this is possible since $\mathrm{res}_U([\theta])$ is the trivial element of $H^1(U, \mathrm{PGL}(2, L))$, but we need to lay our hands on $\overline{\theta}$. Since we have found a representative θ^* of θ which is equal to 1 on U, we write down $\overline{\theta}$ as follows:

$$\overline{\theta}_{\overline{\gamma}} = Y\theta_\gamma \gamma(Y^{-1}) \quad (\gamma \in G)$$

is well defined and has values in $\mathrm{PGL}(2, K'')$. (For this one also needs that $\mathrm{PGL}(2, L)^U$ identifies with $\mathrm{PGL}(2, K'')$.)

Since in fact $\overline{\Gamma}$ is cyclic of order 2, generated by $\overline{\sigma}$, we may recall that θ_σ is (conjugation by) the matrix S, and we will only need to look at

$$\overline{\theta}_{\overline{\sigma}} = Y\theta_\sigma \sigma(Y^{-1}) = YSY^{-1}$$

(note that σ does not move α). But there is a real subtlety, caused by the downstep from $\mathrm{PGL}(2, L)^U$ to $\mathrm{PGL}(2, K'')$. We are still working in PGL, that is, modulo scalar matrices. If we just calculate in the straightforward way, we find (omitting some easy details) that

$$YSY^{-1} = \begin{pmatrix} 0 & \alpha \\ -1/\alpha & 0 \end{pmatrix}.$$

However, we are not allowed to take the right hand side as the GL representative of the looked-for element in PGL, since it is not fixed by U. The element $\alpha^2 = a$ is in K, but α is by construction sent to its negative by τ. A permissible representative is obtained by multiplying the matrix by the scalar α; let $Z = \begin{pmatrix} 0 & a \\ -1 & 0 \end{pmatrix}$ denote the resulting matrix. We now chase the cocycle $\overline{\theta}$ through the connecting homomorphism $\partial : H^1(G/U, \mathrm{PGL}(2, K'')) \to H^2(G/U, K''^{\times})$. To do this, we have to square Z, observe that it is a scalar matrix, and the occurring nonzero scalar in K, modulo norms from K'' to K, is the answer. One readily finds

$$Z^2 = -aI_2.$$

Therefore the form defined by θ is given in the nonabelian part by the quaternion algebra $(-a, K''/K)_K = (-a, b)_K$. Just as for θ', this agrees with the result obtained in [**TT19**] by direct calculation. This completes the proof of Theorem 9.21. □

We recall that Taylor and Truman also showed that $(-a, b)_K$ and $(-1, a)_K$ are equal. The proof uses the fact that $K'K'' = K(\alpha, \beta)$ is embeddable in a Q_8-extension. Our approach does not give a more enlightening proof of this equality. Using Proposition 9.16, it is easy to prove that the two forms H and H' are not isomorphic as Hopf algebras.

We now turn to Hopf-algebra forms defined by *outer cocycles,* again involving nonabelian groups N of order 8. As we will be able to rely on the techniques just developed for inner non-liftable forms, we may be more brief. Note that, once again, $N = S_3$ would not work since it has no outer automorphisms.

We consider as our first candidate $N = D_4$ with generators s, t and relations $s^4 = t^2 = 1$, $tst = s^{-1}$, exactly as before. Using that the subgroup C generated by s is characteristic, one can verify by hand that $\mathrm{Aut}(D_4)$ has order 8. It contains the subgroup $\mathrm{Inn}(D_4)$, whose order is 4 (the center of D_4 is $\{1, s^2\}$), and a representative of the nontrivial class of $\mathrm{Aut}(D_4)$ modulo $\mathrm{Inn}(D_4)$ is the following particular automorphism

$$\varphi_0 : s \mapsto s,\ t \mapsto st.$$

(It is part of the required verifications that this does define an automorphism.) Note that the square of φ_0 is inner, and indeed equal to conjugation by s, and that φ_0^4 is identity.

The Wedderburn decomposition of $\mathbb{Q}[D_4]$ has four factors $\mathbb{Q}$ (they constitute the abelian part, which we will continue to ignore) and one factor $\mathrm{Mat}_2(\mathbb{Q})$ that has come up already. We can view the latter as the quotient $\mathbb{Q}[D_4]/(s^2+1)$; then s corresponds to the matrix $\begin{pmatrix} 0 & 1 \\ -1 & 0 \end{pmatrix}$, and t corresponds to $\begin{pmatrix} 0 & 1 \\ 1 & 0 \end{pmatrix}$.

Let $L/\mathbb{Q}$ be any quartic cyclic extension, with Galois group G generated by γ. We consider the 1-cocycle (= homomorphism) $\theta : G \to \mathrm{Aut}(D_4)$ which is defined by mapping γ to φ_0. Let H' be the $L/\mathbb{Q}$-form determined by this cocycle. We claim: The "nonabelian part" of this form is a division algebra (= skew field). For this it suffices to show that the class of θ in $H^1(G, \mathrm{Aut}(\mathrm{Mat}_2(L)))$ is not trivial.

By the Brauer-Noether theorem, φ_0 is inner, but the point is that it does not come about as conjugation with a group element. With PARI and good luck, we found that φ_0 is conjugation with the matrix

$$C = \begin{pmatrix} 1 & 1 \\ -1 & 1 \end{pmatrix},$$

which corresponds to $1 + s$ (so it does not correspond to any of the eight elements of D_4). We recall that $\mathrm{Aut}(\mathrm{Mat}_2(L)) = \mathrm{PGL}(2, L)$, and we look again at the exact sequence of pointed sets

$$H^1(G, \mathrm{GL}(2, L)) \to H^1(G, \mathrm{PGL}(2, L)) \to H^2(G, L^\times),$$

where the second map is ∂. We intend to show that $\partial[\theta]$ is not trivial as an element of the H^2 term; in fact we will identify it as a generalized quaternion algebra.

The recipe remains the same: Lift θ_γ to $\mathrm{GL}(2, L)$, exponentiate it with the order of γ, and see what we obtain: it must be a scalar multiple of the identity matrix, and this scalar, modulo norms from L, is what we are looking for. As the lift of $\theta_\gamma = \varphi_0$ we take the matrix C. Its fourth power is $-4I_2$. (So there is a small difference with respect to what happened for θ': the lift has infinite order.) Thus the form H' identifies (in the non-abelian part) with the quaternion algebra $(-4, L/K)_K$. As soon as -4 is not a norm from L, this will be nontrivial.

To get a numerical example we proceed as follows. There is an imaginary cyclic quartic field $L/\mathbb{Q}$ ramified only at 2 (and ∞); then all norms from L to $\mathbb{Q}$ are non-negative. Then -4 is a norm from L everywhere except at ∞ and possibly at 2, so the involved skew field E is nontrivial at ∞, possibly at 2, and nowhere else. By the theory of the Brauer group it follows that E must be nontrivial at 2 as well, and indeed must coincide with the usual algebra $\mathbb{H}$ of quaternions.

We note that the corresponding form H_{ab} of $\mathbb{Q}[D_4^{ab}]$ (the abelian part of H_λ, so to speak), is nontrivial as well, since φ_0 exchanges two linear characters of the abelianization D_4^{ab}, which is the noncyclic group of order 4. In fact, $H_{ab} = \mathbb{Q} \times \mathbb{Q} \times L'$ where L' is the quadratic subfield of L.

The example just given seems about as simple as possible. We will now sketch a construction going the other way, using an outer form of $\mathbb{Q}[H_8]$. Let $\mathbb{H}$ denote the non-abelian part of $\mathbb{Q}[H_8]$; this is the usual four-dimensional quaternion algebra with rational coefficients.

The group $N = H_8$ also admits an outer automorphism φ_0. This now is an involution, which exchanges i with j, and sends k to $-k$. One checks that φ_0 is conjugation with $i+j$. Now let $L/\mathbb{Q}$ be a quadratic extension. This leads to an outer $L/\mathbb{Q}$-form H' of $\mathbb{Q}[H_8]$, given by sending the generator of $G = \mathrm{Gal}(L/\mathbb{Q})$ to φ_0. Similarly as above we find using the H^1-terms attached to the short exact sequence $1 \to L^\times \to (L \otimes \mathbb{H})^\times \to \mathrm{Aut}_L(L \otimes \mathbb{H}) \to 1$ that the non-abelian part of this form, call it B, is nontrivial iff $-2 = (i+j)^2$ is not a norm from L. Thus, if we choose $L = \mathbb{Q}(i)$, the form is nontrivial at the prime 2; that is, its completion at 2 is not isomorphic to the completion of $\mathbb{H}$ at 2, and by our knowledge of the local Brauer group this forces B_2 to be a full matrix ring. We claim that locally at all other primes p, the form is trivial. In a nutshell the argument is: at all odd p, we have a form of an unramified quaternion algebra along an unramified quadratic extensions, and all such forms are trivial a priori. In more detail: Let $R \subset \mathbb{H}$ be the $\mathbb{Z}$-span of 1, i, j, k; then for all odd p, R_p is a full matrix ring over $\mathbb{Z}_p$. The given descent datum exhibits B_p as a form of H_p relative to the quadratic extension $L_p/\mathbb{Q}_p$. This form then comes from a form S of the Azumaya algebra R_p, relative to the extension of rings $\mathfrak{O}_{L,p}/\mathbb{Z}_p$. This is a Galois extension of rings, since $L/\mathbb{Q}$ is unramified in p. Hence S/pS is a form of R/pR (a 2-by-2 matrix algebra over $\mathcal{F}_p$) relative to a quadratic extension of $\mathcal{F}_p$; and all such forms are trivial, since there are no (proper) finite skew-fields. Consequently B_p is trivial as well, by lifting of idempotents; recall that all our objects are p-adically complete at the moment.

We have shown that the central simple algebra B "splits" at every prime p; hence it is a 2-by-2 matrix ring over $\mathbb{Q}$ itself. One can check that the abelian part of the form is $\mathbb{Q} \times \mathbb{Q} \times L$. We will see a variant of this construction further below.

No kind of equivalence is to be expected. More precisely, it can happen that a Wedderburn component of a form of $K[N]$ relative to a cyclic extension L/K, defined by sending a generator of $\mathrm{Gal}(L/K)$ to an automorphism φ_0 of N, is trivial as a K-algebra form, even if φ_0 is not inner. One finds an example of this using $K = \mathbb{Q}$, $N = A_4$, and $L/\mathbb{Q}$ a quadratic extension, using an outer automorphism of A_4 that is the restriction of an inner automorphism of A_4. We skip the details.

9.2.4. Groups of order p^3 and outer forms. In this section p will always be an odd prime and N a non-abelian group of order p^3. As is well known, there exist exactly two such groups, one of exponent p and another of exponent p^2. We will only present the results for the former group (they will be summarized in Theorem 9.23 below). It is quite possible to treat the latter group as well, but it turns out that no really new phenomena or techniques are involved, so we prefer to omit this, and leave the details of this, or maybe even generalizations to somewhat larger p-groups, to interested readers.

So let N be the non-abelian group of order p^3 and exponent p, also known as the Heisenberg group. We will use the presentation by generators u, v, t and relations

$$u^p = v^p = t^p = e, \quad uv = vu, \quad u^t = uv, \quad v^t = v.$$

Thus, N is the semidirect product of the elementary abelian group generated by u and v and the cyclic group t, which acts by conjugation as a nontrivial unipotent automorphism. Both the center and the commutator subgroup of N are cyclic of order p and generated by v.

We will look at cyclic extensions L/K and forms of the Hopf algebra $K[N]$ which are "outer forms" in the following sense: the defining cocycle arises by sending a generator of $\mathrm{Gal}(L/K)$ to an outer automorphism φ of N. For this we need to get a hold on outer automorphisms φ of order p. In fact it suffices to know a

Sylow p-subgroup of $\mathrm{Aut}(N)$. Indeed, every φ lies in such a Sylow p-subgroup, and since they are all conjugate (and conjugation in $\mathrm{Aut}(N)$ simply amounts to changing the generators while keeping the presentation), it suffices to have one Sylow p-subgroup under control. As one would expect, the structure of $\mathrm{Aut}(N)$ is known: it is a semidirect product of a 2-dimensional $\mathcal{F}_p$-space and a copy of $\mathrm{GL}(2,p)$ acting on that space. The paper [**Win72**] determines the automorphisms of a considerably larger class of groups, but it takes a little effort to extract what this says for the special case N. The statement is also proved from scratch by Kohl (unpublished manuscript, available on request), and by Nejabati Zenouz as [**NZ19**, Proposition 3.1]; there one finds the reference [**Osi15**], where an analogous result is proved for the Heisenberg group over $\mathbb{Z}$. All we really need here is that $\mathrm{Aut}(N)$ has order $p^3(p-1)^2(p+1)$.

Hence the Sylow group in question, call it P, has order p^3. One immediately sees that $\mathrm{Inn}(N)$ is a subgroup of order p^2 in P, so it will suffice to find just one outer automorphism φ of order p; then P will be generated by $\mathrm{Inn}(N)$ (which is elementary abelian generated by $C(u)$ and $C(t)$) and φ, because φ normalizes $\mathrm{Inn}(N)$.

We define φ by decreeing that it fixes u and v, and by setting $\varphi(t) = ut$. (Note that $C(u)$ has a rather similar description, only $\varphi(t)$ being vt instead of ut.) One has to check a few things. The main thing for φ to be well defined is that $(ut)^p = e$. Indeed $(ut)^p = u(uv)(uv^2)\dots(uv^{p-1})t^p = e$, since p is not 2. To see that φ is not a combination of $C(u)$ and $C(t)$, it suffices to look at the image of t; no combination of $C(u)$ and $C(t)$ is able to produce a factor u in the image of the element t.

So the general element of the Sylow group P has the form $\psi = C(u)^i C(t)^j \varphi^k$, with $0 \le i,j,k < p$, and ψ is inner iff $k = 0$.

In the sequel assume that ζ_p has degree $p-1$ over K (for instance we always can take $K = \mathbb{Q}$), and L/K is cyclic of degree p. We fix a generator σ of $\mathrm{Gal}(L/K)$, and we consider a 1-cocycle $\theta : G \to \mathrm{Aut}_{L-Hopf}(L[N])$, given by sending σ to $\psi \in \mathrm{Aut}(N)$. It is our goal to determine the "non-abelian part" of the form H defined by this cocycle. More precisely speaking, let us consider the Wedderburn decomposition of $K[N]$. This contains as a direct summand the commutative K-algebra $K[N^{ab}]$; the complement A then corresponds to the nonlinear irreducible representations, that is, those representations ρ that do not send v to 1. Consequently, ρ must send $1 + v + \dots + v^{p-1}$ to 0, and we have

$$A = K[N]/(1 + v + \dots + v^{p-1}).$$

We note $K[v]/(1 + v + \dots + v^{p-1}) \cong K(\zeta_p)$ (v goes to ζ_p), and we claim that A is isomorphic to the full matrix algebra $K(\zeta_p)^{p,p}$. This is certainly not new, but we need to write down the details. Consider the matrices (for typographical reasons we shorten ζ_p to ζ)

$$U = \begin{pmatrix} 1 & 0 & 0 & \dots & 0 \\ 0 & \zeta & 0 & \dots & 0 \\ 0 & 0 & \zeta^2 & \dots & 0 \\ & & & \ddots & \\ 0 & 0 & 0 & \dots & \zeta^{p-1} \end{pmatrix}; \qquad T = \begin{pmatrix} 0 & 1 & 0 & \dots & 0 \\ 0 & 0 & 1 & \dots & 0 \\ & & & \ddots & \\ 0 & 0 & 0 & \dots & 1 \\ 1 & 0 & 0 & \dots & 0 \end{pmatrix}.$$

Then it is not hard to check that the map $A \to K(\zeta_p)^{p,p}$ that sends u to U, v to ζ times the identity matrix, and t to T, is well-defined, and an isomorphism. It is clear that every automorphism $\psi \in P$ induces an automorphism ψ_A of the K-algebra A, and actually of A over $K(\zeta)$, because ψ fixes v.

For simplicity, we first treat the case $\psi = \varphi$ (that is, $i = j = 0$ and $k = 1$). The outcome will be that the algebra A_φ descended from A_L via θ (the cocycle which uses φ) is isomorphic to A (in other words, "trivial"). To this end, we use the method of the preceding section. Let φ_A be the automorphism of A_L induced by the automorphism φ of $L[N]$. We exhibit φ_A as the conjugation by a certain matrix $X \in A$; by construction $X^p = \xi I_p$ for some scalar $\xi \in K(\zeta)$, and the obstruction against $A_\varphi \cong A$ is the class of ξ in $H^2(L(\zeta)/K(\zeta), L(\zeta)^\times)$. So the first job is to find X; we know that it exists. Since φ fixes u, we know that X has to commute with U. But the algebra generated by U over $K(\zeta)$ has dimension p, and hence it is its own centralizer. So X is a polynomial in U, and therefore a diagonal matrix. This is all we need to know! Let η be the upper leftmost entry of X. We know a priori that $X^p = \xi I_p$ for some ξ; but then necessarily $\xi = \eta^p$. Hence we get, almost for free, that the cohomology class of ξ in H^2 is trivial, simply because ξ is a p-th power.

This was so easy that one is tempted to have a go at all $\psi \in P$ in the same manner. This is slightly more complicated but still possible. Of course we may omit all ψ that are inner, since then θ is a liftable inner cocycle, the liftability stemming from the fact that N has exponent p. That is, we may take $\psi = C(u)^i C(t)^j \varphi^k$ with $0 \le i, j, k < p$ and $k > 0$, so as to make ψ an outer automorphism. Each ψ is given as conjugation by the matrix $Y = U^i T^j X^k$. We may actually suppose that the upper left corner of X is 1. Now we have to take a closer look. Let $X = \operatorname{diag}(x_0, \dots, x_{p-1})$; so again, $x_0 = 1$. Since now $X^p = I_p$, all x_i must be powers of ζ; we have to make this precise.

LEMMA 9.22. *We have* $x_1 = 1$, $x_2 = \zeta^{-1}$, $x_3 = \zeta^{-3}$, ..., $x_s = \zeta^{-\binom{s}{2}}$.

PROOF. From the construction of X, we have $XT = UTX$. Now XT arises by pushing the diagonal one position to the right, and TX arises by pushing it up one position. That is, the position of the entry x_s in XT is occupied by x_{s+1} in TX. Multiplication by U from the left multiplies row number s by ζ^s (we count from $s = 0$ onward). Hence $\zeta^s x_{s+1} = x_s$, and from this one gets the result by induction over s. □

A one-band matrix is, by definition, a square matrix whose non-zero entries are limited to one diagonal band, that is, to the set of positions where the difference of column index and row index mod p is a fixed number j, called the shifting index. (Diagonal matrices are a special case of this, having $j = 0$.) The matrix Y is then a one-band matrix, with shifting index j. One easily checks the following fact: For any one-band matrix Z with nonzero shifting index, Z^p equals the scalar matrix $\det(Z) I_p$. So the scalar ξ such that $Y^p = \xi I_p$ is simply $\xi = \det(Y)$. This is easy to evaluate. We have $\det(T) = 1$. The determinant of U is ζ to the exponent $0+1+2+(p-1)$, and this is 0 mod p since $p > 2$, so $\det(U) = 1$. Finally we use the preceding lemma to find that $\det(X)$ is ζ to the exponent $-0-0-1-3-\ldots-\binom{p-1}{2}$. One checks that this is again zero modulo p if $p > 3$, and obviously it is -1 for $p = 3$, so $\det(X) = 1$ for $p > 3$ and $\det(X) = 1/\zeta$ for $p = 3$. This gives $\xi = \det(Y) = 1$ for $p > 3$ and $\xi = \det(Y) = \zeta^{-k}$ for $p = 3$, assuming $j > 0$ (that is, Y is not concentrated on the main diagonal). In that latter case $j = 0$, Y is diagonal with powers of ζ on the main diagonal, so $\xi = 1$ no matter what p is. We recall that the obstruction to $A_\varphi \cong A$ is the class $[\xi] \in H^2(L(\zeta)/K(\zeta), L(\zeta)^\times)$, and we repeat that $k \neq 0$ since ψ is assumed to be outer. We sum up what we have proved, even though this entails some further repetition.

THEOREM 9.23. *Let $p > 2$ be prime and N be the nonabelian group of order p^3 and exponent p. Assume K is linearly disjoint with $\mathbb{Q}(\zeta_p)$, and L/K is cyclic of order p. To each outer automorphism ψ of N, we attach an L-form A_ψ of the non-abelian part A of $K[N]$, by mapping a generator of* $\mathrm{Gal}(L/K)$ *to ψ. Then for $p > 3$, A_ψ is always isomorphic to A as an algebra. However if $p = 3$, then for some ψ (exactly those with $j > 0$, see above), A_ψ is not isomorphic to A, and A_ψ is isomorphic to A for some other ψ.*

Recall: If ψ is inner, then A_ψ is always trivial since all inner cocycles are liftable (N has exponent p).

As an aside, let us explain how the form H of $K[N]$ attached to the particular outer automorphism φ arises in a Hopf-Galois extension. In fact [**NZ19**] contains a complete classification of Hopf-Galois situations that involve the given group N and any group G of order p^3 (constituting one of the main results of that PhD thesis), but since we are looking for an example with specific properties, let us resort to an ad hoc construction.

We need a group G of order p^3, a G-Galois extension M/K, and a realization of N as a transitive subgroup of $\mathrm{Perm}(G)$ normalized by $\lambda(G)$, such that the ensuing action of G on N realizes the outer automorphism φ. More precisely, the action of G on N is supposed to factor through a group $\overline{G}$ of order p (corresponding to a cyclic degree p extension L/K contained in M), and a generator of $\overline{G}$ has to act as φ. Most such problems are treated by what has become known as Byott's translation. It suffices to identify G as a subgroup of $\mathrm{Hol}(N)$ with the following properties: The stabilizer of 1_N under G is trivial, and the map $G \subset \mathrm{Hol}(N) \to \mathrm{Aut}(N)$ factors in the desired way: $G \to \overline{G} \to \mathrm{Aut}(N)$, $\overline{G}$ cyclic of order p, and a generator goes to φ.

We consider the following three elements of the holomorph of N (we recall that N is generated by u, v, t): $U = \rho_u$, $V = \rho_v$, and $Z = \rho_t \circ \varphi$. Then $UV = VU$, U and V are of order p, φ commutes with U and V (since φ is identity on u and v), and hence Z normalizes the group generated by U and V. One checks that Z^p is V (for $p = 3$) and identity (for $p > 3$), so it easily follows that G is a nonabelian group of order p^3. We now look at the orbit of 1_N under G. Clearly this orbit contains all products $u^i v^j$, and also t. From this it follows that this orbit must have cardinality p^3, and the stabilizer of 1_N is trivial. (This also gives another proof that G is of order p^3.) The map $\mathrm{Hol}(N) \to \mathrm{Aut}(N)$ simply throws away all ρ_n, $n \in N$, so the resulting map $G \to \mathrm{Aut}(N)$ has image generated by φ, and hence has all desired properties.

9.2.5. Further results. *Cyclic extensions.* — We begin by discussing cyclic extensions L/K of prime power degree p^n. Most of the relevant results come from the paper [**KKTU19b**].

That paper devotes a lot of attention to a discussion of the Hopf-Galois situation, that is, to the question on which type of extensions the Hopf algebras under consideration will act. We have been largely neglecting this aspect, but now we will start out from it, following the lead of [**KKTU19b**].

Let L/K be G-Galois with G cyclic of order p^n (p a fixed odd prime number). Let N be another fixed cyclic group of order p^n. We collect the relevant results in the form of a theorem.

THEOREM 9.24. a) *All Hopf algebras H that make L/K Hopf-Galois are of type N (that is, H must be a L/K-forms of $K[N]$), and there are p^{n-1} inequivalent Hopf-Galois actions on L/K, indexed with $0 \leq s < p^{n-1}$.*

b) *The Hopf algebras H_s that belong to these inequivalent actions are pairwise nonisomorphic as K-Hopf algebras.*
c) *Under the assumptions char$(K) = 0$ and $\zeta_{p^n} \in K$, one has that H_r and H_s are isomorphic as K-algebras if and only if r and s have the same p-adic valuation. In other words, the set $H_0, H_1, H_p, \ldots, H_{p^{n-2}}$ is a full set of representatives without repetition for the isomorphism classes (as K-algebras) of the H_s.*

PROOF. Statement a) is due to Kohl [**Koh98**].

For the remaining statements, the strategy is to describe the Γ-groups $\mathcal{N}_s$ that correspond to the Hopf algebras H_s under the equivalence of categories discussed earlier. Here, as before, Γ denotes the absolute Galois group of K. Then one shows: These Γ-groups are not isomorphic in pairs (which gives b)), but if one forgets the group structure and considers them just as Γ-sets, then $\mathcal{N}_r \cong \mathcal{N}_s$ if and only if r and s have the same p-valuation. For the very interesting details we refer to [**KKTU19b**]. □

This result further confirms the general tendency of Hopf-algebra forms to be "different" in the Hopf-algebra category but "equal" in the category of algebras. We recall that the first such examples that were found concerned $K[N]$ and its form H_λ.

A substantial part of the arguments in [**KKTU19b**] consists in understanding the G-group $\mathcal{N}$ attached to a Hopf algebra $L[N]^G$, where N is a G-group. Contrary to first appearances $\mathcal{N}$ is not simply N. There is a dualizing process involved, and the correct answer is:

$$\mathcal{N} = N^\vee(1),$$

where $N^\vee = \mathrm{Hom}(N, \mathbb{Q}/\mathbb{Z})$ is the (Pontryagin) dual of N, with contragredient G-action as usual, and $-(1)$ denotes the Tate twist. Admittedly we are deviating in terminology from [**KKTU19b**], so let us give some more explanation. The first thing to mention is that $N^\vee(1)$ is written $\widehat{N}$ in that paper. "Tate twist" means that the action of $\gamma \in G$ on an element is multiplied by $\kappa(\gamma)$, where $\kappa(\gamma)$ describes the action of γ on roots of unity ζ in L: $\gamma(\zeta) = \zeta^{\kappa(\gamma)}$. This corresponds precisely to the formula in loc.cit., two lines above Theorem 3.1. The γ^{-1} that appears give the contragredient action on the dual; the γ just after the equality sign gives the Tate twist. Note that it acts on the root of unity $\chi[\gamma^{-1} * \eta]$ (notation from loc.cit.)

The Tate twist has no effect at all if K contains enough roots of unity (it suffices to have $\zeta_e \in K$ where e is the exponent of G). If one assume this, one can show that if N is cyclic as an abstract group, then N and $N^\vee(1) = N^\vee$ are isomorphic as G-groups. If N is not cyclic, this is not to be expected, even with enough roots of unity in K.

This raises the following pretty question. If we make no assumption on roots of unity in K, then can it happen for two G-groups N_1 and N_2, which we assume to be isomorphic as abstract groups, that the G-sets N_1 and N_2 are isomorphic, but the duals $N_1^\vee(1)$ and $N_2^\vee(1)$ are not isomorphic as G-sets?

Indeed this may happen. One nice example, with arithmetic underpinning, is given in [**KKTU19b**, Example 4.12]. Here L is the cyclic real field of degree 9 and conductor 19. One has precisely three Hopf-Galois structures on $L/\mathbb{Q}$, and they are all of cyclic type. The two "nontrivial" Hopf algebras correspond to two G-groups N_1 and N_2, both cyclic of order 9, but with different and nontrivial action of $G = \mathrm{Gal}(L/K)$. It is shown that the G-sets N_1 and N_2 are isomorphic, but not the G-sets $N_1^\vee(1)$ and $N_2^\vee(1)$. (Note the absence of ζ_9 and even ζ_3 in $K = \mathbb{Q}$.)

Let us briefly give an argument, for the reader's convenience. (This is not so much different from [**KKTU19b**], but more pedestrian.) Both N_1 and N_2 are the cyclic group of order 9 with generator η, but the action of $G = \mathrm{Gal}(L/\mathbb{Q})$ is different: a chosen generator $\sigma \in G$ sends $\eta \in N_1$ to η^4 and $\eta \in N_2$ to η^7. In both cases we have three singleton orbits $\{1\}$, $\{\eta^3\}$, $\{\eta^6\}$ under the G-action, and two orbits $\{\eta, \eta^4, \eta^7\}$, $\{\eta^2, \eta^5, \eta^8\}$ of length 3; from which it readily follows that the G-sets N_1 and N_2 are isomorphic. Now we consider the twisted duals. Here the group which acts is bigger than G: we have to take $G \times \mathrm{Gal}(\mathbb{Q}(\zeta_9)/\mathbb{Q})$. The dual of both N_1 and N_2 as an abelian group is cyclic of order 9; fix a generator ω dual to η. Then the G-action is such that in $N_1^\vee(1)$, σ sends ω to ω^7, and in $N_2^\vee(1)$, to ω^4. Let τ be a generator of $\mathrm{Gal}(\mathbb{Q}(\zeta_9/\mathbb{Q}))$, sending ζ_9 to ζ_9^2. Then τ acts on both duals as multiplication by 2. We see that $\sigma\tau^2$ acts trivially on $N_1^\vee(1)$ but nontrivially on $N_2^\vee(1)$. Hence the G-sets $N_1^\vee(1)$ and $N_2^\vee(1)$ cannot be isomorphic.

If one does not insist on a Hopf-Galois context, such examples can be generated in a slightly simpler and more systematic way. In algebraic number theory one considers Tate twists by any integer k; for any G-group N one has $N(k) = N(k-1)(1)$. Now let p be any prime, $L = \mathbb{Q}(\zeta_p)$ and $G = \mathrm{Gal}(L/K)$. Then for any G-group N one has $(N(k))^\vee = N^\vee(-k)$ (since G acts contragrediently on the dual), and therefore, with $\mathbb{Z}/p$ denoting the cyclic order p group with trivial G-action:

$$(\mathbb{Z}/p(k))^\vee(1) = \mathbb{Z}/p(1-k).$$

(This show nicely, by the way, that the functor $N \to N^\vee(1)$ is its own inverse, as it should be. In fact, this functor is simply the description of Cartier duality on the category of Γ-groups.)

The G-group $\mathbb{Z}/p(k)$ is easy to describe. Let us fix a primitive root g modulo p; this corresponds to a generator γ of G. The action of γ on $\mathbb{Z}/p(k)$ is simply multiplication by g^k. This shows: If k is coprime to $p-1$, then the G-set $\mathbb{Z}/p(k)$ has only two orbits, the singleton $\{0\}$ and its complement. It is also easy to see that there are more than two orbits if k is not coprime to $p-1$. This can be made much more precise: apart from $\{0\}$, the orbits are all of length $(p-1)/\gcd(k, p-1)$.

Now it is easy to produce examples: find k_1, k_2 both coprime to $p-1$ such that $\gcd(1-k_1, p-1)$ is not the same as $\gcd(1-k_2, p-1)$. (Then $N_i = \mathbb{Z}/p(k_i)$, $i = 1, 2$, is an example of the required kind.) Take for instance $p = 7$, $k_1 = 1$, $k_2 = 5$. Then $1 - k_1 = 0$ has gcd 6 with $p - 1 = 6$, and $1 - k_2 = -4$ has gcd 2 with 6.

Here is one more example: $p = 13$, $k_1 = 3$, $k_2 = 5$. Then $\gcd(-2, 12) \neq \gcd(-4, 12)$.

Dihedral extensions. — To conclude this section, we discuss dihedral extensions L/K and Hopf forms that come with them. Again, the results come mostly from a paper of Koch, Kohl, Truman and Underwood; this time the reference is [**KKTU19c**]. The overall picture will look a little different from the situation we just described.

As for the case of cyclic p-groups, the authors begin with a thorough discussion of the Hopf-Galois setting. (As they point out, the classification of Hopf-Galois actions goes back to Byott [**Byo04b**].) Fix an odd prime p and let $G = D_p$ with generators r, s and relations $r^p = s^2 = 1$, $srs = r^{-1}$. Fix a G-Galois extension L/K. Then there are $p + 2$ Hopf-Galois actions. One of them is the classical one: $H = K[G]$. The second one is H_λ with the Hopf-Galois action already described in [**GP87**]; it is a form of $K[G]$ as always. The remaining p Hopf-Galois structures are indexed by $c = 0, 1, \ldots, p-1$; the corresponding Hopf algebras H_c are all forms of $K[N]$ where N is the cyclic group of order $2p$, and the action of G on N

defining the form is "always the same". Indeed, $r \in G$ acts trivially, and $s \in G$ acts by inversion. (What differs is the precise definition of the action of H_c on L.) Therefore (see [**KKTU19c**, Proposition 3]) the Hopf algebras H_c are all isomorphic for $0 \leq c < p-1$. They cannot be isomorphic to either $K[G]$ or H_λ, because the latter two are not commutative. We already know that $K[G]$ and H_λ are non-isomorphic as Hopf algebras but isomorphic as algebras. So we have a clear and complete picture: among the $p+2$ Hopf algebras that are relevant in this Hopf-Galois setting, we have three isomorphism classes in the sense of Hopf algebras and just two in the sense of algebras.

We would like to conclude by presenting a very nice explicit example from loc.cit. which exemplifies the "unexpected" algebra isomorphism $K[G] \cong H_\lambda$. Doing so, we will have come full circle, back to the beginning of this section.

We choose $p = 3$, so $G = D_3 = S_3$. Let L be the splitting field of $x^3 - v$ over $\mathbb{Q}$, v not a cube in $\mathbb{Q}$. Let $u = \sqrt[3]{v}$. Then $L = \mathbb{Q}(u, \zeta_3)$ is Galois with group $G = D_3$. With the same choice r, s of generators of D_3 as before, the Galois action is given by the rules

$$r(u) = \zeta_3 u;\ r(\zeta_3) = \zeta_3; \quad s(u) = u;\ s(\zeta_3) = \zeta_3^2.$$

Define the element $b \in L[D_3]$ by

$$b = u \cdot s + u\zeta_3 \cdot sr + u\zeta_3^2 \cdot sr^2.$$

The form H_λ is defined by taking the G-invariants in $L[D_3]$, with G acting on L as just specified, and on $D_3 = G$ by conjugation. One may check without too much trouble that b is fixed under this double action, so b is indeed an element of H_λ. Obviously, b is not zero.

On the other hand a direct computation gives $b^2 = 0$. We sketch how this can be done. It suffices to see that b/u is nilpotent, as u is a central element of $L[D_3]$. (Of course b/u is no longer fixed under G, but no matter.) In the standard representation, s corresponds to the matrix $\begin{pmatrix} 0 & 1 \\ 1 & 0 \end{pmatrix}$, and r corresponds to $\begin{pmatrix} 0 & 1 \\ -1 & -1 \end{pmatrix}$. The element b/u then maps to:

$$\begin{pmatrix} -1-2\zeta_3 & 1-\zeta_3 \\ 2+\zeta_3 & 1+2\zeta_3 \end{pmatrix}.$$

This matrix has trace zero (clear) and determinant zero (one checks that the lower row equals ζ_3 times the upper row). Hence it is nilpotent. The image of b, or of b/u, in the abelian component is zero. Indeed this is easily seen, just replacing r by 1. All this taken together gives $b^2 = 0$.

Since we already know that (as a $\mathbb{Q}$-algebra) $H_\lambda = \mathbb{Q} \times \mathbb{Q} \times A$ with a central simple algebra A of dimension 4, we see that A cannot be a skew field (because this would exclude the existence of nontrivial nilpotents in H_λ), and hence A must be the 2-by-2 matrix ring $\mathrm{Mat}_2(\mathbb{Q})$. On the other hand it is well known that the group ring $\mathbb{Q}[D_3]$ decomposes as $\mathbb{Q} \times \mathbb{Q} \times \mathrm{Mat}_2(\mathbb{Q})$, and hence the algebra H_λ is isomorphic to the group ring $\mathbb{Q}[D_3]$.

For general $p > 3$, the Wedderburn decomposition of $\mathbb{Q}[D_p]$ has again two factors $\mathbb{Q}$, and the non-abelian factor is $\mathrm{Mat}_2(E)$, where E is the real cyclotomic field $\mathbb{Q}(\zeta_p + \zeta_p^{-1})$. In this situation, if we were able to exhibit a nonzero nilpotent element in H_λ, we would learn as before that the non-abelian part of H_λ cannot be a skew field, but this now falls far short of a complete description. (It might be

$\mathrm{Mat}_2(F)$ for any field F of degree $(p-1)/2$, or a direct sum of $(p-1)/2$ copies of $\mathrm{Mat}_2(\mathbb{Q})$, or still something else.)

9.3. Absolutely semisimple forms

In this short section we consider forms in the category of Hopf algebras (not just of algebras). If k is any field, a finite-dimensional k-algebra A is called *absolutely semisimple* (a.s.s. for short), if it is semisimple and all its simple modules remain simple upon base-change from k to its algebraic closure. In equivalent terms, A is a.s.s. if it is semisimple and in all its Wedderburn factors, the skew-field over which the full matrix ring is taken is simply k. That is, the Schur index is always 1, and the center of every Wedderburn factor is k and not a finite nontrivial extension of it. For simplicity we will assume $k = \mathbb{Q}$ in the sequel. A stock example of a.s.s. algebras over $\mathbb{Q}$ is given by the group rings $\mathbb{Q}[S_n]$ of all the symmetric groups. An easy (but isolated) example is $\mathbb{Q}[D_4]$. On the other hand $\mathbb{Q}[H_8]$ is obviously not a.s.s., and the same goes for the group ring of any abelian group which is not 2-elementary.

Pareigis asked the following question [**Par90**, Problem 4.5]: For which finite groups G does there exist a Hopf form H of $\mathbb{Q}[G]$ which is a.s.s. as a $\mathbb{Q}$-algebra? It is not too hard to see that for abelian G, such a form always exists; indeed $\mathbb{Q}[G]$ and $H = \mathbb{Q}^G = \mathrm{Maps}(G, \mathbb{Q})$ become isomorphic over $\mathbb{Q}(\zeta_e)$ where e denotes the exponent of G. The form H is defined, in terms of descent theory, by the cyclotomic action on G. More precisely: The 1-cocycle (in fact a homomorphism) $\mathrm{Gal}(\mathbb{Q}(\zeta_e)/\mathbb{Q}) \to \mathrm{Aut}_{Hopf}(\mathbb{Q}[G]) = \mathrm{Aut}(G)$ which belongs to the form H sends σ_a to multiplication by a on G. Here as usual, σ_a is the automorphism sending ζ_e to ζ_e^a, for all a coprime to e. The purpose of this section is to give an answer to the question for selected nonabelian groups G, and thus convey an idea of what might be going on more generally; a complete treatment of the general case does not seem to be within reach at present. We will at least see that the answer is sometimes yes and sometimes no.

Maybe the simplest class for which the answer is no is the class of nonabelian groups G of odd order p^3 and exponent p^2. Every such group has a presentation by two generators s and t, subject to $s^{p^2} = 1 = t^p$ and the commutation rule $tst^{-1} = s^{1+p}$. The center of G is then cyclic of order p, generated by s^p; and it coincides with the derived group. The abelianized group G^{ab} is elementary of order p^2, generated by $\overline{s}$ and t.

PROPOSITION 9.25. *If G is a nonabelian group of order p^3 as just described, then $\mathbb{Q}[G]$ has no a.s.s. form.*

PROOF. Just suppose $\theta : \mathrm{Gal}(\mathbb{Q}^{alg}/\mathbb{Q}) \to \mathrm{Aut}(G)$ is a 1-cocycle such that the associated Hopf form $H(\theta)$ is a.s.s. By functoriality, θ will then induce a 1-cocycle θ^{ab} with values in $\mathrm{Aut}(G^{ab})$ such that the associated form $H(\theta^{ab})$ is a.s.s., and indeed equal to the abelian part of $H(\theta)$. In other words, $H(\theta^{ab})$ has to be a product of copies of $\mathbb{Q}$. Therefore θ^{ab} must be equivalent to the cyclotomic cocycle described above (now with $e = p^2$). Using the characterization of cohomologous cocycles given at the beginning of the previous section, one sees that in fact θ^{ab} is equal to the cyclotomic cocycle. In other words: For any a coprime to p, θ^{ab} must send σ_a to multiplication by a. It now suffices to prove that for a not congruent to 1 modulo p, no automorphism φ of G exists which induces multiplication by a on G^{ab}.

This is straightforward, but let us give the details. Suppose $\varphi \in \mathrm{Aut}(G)$ is such a lifting. Then we have $\varphi(s) = s^a z$ and $\varphi(t) = t^a w$ for some $z, w \in Z(G)$. (Recall that the center $Z(G)$ is generated by s^p, and coincides with the kernel of $G \to G^{ab}$). We will show that this leads to a contradiction, calculating the term $x := \varphi(s)^{\varphi(t)}$ in two ways (exponential notation means conjugation). First:

$$\begin{aligned} x &= (s^a z)^{t^a w} = (s^a z)^{t^a} \\ &= (s^a)^{t^a} \cdot z = s^{a(1+ap)} \cdot z = s^{a+a^2 p} z. \end{aligned}$$

Second:

$$\begin{aligned} x &= \varphi(s^t) = \varphi(s^{1+p}) \\ &= s^{a(1+p)} z^{1+p} = s^{a+ap} z. \end{aligned}$$

Now since s has order p^2 and a is not congruent to a^2 modulo p by assumption, equating the two terms obtained for x gives a contradiction. □

Now we consider the Heisenberg group G of order p^3 (again p is an odd prime). This group has generators s, t, r and relations as follows: s, t, r have order p, r is central, and $s^t = sr$. The center is generated by r and again it coincides with the derived group; the group G^{ab} is elementary of order p^2 generated by s and t.

PROPOSITION 9.26. *If G is the Heisenberg group of order p^3 as described and if $p \geq 5$, then $\mathbb{Q}[G]$ has no a.s.s. form.*

PROOF. The argument is fairly similar to the preceding one, and so we only sketch it. The first part goes exactly as before: one establishes that the existence of an a.s.s. form would force that for all a coprime to p, the multiplication by a viewed as an automorphism of G^b must lift to an automorphism of G. We fix such an integer a for a moment. Then any putative lifting φ has to satisfy very similar formulas as above: $\varphi(s) = s^a z$ and $\varphi(t) = t^a w$ for some central elements z and w. By calculating $\varphi(s)^{\varphi(t)}$ in two ways and comparing (details omitted), we obtain the equation $\varphi(r) = r^{a^2}$. This is not yet a contradiction; but since p is at least 5, we can pick two values of a whose squares are different modulo p, and this then shows that φ cannot exist. □

We now turn to two interesting groups for which the answer will be affirmative. The first of them is the quaternion group $G = H_8$. We keep the usual notation $\pm i, \pm j, \pm k$ for most of its elements, but the nontrivial central element will be written z instead of -1, to avoid confusion in the group ring. The abelian part of $\mathbb{Q}[H_8]$ is already a.s.s., since H_8^{ab} is 2-elementary. On the other hand, the nonabelian part of the group ring is exactly the standard skew field $\mathbb{H}$ of quaternions with rational (as opposed to real) coefficients. Thus our task, in rough terms, is to find an automorphism of φ which induces the identity on H_8^{ab} (so as not to spoil anything) and does the right thing in the nonabelian part. The nonabelian part of $\mathbb{Q}[H_8]$ is $\mathbb{Q}[H_8]/(z+1)$, and this identifies with the quaternion algebra $\mathbb{H} = (-1,-1)_{\mathbb{Q}}$.

The following construction is a slight twist of what we did near the end of Subsection 9.2.3. Let φ be the automorphism of H_8 given as conjugation by k. It is obvious that φ fixes z and that φ induces the identity modulo z.

Let $\theta : \mathrm{Gal}(\mathbb{Q}(i)/\mathbb{Q}) \to \mathrm{Aut}(H_8) = \mathrm{Aut}_{Hopf}\, \mathbb{Q}[H_8]$ be the 1-cocycle (in fact, homomorphism) that sends complex conjugation to φ. Note that even though φ is inner, the cocycle θ is not inner; we will see the obstruction very soon.

Then the form $H(\theta)$ has trivial abelian part (that is, the abelian part is the same as the abelian part of $\mathbb{Q}[H_8]$), and therefore a.s.s. Therefore we only have to look at the nonabelian part of $H(\theta)$, call it B, which is a form of $\mathbb{H}$. We must show that B is a 2-by-2 matrix ring over $\mathbb{Q}$. The argument will be rather similar to the one given in Subsection 9.2.3. In particular, the reasoning that B_p splits for all odd primes p is exactly the same. Now we look at the 2-adic situation. Note that the quadratic extension is now a different one as in Subsection 9.2.3, and so is the automorphism φ. But we can argue quite similarly that the form B_2 is "trivial" (that is, isomorphic to the original object $\mathbb{H}_2$ — note this is a skew field!) if and only if -1 (the square of k) is a norm in $L_2/\mathbb{Q}_2$. (Recall φ is conjugation by k.) This is not the case; hence by our knowledge of the local Brauer group there is but one possibility: B_2 is nontrivial as a form of H_2, meaning that B_2 is trivial as a central simple algebra. Thus we are done: B is $\mathrm{Mat}_2(\mathbb{Q})$, and we have found the absolutely simple Hopf form of the rational group ring of the quaternion group. (It is also possible to explicitly exhibit a nontrivial zero-divisor in the algebra $\mathbb{Q}(i) \otimes \mathbb{H}$ that is fixed under the automorphism given by conjugation on the first tensor factor and by φ on the second. But we do not want to rely on such explicit arguments.)

The second group is A_5. We will be brief here. From the character table one sees that $\mathbb{Q}[A_5]$ is not a.s.s. More precisely, three of the Wedderburn components are matrix rings over $\mathbb{Q}$, the formats being 1, 4, and 5; but there is one remaining component $\mathrm{Mat}_3(L)$ with $L = \mathbb{Q}(\sqrt{5})$. This corresponds to the fact that there are two (conjugate) irreducible characters of degree 3, each of which takes values in $L \setminus \mathbb{Q}$ on the two distinct conjugacy classes of 5-cycles. Let φ be the automorphism of A_5 given as conjugation by any transposition in S_5 (the choice does not matter). Note that φ is not inner on A_5. Define a cocycle θ by sending the nontrivial automorphism of $L/\mathbb{Q}$ to φ. One can then prove that the Hopf form defined by θ is absolutely semisimple. What happens is the following: the three a.s.s. Wedderburn components remain unaffected (this requires arguments); and in $\mathrm{Mat}_3(L)$, the center L is deformed into the algebra $\mathbb{Q} \times \mathbb{Q}$, which is indeed an $L/\mathbb{Q}$-form of L. We omit all further details; for brevity, and also in the hope that more general and conceptual arguments will be found which make such case-by-case analyses redundant.

CHAPTER 10

Hopf-Galois actions on purely inseparable extensions

While the primary focus of this book is on separable field extensions, it is worth noting that Hopf-Galois theory was originally developed to describe inseparable extensions (where classical Galois theory does not exist). We will give a brief overview of the case where L/K is purely inseparable, including both an early history and a comparison to the separable case.

Suppose L/K is Hopf-Galois with some Hopf algebra H. Then we have an action $L \otimes H \to L$ satisfying certain properties as outlined in Section 1.1. By taking duals we obtain a map $\theta : L \to L \otimes H^*$ such that the map

$$x \otimes y \mapsto (x \otimes 1)\theta(y) : L \otimes L \to L \otimes H^*$$

is an isomorphism (see, e.g., [**Chi00**, Proposition 2.9]). We then say L is an H^*-*Galois object*, and θ is an example of a *coaction.*

Note that coactions are the dual notions of actions, and as not all actions give a Hopf-Galois structure, not all coactions give the structure of an H^*-Galois object. Here, the only coactions we will consider are the ones which give rise to H^*-Galois objects.

Much of the purely inseparable theory relies on coactions, not actions: rather than having L/K be Hopf-Galois with Hopf algebra H, it is advantageous to view L/K as an H^*-Galois object. Furthermore, Hopf-Galois objects can also be described using the language of algebraic geometry. This chapter begins with a brief exposition of the algebraic geometry required for the subsequent sections.

10.1. A little bit of algebraic geometry

Throughout this section, we will assume that all K-algebras are commutative.

The objective here is not to provide a comprehensive overview of algebraic geometry, rather it is to provide the minimum amount required for this chapter. The material presented here is roughly at the level of [**Wat79**], augmented with several examples relevant to our study.

Let A be a K-algebra. Then A gives rise to a set-valued functor $\mathcal{X} = \mathrm{Spec}(A)$ on commutative K-algebras via

$$\mathcal{X}(B) = \mathrm{Hom}_{K\text{-alg}}(A, B)$$

where B is a K-algebra. We say $\mathcal{X}$ is an *affine K-scheme*, and we say that $\mathcal{X}$ is *represented* by A. Additionally, we call $\mathcal{X}(B)$ the *B-valued points* of $\mathcal{X}$.

Affine K-schemes of particular interest to us are $\mathrm{Spec}(H)$ where H is a K-Hopf algebra, and $\mathrm{Spec}(L)$ where L/K is an extension of fields. Note that if A is a monogenic K-algebra, i.e., $A = K(x)$ for some $x \in A$, then the B-valued points of $\mathcal{X} = \mathrm{Spec}(A)$ correspond naturally to a subset of B via

$$f \in \mathcal{X}(B) = \mathrm{Hom}_{K\text{-alg}}(K(x), B) \leftrightarrow f(x) \in B.$$

More generally, if A is generated over K by $x_1, x_2, \ldots, x_n$ then the B-valued points can be viewed as a subset of B^n.

Now $\mathcal{X} = \mathrm{Spec}(A)$ is a functor: if $g : C \to D$ is a K-algebra homomorphism then there is a natural map $\mathcal{X}(g) : \mathcal{X}(C) \to \mathcal{X}(D)$ given by $\mathcal{X}(g)(f) = g \circ f$. Let $\mathcal{Y} = \mathrm{Spec}(B)$ for some K-algebra B. Suppose for each K-algebra C we have a function $F_C : \mathcal{X}(C) \to \mathcal{Y}(C)$ such that

$$\begin{array}{ccc} \mathcal{X}(C) & \xrightarrow{F_C} & \mathcal{Y}(C) \\ {\scriptstyle \mathcal{X}(f)}\downarrow & & \downarrow{\scriptstyle \mathcal{Y}(f)} \\ \mathcal{X}(D) & \xrightarrow{F_D} & \mathcal{Y}(D) \end{array}$$

commutes. Then we say $F : \mathcal{X} \to \mathcal{Y}$ is a *morphism* (or *natural transformation*) of schemes. The collection of affine K-schemes then forms a category.

Let $\mathcal{X} = \mathrm{Spec}(A)$ and $\mathcal{Y} = \mathrm{Spec}(B)$ as above. If $g : A \to B$ is a K-algebra map, then we get an induced morphism of schemes $F : \mathcal{Y} \to \mathcal{X}$ via

$$F_C(f) = f \circ g : A \to C, \; g \in \mathcal{Y}(C).$$

Conversely, any morphism $F : \mathcal{Y} \to \mathcal{X}$ gives a K-algebra homomorphism $g : A \to B$, namely

$$g = F_B(\iota_B),$$

where $\iota_B : B \to B$ is the identity map. This is an application of Yoneda's Lemma [**Wat79**, p. 6].

We will frequently refer to affine schemes simply as "schemes", however the reader should be aware that the term "scheme" is much more general than what is being presented here. It is not necessary for us to consider more general schemes.

EXAMPLE 10.1. Let q be a power of p, and let $K = \mathbb{F}_q((T))$ be the field of formal Laurent series. Let L be a purely inseparable extension of K of degree p^e, say $L = K(x)$ with $x^{p^e} = a \in K$. If $\mathcal{X} = \mathrm{Spec}(L)$ and B is a commutative K-algebra, then

$$\mathcal{X}(B) = \mathrm{Hom}_{K\text{-alg}}(K(x), B) \leftrightarrow \{b \in B \mid b^{p^e} = a\}.$$

Note that if E is a field extension of K then

$$\mathcal{X}(E) = \begin{cases} \{x\} & E \supseteq L \\ \emptyset & E \not\supseteq L \end{cases}.$$

EXAMPLE 10.2. Let $K = \mathbb{Q}$, $L = \mathbb{Q}(\sqrt[3]{2})$, and $\mathcal{X} = \mathrm{Spec}(L)$. Then for any K-algebra B we have

$$\mathcal{X}(B) = \mathrm{Hom}_{\mathbb{Q}\text{-alg}}(\mathbb{Q}(\sqrt[3]{2}), B) \leftrightarrow \{b \in B \mid b^3 = 2\}.$$

For example, if E is a subfield of $\mathbb{C}$ then

$$\mathcal{X}(E) \leftrightarrow E \cap \{\sqrt[3]{2}, \zeta\sqrt[3]{2}, \zeta^2\sqrt[3]{2}\},$$

where $\zeta \in \mathbb{C}$ is a primitive cube root of unity.

Generally, if we fix K^{alg} an algebraic closure of K, then for $\mathcal{X}$ a K-scheme the K^{alg}-valued points of $\mathcal{X}$ are called the *geometric points* of $\mathcal{X}$. In the previous example, the set of geometric points is $\{\sqrt[3]{2}, \zeta\sqrt[3]{2}, \zeta^2\sqrt[3]{2}\}$. By contrast, the scheme in Example 10.1 has a single geometric point, namely x.

REMARK 10.3. Suppose L/K is a separable extension with Galois closure $\widetilde{L}/K$, and let $\mathcal{X} = \operatorname{Spec}(L)$. Let $G = \operatorname{Gal}(\widetilde{L}/K)$ and $G_L = \operatorname{Gal}(\widetilde{L}/L)$. The set $\mathcal{X}(\widetilde{L})$ corresponds naturally with the coset space $\operatorname{Hom}_{K\text{-alg}}(\widetilde{L}, \widetilde{L})/\operatorname{Hom}_{K\text{-alg}}(\widetilde{L}, L) = G/G_L$: clearly, $\sigma \in G$ restricts to a K-algebra map $L \to \widetilde{L}$, and $\sigma_1, \sigma_2 \in G$ restrict to the same map if and only if they agree on L. We will use this fact when we return to Greither-Pareigis theory.

Suppose H is a (commutative) K-Hopf algebra. Then $\mathcal{G} = \operatorname{Spec}(H)$ is a K-scheme; however the coalgebra structure allows us to define a binary operation on $\mathcal{G}(B)$ for any K-algebra B as follows: for $f, g \in \mathcal{G}(B) = \operatorname{Hom}_{K\text{-alg}}(H, B)$ we let

$$(f *_{\mathcal{G}} g)(h) = \operatorname{mult}(f \otimes g)\Delta(h).$$

The identity under this operation is the counit ε, viewing its image in the K-algebra B via the canonical inclusion $K \hookrightarrow B$, and the inverse to f is obtained by composing f with the antipode of H.

The functor $\mathcal{G}$ is called an *affine group scheme*, which we simply call "group scheme" in this text. (Examples of non-affine group schemes include elliptic curves and, more generally, abelian varieties.)

If H_1 and H_2 are K-Hopf algebras, and $\mathcal{G}_1 = \operatorname{Spec}(H_1)$, $\mathcal{G}_2 = \operatorname{Spec}(H_2)$, then a scheme morphism $F : \mathcal{G}_1 \to \mathcal{G}_2$ with the property that F_C is a homomorphism of groups for all commutative K-algebras C corresponds to a K-Hopf algebra map $H_2 \to H_1$ and conversely.

EXAMPLE 10.4. Let $K = \mathbb{F}_q((T))$, and let $H = K[t]/(t^{p^e})$. Then we can make H a Hopf algebra by letting t be *primitive*, i.e., $\Delta(t) = t \otimes 1 + 1 \otimes t$. Let $\mathcal{G} = \operatorname{Spec}(H)$. Then, for any K-algebra B,

$$\mathcal{G}(B) = \operatorname{Hom}_{K\text{-alg}}(H, B) \leftrightarrow \{b \in B \mid b^{p^e} = 0\}.$$

Note that if B is a separable K-algebra then $\mathcal{G}(B) = \{0\}$. For $f, g \in \mathcal{G}(B)$ we have

$$\begin{aligned}(f *_{\mathcal{G}} g)(t) &= \operatorname{mult}(f \otimes g)\Delta(t) \\ &= \operatorname{mult}(f \otimes g)(t \otimes 1 + 1 \otimes t) \\ &= f(t) + g(t).\end{aligned}$$

Identifying $\mathcal{G}(B)$ with a subset of B, the group operation is

$$b *_{\mathcal{G}} c = b + c, \ b, c \in B, \ b^{p^e} = c^{p^e} = 0.$$

The identity element is $0 \in \mathcal{G}(B)$ and the inverse to $b \in \mathcal{G}(B)$ is clearly $-b$.

EXAMPLE 10.5. Let H be the Hopf algebra

$$H = \mathbb{Q}[t]/(t^3 + 3t), \ \Delta(t) = t \otimes 1 + 1 \otimes t + \frac{1}{2}t^2 \otimes t + \frac{1}{2}t \otimes t^2,$$

and let $\mathcal{G} = \operatorname{Spec}(H)$. Then we have

$$\mathcal{G}(B) = \operatorname{Hom}_{\mathbb{Q}\text{-alg}}(H, B) \leftrightarrow \{b \in B \mid b^3 + 3b = 0\},$$

and for $f, g \in \mathcal{G}(B)$, $f(t) = b$, $g(t) = c$ we get

$$\begin{aligned}(f *_{\mathcal{G}} g)(t) &= \operatorname{mult}(f \otimes g)(t \otimes 1 + 1 \otimes t + \frac{1}{2}t^2 \otimes t + \frac{1}{2}t \otimes t^2) \\ &= \operatorname{mult}(b \otimes 1 + 1 \otimes c + \frac{1}{2}b^2 \otimes c + \frac{1}{2}b \otimes c^2) \\ &= b + c + \frac{1}{2}b^2c + \frac{1}{2}bc^2.\end{aligned}$$

Again by identifying maps with their images in B we get $b *_{\mathcal{G}} c = b + c + \frac{1}{2}b^2c + \frac{1}{2}bc^2$. As in the previous example, the identity element is 0 and the inverse to b is $-b$.

In particular, for E a subfield of $\mathbb{C}$ we get $\mathcal{G}(E)$ trivial if $i\sqrt{3} \notin E$, whereas the geometric points are $\mathcal{G}(\mathbb{C}) = \{0, i\sqrt{3}, -i\sqrt{3}\}$. Note that

$$i\sqrt{3} *_{\mathcal{G}} i\sqrt{3} = i\sqrt{3} + i\sqrt{3} + \frac{1}{2}(-3)i\sqrt{3} + \frac{1}{2}i\sqrt{3}(-3) = -i\sqrt{3},$$

so $\mathcal{G}(\mathbb{C}) = \langle i\sqrt{3} \rangle \cong C_3$.

We next discuss the notion of group scheme actions. For what is to follow, it will be clearer to use $\mathcal{N}$ to denote our group scheme, in deliberate parallel to $N \leq \operatorname{Perm}(G)$ in the separable case.

We say a group scheme $\mathcal{N}$ *acts* on a scheme $\mathcal{X}$ if there is a morphism $\mathcal{N} \times \mathcal{X} \to \mathcal{X}$ such that for each K-algebra B the map

$$\mathcal{N}(B) \times \mathcal{X}(B) \to \mathcal{X}(B)$$

is a group action.

For $\eta \in \mathcal{N}(B)$, $b \in \mathcal{X}(B)$ we will denote this action via $(\eta, b) \mapsto \eta[b]$.

EXAMPLE 10.6. Let L and H be the field and Hopf algebra from Examples 10.1 and 10.4 respectively. Let $\mathcal{N} = \operatorname{Spec}(H)$ and $\mathcal{X} = \operatorname{Spec}(L)$. We will, as before, freely identify the B-valued points of $\mathcal{N}$ and $\mathcal{X}$ as subsets of B. For each commutative K-algebra B, define $\mathcal{N}(B) \times \mathcal{X}(B) \to \mathcal{X}(B)$ by

$$\eta[b] = \eta + b,$$

where $\eta, b \in B$ satisfy $\eta^{p^e} = 0$, $b^{p^e} = a$. Note that

$$(\eta[b])^{p^e} = (\eta + b)^{p^e} = \eta^{p^e} + b^{p^e} = 0 + a = a$$

so $\eta[b] \in \mathcal{X}(B)$ as desired. Since $0[b] = b$ and $(\eta_1 + \eta_2)[b] = \eta_1[\eta_2[b]]$ this is clearly a group action for all B. Furthermore, it is easy to verify that the map $\mathcal{N}(B) \times \mathcal{X}(B) \to \mathcal{X}(B)$ is a morphism of schemes.

Many simple examples of K-algebras B fail to produce illustrative group actions, since $\mathcal{N}(E)$ is trivial for E a field. However, let $B = L[y]/(y^{p^e})$. Then

$$\mathcal{N}(B) = \{\eta \in B \mid \eta^{p^e} = 0\} = (y)$$
$$\mathcal{X}(B) = \{b \in B \mid b^{p^e} = a\} = x + (y),$$

and the action relates to coset addition in $B/(y)$.

EXAMPLE 10.7. Let L and H be as in Examples 10.2 and 10.5, and write $\mathcal{N} = \operatorname{Spec}(H)$, $\mathcal{X} = \operatorname{Spec}(L)$. Then $\mathcal{N}(B)$ acts on $\mathcal{X}(B)$ by

$$\eta[b] = b + \frac{1}{2}\eta b + \frac{1}{2}\eta^2 b = \left(1 + \frac{1}{2}\eta + \frac{1}{2}\eta^2\right) b, \ \eta \in \mathcal{N}(B), \ b \in \mathcal{X}(B)$$

as the reader can check. As before, let E be a subfield of $\mathbb{C}$. For the action to be nontrivial we require $\sqrt[3]{2}, i\sqrt{3} \in E$ (note that ζ, a primitive cube root of unity, is in E if and only if $i\sqrt{3} \in E$). In other words, we would need E to contain the Galois closure of L/K. In this case we have

$$\begin{aligned} i\sqrt{3}[\zeta^i \sqrt[3]{2}] &= \left(1 + \frac{1}{2}i\sqrt{3} + \frac{1}{2}(-3)\right)\zeta^i \sqrt[3]{2} \\ &= \left(\frac{-1 - i\sqrt{3}}{2}\right)\zeta^i \sqrt[3]{2} \\ &= \zeta^{i+2}\sqrt[3]{2}, \end{aligned}$$

hence $-i\sqrt{3}[\zeta^i \sqrt[3]{2}] = \zeta^{i+1}\sqrt[3]{2}$.

We claim that in each of the examples above the action is free for each K-algebra B, i.e., the stabilizer of any element in $\mathcal{X}(B)$ is the identity in $\mathcal{G}(B)$. In Example 10.6 this should be clear since the action is addition. For Example 10.7, suppose $\eta[b] = b$. Then

$$\frac{1}{2}\eta b + \frac{1}{2}\eta^2 b = \frac{1}{2}(\eta + \eta^2)b = 0.$$

Since $b^3 = 2$ we know b is not a zero divisor, hence $\eta + \eta^2 = 0$. Thus $\eta = -\eta^2$ and

$$\eta^3 + 3\eta = \eta^3 - 3\eta^2 = \eta^2(\eta - 3) = 0.$$

However, $\eta - 3$ is a unit since

$$(\eta - 3)\left(-\frac{1}{3} - \frac{1}{12}\eta - \frac{1}{36}\eta^2\right) = 1,$$

so $\eta - 3$ is not a zero divisor. Hence, η is trivial and the action is free. Of course, when $\mathcal{X}(B)$ and $\mathcal{N}(B)$ are the same size then the action is transitive as well, for example in the case where $B = \mathbb{Q}(\zeta, \sqrt[3]{2})$ is the Galois closure of L/K.

Each of the examples above gives rise to an isomorphism $\mathcal{N} \times \mathcal{X} \to \mathcal{X} \times \mathcal{X}$ given by $(\eta, b) \mapsto (\eta[b], b)$ for $\eta \in \mathcal{N}(B), b \in \mathcal{X}(B)$.

DEFINITION 10.8. Let $\mathcal{N}$ be a group scheme, acting on a scheme $\mathcal{X}$. If the induced map

$$\mathcal{N}(B) \times \mathcal{X}(B) \to \mathcal{X}(B) \times \mathcal{X}(B),\ (\eta, b) \mapsto (\eta[b], b),\ \eta \in \mathcal{N}(B), b \in \mathcal{X}(B)$$

is an isomorphism for every K-algebra B in which $\mathcal{X}(B) \neq \emptyset$, then we call $\mathcal{X}$ a *principal homogeneous space* (PHS) for $\mathcal{N}$.

A principal homogeneous space of $\mathcal{N}$ is also called a *torsor* for $\mathcal{N}$.

Principal homogeneous spaces are closely connected to Hopf-Galois extensions. Let L/K be a field extension, and let $\mathcal{X} = \operatorname{Spec}(L)$. Let H be a K-Hopf algebra, and let $\mathcal{G} = \operatorname{Spec}(H)$. Then an action $\mathcal{N} \times \mathcal{X} \to \mathcal{X}$ gives rise through duality to a K-algebra map $\gamma : L \to H \otimes L$, and the group scheme action conditions translate to L being an *H-comodule algebra*. Also, the induced map $\mathcal{G} \times \mathcal{X} \to \mathcal{X} \times \mathcal{X}$ above is an isomorphism if and only if the corresponding map

$$L \otimes L \to L \otimes H,\ x \otimes y \mapsto x \otimes \gamma(y)$$

is an isomorphism. Thus

PROPOSITION 10.9. *The K-scheme $\mathcal{X} = \operatorname{Spec}(L)$ is a PHS for $\mathcal{G} = \operatorname{Spec}(H)$ if and only if L is an H-Galois object.*

COROLLARY 10.10. *The K-scheme $\mathcal{X} = \operatorname{Spec}(L)$ is a PHS for $\mathcal{G} = \operatorname{Spec}(H)$ if and only if L/K is H^*-Galois.*

We now wish to re-interpret Greither-Pareigis theory into this algebraic geometry framework. To align the notation with Greither-Pareigis theory we will interchange H and H^* below.

Suppose L/K is separable, and that $\mathcal{X} = \mathrm{Spec}(L)$ is a principal homogeneous space for $\mathcal{N} = \mathrm{Spec}(H^*)$. Let $\widetilde{L}$ be the Galois closure of L/K, $G = \mathrm{Gal}(\widetilde{L}/K)$, $G_L = \mathrm{Gal}(\widetilde{L}/L)$, and consider the group action $\mathcal{N}(\widetilde{L}) \times \mathcal{X}(\widetilde{L}) \to \mathcal{X}(\widetilde{L})$ given by the action of $\mathcal{N}$ on $\mathcal{X}$. As $f : \mathcal{N}(\widetilde{L}) \times \mathcal{X}(\widetilde{L}) \to \mathcal{X}(\widetilde{L}) \times \mathcal{X}(\widetilde{L})$, $f(\eta, x) = (\eta[x], x)$ is an isomorphism, we have $\eta : \mathcal{X}(\widetilde{L}) \to \mathcal{X}(\widetilde{L})$, $\eta(x) = \eta[x]$ is a bijection. Thus we get a map $\alpha : N := \mathcal{N}(\widetilde{L}) \to \mathrm{Perm}(G/G_L)$ using the identification of $\mathcal{X}(\widetilde{L})$ and G/G_L as in Remark 10.3. It is easy to see that α is a homomorphism, and if $\alpha(\eta)$ is trivial then for any $gG_L \in G/G_L$ we have

$$f(\eta, gG_L) = (\eta[gG_L], gG_L) = (gG_L, gG_L) = (1_N[gG_L], gG_L) = f(1_N, gG_L),$$

so since f is an isomorphism we get $\eta = 1_N$. Thus f is an injection, and we may view N as a subgroup of $\mathrm{Perm}(G/G_L)$. Furthermore, this subgroup is regular (that is, free and transitive): note that $|N| = |G/G_L|$ and that each $gG_L \in G/G_L$ has trivial stabilizer.

The regular action described above directly relates to the regular subgroups in the Greither-Pareigis classification. Indeed, keeping the notation above, suppose $N = \{\eta_1, \eta_2, \ldots, \eta_n\} \leq \mathrm{Perm}(G/G_L)$ is regular and normalized by G. Then L/K is Hopf-Galois, with Hopf algebra $H = \widetilde{L}[N]^G$. Note that H is not necessarily commutative, but it is cocommutative. Pick a K-basis $\{h_1, h_2, \ldots, h_n\}$ for H, and let $\{t_1, t_2, \ldots, t_n\}$ be the dual basis in H^*. Write $\eta_j = \sum_{i=1}^n a_{i,j} h_i$, $a_{i,j} \in \widetilde{L}$. Let $\mathcal{N} = \mathrm{Spec}(H^*)$ and $\mathcal{X} = \mathrm{Spec}(\widetilde{L})$. Then

$$\mathcal{N}(\widetilde{L}) = \mathrm{Hom}_{K\text{-alg}}(H^*, \widetilde{L}) \cong \mathrm{Hom}_{\widetilde{L}\text{-alg}}(\widetilde{L} \otimes H^*, \widetilde{L}) \cong \mathrm{Hom}_{\widetilde{L}\text{-alg}}(\widetilde{L}[N]^*, \widetilde{L}) \cong N$$

and

$$\mathcal{X}(\widetilde{L}) = \mathrm{Hom}_{K\text{-alg}}(L, \widetilde{L}) \cong G/G_L.$$

The action of $\mathcal{N}(\widetilde{L})$ on $\mathcal{X}(\widetilde{L})$, induced from the coaction $L \to L \otimes H^*$ extended to $\widetilde{L}$, translates to an action of N on G/G_L via

$$(\eta_j * gG_L)(x) = \sum_{i=1}^n g(h_i \cdot x) t_i(\eta_j) = \sum_{i=1}^n g(h_i \cdot x) a_{i,j}$$

with $\eta_j \in N$, $g \in G$, and $x \in L$. Note that since $h_1 \cdot x \in L$ the above is independent of the choice of coset representative in gG_L. We claim that $\eta_j * 1_G G_L = \eta^{-1}[1_G G_L]$: by regularity it follows that $\eta_j * gG_L = \eta^{-1}[gG_L]$ for all $g \in G$. Indeed, write $h_i = \sum_{j=1}^n b_{i,j} \eta_j$, $b_{i,j} \in \widetilde{L}$, and let $A = [a_{i,j}], B = [b_{i,j}] \in \mathrm{Mat}_n(\widetilde{L})$. Then $A^T B = I$, so

$$a_{1,j} b_{1,k} + a_{2,j} b_{2,k} + \cdots + a_{n,k} b_{n,k} = \delta_{j,k}, \; 1 \leq j, k \leq n.$$

Now for $x \in L$ we have

$$\begin{aligned}(\eta_j * 1_G G_L)(x) &= \sum_{i=1}^{n} (h_i \cdot x) t_i(\eta_j) \\ &= \sum_{i=1}^{n}\sum_{k=1}^{n} \left(b_{i,k}\eta_k^{-1}[1_G G_L]\right)(x) t_i(\eta_j) \\ &= \sum_{i=1}^{n}\sum_{k=1}^{n} (a_{i,j} b_{i,k})\, \eta_k^{-1}[1_G G_L](x) \\ &= \eta_j^{-1}[1_G G_L](x).\end{aligned}$$

Let K^{sep} be the separable closure of K, and let $\Gamma = \mathrm{Gal}(K^{sep}/K)$. For each separable K-algebra A, let $X_A = \mathrm{Hom}_{K\text{-alg}}(A, K^{sep})$. Then Γ acts continuously on X_A via

$$({}^{\gamma}f)(x) = \gamma(f(x)), \gamma \in \Gamma, f \in X_A, x \in A$$

(see, e.g., [**Wat79**, §6.3]). By continuity, this action factors through finite Galois extensions of K, and the assignment $A \mapsto X_A$ is a bijection from separable K-algebras to finite Γ-sets (where the action is continuous).

Return to $N \leq \mathrm{Perm}(G/G_L)$ a regular subgroup, normalized by G, and let $H = \widetilde{L}[N]^G$. Since $\widetilde{L}[N]$ is cocommutative, so is H, hence as before H^* is a commutative K-algebra which is also separable. Thus N corresponds to a finite Γ-set, specifically,

$$\begin{aligned}X_{H^*} &= \mathrm{Hom}_{K\text{-alg}}(H^*, K^{sep}) \cong \mathrm{Hom}_{\widetilde{L}\text{-alg}}(\widetilde{L} \otimes H^*, K^{sep}) \\ &= \mathrm{Hom}_{\widetilde{L}\text{-alg}}(\widetilde{L}[N]^*, K^{sep}) \cong N\end{aligned}$$

and clearly $\mathcal{N}(\widetilde{L}) = \mathcal{N}(K^{sep})$, hence $\mathcal{N}(\widetilde{L}) \cong N$ as well. Under the isomorphism $N \to \mathcal{N}(\widetilde{L})$, the element $\eta \in N$ corresponds to the algebra map $(\widetilde{L} \otimes H)^* \to \widetilde{L}$ obtained by evaluating at η. Then $\mathcal{N}(\widetilde{L}) = \mathrm{Hom}_{K\text{-alg}}(H^*, \widetilde{L}) = \{f_j \mid 1 \leq j \leq n\}$ where $f_j(t_i) = a_{i,j}$.

However,

$${}^{g}\eta_j = {}^{g}\left(\sum_{i=1}^{n} a_{i,j} h_i\right) = \sum_{i=1}^{n} g(a_{i,j}) h_i,$$

so if we pick k such that ${}^{g}\eta_j = \eta_k$ we have $a_{i,k} = g(a_{i,j})$ for all i. The action of G on $\mathcal{N}(\widetilde{L})$ is therefore

$$(g \cdot f_j)(t_i) = g(a_{i,j}) = a_{i,k} = f_k(t_i),$$

in other words $g \cdot f_j = f_k$ if and only if ${}^{g}\eta_j = \eta_k$.

We now revisit the case L/K is purely inseparable and Hopf-Galois for some K-Hopf algebra H. Letting $\mathcal{N} = \mathrm{Spec}(H^*)$ and $\mathcal{X} = \mathrm{Spec}(L)$ as before, we ask: is it possible to look at the K^{sep}-valued points of $\mathcal{N}$ and $\mathcal{X}$ and find a regular embedding $\mathcal{N}(K^{sep}) \hookrightarrow \mathrm{Perm}\left(\mathcal{X}(K^{sep})\right)$? The answer is "no" since $\mathcal{N}(K^{sep})$ is necessarily trivial: the geometric interpretation of Greither-Pareigis theory does not hold, *mutatis mutandis*, in the purely inseparable case. However, return to Example 10.6 above, where $L = K(x)$, $x^{p^e} \in K$, $H = K[t]/(t^{p^e})$ with t primitive, and $\mathcal{N} = \mathrm{Spec}(H) = \mathrm{Spec}(H^*)$ and $\mathcal{X} = \mathrm{Spec}(L)$. For $B = L[y]/(y^{p^e})$ we have

$N_B := \mathcal{N}(B) = (y)$ and $G_B := \mathcal{X}(B) = x + (y)$, and $\mathcal{N}(B)$ acts on $\mathcal{X}(B)$ via addition. We thus have a map $N_B \hookrightarrow \mathrm{Perm}(G_B)$ given by

$$\eta \mapsto (b \mapsto \eta + b),$$

whose image (which, by abuse of notation, we also denote by N_B) is a regular subgroup of $\mathrm{Perm}(G)$.

Thus, $\mathcal{N}(B)$ can be viewed as a regular subgroup of $\mathrm{Perm}(\mathcal{X}(B))$.

REMARK 10.11. Note that, in the work above, $B \cong L \otimes_K L$ via the map $y \mapsto x \otimes 1 - 1 \otimes x$.

10.2. A short history

In this section we assume L/K is a finite field extension, $\mathrm{char}(K) = p > 0$. We will outline some of the Galois-type theories that have been proffered over the years, and discuss the connections with Hopf-Galois theory.

10.2.1. Jacobson's theory of derivations.

Nathan Jacobson [**Jac44**] used derivations as a substitute for the K-algebra automorphisms which lie at the heart of classical Galois theory.

DEFINITION 10.12. A *K-derivation on L* is a K-linear map $D : L \to L$ such that

$$D(xy) = D(x)y + xD(y)$$

for all $x, y \in L$.

Notice that if $D : L \to L$ is a derivation, then

$$D(1) = D(1 \cdot 1) = D(1) \cdot 1 + 1 \cdot D(1) = 2D(1),$$

thus $D(1) = 0$. Hence, by K-linearity, $D(a) = aD(1) = 0$ for all $a \in K$. Generally, let

$$L^D = \{x \in L \mid D(x) = 0\}.$$

Then $K \subseteq L^D \subseteq L$ is a chain of fields, and L^D is called the *field of constants of D*.

Additionally, if we let $D^p = \overbrace{D \circ D \circ \cdots \circ D}^{p \text{ times}}$ then by [**Jac37**, (7)] we have

$$D^p(xy) = D^p(x)y + xD^p(y),$$

and hence D^p is a K-derivation.

Let $\mathcal{D} := \mathcal{D}_K(L)$ denote the set of all K-derivations of L. Then $\mathcal{D}$ has a natural structure of a K-vector space, and

$$L^{\mathcal{D}} = \bigcap_{D \in \mathcal{D}} L^D$$

is a subfield of L containing K. Furthermore, $\mathcal{D}$ carries the structure of a K-Lie algebra via the bracket map

$$[D_1, D_2] = D_1 D_2 - D_2 D_1, \ D_1, D_2 \in \mathcal{D}_K(L);$$

in light of the remarks above, $\mathcal{D}$ is a *restricted p-Lie algebra.* See [**Jac62**, Ch. V] for a treatment of restricted p-Lie algebras and their enveloping algebras, a concept which is mentioned below.

REMARK 10.13. A K-derivation $D : L \to L$ induces a K-algebra map $\varphi : L \to L[t]/(t^2)$ given by

$$\varphi(x) = x + D(x)t, \ x \in L.$$

Conversely, suppose $\varphi : L \to L[t]/(t^2)$ is a K-algebra homomorphism. Then $\varphi(x) \equiv x \pmod t$. Define $D : L \to L$ via

$$\varphi(x) = x + D(x)t.$$

Then D is a derivation. Note, for example,

$$\begin{aligned} \varphi(xy) &= \varphi(x)\varphi(y) \\ xy + D(xy)t &= (x + D(x)t)(y + D(y)t) \\ xy + D(xy)t &= xy + xD(y)t + D(x)yt + D(x)D(y)t^2 \\ D(xy) &= D(x)y + yD(x). \end{aligned}$$

One now sees a connection with classical Galois theory: $x \in L^D$ if and only if $\varphi_D(x) := x + D(x)t = x$. One also sees a connection with algebraic geometry: there is a bijection $\mathcal{D} \to \mathcal{X}(L[t]/(t^2))$, $\mathcal{X} = \operatorname{Spec}(L)$.

EXAMPLE 10.14. Let $L = K(x)$, $x^p \in K$. Define $D : L \to L$ by

$$D(x^i) = ix^{i-1}, \ 1 \le i \le p - 1.$$

Then D is a K-derivation of L, as can be readily checked. Moreover, $\{D\}$ is a K-basis for $\mathcal{D}$, and $L^{\mathcal{D}} = K$.

Suppose now that $L = K(x_1, x_2, \ldots, x_m)$ with $x_i^p \in K$ for all $1 \le i \le m$: we say that L/K has *exponent one*. Then, by [**Jac44**], $\dim_K \mathcal{D}_K(L) = m$, and Jacobson describes a Galois theory as follows. To each intermediate field $K \subseteq F \subseteq L$ we associate the F-vector space $\mathcal{D}_F(L)$; conversely, given a restricted sub-Lie algebra $\mathcal{D}'$ of $\mathcal{D}_K(L)$ the corresponding intermediate field is precisely $L^{\mathcal{D}'}$.

Let us now explain the connection between the theory of derivations and Hopf-Galois theory in the exponent one case. Write $L = K[x_1, \ldots, x_m]$, $x_i^p \in K$ as above. For $1 \le i \le m$, let $D_i = \partial/\partial x_i \in \mathcal{D}_K(L)$. Then $\{D_1, \ldots, D_m\}$ generates $\mathcal{D}_K(L)$ as a K-vector space. Let $H = K[D_1, \ldots, D_m]$. Then H is a K-algebra and $\dim_K H = p^m$ since $D_i^p = 0$ for all $1 \le i \le m$. Moreover, H is a K-Hopf algebra, where each D_i is a primitive element: note that H is the *restricted universal enveloping algebra* (see [**Mon93**, p. 23])) for $\mathcal{D}$. Then H acts on L via these derivations. Since for all $1 \le i \le m$ we have $D_i(1_K) = 0$ and for all $x, y \in L$ we have

$$D_i(xy) = D_i(x)y + xD_i(y) = \operatorname{mult}(D_i(x) \otimes y + x \otimes D_i(y)) = \operatorname{mult} \Delta(D_i)(x \otimes y)$$

we see that L is an H-module algebra.

We claim the K-algebra map $\gamma : L \otimes H \to \operatorname{End}_K(L)$ given by $\gamma(x_i \otimes D_k)(x_j) = x_i D_k(x_j)$ is surjective, hence an isomorphism. Indeed, for $\mathbf{j} = (j_1, j_2, \ldots, j_m)$, $0 \le j_i \le p - 1$ we define $\varepsilon_{\mathbf{j}} : L \to L$ to be the K-linear map with

$$\varepsilon_{\mathbf{j}}(x_1^{i_1} \cdots x_m^{i_m}) = \delta_{\mathbf{i},\mathbf{j}}, \ \mathbf{i} = (i_1, \ldots, i_m).$$

Then $\{\varepsilon_{\mathbf{j}}\}$ is a K-basis for $\operatorname{Hom}_K(L, K)$ and

$$\gamma\left((x_1^{\ell_1} \cdots x_m^{\ell_m}) \prod_{i=1}^{m} \left(\sum_{k=j_i}^{p-1} \frac{1}{(k - j_i)!} x_i^{k-j_i} \otimes D_i^k \right)\right) = x_1^{\ell_1} \cdots x_m^{\ell_m} \varepsilon_{\mathbf{j}},$$

hence γ is an isomorphism.

10.2.2. Higher derivations. Jacobson's Galois theory of derivations is rather limited in scope: it only applies to extensions of the form $K[x_1, \ldots, x_m]/K$ with $x_i^p \in K$ for each i. In [**Swe68**], Sweedler extended the theory to a much larger class of purely inseparable extensions L/K.

DEFINITION 10.15. Let L/K be a finite extension.

(1) The extension L/K is *primitive* if $L = K[x]$ with $x^{p^e} \in K$ for some e: the smallest such e is the *exponent* of the extension.
(2) The extension L/K is *modular* if

$$L = K[x_1] \otimes_K \cdots \otimes_K K[x_m],\ K[x_i]/K \text{ primitive for all } i.$$

Equivalently, L/K is modular if and only if L and $K^{p^{-n}}$ are linearly disjoint over $L \cap K^{p^{-n}}$ for all $n > 0$ [**Swe68**]. Note that separable extensions, which simply require that L and $K^{p^{-1}}$ are disjoint over K, are necessarily modular [**KH75**].

The exponent one extensions described in the previous section are all modular since $K[x_1, \ldots, x_m]$ naturally splits into tensor factors $K[x_i]$ each of exponent one. The notion of a modular extension allows us to consider new yet very basic extensions such as $L = K[x]$ with $x^{p^2} \in K$, $x^p \notin K$. To do so, we first need to generalize the concept of derivation.

DEFINITION 10.16. A *higher derivation of L/K of length n* is a sequence $D = (D_0 := I, D_1, \ldots, D_n)$ of K-linear maps $L \to L$ such that

$$D_m(xy) = \sum_{i=0}^{m} D_i(x) D_{m-i}(y)$$

for all $x, y \in L$. Here, $I : L \to L$ is the identity map.

REMARK 10.17. The definition requires, in particular, that

$$D_1(xy) = D_0(x)D_1(y) + D_1(x)D_0(y) = D_1(x)y + xD_1(y),$$

so D_1 is an ordinary derivation. In fact, one can think of a higher derivation of length 1 as simply an ordinary derivation.

Given a higher derivation $D = (D_0, D_1, \ldots, D_n)$, the field of constants of D is defined in a manner analogous to the ordinary case:

$$L^D = \{x \in L \mid D_i(x) = 0 \text{ for all } 1 \leq i \leq n\}.$$

A higher derivation of length n gives rise to a K-algebra map $\varphi_D : L \to L[t]/(t^{n+1})$ given by

$$\varphi_D(x) = \sum_{i=0}^{n} D_i(x) t^i.$$

In particular, a length one derivation $D = (D_0, D_1)$ corresponds to the map

$$\varphi_D(x) = D_0(x)t^0 + D_1(x)t^1 = x + D_1(x)t \in L[t]/(t^2),$$

recovering the algebra map φ_D coming from an ordinary derivation. The map φ_D is sometimes referred to as an *approximate automorphism of L/K* [**Ger68**] or a *variation of L/K* in $L[t]/(t^{n+1})$ [**Ras71**].

Conversely, any $\varphi : L \to L[t]/(t^{n+1})$ gives a higher derivation in a manner analogous to the ordinary derivation case. Thus the set of higher derivations of length n corresponds bijectively to $\mathcal{X}(L[t]/(t^{n+1}))$, where $\mathcal{X} = \operatorname{Spec}(K)$.

As before, let

$$L^D = \{x \in L \mid \varphi_D(x) = x\} = \{x \in L \mid D_i(x) = 0 \text{ for all } i \neq 0\},$$

and if $\mathcal{D}$ is a set of higher derivations on L/K, its field of constants is

$$L^{\mathcal{D}} = \bigcap_{D \in \mathcal{D}} = L^D.$$

By [**Swe68**, Theorem 1], a (purely inseparable) modular field extension L/K will always have field of constants equal to K for some set $\mathcal{D}$ of higher derivations. In fact, he shows the converse: given any finite purely inseparable extension L/K, if $L^{\mathcal{D}} = K$ for some $\mathcal{D}$ then L/K is modular. Thus any Galois theory rising from derivations, ordinary or higher, can only exist on modular extensions.

EXAMPLE 10.18. Let L/K be a primitive extension of exponent $e \geq 1$, say $L = K[x]$ with $x^{p^e} \in K$. For each $0 \leq m \leq e$ let $D_m : L \to L$ be the K-linear map defined by

$$D_m(x^i) = \binom{i}{m} x^{i-m}$$

with the understanding that $\binom{i}{m} = 0$ for $i < m$. Then

$$D_m(x^{i+j}) = \binom{i+j}{m} x^{i+j-m},$$

while

$$\begin{aligned}\sum_{k=0}^{m} D_k(x^i) D_{m-k}(x^j) &= \sum_{k=0}^{m} \binom{i}{k} x^{i-k} \binom{j}{m-k} x^{j-(m-k)} \\ &= \sum_{k=0}^{m} \binom{i}{k} \binom{j}{m-k} x^{i+j-m} \\ &= \binom{i+j}{m} x^{i+j-m},\end{aligned}$$

the latter equality by Vandermonde's Identity, obtained by comparing the coefficients of $(1+x)^{i+j}$ and $(1+x)^i(1+x)^j$. Thus, $D = (D_0, D_1, \ldots, D_e)$ is a higher derivation of L/K. Furthermore, $L^D = K$.

We can see the connection to Hopf-Galois theory in this case by generalizing the exponent one case.

We close our discussion of higher derivations by pointing out that not all purely inseparable field extensions are modular. Consider $K = \mathbb{F}_p(a, b, c)$ and $L = K[z, xz + y]$, $x^p = a, y^p = b, z^{p^2} = c$. Then for any higher derivation $D = (D_0, D_1, \ldots, D_n)$ it can be shown that $D_i(z^p) = 0$ for all $i > 0$, consequently, $z^p \in L^D$ for all D. Thus the field of constants is not K and L/K is not modular. See [**Swe68**, Example 1.1] for a full treatment of this example. Note that L/L^D is modular; generally, L/K contains a unique minimal intermediate field F for which L/F is modular [**Swe68**, Corollary. 2]. Additionally, [**Swe68**, Theorem 6] proves the existence of a "modular closure" for any L/K.

10.2.3. Heerema's Galois group. In [**Hee71**], Heerema provides a Galois theory for extensions. Heerema's theory has two advantages over the theory of higher derivations. First, it handles both separable and inseparable extensions (Sweedler, in [**Swe68**, §2], also considers general modular extensions, but not in terms of higher derivations), thereby providing a more complete theory in the case L/K is inseparable but not purely so. Second, it uses automorphisms instead of derivations (equivalently, algebra maps into truncated polynomial algebras), and in that sense is closer to classical Galois theory. One noticeable difference between Heerema's Galois group and the classical theory is that the automorphisms are not maps on the field L as we shall presently see.

The perspective in [**Hee71**] is somewhat different from the other theories. Heerema starts with a field L of characteristic $p > 0$ (note that the notation here differs from [**Hee71**] who refers to this field as k) and constructs fixed subfields under his Galois correspondence, thereby constructing the base field (and other intermediate fields, of course).

Let $r \geq 0$ be an integer. Let S be the L-algebra (which depends on r, implicitly) $S = L[t]/(t^{p^r+1})$, and define

$$A = \{\alpha \in \mathrm{Aut}(S) \mid \alpha(t) = t\}$$

where the automorphisms are automorphisms of rings. If we let $A_0 \leq A$ be the subgroup of all α such that $\alpha(x) \equiv x \pmod{t}$ for all $x \in L$ then A_0 is isomorphic to the group of all higher derivations on L of length p^r: assign to a higher derivation $D = (D_0, D_1, \ldots, D_{p^r})$ the automorphism α_D given by

$$\alpha_D(x) = \sum_{i=0}^{p^r} D_i(x)t^i, \; x \in L, \; \alpha_D(t) = t.$$

Note that $\alpha_D(x) \equiv D_0(x) \equiv x \pmod{t}$ for all $x \in L$ so $\alpha_D \in A_0$ as desired.

Now suppose G is a subgroup of A. We then let

$$K = L^G = \{x \in L \mid \alpha(x) = x \text{ for all } \alpha \in G\}.$$

Then L is a normal extension of K such that $L^{p^{r+1}}/K$ is separable; furthermore, L decomposes into a compositum $L = K_1K_2$ where K_1/K is modular and K_2/K is both normal and separable. Furthermore, define

$$\begin{aligned} G_0 &= \{\alpha \in G \mid \alpha(L) \subseteq L\}, \\ G_L &= \{\alpha \in G \mid \alpha(x) \equiv x \pmod{t} \text{ for all } x \in L\}. \end{aligned}$$

Then we get $K_1 = L^{G_L}$ and $K_2 = L^{G_0}$. One can see that G_L captures the classical Galois theory of field automorphisms, while G_0 captures the purely inseparable theory of higher derivations. For proofs of all of the facts above, see [**Hee71**, Theorem 3.1].

EXAMPLE 10.19. Let $K = \mathbb{F}_q((T))$, and let $L = K(x, y)$ be an extension of K of degree p^2 with $x^p \in K$, $y^p - y \in K$. Let $\alpha, \beta : L[t]/(t^{p^2}) \to L[t]/(t^{p^2})$ be K-algebra maps with

$$\begin{aligned} &\alpha(x) = x && \alpha(y) = y + 1 && \alpha(t) = t \\ &\beta(x) = x + t^p && \beta(y) = y && \beta(t) = t. \end{aligned}$$

Then $G := \langle \alpha, \beta \rangle$ is an elementary abelian group contained in A (here we have chosen $r = 0$). Clearly, $L^G = K$. With the notation above, we have $G_0 = \langle \alpha \rangle$ and

$G_L = \langle \beta \rangle$. Thus $K_1 = L^{\langle \beta \rangle} = K[y]$, a Galois extension of K; and $K_2 = L^{\langle \alpha \rangle} = K(x)$ a purely inseparable modular extension of K.

In 2018, Battiston ([**Bat18**]) created a descent theory using (a slight variation of) the Heerema-Galois group. In the example L/K above, the Battiston's Heerema-Galois group is $\mathrm{Aut}_{K[y]}(L[y])$. There appear to be connections between the Galois descent implicit in Greither-Pareigis theory in the separable case and the descent theory of Battiston in the inseparable case.

10.2.4. Chase's truncated automorphism scheme. Another approach to studying purely inseparable extensions via automorphisms appears in [**Cha76**]. Chase's approach uses the notion of a "truncated automorphism scheme" which we now describe. As in the previous sections, we adjust the notations of cited works to fit the conventions of this book, and suppose L/K is a purely inseparable extension.

We say a K-algebra A is a *truncated polynomial algebra* if it is of the form

$$A = K[t_1, \cdots t_r]/(t_1^{n_1}, \cdots, t_r^{n_r})$$

for some positive integers $n_1, \ldots, n_r$. Then A is local, with maximal ideal $(t_1, \ldots, t_r)$, and is evidently a finite dimensional K-vector space. We call the affine K-scheme $\mathrm{Spec}(A)$ *truncated* as well. Note that the set of L-valued points of $\mathrm{Spec}(A)$ is simply $\{0\}$ for L a field extension of K; however one can get B-valued points when B is also a truncated polynomial algebra.

DEFINITION 10.20. The functor $G_{\mathrm{tr}}(L/K)$ which assigns to each truncated polynomial algebra A the group $\mathrm{Aut}_{A\text{-alg}}(L \otimes A)$ is called the *truncated automorphism scheme* for L/K.

The notion of the truncated automorphism scheme was first introduced in [**Cha72**] in the case where L/K is purely inseparable and modular.

The truncated automorphism scheme $G_{\mathrm{tr}}(L/K)$ acts on $\mathcal{X} = \mathrm{Spec}(L)$. This action has a universal property, namely if G is any truncated K-group scheme acting on L then there is a unique map $G \to G_{\mathrm{tr}}(L/K)$ of K-group schemes preserving the action.

Chase's main theorem of his truncated automorphism scheme action is the following.

THEOREM 10.21. *Let L/K be a finite, purely inseparable extension. Then the following are equivalent.*

(1) *L/K is modular.*
(2) *$L \otimes L$ is a truncated polynomial algebra.*
(3) *$L^{G_{\mathrm{tr}}(L/K)} = K$.*
(4) *$L^G = K$ for some truncated K-group scheme G acting on L/K.*

PROOF. See [**Cha76**, Theorem 2.2]. □

The Galois theory of Chase is then the content of [**Cha76**, Theorem 2.3]. Assume that L/K is modular. If F is an intermediate field of L/K with the property that L/F is modular, then $G_{\mathrm{tr}}(L/K)$ has a unique maximal subgroup scheme $G_{\mathrm{tr}}(L/F/K)$ such that $F \subseteq K^{G_{\mathrm{tr}}(L/F/K)}$. This subgroup scheme is defined, for A a truncated polynomial algebra, by

$$G_{\mathrm{tr}}(L/F/K)(A) = \mathrm{Aut}_{F \otimes A\text{-alg}}(L \otimes A).$$

Furthermore, one has $F \subseteq K^{G_{\mathrm{tr}}(L/F/K)}$ if and only if L/F is purely inseparable and modular; if so, $G_{\mathrm{tr}}(L/F/K)$ has order $[L : F]^{[L:K]}$. In the case $F = K$ we see that

$L^{G_{\mathrm{tr}}(L/K)} = K$ if and only if L/K is modular, furthermore $\dim_k G_{\mathrm{tr}}(L/K) = [L : K]^{[L:K]}$. In particular, note that the analogue of the Galois group is much bigger than the degree of the field extension.

EXAMPLE 10.22. Let L/K be a simple extension of exponent e, say $L = K(x)$, $x^{p^e} = b \in K$. We shall explicitly compute $G_{\mathrm{tr}}(L/K)$. Let A be a truncated polynomial algebra, then $G_{\mathrm{tr}}(L/K)(A) = \mathrm{Aut}_{A\text{-alg}}(L \otimes A)$. As $L \otimes A$ is generated as an A-algebra by $x \otimes 1$, any $\varphi \in G_{\mathrm{tr}}(L/K)(A)$ is determined by $\varphi(x \otimes 1)$. Write

$$\varphi(x \otimes 1) = x \otimes 1 + \sum_{i=0}^{p^e-1} x^i \otimes a_i$$

for some choice $a_0, \ldots, a_{p^e-1} \in A$. Then

$$(\varphi(x \otimes 1))^{p^e} = \varphi(x^{p^e} \otimes 1) = \varphi(b \otimes 1) = b \otimes 1,$$

while

$$\begin{aligned}(\varphi(x \otimes 1))^{p^e} &= x^{p^e} \otimes 1 + \sum_{i=0}^{p^e-1} x^{ip^e} \otimes a_i^{p^e} \\ &= b \otimes 1 + \sum_{i=0}^{p^e-1} b^i \otimes a_i^{p^e} \\ &= b \otimes 1 + 1 \otimes \left(\sum_{i=0}^{p^e-1} b^i a_i^{p^e} \right),\end{aligned}$$

hence the sum on the right is zero. Thus, selecting $\varphi \in G_{\mathrm{tr}}(L/K)(A)$ is equivalent to picking $a_0, \ldots, a_{p^e-1} \in A$ such that $a_0^{p^e} + b a_1^{p^e} + \cdots + b^{p^e-1} a_{p^e-1}^{p^e} = 0$, hence $G_{\mathrm{tr}}(L/K)$ is represented by

$$K[t_0, \ldots, t_{p^e-1}]/(t_0^{p^e} + b t_1^{p^e} + \cdots + b^{p^e-1} t_{p^e-1}^{p^e}).$$

Of course, it is desirable to replace G_{tr} with something smaller — a K-group scheme G of order $[L : K]$. Given such a G, in [**Cha76**, Theorem 5.3] it is shown that the truncated K-group scheme $G^{\#}$ which can be defined by

$$G^{\#}(A) = G(L \otimes A)$$

for A a truncated polynomial K-algebra, is isomorphic to $G_{\mathrm{tr}}(L/K)$ as K-group schemes. One sees a parallel to Hopf-Galois theory in the separable case, where any K-Hopf algebra H for which L/K is Hopf-Galois has the property $\widetilde{L} \otimes H \cong \widetilde{L}[\mathrm{Gal}(\widetilde{L}/K)]$ where $\widetilde{L}$ is the Galois closure.

EXAMPLE 10.23. Return to the situation in Example 10.22. Define the truncated polynomial K-group scheme G by

$$G(A) = \{\varphi \in G_{\mathrm{tr}}(L/K)(A) \mid \varphi(x \otimes 1) = x \otimes 1 + 1 \otimes c \text{ for some } c \in A \text{ with } c^{p^e} = 0\}.$$

Clearly, $G(A) \subset G_{\mathrm{tr}}(L/K)$. Let us denote the element of $G(A)$ sending $x \otimes 1$ to $x \otimes 1 + 1 \otimes c$ by φ_c. Then

$$\varphi_c \varphi_d(x \otimes 1) = \varphi_c(x \otimes 1 + 1 \otimes d) = x \otimes 1 + 1 \otimes c + 1 \otimes d = x \otimes 1 + 1 \otimes (c + d),$$

hence $\varphi_c \varphi_d = \varphi_{c+d}$. Then $G = \mathrm{Spec}(H)$, $H = K[t]/(t^{p^e})$, $\Delta(t) = t \otimes 1 + 1 \otimes t$ and G is the eth Frobenius kernel of the *additive group scheme* $\mathbb{G}_a$ (defined by $\mathbb{G}_a(A) =$

$(A,+))$. The action of G on L/K translates to a K-algebra map $\alpha : L \to L \otimes H$ given by $\alpha(x) = x \otimes 1 + 1 \otimes t$. Thus,

$$L^G = \{y \in L \mid \alpha(y) = y \otimes 1\} = K$$

as desired.

EXAMPLE 10.24. Let L/K be as above, only now we specialize to the case $e = 2$. For $c \in A$, $c^{p^2} = 0$, let $\varphi_c \in G_{\mathrm{tr}}(L/K)(A)$ be given by

$$\varphi_c(x \otimes 1) = x \otimes 1 + 1 \otimes c - \sum_{i=1}^{p-1} \frac{1}{i!(p-i)!} x^{pi} \otimes c^{p(p-i)}.$$

Note that $\varphi_c(x^p \otimes 1) = x^p \otimes 1 + 1 \otimes c^p$. Then, after considerable calculation, we get

$$\varphi_c \varphi_d = \varphi_a, \ a = c + d - \sum_{i=1}^{p-1} \frac{1}{i!(p-i)!} c^{pi} d^{p(p-i)},$$

so if we let $G'(A) = \{\varphi_a \mid a \in A, \ a^{p^2} = 0\}$ then $G'(A) = \mathrm{Spec}(H')$ with $H' = K[t]/(t^{p^2})$ and

$$\Delta(t) = t \otimes 1 + 1 \otimes t - \sum_{i=1}^{p-1} \frac{1}{i!(p-i)!} t^{pi} \otimes t^{p(p-i)}.$$

The coaction $\alpha' : L \to L \otimes H$ is given by

$$\alpha'(x) = x \otimes 1 + 1 \otimes t - \sum_{i=1}^{p-1} \frac{1}{i!(p-i)!} x^{pi} \otimes t^{p(p-i)}.$$

Notice that the G' constructed in Example 10.24 is different from the G' in Example 10.23 in the case $e = 2$. That an extension could be Hopf-Galois for multiple Hopf algebras was evident to Chase, who wrote in [**Cha76**] "Scrutiny of the simplest examples shows that a modular extension can be a PHS for many different truncated $[K]$-group schemes G". (We will explore this in greater detail in Section 10.3). In [**Cha76**, Theorem 5.3] he then shows that if G and G' are any two group schemes for which L is a PHS then $G^\# \cong (G')^\#$.

In fact, any finite modular extension L/K is Hopf-Galois for some Hopf algebra H. Since L/K is modular we can write $L = K(x_1, \ldots, x_n)$ with $x_i^{p^{e_i}} \in K, x_i^{p^{e_i - 1}} \notin K$, for each i. Let $H = K[t_1, \ldots, t_n]/(t_1^{p^{e_1}}, \ldots, t_n^{p^{e_n}})$, and make H into a Hopf algebra by letting each t_i be primitive. Then L/K is an H-Galois object via the map $x_i \mapsto x_i \otimes 1 + 1 \otimes x_i$, and hence L/K is an H^*-Galois extension.

Chase's theory can be related to the theory of derivations. The action of $G_{\mathrm{tr}}(L/K)$ on L/K induces a isomorphism of restricted Lie algebras $\mathrm{Lie}(G_{\mathrm{tr}}(L/K)) \to \mathcal{D}_K(L)$, and under this isomorphism the image of $\mathrm{Lie}(G_{\mathrm{tr}}(L/K_1/K))$ is $\mathcal{D}_{K_1}(L)$.

10.3. Hopf-Galois structures on modular extensions

The purpose of this section is to expand upon Chase's comment concerning the plethora of Hopf-Galois structures on a modular extension. Even the simplest example exhibits differences with the separable theory.

EXAMPLE 10.25. Let L/K be a purely inseparable extension of degree p^e, $p > 2$. Write $L = K[x]$ with $x^{p^e} = b \in K$, $x^{p^{e-1}} \notin K$. Let $H = K[t]/(t^{p^e})$ with t primitive and define a coaction θ_x of H on L via

$$\theta_x(x) = x \otimes 1 + 1 \otimes t$$

as Chase does in [**Cha76**, Proposition 5.2]. This makes L/K an H-Galois object, but the exact coaction depends on the choice of x. For example, let $y = x^s$, $p \nmid s$. Then $y^{p^e} = (y^{p^e})^s = b^s \in K$ and we can define

$$\theta_y(y) = y \otimes 1 + 1 \otimes t$$

to be the coaction which makes L/K and H-Galois object and L/K is H^*-Galois. Note that

$$\begin{aligned}\theta_x(y) &= (\theta_x(x))^s \\ &= \sum_{i=0}^{s} \binom{s}{i} x^i \otimes t^{s-i} \\ &= y \otimes 1 + 1 \otimes t + \sum_{i=1}^{s-1} \binom{s}{i} x^i \otimes t^{s-i} \\ &\neq \theta_y(y),\end{aligned}$$

hence there are an infinite number of Hopf-Galois structures on L/K all arising from the same Hopf algebra. While in the separable case a Hopf algebra can act in multiple ways on a field extension (see [**Chi13**], [**CRV15b**], and in particular [**KKTU19b**]), it follows from Greither-Pareigis theory that the total number of Hopf-Galois structures must be finite.

We will now show that modular extensions L/K of exponent $e > 1$ have an infinite number of non-isomorphic Hopf algebras H which make L/K Hopf-Galois. The construction, and further details, can be found in [**Koc14**]. As L is the tensor product of primitive extensions, it suffices to consider the case $L = K(x)$, $x^{p^e} = a \in K$, $x^{p^{e-1}} \notin K$. The description of the Hopf algebras is facilitated by a short discussion of Witt vectors.

For each $d \geq 0$ we define a polynomial w_d in $d+1$ indeterminates $Z_0, Z_1, \ldots, Z_d$ by

$$w_d(Z_0, Z_1, \ldots, Z_d) = Z_0^{p^d} + pZ_1^{p^{d-1}} + \cdots + p^d Z_d$$

which we call *Witt polynomials.* Using these we define, for each $d \geq 0$, a polynomial S_d in $2(d+1)$ indeterminates $X_0, X_1, \ldots, X_d; Y_0, Y_1, \ldots, Y_d$ recursively as follows:

$$w_d(S_0, \ldots, S_d) = w_d(X_0, \ldots, X_d) + w_d(Y_0, \ldots, Y_d).$$

For example, we have $S_0(X_0; Y_0) = X_0 + Y_0$, and

$$S_1((X_0, X_1); (Y_0, Y_1)) = X_1 + Y_1 - \sum_{i=1}^{p-1} \frac{(p-1)!}{i!(p-i)!} X_0^i Y_0^{p-i}.$$

The S_d, whose coefficients are all integers, are typically used in the description of Witt vectors–see, e.g., [**Jac75**].

Now pick $0 < r < e$ and $a \in K^\times$. Let $H = H_{e,r,a} = K[t]/(t^{p^e})$ with

$$\Delta(t) = S_d((a_d t^{p^{dr}} \otimes 1, \ldots, a_1 t^{p^r} \otimes 1, t \otimes 1); (1 \otimes a_d t^{p^{dr}}, \ldots, 1 \otimes a_1 t^{p^r}, 1 \otimes t))$$

where $d = \lceil n/r \rceil - 1$, and $\{a_i \mid i \in \mathbb{Z}^+\}$ are given recursively by

$$a_1 = a^{1/p} \in K^{1/p},\ a_{i+1} = a^{1/p} a_i^{p^r}.$$

Then H is a Hopf algebra which has a coaction on L via

$$\theta(x) = S_d((a_d x^{p^{dr}} \otimes 1, \ldots, a_1 x^{p^r} \otimes 1, x \otimes 1); (1 \otimes a_d t^{p^{dr}}, \ldots, 1 \otimes a_1 t^{p^r}, 1 \otimes t)).$$

(Given that the $a_i \notin K$ in general, it is not immediately clear that $\theta(x) \in L \otimes H$; this follows by [**Koc14**, Lemma 2.1].) This makes L/K an H-Galois object.

In the language of schemes, with $\mathcal{X} = \operatorname{Spec}(L)$ and $\mathcal{N} = \operatorname{Spec}(H)$ as before, the B-valued points of $\mathcal{X}$ (resp. $\mathcal{N}$) are the elements $b \in B$ with $b^{p^e} = a$ (resp. $b^{p^e} = 0$). The group operation is given by

$$\eta_1 *_{\mathcal{N}} \eta_2 = S_d((a_d \eta_1^{p^{dr}}, \ldots, a_1 \eta_1^{p^r}, \eta_1); (a_d \eta_2^{p^{dr}}, \ldots, a_1 \eta_2^{p^r}, \eta_2)),\ \eta_1, \eta_2 \in \mathcal{N}(B)$$

and the action on $\mathcal{X}(B)$ is

$$\eta[b] = S_d((a_d \eta^{p^{dr}}, \ldots, a_1 \eta^{p^r}, \eta); (a_d b^{p^{dr}}, \ldots, a_1 b^{p^r}, b)),\ \eta_1 \in \mathcal{N}(B),\ b \in \mathcal{X}(B).$$

EXAMPLE 10.26. Let $L = K[x]$ with $x^{p^2} \notin K$, $x^{p^3} = b \in K$. Here $e = 3$ and, along with the isomorphic Hopf algebras in Example 10.25 the construction above yields two families of Hopf-Galois extensions, parameterized by a fixed choice $a \in K^{1/p}$, corresponding to the two choices of $0 < r < 3$.

The case $r = 1$: Then $d = 1$ and we have $a_1 = a$. The comultiplication in the Hopf algebra $H_{3,1,a} = H[t]/(t^{p^3})$ is

$$\begin{aligned}\Delta(t) &= S_1((a^{1/p} t^{p^2} \otimes 1, t \otimes 1); (1 \otimes a^{1/p} t^{p^2})) \\ &= t \otimes 1 + 1 \otimes t - \sum_{i=1}^{p-1} a \frac{(p-1)!}{i!(p-i)!} t^{p^2 i} \otimes t^{p^2(p-i)}\end{aligned}$$

and the coaction $\theta : L \to L \otimes H$ is given by

$$\begin{aligned}\theta(x) &= S_1((a^{1/p} x^{p^2} \otimes 1, x \otimes 1); (1 \otimes a^{1/p} t^{p^2})) \\ &= x \otimes 1 + 1 \otimes t - \sum_{i=1}^{p-1} a \frac{(p-1)!}{i!(p-i)!} x^{p^2 i} \otimes t^{p^2(p-i)}.\end{aligned}$$

The case $r = 2$: Then $d = 0$ and, regardless of a, we get

$$\Delta(t) = t \otimes 1 + 1 \otimes t,\ \theta(x) = x \otimes 1 + 1 \otimes t.$$

Thus, we recover the 2-dimensional analogue of Example 10.25.

Different choices of a may give isomorphic Hopf algebras: by [**Koc14**, Proposition 3.1] we know $H_{e,r,a} \cong H_{e',r',a'}$ if and only if $e = e', r = r'$ and $a/a' \in K^{p^{e+1}-1}$. However, there remain an infinite number of choices of a which generate non-isomorphic Hopf algebras.

We conclude by remarking that [**Koc14**, Proposition 5.6] gives an explicit construction of $H^*_{e,e-1,a}$, and hence the Hopf algebras which act on L/K. For general r this is a messy computation.

Stark differences between the separable and purely inseparable case are visible here.

In the case L/K is separable and Hopf-Galois with Hopf algebra H, then $L \otimes L \cong L^n$, $n = [L : K]$ is a Hopf-Galois extension of $L \otimes K \cong L$ with Hopf algebra $L \otimes H$. But then $L \otimes H \cong L[N]$ for N some regular subgroup of $\operatorname{Perm}(G)$ necessarily normalized by G to allow for descent to L/K. One can see that base change to L simplifies matters here.

On the other hand, if L/K is purely inseparable and modular, hence Hopf-Galois for some Hopf algebra H, then $L \otimes L$ is a truncated polynomial L-algebra which is a Hopf-Galois extension over L with Hopf algebra $L \otimes H$, which is also a truncated polynomial L-algebra. If we instead consider the dual $(L \otimes H)^*$, then a coaction $L \otimes L \to (L \otimes L) \otimes (L \otimes H)^*$ is equivalent to endowing $L \otimes L$ with a comultiplication which makes $L \otimes L$ an L-Hopf algebra. The previous example (with L in place of K) demonstrates that one can often put an infinite number of Hopf algebra structures on an algebra, hence base changing to L does not effectively simplify the problem.

Part II: Hopf-Galois Module Theory

CHAPTER 11

Hopf-Galois module theory

This chapter surveys some of the applications of Hopf-Galois theory to questions of integral module structure in extensions of local or global fields, with particular emphasis on developments since the publication of [**Chi00**].

We begin by expanding a little on the summary of classical Galois module theory given in Chapter 1. If L/K is a G-Galois extension of fields then by the Normal Basis Theorem L is a free module of rank one over the group algebra $K[G]$. In the case that L/K is an extension of local or global fields we can investigate integral analogues of this result, with the most natural approach being to study the structure of the ring of integers (or valuation ring) $\mathfrak{O}_L$ as a module over the integral group ring $\mathfrak{O}_K[G]$. Noether's Theorem [**Noe32**] (see also [**Frö83**, Theorem 3]) states that $\mathfrak{O}_L$ is a locally free $\mathfrak{O}_K[G]$-module if and only if L/K is at most tamely ramified. In the local case this is the desired analogue of the Normal Basis Theorem; in the global case it is a necessary condition for $\mathfrak{O}_L$ to be a free $\mathfrak{O}_K[G]$-module.

The integral group ring $\mathfrak{O}_K[G]$ is an example of an $\mathfrak{O}_K$*-order* in $K[G]$: it is a finitely generated projective $\mathfrak{O}_K$-submodule of $K[G]$ that is closed under multiplication, and we have $K \otimes_{\mathfrak{O}_K} \mathfrak{O}_K[G] = K[G]$. One approach to studying wildly ramified extensions is to replace the integral group ring with a larger order in $K[G]$ called the *associated order* of $\mathfrak{O}_L$ in $K[G]$:

$$\mathfrak{A}_{K[G]} = \{z \in K[G] \mid z \cdot x \in \mathfrak{O}_L \text{ for all } x \in \mathfrak{O}_L\}.$$

By construction, this is the largest subring of $K[G]$ for which $\mathfrak{O}_L$ is a module. If $\mathfrak{O}_L$ is a free $\mathfrak{A}_{K[G]}$-module then it must have rank one, and $\mathfrak{A}_{K[G]}$ is the only $\mathfrak{O}_K$-order in $K[G]$ over which $\mathfrak{O}_L$ can possibly be free [**Chi00**, Proposition 12.5]. In general, we have $\mathfrak{O}_K[G] \subseteq \mathfrak{A}_{K[G]}$, and Noether's Theorem implies that we have equality if and only if L/K is at most tamely ramified. The use of associated orders was pioneered by Leopoldt [**Leo59**], who proved that if $L/\mathbb{Q}$ is an abelian extension then $\mathfrak{O}_L$ is a free $\mathfrak{A}_{\mathbb{Q}[G]}$-module. This theorem implies the corresponding result for abelian extensions of $\mathbb{Q}_p$, but it does not generalize easily to other base fields, or to nonabelian Galois groups.

Hopf-Galois theory broadens the scope of these investigations considerably. If L/K is an H-Galois extension of local or global fields then L is a free H-module of rank one [**Chi00**, Theorem 2.16], and we may study $\mathfrak{O}_L$ as a module over its associated order in H:

$$\mathfrak{A}_H = \{z \in H \mid z \cdot x \in \mathfrak{O}_L \text{ for all } x \in \mathfrak{O}_L\}.$$

If L/K admits a number of Hopf-Galois structures then we can compare the structure of $\mathfrak{O}_L$ as a module over its various associated orders. Some of the most striking results in this direction arise in the case in which L/K is already Galois in the classical sense. As mentioned in Chapter 1, the publication of [**Chi00**] was largely motivated by results of Byott [**Byo00**], who exhibited a family of G-Galois extensions of p-adic fields L/K for which $\mathfrak{O}_L$ is not a free $\mathfrak{A}_{K[G]}$-module but is a free $\mathfrak{A}_H$-module for some other Hopf algebra H giving a Hopf-Galois structure on the extension. Hopf-Galois theory can also provide contexts in which we can ask module-theoretic questions about the structure of rings of integers in extensions of local or global fields that are not Galois in the classical sense.

More generally, if L/K is an H-Galois extension then we can study the structure of each fractional ideal $\mathfrak{B}$ of L as a module over its associated order in H:

$$\mathfrak{A}_H(\mathfrak{B}) = \{z \in H \mid z \cdot x \in \mathfrak{B} \text{ for all } x \in \mathfrak{B}\}.$$

In order to streamline the exposition of the results in this chapter, we have chosen to focus on the study of the ring of integers $\mathfrak{O}_L$. A number of the results we present can be generalized to more general fractional ideals: we highlight where this is the case and refer the reader to the original papers for details of these generalizations.

In Section 11.1 we study generalized normal basis generators in Hopf-Galois extensions. We then turn to the study of associated orders in Hopf-Galois extensions. Section 11.2 concerns Childs' theorem, which is the flagship result in this area: if L/K is a finite H-Galois extension of p-adic fields and $\mathfrak{A}_H$ is a Hopf order in H then $\mathfrak{O}_L$ is a free $\mathfrak{A}_H$-module. We summarize results concerning the application of this result to questions of integral Hopf-Galois module structure in Galois extensions of p-adic fields of degree p or p^2. We then present some tools for studying associated orders that are not Hopf, including subextension techniques based upon the results of Section 8.1. In Section 11.5 we focus on extensions of p-adic fields that are at most tamely ramified: the evidence to date leads us to conjecture that for these extensions each fractional ideal is free over its associated order in every Hopf-Galois structure admitted by the extension. Finally, in Section 11.6 we survey the application of Hopf-Galois theory to questions of integral module structure in extensions of global fields, including some non-normal extensions.

We will always assume that L/K denotes a finite separable extension of fields with Galois closure $\widetilde{L}$, and write $G = \mathrm{Gal}(\widetilde{L}/K)$, $G_L = \mathrm{Gal}(\widetilde{L}/L)$, and $X = G/G_L$. In particular: if L/K is a Galois extension then $G = \mathrm{Gal}(L/K)$, $G_L = \{1\}$, and $X = G$.

11.1. The Normal Basis Theorem for Hopf-Galois structures

The Hopf-Galois analogue of the Normal Basis Theorem is that if L/K is H-Galois then L is a free H-module of rank one [**Chi00**, Theorem 2.16]. An element $x \in L$ with the property that $L = H \cdot x$ is necessarily a free generator of L as an H-module: we shall call such an element a *normal basis generator for L with respect to H*.

11.1.1. Normal basis generators for separable Hopf-Galois extensions. By Greither-Pareigis theory each Hopf algebra giving a Hopf-Galois structure on L/K has the form $H = \widetilde{L}[N]^G$ for N some regular subgroup of $\mathrm{Perm}(X)$ normalized by the image of G under the left translation map $\lambda : G \to \mathrm{Perm}(G)$. In this subsection we give a concrete criterion, in terms of G and N, for an element $x \in L$ to be a normal basis generator for L with respect to H. These results are an amalgamation of results in [**Tru16a**] and [**Tru18b**].

Recall from [**GP87**] that the action of $\widetilde{L}[N]^G$ on L is obtained via descent from the action of $\widetilde{L}[N]$ on $M = \mathrm{Map}(X, \widetilde{L})$. The $\widetilde{L}$-algebra M has a basis of mutually orthogonal idempotents $\{u_{\overline{g}} \mid \overline{g} \in X\}$, where $\overline{g}$ denotes the left coset $gG_L \in X$ and

$$u_{\overline{g}}(\overline{h}) = \delta_{\overline{g},\overline{h}} \text{ for all } \overline{g}, \overline{h} \in X.$$

The group N permutes these idempotents by acting on the subscripts:

$$\eta \cdot u_{\overline{g}} = u_{\eta(\overline{g})} \text{ for all } \eta \in N \text{ and } \overline{g} \in X,$$

and extending this action to $\widetilde{L}[N]$ makes M into an $\widetilde{L}[N]$-Galois extension of $\widetilde{L}$. By Galois descent we obtain that $\widetilde{L}[N]^G$ is a K-Hopf algebra and that M^G is an $\widetilde{L}[N]^G$-Galois extension of K. Finally, we may identify L with the fixed ring M^G via the K-algebra isomorphism $L \xrightarrow{\sim} M^G$ defined by

$$x \mapsto f_x = \sum_{\overline{g} \in X} \overline{g}(x) u_{\overline{g}} \text{ for all } x \in L.$$

LEMMA 11.1. *An element $f \in M^G$ is a free $\widetilde{L}[N]^G$-generator of M^G if and only if it is a free $\widetilde{L}[N]$-generator of M.*

PROOF. If $M^G = \widetilde{L}[N]^G \cdot f$ for some $f \in M^G$ then $\widetilde{L} \otimes_K M^G = \widetilde{L} \otimes_K (\widetilde{L}[N]^G \cdot f)$. By Galois descent, we have $\widetilde{L} \otimes_K M^G = M$ and $\widetilde{L} \otimes_K \widetilde{L}[N]^G = \widetilde{L}[N]$, so we obtain $M = \widetilde{L}[N] \cdot f$. Conversely, if $M = \widetilde{L}[N] \cdot f$ for some $f \in M^G$ then we have $M^G = (\widetilde{L}[N] \cdot f)^G = \widetilde{L}[N]^G \cdot f$. □

LEMMA 11.2. *For $x \in L$, the element f_x is an $\widetilde{L}[N]$-generator of M if and only if the matrix*

$$T_N(x) = (\eta(\overline{g})[x]) \text{ (where } \eta \in N \text{ and } \overline{g} \in X\text{)}$$

is nonsingular.

Strictly speaking, the definition of $T_N(x)$ depends upon orderings of the group N and the set X. However, the question of whether it is nonsingular does not.

PROOF. The set $\{u_{\overline{g}} \mid \overline{g} \in X\}$ is an $\widetilde{L}$-basis of M. For $x \in L$ and $\eta \in N$, we have

$$\begin{aligned} \eta \cdot f_x &= \eta \cdot \left(\sum_{\overline{g} \in X} \overline{g}(x) u_{\overline{g}} \right) \\ &= \sum_{\overline{g} \in X} \overline{g}(x) u_{\eta(\overline{g})} \\ &= \sum_{\overline{g} \in X} \eta^{-1}(\overline{g})[x] u_{\overline{g}}. \end{aligned}$$

Therefore the transition matrix from the set $\{u_{\overline{g}} \mid \overline{g} \in X\}$ to the set $\{\eta \cdot f_x \mid \eta \in N\}$ is row equivalent to the matrix $T_N(x)$ defined above, and so f_x is an $\widetilde{L}[N]$-generator of M if and only if this matrix is nonsingular. □

Combining Lemmas 11.1 and 11.2 we obtain:

PROPOSITION 11.3. *An element $x \in L$ is a normal basis generator of L with respect to H if and only if the matrix*

$$T_N(x) = (\eta(\overline{g})[x]) \text{ (where } \eta \in N \text{ and } \overline{g} \in X)$$

is nonsingular.

This result is useful for relating normal basis generators with respect to different Hopf-Galois structures on a given extension: see for example Proposition 11.15.

11.1.2. Valuation criteria for normal basis generators. For extensions of local fields L/K it is sometimes possible to detect normal basis generators in terms of the normalized valuation $v_L : L \to \mathbb{Z} \cup \{\infty\}$. We say that an H-Galois extension of local fields L/K has a *valuation criterion with respect to H* if there is an integer b such that every element $x \in L$ satisfying $v_L(x) = b$ is a normal basis generator of L with respect to H. If such an integer b exists, then all integers that are congruent to b modulo the ramification index of L/K satisfy the same property. Valuation criteria for normal basis generators are an essential component in the development of *scaffolds*: see Chapter 15

The first results on valuation criteria for normal basis generators in Galois extensions of local fields were due to Byott and Elder [**BE07**], who studied elementary abelian totally ramified p-extensions of p-adic fields. In [**Tho08**] Thomas proved an analogous result for arbitrary totally ramified abelian p-extensions of local fields in characteristic p. In both of these cases, the valuation criterion was described in terms of the largest ramification break associated with L/K (see [**Ser62**] or [**Chi00**, §23]). In [**Eld10**] Elder reformulated the existing results on the valuation criterion in terms of the valuation of the relative different $\mathcal{D}_{L/K}$ and resolved the question for local fields of characteristic p:

THEOREM 11.4. *Let L/K be a Galois extension of complete local fields of characteristic p with perfect residue fields. If L/K is a totally ramified p-extension then L/K has a valuation criterion with respect to $K[G]$, with $b \equiv -\mathcal{D}_{L/K} - 1 \pmod{[L : K]}$, and given $i \not\equiv b \pmod{[L : K]}$ there exists $x_i \in L$ that satisfies $v_L(x_i) = i$ but is not a normal basis generator for L with respect to $K[G]$.*

If L/K is not a totally ramified p-extension then given any $i \in \mathbb{Z}$ there exists $x_i \in L$ that satisfies $v_L(x_i) = i$ but is not a normal basis generator for L with respect to $K[G]$.

The most comprehensive results on the Galois case are due to de Smit, Florence, and Thomas [**dSFT12**]. They note that a necessary condition for an element $x \in L$ to be a normal basis generator for L with respect to $K[G]$ is that $\mathrm{Tr}_{L/K}(x) \neq 0$, and that every element of L of valuation b satisfies $\mathrm{Tr}_{L/K}(x) \neq 0$ if and only if L/K is totally ramified and $b \equiv -\mathcal{D}_{L/K} - 1$ [**dSFT12**, Proposition 1.1]. Using this, they prove [**dSFT12**, Proposition 1.2]:

PROPOSITION 11.5. *A Galois extension of local fields L/K with residue characteristic p has a valuation criterion for normal basis generators with respect to $K[G]$ if and only if*

(1) *L/K is a totally ramified p-extension;*
(2) *Every non-zero $K[G]$-submodule of L contains an element of valuation 0.*

In the case that K has characteristic p condition (1) of Proposition 11.5 implies condition (2), and so Proposition 11.5 implies Theorem 11.4 above. In the case that

K has characteristic 0, de Smit, Florence, and Thomas identify precisely the abelian extensions of K that have a valuation criterion for normal basis generators with respect to $K[G]$. If L/K is a totally ramified Kummer p-extension whose Galois group G has exponent $m = p^s$ then this occurs if and only if L is obtained by adjoining m-th roots of units of $\mathfrak{O}_K$; in this case we say that L is a *unit root Kummer extension* of K. More generally [**dSFT12**, Theorem 1.4]:

THEOREM 11.6. *Let L/K be a totally ramified abelian extension of local fields whose degree is a power of the residue characteristic p, let $m = p^s$ be the exponent of G, and let $r \mid m$ be the number of m-th roots of unity inside K. If $p = 2$ and $8 \mid m$, assume that $r \neq 2$. Then L/K has a valuation criterion for normal basis generators with respect to $K[G]$ if and only if every cyclic subextension F/K of L/K of degree r is a unit root Kummer extension.*

In [**Byo11**] Byott generalized the results of [**Eld10**] to give a Hopf-Galois valuation criterion in characteristic p.

THEOREM 11.7. *Let L/K be a separable H-Galois extension of local fields of characteristic p. If L/K is a totally ramified p-extension then L/K has a valuation criterion with respect to H, with $b \equiv -\mathcal{D}_{L/K} - 1 \pmod{[L:K]}$, and given $i \not\equiv b \pmod{[L:K]}$ there exists $x_i \in L$ that satisfies $v_L(x_i) = i$ but is not a normal basis generator for L/K with respect to H.*

If L/K is not a totally ramified p-extension then given any $i \in \mathbb{Z}$ there exists $x_i \in L$ that satisfies $v_L(x_i) = i$ but is not a normal basis generator for L/K with respect to H.

In the case that L/K is a totally ramified p-extension the strategy of the proof is to show that an element $t \in L$ is a normal basis generator for L with respect to H if and only if $\mathrm{Tr}_{L/K}(x) \neq 0$ [**Byo11**, Lemma 2.3]; a consequence this is that in this case the set of normal basis generators is the same for all Hopf-Galois structures on L/K.

At the time of writing there is no known analogue of Theorem 11.7 for Hopf-Galois extensions of p-adic fields.

11.2. Hopf orders and Childs' theorem

A *Hopf order* in a K-Hopf algebra H is an $\mathfrak{O}_K$-order in H which is also an $\mathfrak{O}_K$-Hopf algebra with operations inherited from those on H [**Chi00**, Definition 5.1]. The question of determining all Hopf orders in a given Hopf algebra (in particular, in a given group algebra) is a topic of long-standing interest: Chapter 12 of this book surveys some of the extensive work in this area.

The importance of Hopf orders in Hopf-Galois module theory is due to Childs' Theorem:

THEOREM 11.8 (Childs). *Let L/K be an H-Galois extension of p-adic fields and suppose that $\mathfrak{A}_H$ is a Hopf order in H. Then $\mathfrak{O}_L$ is a free $\mathfrak{A}_H$-module.*

Childs' Theorem is proved in detail in [**Chi00**, Chapter 3]; we summarize the proof below.

To help illuminate this summary, we consider first a Galois extension of p-adic fields L/K that is at most tamely ramified, and reinterpret one half of the standard modern proof of Noether's Theorem [**Frö83**, Theorem 3] using the fact that $\mathfrak{O}_K[G]$ is a Hopf order in $K[G]$ [**Chi00**, §5]. Since L/K is at most tamely ramified there exists an element $t \in \mathfrak{O}_L$ such that $\mathrm{Tr}_{L/K}(t) = 1$; we use this element to define a map $\nu : \mathfrak{O}_L \to \mathfrak{O}_K[G] \otimes_{\mathfrak{O}_K} \mathfrak{O}_L$ by

$$\nu(x) = \sum_{g \in G} g \otimes tg^{-1}(x),$$

and show that ν is a section to the $\mathfrak{O}_K[G]$-linear multiplication map $\mu : \mathfrak{O}_K[G] \otimes_{\mathfrak{O}_K} \mathfrak{O}_L \to \mathfrak{O}_L$. Hence $\mathfrak{O}_L$ is a projective $\mathfrak{O}_K[G]$-module, and since $K \otimes_{\mathfrak{O}_K} \mathfrak{O}_L = L$ is a free $K \otimes_{\mathfrak{O}_K} \mathfrak{O}_K[G] = K[G]$-module, a theorem of Swan [**Swa60**] implies that $\mathfrak{O}_L$ is a free $\mathfrak{O}_K[G]$-module.

In the language of Hopf algebras, the element t satisfies $\theta \cdot t = 1$, where $\theta = \sum_{g \in G} g$ is a left integral of $\mathfrak{O}_K[G]$, and since $\Delta(g) = g \otimes g$ and $S(g) = g^{-1}$ for all $g \in G$ the definition of the map ν can be rewritten in Sweedler notation as

$$\nu(x) = \sum_{(\theta)} \theta_{(1)} \otimes tS(\theta_{(2)})(x).$$

With this formulation in mind, we return to the case in which L/K is an H-Galois extension of p-adic fields (not necessarily Galois or tamely ramified).

SKETCH PROOF OF CHILDS' THEOREM 11.8. The assumption that $\mathfrak{A}_H$ is a Hopf order in H implies that $\mathfrak{O}_L$ is an $\mathfrak{A}_H$*-tame extension of* $\mathfrak{O}_K$. That is: $\mathfrak{O}_L$ is a faithful $\mathfrak{A}_H$-module, $\mathfrak{A}_H$ has the same $\mathfrak{O}_K$-rank as $\mathfrak{O}_L$, and there exists a left integral $\theta \in \mathfrak{A}_H$ and an element $t \in \mathfrak{O}_L$ such that $\theta \cdot t = 1$ [**Chi00**, Theorem 13.3]. Define a map $\nu : \mathfrak{O}_L \to \mathfrak{A}_H \otimes_{\mathfrak{O}_K} \mathfrak{O}_L$ by

$$\nu(x) = \sum_{(\theta)} \theta_{(1)} \otimes tS(\theta_{(2)})(x).$$

Then ν is a section to the $\mathfrak{A}_H$-module multiplication map $\mu : \mathfrak{A}_H \otimes_{\mathfrak{O}_K} \mathfrak{O}_L \to \mathfrak{O}_L$ [**Chi00**, Theorem 13.5], and so $\mathfrak{O}_L$ is a projective $\mathfrak{A}_H$-module. To complete the proof, we appeal to a generalization of Swan's Theorem due to Schneider [**Sch77**]: since $\mathfrak{O}_L$ is a projective $\mathfrak{A}_H$-module and $K \otimes_{\mathfrak{O}_K} \mathfrak{O}_L$ is a free $K \otimes_{\mathfrak{O}_K} \mathfrak{A}_H$-module, $\mathfrak{O}_L$ is a free $\mathfrak{A}_H$-module. □

We remark that if $\mathfrak{O}_L$ is an $\mathfrak{A}$-tame extension of $\mathfrak{O}_K$ for some Hopf order $\mathfrak{A}$ in H, then the argument above shows that $\mathfrak{O}_L$ is a free $\mathfrak{A}$-module, and so necessarily $\mathfrak{A} = \mathfrak{A}_H$: see [**Chi00**, Proposition 12.5].

If $\mathfrak{A}_H$ is a Hopf order in H then we might study the more stringent question of whether $\mathfrak{O}_L$ is an $\mathfrak{A}_H$-Galois extension of $\mathfrak{O}_K$. The relationship between the notions of $\mathfrak{A}_H$-Galois, $\mathfrak{A}_H$-free, and $\mathfrak{A}_H$-tame is explored in [**Chi00**, §14]. In particular, if $\mathfrak{A}_H$ is a local Hopf order in H then these three conditions are equivalent, by work of Childs and Hurley: see [**Chi00**, Theorem 14.7] or [**CH86**]. Furthermore, in this case an element $t \in \mathfrak{O}_L$ is a free generator of $\mathfrak{O}_L$ as an $\mathfrak{A}_H$-module if and only if there exists a left integral θ of $\mathfrak{A}_H$ such that $\theta \cdot t = 1$; this idea has been exploited in, for example, [**Tru12**],[**Tru16b**], and [**Tay20**].

Conversely, given a Hopf algebra H we might seek to determine the Hopf orders $\mathfrak{H} \subset H$ that are *realizable* in the sense that there exists an H-Galois extension L/K such that $\mathfrak{A}_H = \mathfrak{H}$ or, more stringently, such that $\mathfrak{O}_L$ is an $\mathfrak{H}$-Galois extension of $\mathfrak{O}_K$. Chapter 12 contains a thorough discussion of these ideas in the case that H is a group algebra; we highlight some of these results below, along with some examples for Hopf algebras that are not group algebras.

The converse of Child's Theorem is false: there are many examples of H-Galois extension of p-adic fields for which $\mathfrak{O}_L$ is a free $\mathfrak{A}_H$-module but $\mathfrak{A}_H$ is not a Hopf order in H. Elsewhere in this chapter we present tools for analyzing rings of integers, and making comparisons between different Hopf-Galois structures on a given extension, in these cases.

11.2.1. Galois extensions of degree p. A Galois extension L/K of p-adic fields of degree p admits only the classical Hopf-Galois structure, given by $K[G]$. The classification of the $\mathfrak{O}_K$-Hopf orders in $K[G]$ is known: see [**TO70**], [**Chi00**, Chapter 4], or Theorem 12.42 of this book. Writing $e = v_K(p)$, $e' = e/(p-1)$, and $G = \langle g \rangle$, the Hopf orders in $K[G]$ are the *Larson orders*

$$\mathfrak{H}_i = \mathfrak{O}_K \left[\frac{g-1}{\pi_K^i} \right] \text{ for } 0 \leq i \leq e'.$$

The extension L/K is either unramified or totally ramified, and has a single ramification break b.

First suppose that L/K is unramified. Then $b = -1$, and Noether's Theorem implies that $\mathfrak{A}_{K[G]} = \mathfrak{H}_0 = \mathfrak{O}_K[G]$ and that $\mathfrak{O}_L$ is a free $\mathfrak{O}_K[G]$-module. In fact, in this case $\mathfrak{O}_L/\mathfrak{O}_K$ is a Galois extension of rings in the sense of [**CHR65**], and so $\mathfrak{O}_L$ is an $\mathfrak{O}_K[G]$-Galois extension of $\mathfrak{O}_K$.

Now suppose that L/K is totally ramified. Then either $b < pe'$ and p does not divide b, or $b = pe'$.

The question of whether $\mathfrak{O}_L$ is an $\mathfrak{A}_{K[G]}$-Galois extension of $\mathfrak{O}_K$ was addressed in [**Chi87**] (see also [**Chi00**, Theorem 24.1, Corollary 24.15, and Proposition 30.7]).

THEOREM 11.9. *The associated order $\mathfrak{A}_{K[G]}$ is a Hopf order in $K[G]$ if and only if $b \equiv -1 \pmod p$ or $b = pe'$.*

- *If $b \equiv -1 \pmod p$ then, writing $b = pi - 1$ for some $i \leq e'$, we have $\mathfrak{A}_{K[G]} = \mathfrak{H}_i$. In this case $\mathfrak{O}_L$ is an $\mathfrak{A}_{K[G]}$-Galois extension of $\mathfrak{O}_K$.*
- *If $b = pe'$ then $\mathfrak{A}_{K[G]} = \mathfrak{H}_{e'}$. In this case $\mathfrak{O}_L$ is a free $\mathfrak{A}_{K[G]}$-module, but not an $\mathfrak{A}_{K[G]}$-Galois extension of $\mathfrak{O}_K$.*

It can be shown (see Subsection 12.11.1) that each Hopf order $\mathfrak{H}_i \subset K[G]$ is realizable in the strong sense: there exists a Galois extension L/K such that $\mathfrak{O}_L$ is an $\mathfrak{H}_i$-Galois extension of $\mathfrak{O}_K$.

More generally, the question of whether $\mathfrak{O}_L$ is free over $\mathfrak{A}_{K[G]}$ was resolved by Ferton [**Fer74**]:

THEOREM 11.10. *Write $b = qp + r$ with $0 \leq r < p$.*

- *If $b < pe' - 1$ then $\mathfrak{O}_L$ is a free $\mathfrak{A}_{K[G]}$-module if and only if r divides $p-1$.*
- *If $pe' - 1 \leq b \leq pe'$ then $\mathfrak{O}_L$ is a free $\mathfrak{A}_{K[G]}$-module if and only if the period of the continued fraction expansion of b/p is at most 4.*

A natural class of extensions to which the results of this subsection apply are Kummer extensions of p-adic fields of degree p: a detailed analysis of the ramification of these is given in [**Chi00**, §24].

11.2.2. Galois extensions of degree p^2. In [**Byo00**] Byott exhibited examples of Galois extensions of p-adic fields L/K for which $\mathfrak{O}_L$ is not a free $\mathfrak{A}_{K[G]}$-module, but is a free $\mathfrak{A}_H$-module for some Hopf algebra H giving a nonclassical Hopf-Galois structure on the extension; the climax of [**Chi00**] is a presentation and discussion of these results. In these examples $\mathfrak{A}_H$ is a Hopf order, but $\mathfrak{A}_{K[G]}$ is not (we reiterate, however, that the failure of $\mathfrak{A}_{K[G]}$ to be a Hopf order does not in general imply that $\mathfrak{O}_L$ is not a free $\mathfrak{A}_{K[G]}$-module). Byott's construction employs the theory of Lubin-Tate formal groups, and the smallest examples produced are elementary abelian extensions of degree p^2.

In this subsection, we assume that K is a p-adic field containing a primitive p-th root of unity ζ, and consider totally ramified Galois extensions L/K of degree p^2. Building upon a partial result in [**Chi96**], in [**Byo02**] Byott analyzed in detail the Hopf-Galois module structure of such extensions. We summarize this analysis in the case that that p is odd, referring to [**Byo02**] for details of the exceptional case $p = 2$. As above, let $e = v_K(p)$ and $e' = e/(p-1)$; since $\zeta \in K$, we have $e' \in \mathbb{Z}$.

The extension L/K admits p Hopf-Galois structures if its Galois group G is cyclic, and p^2 if G is elementary abelian [**Byo02**, Theorem 2.5]. The nonclassical Hopf-Galois structures correspond bijectively with pairs (T, d), where T is a subgroup of G of order p and $d \in \{1, \ldots, p-1\}$ is an integer; it is possible to include the classical Hopf-Galois structure in this classification by allowing $d = 0$, which gives the classical Hopf-Galois structure regardless of the choice of T.

A pair (T, d) determines permutations $\rho, \eta \in \mathrm{Perm}(G)$ and a subgroup $N_{T,d} = \langle \rho, \eta \rangle$ that is regular and normalized by $\lambda(G)$. A consequence of assuming p to be odd is that $N_{T,d} \cong G$; describing it in terms of two generators allows for simultaneous treatment of the cyclic and elementary abelian cases. The Hopf algebra giving the Hopf-Galois structure corresponding to (T, d) is $H_{T,d} = L[N_{T,d}]^G$.

Following [**Byo02**, Lemma 2.10] we can give a simple description of $H_{T,d}$ as a K-algebra. First note that the K-algebra $K[\rho]$ has a basis of mutually orthogonal idempotents

$$e_s = \frac{1}{p} \sum_{k=0}^{p-1} \zeta^{-ks} \rho^k \text{ for } 0 \leq s \leq p-1.$$

Next let $M = L^T$, write $M = K(v)$ for some element v satisfying $v^p \in K$ and a further condition depending on d (see equation (2.9) in loc. cit), and define

$$a_v = \sum_{s=0}^{p-1} v^s e_s.$$

Then

$$H_{T,d} = K[\rho, a_v \eta].$$

With this notation to hand, we can present Byott's description of the Hopf orders in $H_{T,d}$. The Hopf orders in $K[G]$ were classified by Byott [**Byo93**], Greither [**Gre92**], and Underwood [**Und94**] (see Section 12.6), and the construction of *Greither orders* (see Subsection 12.6.2) was extended to nonclassical Hopf-Galois structures of type C_{p^2} by Childs [**Chi96**]. In [**Byo02**, Theorem 4.5] it is shown that the Hopf orders $\mathfrak{H}$ in $H_{T,d}$ are the orders

$$\mathfrak{H}_{i,j,v} = \mathfrak{O}_K \left[\frac{\rho - 1}{\pi_K^i}, \frac{a_v \eta - 1}{\pi_K^j} \right],$$

where $0 \leq i, j \leq e'$ and v is required to satisfy certain congruence conditions, depending upon whether $N_{T,d}$ is cyclic or elementary abelian, and upon the relationship between the values of i and j. See loc. cit. for details.

Now let $b_1 \leq b_2$ be the ramification breaks of L/K. A necessary condition for $\mathfrak{O}_L$ to be a Hopf-Galois extension of $\mathfrak{O}_K$ with respect to the Hopf-Galois structure given by $H_{T,d}$ is that $\mathfrak{O}_L$ should be a Hopf-Galois extension of $\mathfrak{O}_M$ with respect to the unique classical Hopf-Galois structure on L/M and that $\mathfrak{O}_M$ should be a Hopf-Galois extension of $\mathfrak{O}_K$ with respect to the unique classical Hopf-Galois structure on M/K. Applying the results of the previous subsection, we see that we require

$b_1 = pj - 1$ and $b_2 = p^2 i - 1$ for some i, j satisfying $0 \leq i, j \leq e'$, and $j \leq pi$ [**Byo02**, Lemma 5.3].

To obtain a criterion we write $L = M(x)$ for a certain $x \in L$ with $x^p \in M$; studying the Galois action on x yields an element $\beta \in M$ satisfying certain congruence conditions [**Byo02**, Lemma 5.5]. We then have [**Byo02**, Lemma 6.1]:

THEOREM 11.11. *Suppose that $b_1 = pj - 1$ and $b_2 = p^2 i - 1$ with $j \leq pi$. Then $\mathfrak{O}_L$ is an $\mathfrak{H}_{i,j,v}$-Galois extension of $\mathfrak{O}_K$ if and only if $v\beta \equiv 1 \pmod{\pi_K^{e'-i+j}\mathfrak{O}_M}$.*

The interplay between the values of i, j, and d is studied in [**Byo02**, §8]. In particular, if $\mathfrak{O}_L$ is an $\mathfrak{H}_{i,j,v}$-Galois extension of $\mathfrak{O}_K$ then either $p \mid j$ or $j \equiv 1 \pmod{p}$. In the square $0 \leq i, j \leq e'$, the pairs (i, j) where the associated order in the classical Hopf Galois structure (given by $K[G]$) is a Hopf order take up half of the possible points. By contrast, pairs (i, j) corresponding to the cases in which the associated order in $K[G]$ is not a Hopf order but the associated order in some non-classical Hopf Galois structure is a Hopf order consist of a single point together with points on a single line segment. Thus the phenomenon of [**Byo00**], in which a nonclassical Hopf-Galois structure gives a "better" description of $\mathfrak{O}_L$ than the classical Hopf-Galois structure, is rare.

Since L/K admits multiple Hopf-Galois structures, we can make comparisons between the description of $\mathfrak{O}_L$ afforded by each of them. From [**Byo02**, Lemma 9.1] we have

LEMMA 11.12. *Suppose that $b_1 = pj - 1$ and $b_2 = p^2 i - 1$ with $j \leq pi$ and that $\mathfrak{O}_L$ is a Hopf-Galois extension of $\mathfrak{O}_K$ with respect to $H_{T,d}$.*

- *If $i \geq 2j$ then $\mathfrak{O}_L$ is a Hopf-Galois extension of $\mathfrak{O}_K$ with respect to $H_{T,d'}$ for all $d' \in 0, 1, \ldots, p-1$.*
- *If $i < 2j$ then $\mathfrak{O}_L$ is a Hopf-Galois extension of $\mathfrak{O}_K$ with respect to a unique Hopf-Galois structure on L/K.*

The complete description of the behaviour of $\mathfrak{O}_L$ with respect to the various Hopf-Galois structures admitted by L/K is given in [**Byo02**, §9].

Finally: in contrast to the degree p case, not every Hopf order in each Hopf algebra $H_{T,d}$ is realizable. For the case that G is cyclic, questions of realizability for Greither orders in $K[G]$ are studied in Subsection 12.11.2. From [**Byo02**, Theorem 7.1] we have

THEOREM 11.13. *Let M be an extension of K of degree p with ramification break $b = pj - 1$, let i be an integer with $j \leq pi$, and let $\mathfrak{H}_{i,j,v}$ be a Hopf order as described above, with $v \in M$. Then there exists a totally ramified Galois extension L/K of degree p^2 such that $M \subset L$ and $\mathfrak{O}_L$ is $\mathfrak{H}_{i,j,v}$-Galois if and only if $v_K(v^p - 1) = p(e' - i) + j$.*

The more general question of determining criteria for $\mathfrak{O}_L$ to be free over its associated order in $K[G]$ is not well understood; the analogous question for non-classical Hopf-Galois structures admitted by L/K even less so. In general, the ramification breaks b_1, b_2 do not convey sufficient information.

If L/K is a cyclic Kummer extension, $b_1 < pe'$, and $p \nmid b_1$, then work of Miyata [**Miy98**] and Byott [**Byo08**] implies that $\mathfrak{O}_L$ is a free $\mathfrak{A}_{K[G]}$ module if and only if the principal remainder of b_1 modulo p^2 divides $p^2 - 1$, generalizing the first part of Theorem 11.10. If G is cyclic, $p \nmid e$, and $b_i \geq p^i e' - 1$ for each i then criteria for $\mathfrak{O}_L$ to be a free $\mathfrak{A}_{K[G]}$-module follow from work of Bertrandias [**Ber79**], generalizing the second part of Theorem 11.10. In fact, the cited works of Byott and Bertrandias study cyclic extension of arbitrary prime power degree. The elementary abelian case was studied by Miyata in [**Miy07**].

The study of the integral Galois module structure of totally ramified Galois extensions L/K of degree p^2, where L, K are local fields of characteristic p, played a large part in the creation and evolution of *Galois scaffolds*: see [**Eld09**], [**BE13**], [**BE14**], for example. Galois scaffolds are now just one part of a more general theory of *scaffolds*, another being the theory of *Hopf-Galois scaffolds*: see Chapter 15 of this book.

11.3. Associated orders for opposite Hopf-Galois structures

In this section we suppose that L/K is a separable extension of local or global fields. We have seen in Chapter 8 that if N is a regular subgroup of $\mathrm{Perm}(X)$ that is normalized by $\lambda(G)$ then so too is the opposite subgroup $N^{opp} = \mathrm{Cent}_{\mathrm{Perm}(X)}(N)$. Both of these subgroups correspond to Hopf-Galois structures on L/K, with Hopf algebras $H = \widetilde{L}[N]^G$ and $H^{opp} = \widetilde{L}[N^{opp}]^G$ respectively. We call the Hopf-Galois structure given by H^{opp} on L/K *opposite* of the one given by H.

In this section we present results from [**Tru16a**] and [**Tru18b**] concerning questions of integral module structure in L with respect to H and H^{opp}. The results there apply to any fractional ideal of L; the arguments presented here generalize, mutatis mutandis, to that case. The main theorem is the following [**Tru18b**, Theorem 1.2]:

THEOREM 11.14. *The ring of integers $\mathfrak{O}_L$ is a free $\mathfrak{A}_H$-module if and only if it is a free $\mathfrak{A}_{H^{opp}}$-module.*

The first step is to use Theorem 11.3 to show that the normal basis generators with respect to H and H^{opp} are the same:

PROPOSITION 11.15. *An element $x \in L$ generates L as a free H-module if and only if it generates L as a free H^{opp}-module.*

PROOF. By Lemmas 11.1 and 11.2 it is sufficient to show that $T_N(x)$ is nonsingular if and only if $T_{N^{opp}}(x)$ is nonsingular. Enumerating the elements of X as $\overline{g_j}$ we define $\eta_j \in N$ to be the unique element such that $\eta_j(\overline{g_1}) = \overline{g_j}$ and define $\eta'_j \in N^{opp}$ analogously. We then have:

$$\begin{aligned}
\det(T_N(x)) &= \det\left(\eta_i(\overline{g_j})[x]\right) \\
&= \det\left((\eta_i(\eta'_j(\overline{g_1}))[x])\right) \\
&= \det\left((\eta'_j(\eta_i(\overline{g_1}))[x])\right) \text{ since } N, N^{opp} \text{ commute inside } \mathrm{Perm}(X) \\
&= \det\left((\eta'_j(\overline{g_i})[x])\right) \\
&= \det(T_{N^{opp}}(x)^T)
\end{aligned}$$

Therefore $\det(T_N(x)) = \det(T_{N^{opp}}(x))$, and so $T_N(x)$ is nonsingular if and only if $T_{N^{opp}}(x)$ is nonsingular, which completes the proof. $\square$

Next we note that the fact that N and N^{opp} commute inside $\mathrm{Perm}(X)$ is reflected in the actions of H and H^{opp} on L. Recall that the actions of H and H^{opp} on L are obtained via Galois descent from the actions of $\widetilde{L}[N]$ and $\widetilde{L}[N^{opp}]$ on $\mathrm{Map}(X, \widetilde{L})$ (see discussion at the start of Subsection 11.1.1). Since N and N^{opp} commute inside $\mathrm{Perm}(X)$ we have

$$z \cdot (z' \cdot f) = z' \cdot (z \cdot f) \text{ for all } z \in \widetilde{L}[N],\ z' \in \widetilde{L}[N^{opp}],\ f \in \mathrm{Map}(X, \widetilde{L}),$$

and so

$$z \cdot (z' \cdot x) = z' \cdot (z \cdot x) \text{ for all } z \in H,\ z' \in H^{opp},\ x \in L. \tag{11.1}$$

Now we have:

PROOF OF THEOREM 11.14. Suppose that $x \in \mathfrak{O}_L$ is a free generator of $\mathfrak{O}_L$ as an $\mathfrak{A}_H$-module. Then x is a free generator of L as an H-module, and so by Proposition 11.15 it is a free generator of L as an H^{opp}-module. Therefore for each $a \in \mathfrak{A}_H$ we may define $z_a \in H^{opp}$ by $z_a \cdot x = a \cdot x$. We claim that

$$\mathfrak{A}_{H^{opp}} = \{z_a \mid a \in \mathfrak{A}_H\}.$$

First we show that $z_a \in \mathfrak{A}_{H^{opp}}$ for each $a \in \mathfrak{A}_H$. Let $a \in \mathfrak{A}_H$ and $y \in \mathfrak{O}_L$. Since $\mathfrak{O}_L$ is a free $\mathfrak{A}_H$-module, there exists a unique element $b \in \mathfrak{A}_H$ such that $y = b \cdot x$. Now we have:

$$\begin{aligned} z_a \cdot y &= z_a \cdot (b \cdot x) \\ &= b \cdot (z_a \cdot x) \text{ by (11.1)} \\ &= b \cdot (a \cdot x), \end{aligned}$$

and this lies in $\mathfrak{O}_L$ since $a, b \in \mathfrak{A}_H$. Therefore $z_a \in \mathfrak{A}_{H^{opp}}$.

On the other hand, if $z \in \mathfrak{A}_{H^{opp}}$ then $z \cdot x = z_a \cdot x$ for some $a \in \mathfrak{A}_H$, and this implies that $z = z_a$ because x is a free generator of L as an H^{opp}-module. Therefore $\mathfrak{A}_{H^{opp}} = \{z_a \mid a \in \mathfrak{A}_H\}$, as claimed.

To prove the converse statement, we note that $(N^{opp})^{opp} = N$, and so the Hopf-Galois structure given by $(H^{opp})^{opp}$ on L/K coincides with that given by H; we may therefore apply the argument above with the roles of H, H^{opp} interchanged. □

The prototypical example of a pair of opposite Hopf-Galois structures is the action of the classical and canonical nonclassical Hopf-Galois structure on a non-abelian Galois extension; in this context Theorem 11.14 implies that $\mathfrak{O}_L$ is free over its associated order in H_λ if and only if it is free over its associated order in $K[G]$. This case of the theorem was established first in [**Tru16a**], which also includes explicit formulas for the elements z_a in this case. Theorem 11.14 has some interesting consequences in this context. For example: if L/K is a Galois extension of local fields and is at most tamely ramified then $\mathfrak{O}_L$ is a free $\mathfrak{O}_K[G]$-module by Noether's Theorem, and so $\mathfrak{O}_L$ is also free over its associated order in H_λ. If L/K is a Galois extension of local fields and is weakly ramified then $\mathfrak{O}_L$ is free over its associated order in $K[G]$ by a theorem of Johnston [**Joh15**, Theorem 1.2], and so $\mathfrak{O}_L$ is also free over its associated order in H_λ. If $L/\mathbb{Q}$ is a tamely ramified Galois extension and $[L : \mathbb{Q}]$ is not divisible by 4, then $\mathfrak{O}_L$ is free over $\mathbb{Z}[G]$ by M. J. Taylor's Theorem (see Section 11.6), and so $\mathfrak{O}_L$ is free over its associated order in H_λ.

11.4. Subextension techniques

In this section we suppose that L/K is a Galois extension of local or global fields. We describe two useful results for relating the integral Galois module structure of the extension L/K with that of certain subextensions, and use the results of Chapter 8 to generalize these results to the Hopf-Galois context.

Firstly, suppose that J is a normal subgroup of G and let $\nu : K[G] \to K[G/J]$ be the natural surjective homomorphism of K-algebras. Writing $F = L^J$, we have $z \cdot x = \nu(z) \cdot x$ for all $x \in F$. Now let $\theta_J = \sum_{j \in J} j \in K[G]$; then we have

$\theta_J(x) = \mathrm{Tr}_{L/F}(x)$ for all $x \in L$, and $\theta_J z = z\theta_J$ for all $z \in K[G]$ since J is normal in G. It follows that

$$\theta_J \cdot (z \cdot x) = \nu(z) \cdot (\theta_J \cdot x) \text{ for all } z \in K[G] \text{ and all } x \in L.$$

Hence, if x is a normal basis generator for L with respect to $K[G]$, then we have

$$F = \theta_J(L) = \theta_J(K[G] \cdot x) = \nu(K[G]) \cdot \theta_J(x) = K[G/J] \cdot \theta_J(x),$$

so $\theta_J(x)$ is a normal basis generator for F with respect to $K[G/J]$.

An integral analogue of this result appears in [**BL96**, Lemma 6]: let $\mathfrak{A}_{L/K}$ denote the associated order of $\mathfrak{O}_L$ in $K[G]$, and let $\mathfrak{A}_{F/K}$ denote the associated order of $\mathfrak{O}_F$ in $K[G/J]$. Suppose that $\mathfrak{O}_L$ is a free $\mathfrak{A}_{L/K}$-module, with generator x, say, and suppose in addition that L/F is at most tamely ramified. Then the trace map $\mathrm{Tr}_{L/F} : \mathfrak{O}_L \to \mathfrak{O}_F$ is surjective, and we have

$$\mathfrak{O}_F = \mathrm{Tr}_{L/F}(\mathfrak{O}_L) = \theta_J(\mathfrak{O}_L) = \theta_J(\mathfrak{A}_{L/K} \cdot x) = \nu(\mathfrak{A}_{L/K}) \cdot \theta_J(x).$$

Hence $\mathfrak{O}_F$ is a free module over the $\mathfrak{O}_K$-order $\nu(\mathfrak{A}_{L/K})$ in $K[G/J]$, so $\nu(\mathfrak{A}_{L/K}) = \mathfrak{A}_{F/K}$ and $\mathfrak{O}_F$ is a free $\mathfrak{A}_{F/K}$-module.

Hopf-Galois analogues of these results appear in [**Tay20**] (see also [**TT20**]). Suppose that L/K is H-Galois with $H = L[N]^G$ for some regular subgroup N of $\mathrm{Perm}(G)$ normalized by $\lambda(G)$, and that P is a normal subgroup of N that is normalized by $\lambda(G)$. Let $\nu : L[N] \to L[N/P]$ be the natural surjective homomorphism of L-algebras.

LEMMA 11.16. *We have $\nu({}^g z) = {}^g\nu(z)$ for all $z \in L[N]$ and all $g \in G$, and so ν restricts to a K-algebra homomorphism $\nu : L[N]^G \to L[N/P]^G$.*

PROOF. It is sufficient to consider the case $z = c\eta$ with $c \in L$ and $\eta \in N$. In this case, for each $g \in G$ we have

$$\nu({}^g(c\eta)) = \nu(g(c)\,{}^g\eta) = g(c)({}^g\eta)P,$$

and

$${}^g\nu(c\eta) = {}^g(c\eta P) = g(c)({}^g\eta)P,$$

by the description of the action of G on $L[N/P]$ given in Proposition 8.6. Therefore ν is G-equivariant, and so maps fixed points to fixed points. □

As in Section 8.1 of Chapter 8, let $H_P = L[P]^G$ be the Hopf subalgebra of $H = L[N]^G$ arising from P. Let $F = L^{H_P}$, and let $J = \mathrm{Gal}(L/F)$.

LEMMA 11.17. *Let $z \in L[N]^G$. Then for all $x \in F$ we have*

$$z \cdot x = \nu(z) \cdot x.$$

PROOF. Let $\eta_1, \ldots, \eta_r$ be a set of coset representatives for P in N, and for each i let $g_i = \eta_i^{-1}[1_G]$. Recall from Proposition 8.5 that P acts regularly on J; hence the g_i are a set of coset representatives for J in G. Write

$$z = \sum_{i=1}^{r} \sum_{\pi \in P} c_{i,\pi} \pi \eta_i \text{ with } c_{i,j} \in L.$$

Now for $x \in L$ we have

$$\begin{aligned} z \cdot x &= \left(\sum_{i=1}^{r} \sum_{\pi \in P} c_{i,\pi} \pi \eta_i \right) \cdot x \\ &= \sum_{i=1}^{r} \sum_{\pi \in P} c_{i,\pi} \eta_i^{-1} \pi^{-1} [1_G](x) \\ &= \sum_{i=1}^{r} \sum_{\pi \in P} c_{i,\pi} g_i j_{i,\pi}(x) \text{ for some } j_{i,\pi} \in J \\ &= \sum_{i=1}^{r} \sum_{\pi \in P} c_{i,\pi} g_i(x) \text{ since } x \in F \\ &= \left(\sum_{i=1}^{r} \sum_{\pi \in P} c_{i,\pi} \eta_i P \right) \cdot x \\ &= \nu(z) \cdot x. \end{aligned}$$

□

Now let $\theta_P = \sum_{\pi \in P} \pi \in L[N]$; then $\theta_P z = z \theta_P$ for all $z \in L[N]$, since P is normal in N. Since P is normalized by $\lambda(G)$, we have $\theta_P \in L[P]^G$, and for $x \in L$ we have

$$\theta_P \cdot x = \sum_{\pi \in P} \pi^{-1}[1_G](x) = \sum_{j \in J} j(x) = \mathrm{Tr}_{L/F}(x).$$

Hence, using Lemma 11.17, we have

$$\theta_P \cdot (z \cdot x) = z \cdot (\theta_P \cdot x) = \nu(z) \cdot (\theta_P \cdot x) \text{ for all } z \in L[N]^G \text{ and all } x \in L.$$

In particular, if $x \in L$ is a normal basis generator for L with respect to $L[N]^G$ then

$$F = \theta_P(L) = \theta_P(L[N]^G \cdot x) = \nu(L[N]^G) \cdot \theta_P(x) = L[N/P]^G \cdot \theta_P(x),$$

so $\theta_P(x)$ is a normal basis generator for F with respect to $L[N]^G$.

As in the classical case, we can formulate an integral analogue:

PROPOSITION 11.18. *Let $\mathfrak{A}_{L/K}$ denote the associated order of $\mathfrak{O}_L$ in $L[N]^G$, and let $\mathfrak{A}_{F/K}$ denote the associated order of $\mathfrak{O}_F$ in $L[N/P]^G$. Suppose that $\mathfrak{O}_L$ is a free $\mathfrak{A}_{L/K}$-module and that L/F is at most tamely ramified. Then $\mathfrak{A}_{F/K} = \nu(\mathfrak{A}_{L/K})$ and $\mathfrak{O}_F$ is a free $\mathfrak{A}_{F/K}$-module.*

PROOF. Let $x \in \mathfrak{O}_L$ be a free generator of $\mathfrak{O}_L$ as an $\mathfrak{A}_{L/K}$-module. Then we have

$$\begin{aligned} \mathfrak{O}_F &= \mathrm{Tr}_{L/F}(\mathfrak{O}_L) \\ &= \theta_P\left(\mathfrak{A}_{L/K} \cdot x\right) \\ &= \nu(\mathfrak{A}_{L/K}) \cdot (\theta_P \cdot x) \\ &= \nu(\mathfrak{A}_{L/K}) \cdot \mathrm{Tr}_{L/F}(x). \end{aligned}$$

Therefore $\mathfrak{O}_F$ is a free module over the $\mathfrak{O}_K$-order $\nu(\mathfrak{A}_{L/K})$ in $L[N/P]^G$, so $\mathfrak{A}_{F/K} = \nu(\mathfrak{A}_{L/K})$ and $\mathfrak{O}_F$ is a free $\mathfrak{A}_{F/K}$-module. □

As noted in Chapter 8, an interesting feature of the Hopf-Galois correspondence for a Galois extension L/K is that subextensions F/K that are not Galois may be Hopf-Galois with respect to an appropriate quotient Hopf algebra. Proposition 11.18 allows us to study the integral Hopf-Galois module structure of such subextensions.

We now consider a partial converse to Proposition 11.18 which shows that, under certain circumstances, the integral Hopf-Galois module structure of L/K is determined by that of certain subextensions.

First we describe the classical results we will generalize. Suppose that $G = G_1 \times G_2$ for some subgroups G_1, G_2; then $K[G] \cong K[G_1] \otimes_K K[G_2]$. For each i, let $F_i = L^{G_i}$; then F_1/K is a Galois extension with Galois group G_2, and F_2/K is a Galois extension with Galois group G_1. Furthermore, F_1/K and F_2/K are linearly disjoint and $L = F_1F_2$. It is not hard to see that if x_1 is a normal basis generator of F_1 with respect to $K[G_2]$ and x_2 is a normal basis generator of F_2 with respect to $K[G_1]$ then x_1x_2 is a normal basis generator of L with respect to $K[G]$.

An integral analogue of this result appears in [**BL96**, Lemma 5]: we say that F_1, F_2 are *arithmetically disjoint* over K if they are linearly disjoint over K and their discriminants $\mathfrak{d}_{F_1/K}$ and $\mathfrak{d}_{F_2/K}$ are coprime. In this case we have [**FT93**, Chapter III, (2.13)]

$$\mathfrak{O}_L = \mathfrak{O}_{F_1}\mathfrak{O}_{F_2} \cong \mathfrak{O}_{F_1} \otimes_{\mathfrak{O}_K} \mathfrak{O}_{F_2}.$$

Now let $\mathfrak{A}_1$ denote the associated order of $\mathfrak{O}_{F_1}$ in $K[G_2]$, let $\mathfrak{A}_2$ denote the associated order of $\mathfrak{O}_{F_2}$ in $K[G_1]$, and let $\mathfrak{A}$ denote the associated order of $\mathfrak{O}_L$ in $K[G]$. Then $\mathfrak{A} = \mathfrak{A}_1 \otimes_{\mathfrak{O}_K} \mathfrak{A}_2$ and if x_i is a free generator of $\mathfrak{O}_{F_i}$ as an $\mathfrak{A}_i$-module for each i then x_1x_2 is a free generator of $\mathfrak{O}_L$ as an $\mathfrak{A}$-module.

We now derive Hopf-Galois analogues of these results; essentially these are slight reinterpretations of results of Gill Muñoz and Rio [**GR21**] (see also [**TT20**]). Suppose that L/K is H-Galois with $H = L[N]^G$ for some regular subgroup N of $\mathrm{Perm}(G)$ normalized by $\lambda(G)$ and that $N = N_1 \times N_2$ for N_1, N_2 of N that are themselves normalized by $\lambda(G)$. For each i let $H_i = L[N_i]^G$ be the Hopf subalgebra of H arising from N_i.

LEMMA 11.19. *We have* $L[N]^G = L[N_1]^G \otimes_K L[N_2]^G$.

PROOF. By Galois descent we have identifications $L \otimes_K L[N]^G = L[N]$ and $L \otimes_K L[N_i]^G = L[N_i]$ for each i, and since $N = N_1 \times N_2$ we have $L[N] = L[N_1] \otimes_L L[N_2]$. Hence

$$\begin{aligned} L \otimes_K L[N]^G &= L[N] \\ &= L[N_1] \otimes_L L[N_2] \\ &= (L \otimes_K L[N_1]^G) \otimes_L (L \otimes_K L[N_2]^G) \\ &= L \otimes_K (L[N_1]^G \otimes_K L[N_2]^G). \end{aligned}$$

By faithful flatness this yields an identification

$$L[N]^G = L[N_1]^G \otimes_K L[N_2]^G.$$

□

Now for each i let $F_i = L^{H_i}$, and recall from Corollary 7.6 that $[L : F_i] = \dim_K(H_i) = |N_i|$, so that $[F_i : K] = |N|/|N_i|$. We will suppose that F_1 and F_2 are linearly disjoint over K; we then have $F_1F_2 \subseteq L$ and

$$[F_1F_2 : K] = [F_1 : K][F_2 : K] = \frac{|N|}{|N_1|}\frac{|N|}{|N_2|} = |N| = [L : K],$$

so that $L = F_1F_2 \cong F_1 \otimes_K F_2$. We note that neither of the extensions F_i/K is necessarily a Galois extension; if either is, then the fact that $F_1 \cap F_2 = K$ implies that F_1 and F_2 are linearly disjoint over K. In any case, by Proposition 8.2 the extension F_1/K is $L[N_2]^G$-Galois and the extension F_2/K is $L[N_1]^G$-Galois. By Lemma 11.19, we see that if x_1 is a normal basis generator of F_1 with respect to $L[N_2]^G$ and x_2 is a normal basis generator of F_2 with respect to $L[N_1]^G$ then x_1x_2 is a normal basis generator of L with respect to $L[N]^G$.

As in the classical case, we can formulate an integral analogue, whose proof is immediate:

PROPOSITION 11.20. *Suppose that F_1/K and F_2/K are arithmetically disjoint, let $\mathfrak{A}_1$ denote the associated order of $\mathfrak{O}_{F_1}$ in $L[N_2]^G$, let $\mathfrak{A}_2$ denote the associated order of $\mathfrak{O}_{F_2}$ in $L[N_1]^G$, and let $\mathfrak{A}$ denote the associated order of $\mathfrak{O}_L$ in $L[N]^G$. Then $\mathfrak{A} = \mathfrak{A}_1 \otimes_{\mathfrak{O}_K} \mathfrak{A}_2$, and if x_i is a free generator of $\mathfrak{O}_{F_i}$ as an $\mathfrak{A}_i$-module for each i then x_1x_2 is a free generator of $\mathfrak{O}_L$ as an $\mathfrak{A}$-module.*

11.5. Tamely ramified extensions

If L/K is a Galois extension of p-adic fields which is at most tamely ramified (henceforth, "tame"), then Noether's Theorem implies that $\mathfrak{O}_L$ is a free $\mathfrak{O}_K[G]$-module. We therefore cannot achieve a "better" description of $\mathfrak{O}_L$ by studying it relative to a different Hopf-Galois structure on the extension. However, in this section we show that the assumption of tame ramification is sufficient to ensure that $\mathfrak{O}_L$ is free over its associated order in many (and we conjecture all) Hopf-Galois structures admitted by the extension. Furthermore, these results encompass many non-normal extensions in addition to Galois extensions, and pave the way for investigations into the integral Hopf-Galois module structure of extensions of number fields, which we summarize in the final section of this chapter.

We suppose that L/K is an extension of p-adic fields which is H-Galois with $H = \widetilde{L}[N]^G$ for some regular subgroup N of $\mathrm{Perm}(X)$ normalized by $\lambda(G)$.

The $\mathfrak{O}_K$-order $\mathfrak{O}_{\widetilde{L}}[N]^G$ is then a formal analogue of the integral group ring in the Galois case. Indeed, if L/K is a Galois extension (so that $\widetilde{L} = L$ and $G_L = \{1\}$) then the classical Hopf-Galois structure on L/K corresponds to the regular subgroup $N = \rho(G)$ of $\mathrm{Perm}(G)$, and in this case have $L[N]^G = L[\rho(G)]^G = L^G[\rho(G)] = K[\rho(G)]$ and $\mathfrak{O}_L[N]^G = \mathfrak{O}_K[\rho(G)]$. The analogy encompasses the action of $\mathfrak{O}_{\widetilde{L}}[N]^G$ on L: (see [**Tru18a**, Proposition 2.1])

LEMMA 11.21. *We have $\mathfrak{O}_{\widetilde{L}}[N]^G \subseteq \mathfrak{A}_H$.*

PROOF. Let $x \in \mathfrak{O}_L$ and $z \in \mathfrak{O}_{\widetilde{L}}[N]^G$, and write $z = \sum_{n \in N} c_n n$ with $c_n \in \mathfrak{O}_{\widetilde{L}}$. Since L/K is $\widetilde{L}[N]^G$-Galois we have

$$z \cdot x = \sum_{\eta \in N} c_\eta \eta^{-1}(\overline{1})[x] \in L.$$

On the other hand, for each $\eta \in N$ any group element representing $\eta^{-1}(\overline{1})$ is a Galois automorphism of $\widetilde{L}$, so $\eta^{-1}(\overline{1})[x] \in \mathfrak{O}_{\widetilde{L}}$ for each $\eta \in N$. Therefore $z \cdot x \in \mathfrak{O}_{\widetilde{L}} \cap L = \mathfrak{O}_L$, whence $z \in \mathfrak{A}_H$. □

In the Galois case, the integral group ring $\mathfrak{O}_K[G]$ is a Hopf order in $K[G]$, the trace element $\theta = \sum_{g \in G} g$ is a left integral of $\mathfrak{O}_K[G]$, and the assumption that L/K is at most tamely ramified implies that there is an element $t \in \mathfrak{O}_L$ such that

$\theta \cdot t = 1$; thus in the language of Section 11.2 $\mathfrak{O}_L$ is an $\mathfrak{O}_K[G]$-tame extension of $\mathfrak{O}_K$.

In the Hopf-Galois case, the element $\theta = \sum_{\eta \in N} \eta$ lies in $\mathfrak{O}_{\widetilde{L}}[N]^G$, and for all $x \in \mathfrak{O}_L$ we have

$$\theta \cdot x = \sum_{\eta \in N} \eta^{-1}[\overline{1}](x) = \sum_{\overline{g} \in X} \overline{g}(x) = \mathrm{Tr}_{L/K}(x).$$

The element θ is a left and right integral of the Hopf algebra $\widetilde{L}[N]^G$, but this language does not apply to $\mathfrak{O}_{\widetilde{L}}[N]^G$, since this need not be a Hopf order in $\widetilde{L}[N]^G$ (we obtain a criterion later). For the same reason, it is not appropriate to say that $\mathfrak{O}_L$ is an $\mathfrak{O}_{\widetilde{L}}[N]^G$-tame extension of $\mathfrak{O}_K$.

If L/K is wildly ramified then $\mathrm{Tr}_{L/K}(\mathfrak{O}_L) \subseteq \mathfrak{p}_K$, so $\theta \cdot x \in \mathfrak{p}_K$ for all $x \in \mathfrak{O}_L$, and so the element θ/π_K lies in $\mathfrak{A}_H$ but not in $\mathfrak{O}_{\widetilde{L}}[N]^G$. Therefore in this case we have $\mathfrak{O}_{\widetilde{L}}[N]^G \subsetneq \mathfrak{A}_H$. We shall see that if L/K is tame then there are many cases in which $\mathfrak{O}_{\widetilde{L}}[N]^G = \mathfrak{A}_H$.

In the remainder of this section we prove the following result, at the time of writing the most general known concerning the integral Hopf-Galois module structure of tame extensions of p-adic fields. Recall that L/K is called *almost classically Galois* if G_L has a normal complement in G. By an abuse of language, we consider Galois extensions as almost classically Galois extensions in which G_L is trivial.

THEOREM 11.22. *Suppose that L/K is tame and almost classically Galois and that N is abelian. Then $\mathfrak{A}_H = \mathfrak{O}_{\widetilde{L}}[N]^G$ and $\mathfrak{O}_L$ is a free $\mathfrak{A}_H$-module.*

Theorem 11.22 first appeared in [**Tru18a**, Theorem 5.3], and in fact remains true if the ring of integers $\mathfrak{O}_L$ is replaced by any fractional ideal $\mathfrak{B}$ of L. However, the proof given in loc. cit. is rather intricate and involved; by specializing to the case $\mathfrak{B} = \mathfrak{O}_L$ we are able to apply the subextension techniques of Section 11.4 to give a more streamlined proof.

We shall see that if N is not abelian then it is possible for $\mathfrak{A}_H$ to be strictly larger than $\mathfrak{O}_{\widetilde{L}}[N]^G$, even if L/K is tame. However, we conjecture that the second conclusion of Theorem 11.22 is valid more generally:

CONJECTURE 11.23. *If L/K is tame then $\mathfrak{O}_L$ is a free $\mathfrak{A}_H$-module.*

We now assemble the results required to prove Theorem 11.22. We begin with the extreme case of unramified extensions; since such extensions are automatically Galois we have $\widetilde{L} = L$, $G_L = \{1\}$, and $X = G$ in this case.

PROPOSITION 11.24. *Suppose that L/K is unramified. Then $\mathfrak{A}_H = \mathfrak{O}_L[N]^G$ and $\mathfrak{O}_L$ is a free $\mathfrak{A}_H$-module.*

PROOF. Since L/K is unramified the extension of commutative rings $\mathfrak{O}_L/\mathfrak{O}_K$ is a Galois extension in the sense of [**CHR65**]. By Galois descent, the $\mathfrak{O}_L$-Hopf order $\mathfrak{O}_L[N]$ in $L[N]$ descends to an $\mathfrak{O}_K$-Hopf order $\mathfrak{O}_L[N]^G$ in $L[N]^G$. The element θ is then a left integral of $\mathfrak{O}_L[N]^G$, and since L/K is unramified there exists $t \in \mathfrak{O}_L$ such that $\theta \cdot t = 1$. Therefore $\mathfrak{O}_L$ is an $\mathfrak{O}_L[N]^G$-tame extension of $\mathfrak{O}_K$, and so $\mathfrak{O}_L$ is a free $\mathfrak{O}_L[N]^G$-module (see Section 11.2). □

In fact, if L/K is unramified then $\mathfrak{O}_L[N]^G$ gives a Hopf-Galois structure on the extension of commutative rings $\mathfrak{O}_L/\mathfrak{O}_K$: see [**Tru13**, Theorem 1.3]. In general,

$\mathfrak{O}_{\widetilde{L}}[N]^G$ is a Hopf order in $\widetilde{L}[N]^G$ if and only if the inertia subgroup of G is contained in the kernel of the action of G on N: see [**BB99**, Proposition 4.6].

If L/K is a Galois extension of p-adic fields and p does not divide $[L : K]$ then L/K is certainly tame. Furthermore, $\mathfrak{O}_K[G]$ is a maximal order in $K[G]$ [**CR90**, Proposition 27.1], and the fact that $K \otimes_{\mathfrak{O}_K} \mathfrak{O}_L = L$ is a free $K \otimes_{\mathfrak{O}_K} \mathfrak{O}_K[G] = K[G]$-module implies that $\mathfrak{O}_L$ is free over $\mathfrak{O}_K[G]$ [**Rei75**, Theorem 18.10]: this is a special case of Noether's Theorem. The following is a partial generalization of these ideas to separable Hopf-Galois extensions [**Tru11**, Theorem 4.4]:

PROPOSITION 11.25. *Suppose that N is abelian and that $p \nmid [L : K]$. Then $\mathfrak{A}_H = \mathfrak{O}_{\widetilde{L}}[N]^G$ and $\mathfrak{O}_L$ is a free $\mathfrak{A}_H$-module.*

PROOF. The Hopf algebra $H = \widetilde{L}[N]^G$ is a separable K-algebra [**Tru18b**, Lemma 4.2], and is commutative because N is abelian. It therefore contains a unique maximal $\mathfrak{O}_K$-order $\mathfrak{M}$, which is the integral closure of $\mathfrak{O}_K$ in H [**CR90**, Proposition 26.10]. If $z \in \mathfrak{M}$ then z is integral over $\mathfrak{O}_K$ in H, so z is integral over $\mathfrak{O}_{\widetilde{L}}$ in $\widetilde{L}[N]$, so z lies in the unique maximal order in $\widetilde{L}[N]$, which is $\mathfrak{O}_{\widetilde{L}}[N]$ [**CR90**, Proposition 27.1]; hence $z \in \mathfrak{O}_{\widetilde{L}}[N] \cap \widetilde{L}[N]^G = \mathfrak{O}_{\widetilde{L}}[N]^G$. Therefore $\mathfrak{M} = \mathfrak{O}_{\widetilde{L}}[N]^G$, and so $\mathfrak{O}_{\widetilde{L}}[N]^G = \mathfrak{A}_H$ by Lemma 11.21. Since $\mathfrak{A}_H$ is a maximal order and $K \otimes_{\mathfrak{O}_K} \mathfrak{O}_L = L$ is a free $K \otimes_{\mathfrak{O}_K} \mathfrak{A}_H = H$-module, [**Rei75**, Theorem 18.10] implies that $\mathfrak{O}_L$ is a free $\mathfrak{A}_H$-module. □

Note that we could not appeal to Schneider's Theorem [**Sch77**] in the final step of the proof of Proposition 11.25, since $\mathfrak{O}_{\widetilde{L}}[N]^G$ need not be a Hopf order in $\widetilde{L}[N]^G$.

The first conclusion of Proposition 11.25 need not hold if the assumption that N is abelian is removed:

EXAMPLE 11.26. Let $p \equiv 2 \pmod 3$, let $K = \mathbb{Q}_p$, and let L be the splitting field of $x^3 - p$ over K. Then $L = K(a, \zeta)$, where $a^3 = p$ and ζ is a primitive cube root of unity, and L/K is Galois with group $G \cong D_3$. The extension L/K cannot be unramified, since it has nonabelian Galois group, and it cannot be totally ramified, since the subextension $K(\zeta)/K$ is nontrivial and unramified. Therefore the inertia subgroup must be the unique nontrivial proper normal subgroup of G, which has order 3, and so L/K is tamely ramified. Since G is nonabelian L/K admits a canonical nonclassical Hopf-Galois structure, with Hopf algebra $H_\lambda = L[\lambda(G)]^G$. We will show that the associated order $\mathfrak{A}_\lambda$ of $\mathfrak{O}_L$ in H_λ strictly contains $\mathfrak{O}_L[\lambda(G)]^G$. We present G as

$$G = \langle r, s \mid r^3 = s^2 = 1, srs = r^{-1} \rangle,$$

where

$$\begin{aligned} r(a) &= \zeta a, & r(\zeta) &= \zeta, \\ s(a) &= a, & s(\zeta) &= \zeta^{-1}, \end{aligned}$$

and recall that G acts on $L[\lambda(G)]$ by acting on L as Galois automorphisms and acting on $\lambda(G)$ by conjugation via the left regular embedding $\lambda : G \to \mathrm{Perm}(G)$. We may verify that the element

$$z = a^2\lambda(s) + \zeta^2 a^2 \lambda(r^{-1}s) + \zeta a^2 \lambda(rs) \in \mathfrak{O}_L[\lambda(G)]$$

is fixed by every element of G, and so actually lies in $\mathfrak{O}_L[\lambda(G)]^G$. Therefore it acts on elements of L according to the formula given in the theorem of Greither and Pareigis. It is sufficient to consider its action on elements of the form $\zeta^i a^j$ for

$i = 1, 2$ and $j = 0, 1, 2$, since these form an $\mathfrak{O}_K$-basis of $\mathfrak{O}_L$. For such an element, we have:

$$\begin{aligned} z \cdot (\zeta^i a^j) &= a^2(\lambda(s) + \zeta^2\lambda(r^{-1}s) + \zeta\lambda(rs)) \cdot (\zeta^i a^j) \\ &= a^2(\lambda(s)^{-1}[1_G] + \zeta^2\lambda(r^{-1}s)^{-1}[1_G] + \zeta\lambda(rs)^{-1}[1_G])(\zeta^i a^j) \\ &= a^2(s + \zeta^2 sr + \zeta sr^2)(\zeta^i a^j) \\ &= a^{2+j}\zeta^{-i}(1 + \zeta^{2(1+j)} + \zeta^{(1+j)}) \\ &= \begin{cases} 3p\zeta^{-i}a & \text{if } j = 2 \\ 0 & \text{otherwise} \end{cases} \\ &\in p\mathfrak{O}_L \end{aligned}$$

Hence $z \cdot x \in p\mathfrak{O}_L$ for all $x \in \mathfrak{O}_L$, and so z/p lies in $\mathfrak{A}_\lambda$ but not in $\mathfrak{O}_L[\lambda(G)]^G$. Therefore $\mathfrak{A}_\lambda \supsetneq \mathfrak{O}_L[\lambda(G)]^G$. However, since L/K is tamely ramified, $\mathfrak{O}_L$ is a free module over its associated order $\mathfrak{O}_K[G]$ in $K[G]$, and so $\mathfrak{O}_L$ is a free module over $\mathfrak{A}_\lambda$, by Theorem 11.14.

However, we conjecture that if $p \nmid [L : K]$ then $\mathfrak{A}_H$ is a maximal order in H; if this were true then, applying [**Rei75**, Theorem 18.10] as in the proof of Proposition 11.25, it would follow that $\mathfrak{O}_L$ is a free $\mathfrak{A}_H$-module.

We can combine Propositions 11.24 and 11.25 using the subextension techniques of Section 11.4 to give a proof of Theorem 11.22 in the Galois case:

THEOREM 11.27. *Suppose that L/K is a tame Galois extension and that N is abelian. Then $\mathfrak{A}_H = \mathfrak{O}_L[N]^G$ and $\mathfrak{O}_L$ is a free $\mathfrak{A}_H$-module.*

PROOF. Write $|N| = p^r m$ with $(p, m) = 1$, and write $N = A \times B$ with $|A| = m$ and $|B| = p^r$, as in Theorem 8.22. Since A, B are the unique subgroups of N of their respective orders, each is normalized by $\lambda(G)$; hence we obtain Hopf subalgebras $H_A = L[A]^G$ and $H_B = L[B]^G$ of H, and $H = H_A \otimes_K H_B$ by Lemma 11.19.

Let $F_A = L^{H_A}$ and $F_B = L^{H_B}$.

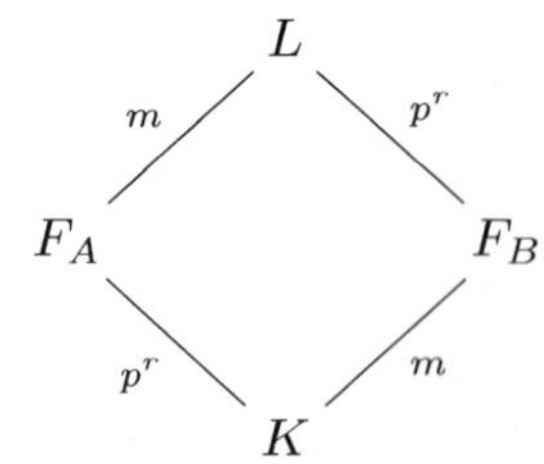

Since $[F_A : K] = p^r$ and L/K is tamely ramified, F_A/K is unramified, hence Galois. Since in addition we have $F_A \cap F_B = K$, we see that $L = F_A F_B$. The fact that F_A/K is unramified also implies that F_A and F_B are arithmetically disjoint over K, so Proposition 11.20 applies. The extension F_A/K is Hopf-Galois with respect to H_B and is unramified, so $\mathfrak{O}_{F_A}$ is a free $\mathfrak{O}_L[B]^G$-module, by Proposition 11.24. The extension F_B/K is Hopf-Galois with respect to H_A and has degree prime to p, so $\mathfrak{O}_{F_B}$ is a free $\mathfrak{O}_L[A]^G$-module, by Proposition 11.25. Therefore by Proposition 11.20 $\mathfrak{O}_L$ is a free $\mathfrak{O}_L[N]^G$-module. □

Finally, we complete the proof of Theorem 11.22 by using the theory of induced Hopf-Galois structures (see Section 8.3) to reduce to the Galois case, then applying Theorem 11.27 and Proposition 11.18:

PROOF OF THEOREM 11.22. Since L/K is tame, the extension $\widetilde{L}/L$ is unramified (see [**Tru18a**, Proposition 5.2]), and so G_L is cyclic. Since L/K is an almost classically Galois extension, G_L has a normal complement in G, and so by Theorem 8.18 we may induce a Hopf-Galois structure on $\widetilde{L}/K$ from the given Hopf-Galois structure on L/K (corresponding to the regular subgroup N of $\mathrm{Perm}(X)$) and the classical Hopf-Galois structure on $\widetilde{L}/L$ (corresponding to the regular subgroup $\rho(G_L)$ of $\mathrm{Perm}(G_L)$). The regular subgroup of $\mathrm{Perm}(G)$ corresponding to this Hopf-Galois structure is $M \cong N \times G_L$; in particular, M is abelian. By Theorem 11.27, $\mathfrak{O}_{\widetilde{L}}$ is a free $\mathfrak{O}_{\widetilde{L}}[M]^G$-module. By Lemma 11.16 that there is a K-algebra homomorphism $\nu : \widetilde{L}[M]^G \to \widetilde{L}[N]^G$ that maps $\mathfrak{O}_{\widetilde{L}}[M]^G$ to $\mathfrak{O}_{\widetilde{L}}[N]^G$. Since $\widetilde{L}/L$ is unramified, we may apply Proposition 11.18 to conclude that $\mathfrak{O}_L$ is a free module over $\nu(\mathfrak{O}_{\widetilde{L}}[M]^G) = \mathfrak{O}_{\widetilde{L}}[N]^G$. □

11.6. Extensions of number fields

At the time of writing [**Chi00**] nothing was known about the Hopf-Galois module structure of rings of integers in extensions of number fields. In this section we survey some of the results that have now been established in this area, which is still in its infancy. These results concern extensions that are at most tamely ramified ("tame"); we therefore begin with a brief discussion of some aspects of classical tame Galois module theory (see [**Frö83**] for a much more thorough survey).

If L/K is a tame Galois extension of number fields then Noether's Theorem implies that $\mathfrak{O}_L$ is a *locally free* $\mathfrak{O}_K[G]$-module; that is: for each prime ideal $\mathfrak{p}$ of $\mathfrak{O}_K$ the completed ring of integers $\mathfrak{O}_{L,\mathfrak{p}} = \mathfrak{O}_K \otimes_{\mathfrak{O}_K} \mathfrak{O}_L$ is a free $\mathfrak{O}_{K,\mathfrak{p}}[G]$-module. We seek conditions for $\mathfrak{O}_L$ to be a free $\mathfrak{O}_K[G]$-module, or study the isomorphism classes of locally free $\mathfrak{O}_K[G]$-modules realizable by rings of integers in tame extensions. The first theorem in this direction is the Hilbert-Speiser Theorem [**Hil65**]: if $L/\mathbb{Q}$ is a tame abelian extension then $\mathfrak{O}_L$ is a free $\mathbb{Z}[G]$-module.

The Galois module structure of tame nonabelian extensions of $\mathbb{Q}$ was intensively studied in the 1970s, beginning with Martinet's discovery of a class of tame extensions $L/\mathbb{Q}$ with Galois group $G \cong H_8$ (the quaternion group of order 8) for which $\mathfrak{O}_L$ is not a free $\mathbb{Z}[G]$-module [**Mar71**], and culminating with M. J. Taylor's proof [**Tay81**] of Fröhlich's conjecture that, in general, the isomorphism class of $\mathfrak{O}_L$ as a $\mathbb{Z}[G]$-module is governed by the behaviour of the extended Artin L-functions attached to the irreducible symplectic characters of G.

It is also natural to study the consequences of varying the base field in the Hilbert-Speiser Theorem. In [**GRRS99**] it was shown that, given a number field $K \neq \mathbb{Q}$, there exists a prime number p and a tame Galois extension L/K of degree p such that $\mathfrak{O}_L$ is not a free $\mathfrak{O}_K[G]$-module; thus $\mathbb{Q}$ is the only number field for which every tame abelian extension has a normal integral basis. Gómez Ayala [**GA94**] studied tame Kummer extensions of number fields L/K of prime degree, and determined a criterion for $\mathfrak{O}_L$ to be a free $\mathfrak{O}_K[G]$-module in this case; this result was generalized to an arbitrary tame Kummer extension in [**DCR13**].

Now we turn to questions of Hopf-Galois module structure. We shall assume that L/K is an extension of number fields that is $\widetilde{L}[N]^G$-Galois; note that we do not make a standing assumption that L/K is a Galois extension. The following generalizations of Propositions 11.24 and 11.25 appear in [**Tru11**], and allow us to determine the local structure of $\mathfrak{O}_L$ over its associated order in $\widetilde{L}[N]^G$ in some cases.

PROPOSITION 11.28. *Let $\mathfrak{p}$ be a prime ideal of $\mathfrak{O}_K$ that is unramified in $\widetilde{L}$. Then $\mathfrak{O}_{L,\mathfrak{p}}$ is a free $\mathfrak{O}_{\widetilde{L},\mathfrak{p}}[N]^G$-module.*

PROPOSITION 11.29. *Suppose that N is abelian, and let $\mathfrak{p}$ be a prime ideal of $\mathfrak{O}_K$ lying above a prime number that does not divide $[L:K]$. Then $\mathfrak{O}_{L,\mathfrak{p}}$ is a free $\mathfrak{O}_{\widetilde{L},\mathfrak{p}}[N]^G$-module.*

A Galois extension of number fields is called *domestic* if no prime ideal of $\mathfrak{O}_K$ lying above a prime number that divides $[L:K]$ is ramified in L.

PROPOSITION 11.30. *Suppose that L/K is domestic and that N is abelian. Then $\mathfrak{O}_L$ is a locally free $\mathfrak{O}_L[N]^G$-module.*

If L/K is a Galois extension of number fields of prime power degree p^n, then it is tame if and only if no prime ideal of $\mathfrak{O}_K$ lying above p is ramified in L, which occurs if and only if it is domestic; Proposition 11.30 therefore implies that if L/K is a tame extension of number fields of prime power degree which is Hopf-Galois with respect to a commutative Hopf algebra $H = L[N]^G$ then $\mathfrak{O}_L$ is a locally free $\mathfrak{O}_L[N]^G$-module [**Tru11**, Corollary 5.10].

Building on this result, [**Tru12**] and [**Tru16b**] study the Hopf-Galois module structure of rings of algebraic integers in tame Kummer extensions L/K of number fields with Galois group isomorphic to $C_p \times C_p$, for some prime number p. In these cases all of the Hopf-Galois structures are of abelian type (see [**Byo02**]), and so Proposition 11.30 applies. Using detailed local information and idèlic descriptions of locally free class groups, necessary and sufficient conditions are obtained for $\mathfrak{O}_L$ to be free over its associated order in each Hopf-Galois structure. In general these conditions take the form of certain ideals having trivial class in certain ray class groups. However, the results for $p = 2$ and $K = \mathbb{Q}$ are more concrete. The tame biquadratic extensions of $\mathbb{Q}$ are precisely the extensions of the form $L = \mathbb{Q}(\alpha, \beta)$, with $\alpha^2 = a$ and $\beta^2 = b$ integers that are congruent to 1 modulo 4. The extension $L/\mathbb{Q}$ admits four Hopf-Galois structures: the classical structure (of elementary abelian type) and three nonclassical structures, each of cyclic type and each corresponding to a quadratic subfield of $L/\mathbb{Q}$ [**Byo02**, Theorem 2.5]. From [**Tru12**, Section 6] we have

PROPOSITION 11.31. *Let $u, v \in \{\alpha, \beta, \alpha\beta\}$ with $u \neq v$, and consider the equations*

$$x^2 + v^2 y^2 = \pm 2 \gcd(u^2, v^2). \tag{11.2}$$

A necessary condition for the ring of integers $\mathfrak{O}_L$ to be free over its associated order in the Hopf-Galois structure corresponding to the quadratic subfield $\mathbb{Q}(v)$ is that one of the equations (11.2) *has an integer solution. A sufficient condition for $\mathfrak{O}_L$ to be free over this associated order is that there exist integers $x \equiv y \equiv 1 \pmod 8$ satisfying one of the equations* (11.2).

In particular, Proposition 11.31 implies that if $\mathbb{Q}(v)$ is real then $\mathfrak{O}_L$ is not free over its associated order in the Hopf-Galois structure corresponding to $\mathbb{Q}(v)$, and so it is impossible for $\mathfrak{O}_L$ to be simultaneously free over its associated orders in all of the Hopf-Galois structures on the extension $L/\mathbb{Q}$. In [**Tru12**, Section 7] examples are constructed in which $\mathfrak{O}_L$ is free over its associated order in precisely one, two, or three of these Hopf-Galois structures.

Next we consider *quaternion fields*: Galois extensions $L/\mathbb{Q}$ with $G \cong H_8$. As mentioned above, tame quaternion fields have been important in the history of Galois module theory, because they provided the first examples of tame Galois extensions of $\mathbb{Q}$ for which $\mathfrak{O}_L$ is not a free $\mathbb{Z}[G]$-module [**Mar71**]. In [**Tay20**] the techniques of Hopf-Galois theory are applied to tame quaternion fields in the hope that a different Hopf-Galois structure may give a more satisfactory description of the ring of algebraic integers in the case that it is poorly described by classical techniques. Whilst this hope has not yet been realized, much has been learned about the Hopf-Galois module structure of these extensions.

Quaternion fields admit Hopf-Galois structures of all types of order 8 [**SV18**, Table A.1]; the corresponding regular subgroups are determined in [**TT19**], which also studies certain isomorphism problems for the Hopf algebras that appear (see Subsection 9.2.3).

One general result is the following [**Tay20**, Proposition 4.3.1]:

PROPOSITION 11.32. *Let $L/\mathbb{Q}$ be a tame quaternion field. Then $\mathfrak{O}_L$ is locally free, but not free, over its associated order in each of the Hopf-Galois structures of cyclic type.*

PROOF. Suppose that N is a cyclic regular subgroup of $\mathrm{Perm}(G)$ that is normalized by $\lambda(G)$, so that $H = L[N]^G$ gives a Hopf-Galois structure of cyclic type on $L/\mathbb{Q}$. Since N is abelian, the discussion following Proposition 11.30 implies that $\mathfrak{O}_L$ is a locally free $\mathfrak{O}_L[N]^G$-module.

As in Example 8.8, let P be the unique subgroup of N of order 2; then P is normal in N and normalized by $\lambda(G)$, and the corresponding fixed field K is the unique biquadratic subfield of $L/\mathbb{Q}$. We note that, since K embeds into a quaternion field, it is totally real: see [**JY88**, Proposition I.2.8]. By Theorem 8.2 we obtain a quotient Hopf-Galois structure of cyclic type on $K/\mathbb{Q}$, with Hopf algebra $L[N/P]^G$, say. By Proposition 11.18, if $\mathfrak{O}_L$ were a free $\mathfrak{O}_L[N]^G$-module then $\mathfrak{O}_K$ would be a free $\mathfrak{O}_L[N/P]^G$-module. Since K is real, this is impossible: see the discussion following Proposition 11.31. Therefore $\mathfrak{O}_L$ is not a free $\mathfrak{O}_L[N]^G$-module. □

Some of the other results of [**Tay20**] apply only to a special class of tame quaternion fields, following a construction due to Fujisaki [**Fuj90**]. In [**Tay20**] these are called *tame Fujisaki extensions*. From [**Tay20**, Theorem 6.0.1] we have

THEOREM 11.33. *Let $L/\mathbb{Q}$ be a tame Fujisaki extension. Then $\mathfrak{O}_L$ is locally free over its associated order in each Hopf-Galois structure admitted by the extension.*

For the Hopf-Galois structures of abelian type we may apply Proposition 11.30. The extension $L/\mathbb{Q}$ admits precisely two Hopf-Galois structures of quaternion type: the classical and canonical nonclassical Hopf-Galois structures. Since $L/\mathbb{Q}$ is tame, $\mathfrak{O}_L$ is certainly a locally free $\mathbb{Z}[G]$-module, and local freeness over its associated order in the canonical nonclassical structure follows by Theorem 11.14. There are 6 Hopf-Galois structures of dihedral type, which can be arranged into 3 pairs using concept of opposite Hopf-Galois structures (see Section 11.3). Local freeness of $\mathfrak{O}_L$ over its associated order in these Hopf-Galois structures is established via a combination of explicit computation and the application of Theorem 11.14: see [**Tay20**, Chapter 6]. We note that in the case $N = \lambda(G)$ (corresponding to the canonical nonclassical Hopf-Galois structure) or in 3 of the cases in which N is dihedral, it is possible for the associated order of $\mathfrak{O}_L$ to be strictly larger than $\mathfrak{O}_L[N]^G$.

It is also shown in [**Tay20**, Chapter 7] that if $L/\mathbb{Q}$ is a tame Fujisaki extension then the structure of $\mathfrak{O}_L$ as a module over its associated order in the Hopf-Galois structures of dihedral type is closely linked to its structure over $\mathbb{Z}[G]$, its associated order in the classical Hopf-Galois structure. We refer to loc. cit. for the intricate details.

Finally we consider the application of Hopf-Galois theory to the study of the ring of algebraic integers in a non-normal extension of number fields. Let L/K be a radical extension of number fields of prime degree p (that is: $L = K(\omega)$ with $\omega^p \in K$ but $\omega \notin K$). Then L/K admits a unique Hopf-Galois structure: if K contains a primitive pth root of unity then L/K is a Galois extension, and this structure is the classical one; if K does not contain a primitive pth root of unity then it is a non-normal extension, and its unique Hopf-Galois structure is described in, for example, [**Chi89**].

If we suppose in addition that L/K is tame then, in the Galois case, the ring of algebraic integers $\mathfrak{O}_L$ is locally free over $\mathfrak{O}_K[G]$ by Noether's Theorem, and Gómez Ayala [**GA94**] has determined a criterion for $\mathfrak{O}_L$ to be a free $\mathfrak{O}_K[G]$-module. In the non-normal case, we have [**Tru20**, Theorem 5.5]:

THEOREM 11.34. *Suppose that p is unramified in K. Let H give the unique Hopf-Galois structure on L/K, and let $\mathfrak{A}_H$ denote the associated order of $\mathfrak{O}_L$ in H. Then:*

(1) *$\mathfrak{O}_L$ is a locally free $\mathfrak{A}_H$-module;*
(2) *$\mathfrak{O}_L$ is a free $\mathfrak{A}_H$-module if and only if there exists $\beta \in \mathfrak{O}_L$ such that*
   (a) *$L = K(\beta)$;*
   (b) *$b = \beta^p \in \mathfrak{O}_K$;*
   (c) *for $j = 0, 1, \dots, p-1$ the ideals*

$$\mathfrak{b}_j = \prod_{\mathfrak{p}} \mathfrak{p}^{\lfloor v_{\mathfrak{p}}(b^j)/p \rfloor}$$

are principal, with generators b_j such that

$$\sum_{j=0}^{p-1} \frac{\beta^j}{b_j} \equiv 0 \pmod{p\mathfrak{O}_L}.$$

Furthermore, in this case the element

$$\frac{1}{p}\sum_{j=0}^{p-1} \frac{\beta^j}{b_j}$$

is a free generator of $\mathfrak{O}_L$ as an $\mathfrak{A}_H$-module.

To prove part (1) of Theorem 11.34 we note first that the Hopf algebra giving the unique Hopf-Galois structure on L/K is commutative, and so Proposition 11.29 implies that if $\mathfrak{p}$ is a prime ideal of $\mathfrak{O}_K$ that does not lie above p then $\mathfrak{O}_{L,\mathfrak{p}}$ is a free $\mathfrak{A}_{H,\mathfrak{p}}$-module. If $\mathfrak{p}$ does lie above p then we can show explicitly that $\mathfrak{O}_{L,\mathfrak{p}}$ is a free $\mathfrak{A}_{H,\mathfrak{p}}$-module: see [**Tru20**, Theorem 5.1]. Note that since L/K is not a Galois extension the prime ideals of $\mathfrak{O}_K$ lying above p are not necessarily unramified in L, so we cannot apply Proposition 11.28.

To derive the criteria for global freeness expressed in part (2) of Theorem 11.34 we employ a result of Bley and Johnston [**BJ08**, Proposition 2.1], which in this case implies that $\mathfrak{O}_L$ is a free $\mathfrak{A}_H$ module if and only if the set $\mathfrak{M}_H\mathfrak{O}_L$, where $\mathfrak{M}_H$ is the unique maximal order in H, is a free $\mathfrak{M}_H$ module with a generator lying in $\mathfrak{O}_L$. To obtain criteria for this to occur we use an idèlic description of the locally free class group of $\mathfrak{M}_H$ due to Fröhlich (see [**Tru20**, Section 5]); this is facilitated by the fact that we have $H \cong K^p$ as K-algebras [**Tru20**, Proposition 4.3]. We note a strong parallel with the situation in which L/K is a Galois extension: in this case $\mathfrak{O}_L$ is locally free over $\mathfrak{O}_K[G]$ by Noether's theorem, and the result of Bley and Johnston cited above implies that $\mathfrak{O}_L$ is a free $\mathfrak{O}_K[G]$-module if and only if $\mathfrak{M}_{K[G]}\mathfrak{O}_L$ is a free $\mathfrak{M}_{K[G]}$ module with a generator lying in $\mathfrak{O}_L$ (this time $\mathfrak{M}_{K[G]}$ denotes the unique maximal order in $K[G]$). Since we also have $K[G] \cong K^p$ as K-algebras in this case, the study of $\mathfrak{M}_{K[G]}\mathfrak{O}_L$ as an $\mathfrak{M}_{K[G]}$-module closely resembles the study of $\mathfrak{M}_H\mathfrak{O}_L$ as an $\mathfrak{M}_H$-module in the non-normal case. Ultimately, we find that the criteria for global freeness expressed in part (2) of Theorem 11.34 are identical to those determined in [**GA94**] for the Galois case. For full details, we refer the reader to [**Tru20**, Section 6].

CHAPTER 12

Hopf orders in group rings

12.1. Hopf orders and Galois module theory

Let K be a field that is complete with respect to a discrete valuation $\nu : K \to \mathbb{Z} \cup \{\infty\}$. Let $\mathfrak{O}_K = \{x \in K \mid \nu(x) \geq 0\}$ denote its valuation ring, with group of units $\mathfrak{O}_K^\times = \{x \in \mathfrak{O}_K \mid \nu(x) = 0\}$ and unique maximal ideal $\mathfrak{p} = \{x \in \mathfrak{O}_K \mid \nu(x) \geq 1\}$, and let π denote a uniformizing parameter of K (so that $\nu(\pi) = 1$). Assume that the residue field $k = \mathfrak{O}_K/\mathfrak{p}$ is finite, with characteristic p. Then K is either a finite extension of the p-adic rationals $\mathbb{Q}_p$ (a p-adic field) or the field of formal Laurent series $\mathbb{F}_q((T))$ over a finite field $\mathbb{F}_q$, where q is a power of p.

Now let L be a G-Galois extension of K, with valuation ring $\mathfrak{O}_L$. Then, as discussed in the overview of Galois module theory in Chapter 11, $\mathfrak{O}_L$ is a module over its associated order $\mathfrak{A}$ in the Hopf algebra $K[G]$, and we seek criteria for $\mathfrak{O}_L$ to be a free $\mathfrak{A}$-module. If $\mathrm{char}(K) = 0$ then, by Childs' generalization of Noether's theorem (Theorem 11.8), a sufficient condition for this to occur is that $\mathfrak{A}$ is a *Hopf order* in $K[G]$:

DEFINITION 12.1. An $\mathfrak{O}_K$-order in $K[G]$ that is also an $\mathfrak{O}_K$-Hopf algebra with operations inherited from those on $K[G]$ is called an *$\mathfrak{O}_K$-Hopf order* in $K[G]$.

To show that an $\mathfrak{O}_K$-order H in $K[G]$ is an $\mathfrak{O}_K$-Hopf order, it suffices to show that the comultiplication on $K[G]$ restricts to H: see [**Und15**, Proposition 3.4.5].

It is not hard to show that the integral group ring $\mathfrak{O}_K[G]$ is a Hopf order in $K[G]$, and the proof of the "if" part of Noether's theorem can be interpreted in this Hopf-theoretic framework: see Section 11.2. Childs' theorem generalizes this approach to the case of wild extensions: the strategy is to determine $\mathfrak{A}$ and check to see whether it is an $\mathfrak{O}_K$-Hopf order in $K[G]$. This requires, of course, good knowledge of the structure of $\mathfrak{O}_K$-Hopf orders in $K[G]$ for G a finite group.

Childs also generalizes the "only if" part of Noether's theorem, again assuming that $\mathfrak{A}$ is a Hopf order. In this case $\mathfrak{O}_L$ is an *$\mathfrak{A}$-extension of* $\mathfrak{O}_K$: that is, $\mathfrak{O}_L$ is an $\mathfrak{A}$-module algebra and the fixed ring

$$\{x \in \mathfrak{O}_L \mid a \cdot x = \varepsilon(a)x \ \forall a \in \mathfrak{A}\}$$

is equal to $\mathfrak{O}_K$. Childs shows that there is a left integral θ of $\mathfrak{A}$ such that $\theta \cdot \mathfrak{O}_L = \mathfrak{O}_K$ [**Chi00**, Theorem 13.3]. That is, $\mathfrak{O}_L$ is a *tame* *$\mathfrak{A}$-extension of* $\mathfrak{O}_K$ (see Section 11.2). In this language, a G-Galois extension L/K is at most tamely ramified if and only if $\mathfrak{O}_L$ is a tame $\mathfrak{O}_K[G]$-extension of $\mathfrak{O}_K$.

EXAMPLE 12.2. Let $K = \mathbb{Q}_p(\zeta_p)$ and let π be the uniformizing parameter for K. Then the polynomial $x^p - \pi$ is irreducible over K by Eisenstein's criterion. Let $L = K(z)$, where z is a root of $x^p - \pi$. Then L/K is Galois with group $C_p = \langle g \rangle$; the Galois action is given as $g^i(z) = \zeta_p^i z$. Since $\pi = z^p$, L/K is wild.

We have $\mathfrak{O}_L = \mathfrak{O}_K[z]$. The associated order $\mathfrak{A}$ contains the elements $\{e_i\}$, $0 \le i \le p-1$, where

$$e_i = \frac{1}{p}\sum_{j=0}^{p-1} \zeta_p^{-ij} g^j.$$

Thus $\mathfrak{A}$ is the maximal integral order in $K[C_p]$. Arguing via duality (see Section 12.2 below) we can show that $\mathfrak{A}$ is an $\mathfrak{O}_K$-Hopf order in $K[C_p]$. By the results of Childs cited above, $\mathfrak{O}_L$ is a free rank one $\mathfrak{A}$-module and a tame $\mathfrak{A}$-extension of $\mathfrak{O}_K$.

We might also ask: given a Hopf order H in a group algebra $K[G]$ (for some finite group G) does there exist a G-Galois extension L/K such that $\mathfrak{A} = H$?

DEFINITION 12.3. An $\mathfrak{O}_K$-Hopf order H in $K[G]$ is *realizable* if there exists a G-Galois extension L/K whose associated order $\mathfrak{A}$ in $K[G]$ is H.

Realizability is of central importance in Galois module theory. For if H is realizable, then there exists a G-Galois extension L/K for which $H = \mathfrak{A}$. Thus $\mathfrak{A}$ is a Hopf order and so by Childs' Theorem (Theorem 11.8) $\mathfrak{O}_L$ is a free rank one $\mathfrak{A}$-module; we have characterized $\mathfrak{O}_L$ as a Galois module. Moreover, $\mathfrak{O}_L$ is a tame $\mathfrak{A}$-extension of $\mathfrak{O}_K$.

There is also a narrower definition of realizability:

DEFINITION 12.4. An $\mathfrak{O}_K$-Hopf order H in $K[G]$ is *realizable as a Galois group* if there exists a G-Galois extension L/K for which the valuation ring $\mathfrak{O}_L$ of L is an H-Galois extension of $\mathfrak{O}_K$.

If H is realizable as a Galois group then it is realizable in the sense of Definition 12.3. For if L/K is a G-Galois extension for which $\mathfrak{O}_L$ is an H-Galois extension of $\mathfrak{O}_K$ then $\mathfrak{O}_L$ is a free H-module of rank one, so $H = \mathfrak{A}$ (by [**Chi00**, Proposition 12.5]) and $\mathfrak{O}_L$ is a tame $\mathfrak{A}$-extension of $\mathfrak{O}_K$. However, it is possible for $\mathfrak{A}$ to be a Hopf order in $K[G]$ (so that that $\mathfrak{O}_L$ is a free rank one $\mathfrak{A}$-module and a tame $\mathfrak{A}$-extension of $\mathfrak{O}_K$) without $\mathfrak{O}_L$ being an $\mathfrak{A}$-Galois extension of $\mathfrak{O}_K$. To explain this, we first need to generalize the notion of the discriminant of $\mathfrak{O}_L$.

Recall that a *(left) integral* of H is an element $\theta \in H$ that satisfies

$$h\theta = \varepsilon(h)\theta,$$

for all $h \in H$. The set of integrals of H forms an ideal of H, denoted by I_H (see [**Chi00**, §3]), and by [**Und11**, Proposition 4.3.3], there exists a generating integral Λ for I_H. Suppose that $\mathfrak{O}_L$ is an H-extension of $\mathfrak{O}_K$, and let $\{x_1, x_2, \ldots, x_n\}$ denote a basis for $\mathfrak{O}_L$ over $\mathfrak{O}_K$. Let M denote the $n \times n$ matrix whose i, jth entry is

$$m_{i,j} = \Lambda \cdot (x_i x_j).$$

The *discriminant* of $\mathfrak{O}_L$ over $\mathfrak{O}_K$ with respect to H is the ideal of $\mathfrak{O}_K$ defined by

$$\det(M)\mathfrak{O}_K.$$

The discriminant is independent of the choice of $\mathfrak{O}_K$-basis for $\mathfrak{O}_L$. We denote the discriminant with respect to H by $\mathrm{disc}_H(\mathfrak{O}_L/\mathfrak{O}_K)$.

PROPOSITION 12.5. *Let K be a p-adic field and let L/K be a G-Galois extension. Let H be an $\mathfrak{O}_K$-Hopf order in $K[G]$, and suppose that $\mathfrak{O}_L$ is an H-extension of $\mathfrak{O}_K$. Then $\mathfrak{O}_L$ is an H-Galois extension of $\mathfrak{O}_K$ if and only if* $\mathrm{disc}_H(\mathfrak{O}_L/\mathfrak{O}_K) = \mathfrak{O}_K$.

PROOF. See [**Und11**, Proposition 10.3.8]. □

Now we give an example of an extension L/K for which $\mathfrak{A}$ is a Hopf order in $K[G]$ (so $\mathfrak{O}_L$ is a free rank one $\mathfrak{A}$-module and a tame $\mathfrak{A}$-extension of $\mathfrak{O}_K$) but $\mathfrak{O}_L$ is not an $\mathfrak{A}$-Galois extension of $\mathfrak{O}_K$.

EXAMPLE 12.6. Let $K = \mathbb{Q}_p(\zeta_p)$, where ζ_p is a primitive pth root of unity, let π be a uniformizing parameter for $\mathfrak{O}_K$. Then the polynomial $x^p - \pi$ is irreducible over K by Eisenstein's criterion. Let $L = K(z)$, where z is a root of $x^p - \pi$. Then L/K is Galois with group $C_p = \langle g \rangle$; the Galois action is given as $g^i(z) = \zeta_p^i z$. Since $\pi = z^p$, L/K is wild. We have $\mathfrak{O}_L = \mathfrak{O}_K[z]$.

As stated in Example 12.2, the associated order $\mathfrak{A}$ is an $\mathfrak{O}_K$-Hopf order in $K[C_p]$. By results of Childs discussed above, $\mathfrak{O}_L$ is isomorphic to $\mathfrak{A}$ as $\mathfrak{A}$-modules and $\mathfrak{O}_L$ is a tame $\mathfrak{A}$-extension of $\mathfrak{O}_K$.

Note that

$$\Lambda = \frac{1}{p} \sum_{g \in C_p} g$$

is a generating integral for $\mathfrak{A}$. We compute $\mathrm{disc}_{\mathfrak{A}}(\mathfrak{O}_K[z]/\mathfrak{O}_K)$. We first compute $\mathrm{disc}(\mathfrak{O}_K[z]/\mathfrak{O}_K)$. With respect to the basis $\{1, z, z^2, \ldots, z^{p-1}\}$, we have by [**Lan65**, p. 348],

$$\mathrm{disc}(\mathfrak{O}_K[z]/\mathfrak{O}_K) = (-1)^{p(p-1)/2} \mathrm{N}_{L/K}(pz^{p-1})\mathfrak{O}_K = p^p \pi^{p-1} \mathfrak{O}_K.$$

Thus

$$\begin{aligned} p^p \pi^{p-1} \mathfrak{O}_K &= \det(\mathrm{Tr}_{L/K}(z^a z^b))\mathfrak{O}_K \\ &= p^p \det(\mathrm{Tr}_{L/K}(z^a z^b)/p)\mathfrak{O}_K \\ &= p^p \det(\Lambda(z^a z^b))\mathfrak{O}_K. \end{aligned}$$

Thus,

$$\mathrm{disc}_{\mathfrak{A}}(\mathfrak{O}_L/\mathfrak{O}_K) = \pi^{p-1}\mathfrak{O}_K \neq \mathfrak{O}_K.$$

Thus by Proposition 12.5, $\mathfrak{O}_L$ is not a $\mathfrak{A}$-Galois extension of $\mathfrak{O}_K$.

In this chapter we shall construct and classify classes of Hopf orders in group algebras over K and decide which of those are realizable or not (in the sense of Definition 12.4).

12.2. Dual Hopf orders

If H is an $\mathfrak{O}_K$-Hopf algebra, free of finite rank over $\mathfrak{O}_K$, then the linear dual $H^* = \mathrm{Hom}_{\mathfrak{O}_K}(H, \mathfrak{O}_K)$ is an $\mathfrak{O}_K$-Hopf algebra with structure maps induced from those on H by duality (see [**Und11**, Proposition 4.1.7] or [**Und15**, Proposition 3.1.12]).

If K is a field and G is a finite group then the group algebra $K[G]$ is a K-Hopf algebra with comultiplication $\Delta : K[G] \to K[G] \otimes_K K[G]$ defined as $g \mapsto g \otimes g$, counit $\varepsilon : K[G] \to K$ given as $g \mapsto 1$, and antipode $S : K[G] \to K[G]$ defined as $g \mapsto g^{-1}$, for $g \in G$.

Let $\{e_g\}_{g \in G}$ be the basis for $K[G]^*$ that is dual to the basis $\{g\}_{g \in G}$ for $K[G]$, i.e., $\langle e_g, h \rangle = \delta_{g,h}$ where

$$\langle -, - \rangle : K[G]^* \times K[G] \to K$$

is the duality pairing.

By [**Und11**, Proposition 4.1.7] (or [**Und15**, Proposition 3.1.12]), $K[G]^*$ is a K-Hopf algebra. Multiplication on $K[G]^*$ is defined through duality:

$$\begin{aligned}(e_g e_h)(s) &= m_{K[G]}(e_g \otimes e_h)\Delta_{K[G]}(s)\\ &= m_{K[G]}(e_g \otimes e_h)(s \otimes s)\\ &= e_g(s)e_h(s)\\ &= \delta_{g,s}\delta_{h,s}\\ &= \delta_{g,h}e_g(s)\end{aligned}$$

for $g, h, s \in G$. Thus $\{e_g\}_{g\in G}$ is a set of mutually orthogonal idempotents and as a K-algebra $K[G]^*$ decomposes as

$$K[G]^* = \bigoplus_{g\in G} Ke_g.$$

The unit map $i_{K[G]^*} : K \to K[G]^*$ is defined through duality:

$$i_{K[G]^*}(r)(g) = r(\varepsilon_{K[G]}(g)) = r,$$

for $r \in K$, $g \in G$.

The Hopf algebra structure on $K[G]^*$ is also defined through duality. Comultiplication is given as

$$\begin{aligned}\Delta_{K[G]^*}(e_g)(s \otimes t) &= e_g(m_{K[G]}(s \otimes t))\\ &= e_g(st)\\ &= \Big(\sum_{\substack{a,b\in G,\\ g=ab}} e_a \otimes e_b\Big)(s \otimes t)\end{aligned}$$

for $g, s, t \in G$. Thus

$$\Delta_{K[G]^*}(e_g) = \sum_{\substack{a,b\in G,\\ g=ab}} e_a \otimes e_b.$$

The counit map is given as

$$\begin{aligned}\varepsilon_{K[G]^*}(e_g)(r) &= e_g(i_{K[G]}(r))\\ &= e_g(1)r\end{aligned}$$

for $g, r \in G$, thus

$$\varepsilon_{K[G]^*}(e_g) = e_g(1).$$

Finally, the antipode is given by

$$\begin{aligned}S_{K[G]^*}(e_g)(t) &= e_g(S_{K[G]}(t))\\ &= e_g(t^{-1})\\ &= e_{g^{-1}}(t)\end{aligned}$$

for $g, t \in G$, thus

$$S_{K[G]^*}(e_g) = e_{g^{-1}}.$$

PROPOSITION 12.7. *Let $G = C_{p^n}$ denote the cyclic group of order p^n, $n \geq 1$, and suppose that K contains a primitive p^nth root of unity, ζ_{p^n}. Then $K[C_{p^n}]^* \cong K[C_{p^n}]$ as K-Hopf algebras.*

PROOF. Let $\{e_i\}_{i=0}^{p^n-1}$ be the basis for $K[C_{p^n}]^*$ dual to the basis $\{g^i\}_{i=0}^{p^n-1}$ for $K[C_{p^n}]$. There is a K-Hopf algebra isomorphism $\psi : K[C_{p^n}]^* \to K[C_{p^n}]$ defined as

$$\psi(e_i) = \frac{1}{p^n} \sum_{j-0}^{p^n-1} \zeta_{p^n}^{-ij} g^j$$

for $0 \le i \le p^n - 1$. □

REMARK 12.8. Suppose K contains a primitive p^nth root of unity, ζ_{p^n}. Let $\widehat{C}_{p^n}$ denote the group of characters of C_{p^n}. Then $\widehat{C}_{p^n}$ is cyclic of order p^n, generated by the character $\gamma : C_{p^n} \to K$; we have $C_{p^n} = \langle \gamma \rangle = \{\gamma^i\}$, $0 \le i \le p^n - 1$, where $\gamma^i(g^j) = \zeta_{p^n}^{ij}$, $0 \le j \le p^n - 1$; $\widehat{C}_{p^n} \cong C_{p^n}$, the isomorphism being given as $\gamma \mapsto g$. $K[\widehat{C}_{p^n}] \cong K[C_{p^n}]$ as K-Hopf algebras through the map $\gamma \mapsto g$.

We make the identification

$$K[C_{p^n}]^* := K[\widehat{C}_{p^n}],$$

where e_i, $0 \le i \le p^n - 1$, is identified with the element $\frac{1}{p^n} \sum_{j=0}^{p^n-1} \zeta_{p^n}^{-ij} \gamma^j$ in $K[\widehat{C}_{p^n}]$.

The linear dual $H^* = \mathrm{Hom}_{\mathfrak{O}_K}(H, \mathfrak{O}_K)$ can be defined as

$$H^* = \{x \in K[G]^* \mid \langle x, H \rangle \subseteq \mathfrak{O}_K\}.$$

PROPOSITION 12.9. *Let H be an $\mathfrak{O}_K$-Hopf order in $K[G]$, G a finite group. Then H^* is an $\mathfrak{O}_K$-Hopf order in $K[G]^*$.*

PROOF. By [**Und15**, Proposition 3.1.12], $K[G]^*$ is a finite dimensional Hopf algebra over K. We check that H^* is an $\mathfrak{O}_K$-order in $K[G]^*$. By [**Und11**, Proposition 4.1.7], H^* is an $\mathfrak{O}_K$-Hopf algebra with multiplication induced from that of $K[G]^*$ since $\Delta_{K[G]}(H) \subseteq H \otimes_{\mathfrak{O}_K} H$. Thus H^* is an $\mathfrak{O}_K$-subalgebra of $K[G]^*$.

In addition, H^* is an $\mathfrak{O}_K$-submodule $K[G]^*$, free of rank $|G|$ over $\mathfrak{O}_K$, and

$$K \otimes_{\mathfrak{O}_K} H^* \cong (K \otimes_{\mathfrak{O}_K} H)^* \cong K[G]^*$$

as K-algebras. Thus H^* is an $\mathfrak{O}_K$-order in $K[G]^*$.

Finally, the maps $\Delta_{K[G]^*}$, $\varepsilon_{K[G]^*}$ and $S_{K[G]^*}$ induce on H^* its structure as an $\mathfrak{O}_K$-Hopf algebra. Thus H^* is an $\mathfrak{O}_K$-Hopf order in $K[G]^*$. □

We stated earlier that the integral group ring $\mathfrak{O}_K[G]$ is an example of an $\mathfrak{O}_K$-Hopf order in $K[G]$. In fact:

PROPOSITION 12.10. *Let H be an $\mathfrak{O}_K$-Hopf order in $K[G]$. Then $\mathfrak{O}_K[G] \subseteq H$.*

PROOF. The group ring $\mathfrak{O}_K[G]$ is an $\mathfrak{O}_K$-Hopf order in $K[G]$. Thus by Proposition 12.9, $\mathfrak{O}_K[G]^*$ is an $\mathfrak{O}_K$-Hopf order in $K[G]^*$. We have

$$\mathfrak{O}_K[G]^* = \bigoplus_{g \in G} \mathfrak{O}_K e_g,$$

where $\{e_g\}_{g \in G}$ is the basis for $\mathfrak{O}_K[G]^*$ dual to the basis $\{g\}_{g \in G}$ for $\mathfrak{O}_K[G]$.

We claim that $\mathfrak{O}_K[G]^*$ is the maximal $\mathfrak{O}_K$-Hopf order in $K[G]^*$. To see this, let $\alpha = \sum_{g \in G} r_g e_g \in K[G]^*$ with $r_g \in K \setminus \mathfrak{O}_K$ for some $g \in G$. Then α cannot be integral over $\mathfrak{O}_K$. Thus $\mathfrak{O}_K[G]^*$ is maximal. Consequently, $H^* \subseteq \mathfrak{O}_K[G]^*$ and so $\mathfrak{O}_K[G] \subseteq H$. □

Proposition 12.10 is also proved in [**Lar67**, Theorem], [**Und11**, Proposition 4.4.3] without using duality.

12.2.1. Galois H^*-objects and realizability. We have seen that if H is an $\mathfrak{O}_K$-Hopf algebra, free of finite rank over $\mathfrak{O}_K$, then the linear dual $H^* = \mathrm{Hom}_{\mathfrak{O}_K}(H, \mathfrak{O}_K)$ is an $\mathfrak{O}_K$-Hopf algebra. Let $\Delta = \Delta_{H^*}$ and $\varepsilon = \varepsilon_{H^*}$ denote the comultiplication and counit map of H^*. An $\mathfrak{O}_K$-module A is called a *right H^*-comodule* if there exists an $\mathfrak{O}_K$-linear map $\alpha : A \to A \otimes_{\mathfrak{O}_K} H^*$ for which

(i) $(\alpha \otimes Id)\alpha = (Id \otimes \Delta)\alpha$,
(ii) $(Id \otimes \varepsilon)\alpha(b) = b \otimes 1$, $\forall b \in A$.

We adapt the Sweedler notation to write

$$\alpha(b) = \sum_{(b)} b_{(1)} \otimes b_{(2)},$$

for $b, b_{(1)} \in A$, $b_{(2)} \in H^*$.

Now suppose that A is an $\mathfrak{O}_K$-algebra with multiplication $m_A : A \otimes_{\mathfrak{O}_K} A \to A$ and unit map $i_A : \mathfrak{O}_K \to A$ that is also a (right) H^*-comodule. Then $A \otimes_{\mathfrak{O}_K} A$ is an H^*-comodule via

$$\alpha : A \otimes A \to A \otimes A \otimes H^*$$

defined as $b \otimes c \mapsto \sum_{(b),(c)} b_{(1)} \otimes c_{(1)} \otimes b_{(2)}c_{(2)}$ and $\mathfrak{O}_K$ is an H^*-comodule via $\alpha(r) = r \otimes 1_{H^*}$. We say that A is an *H^*-comodule algebra* if m_A, i_A are H^*-comodule homomorphisms, and that A is a *Galois H^*-object* if in addition the map

$$\gamma : A \otimes A \to A \otimes H^*$$

defined by $\gamma(b \otimes c) = (b \otimes 1)\alpha(c)$ is an isomorphism. (See Chapter 10.)

These concepts appeared in Section 10.1. Similarly to Proposition 10.9 and Corollary 10.10, we have

PROPOSITION 12.11. *Let H be an $\mathfrak{O}_K$-Hopf algebra that is free of finite rank over $\mathfrak{O}_K$. Then A is an H-Galois extension of $\mathfrak{O}_K$ if and only if A is a Galois H^*-object.*

PROOF. See [**Chi00**, (2.10) Corollary]. □

We can infer the realizability (in the sense of Definition 12.4) of $\mathfrak{O}_K$-Hopf orders though the use of short exact sequences.

Let G be a finite abelian group with $G' \leq G$. The short exact sequence of groups

$$1 \to G' \to G \to G'' \to 1,$$

$G'' \cong G/G'$, induces a short exact sequence of K-Hopf algebras

$$K \to K[G'] \to K[G] \to K[G''] \to K$$

[**Chi00**, §4].

Let H be an $\mathfrak{O}_K$-Hopf order in $K[G]$. Then there is a short exact sequence of $\mathfrak{O}_K$-Hopf orders

$$\mathfrak{O}_K \to H' \to H \to H'' \to \mathfrak{O}_K,$$

dualizing as

$$\mathfrak{O}_K \to J' \to J \to J'' \to \mathfrak{O}_K,$$

where $J' = (H'')^*$, $J = H^*$ and $J'' = (H')^*$.

PROPOSITION 12.12. *Suppose A is a Galois J-object with J-comodule algebra map $\alpha : A \to A \otimes J$. Then*

$$A_{J'} = \{x \in A \mid \alpha(x) \in A \otimes_{\mathfrak{O}_K} J' \subseteq A \otimes_{\mathfrak{O}_K} J\}$$

is a Galois J'-object and A is a Galois $(A_{J'} \otimes_{\mathfrak{O}_K} J'')$-object.

PROOF. See [**Gre92**, Part II, Lemma 1.6]. □

COROLLARY 12.13. *Suppose that A is an H-Galois extension of $\mathfrak{O}_K$. Then $A_{J'}$ is an H''-Galois extension of $\mathfrak{O}_K$ and A is an H'-Galois extension of $A_{J'}$.*

PROOF. This follows from Proposition 12.11. □

Corollary 12.13 generalizes the classical Fundamental Theorem of Galois Theory in the following sense. Suppose H is realizable as a Galois group via the Galois extension L/K with group G. Then $\mathfrak{O}_L$ is an H-Galois extension of $\mathfrak{O}_K$. Thus $(\mathfrak{O}_L)_{J'}$ is a Galois H''-extension of $\mathfrak{O}_K$. In fact, $(\mathfrak{O}_L)_{J'}$ is the ring of integers of the fixed field L' of $G' \subseteq G$. In other words, the Hopf quotient

$$H/(H')^+H \cong H''$$

is realizable as a Galois group via the extension L'/K with group $G/G' \cong G''$.

Moreover, $\mathfrak{O}_L$ is a Galois H'-extension of $(\mathfrak{O}_L)_{J'}$ and the Hopf kernel H' is realizable as a Galois group via the Galois extension L/L' with group G'.

Thus:

COROLLARY 12.14. *If H is realizable as a Galois group, then H'' and H' are realizable as Galois groups.*

12.3. Byott's theorem on realizability

In [**Byo04c**, Theorem 6.1] Byott provided a convenient criterion for realizability.

Byott's result requires that certain Hopf algebras are local rings. So we review a general method for showing that a commutative $\mathfrak{O}_K$-Hopf algebra is local.

Let A be a commutative ring. The *nilradical* of A, denoted as $\mathrm{nil}(A)$, is the ideal consisting of all of the nilpotent elements of A.

PROPOSITION 12.15. *Let*

$$\mathfrak{O}_K \to H' \to H \overset{s}{\to} H'' \to \mathfrak{O}_K$$

be a short exact sequence of finite (finitely generated, projective as $\mathfrak{O}_K$-modules) commutative $\mathfrak{O}_K$-Hopf algebras. Then H is local if and only if H' and H'' are local.

PROOF. We may consider the situation over $k = \mathfrak{O}_K/\mathfrak{p}$.

Suppose that $kH' = k \otimes_{\mathfrak{O}_K} H'$ and $kH'' = k \otimes_{\mathfrak{O}_K} H''$ are local. We show that $kH = k \otimes_{\mathfrak{O}_K} H$ is local. Then $\ker(\varepsilon_{kH'})$ is the unique maximal ideal of kH' and $\ker(\varepsilon_{kH''})$ is the unique maximal ideal of kH''. By [**Und11**, Proposition 1.3.2], $\mathrm{nil}(kH')$ is the intersection of all prime ideals of kH'. But each prime ideal of kH' is maximal, hence $\mathrm{nil}(kH') = \ker(\varepsilon_{kH'})$. Likewise $\mathrm{nil}(kH'') = \ker(\varepsilon_{kH''})$.

We always have $\mathrm{nil}(kH) \subseteq \ker(\varepsilon_{kH})$. Let $x \in \ker(\varepsilon_{kH})$. Then $s(x) \in \mathrm{nil}(kH'')$ and so, $(s(x))^n = s(x^n) = 0$ for some $n > 0$. Thus $x^n \in \ker(\varepsilon_{kH'})kH = \mathrm{nil}(kH')kH$. Thus $x^n \in \mathrm{nil}(kH)$. And so, $\mathrm{nil}(kH) \subseteq \ker(\varepsilon_{kH})$, which implies that kH is local. It follows that H is local.

Conversely, suppose that kH is local, so that $\ker(\varepsilon_{kH}) = \mathrm{nil}(kH)$. Then $\ker(\varepsilon_{kH'}) = \mathrm{nil}(kH')$ and $\ker(\varepsilon_{kH''}) = \mathrm{nil}(kH'')$ and so, KH' and kH'' are local. Thus H' and H'' are local. □

THEOREM 12.16 (Byott). *Let K be a p-adic field, let G be an abelian group of order p^n, $n \geq 1$. Let H be an $\mathfrak{O}_K$-Hopf order in $K[G]$ and suppose that H^* is a local ring. Then H is realizable as a Galois group if and only if H^* is a monogenic $\mathfrak{O}_K$-algebra.*

PROOF. Suppose H^* is monogenic of rank $q = p^n$ over $\mathfrak{O}_K$. Set $m = q - 1$. Let $\overline{x} = (x_1, x_2, \cdots, x_m)$, x_i indeterminate. Set $q_1 = q$, $q_i = 1$, $2 \le i \le m$. Let $\mathrm{wt} : \mathfrak{O}_K[[\overline{x}]] \to \mathbb{Z} \cup \{\infty\}$ denote the weight function determined by the integers $q_1, q_2, \ldots, q_m$ [**Byo04c**, Definition 4.5].

By Byott's Embedding Theorem ([**Byo04c**, Theorem 5.4]) there exists an isogeny

$$\varphi : F \to G$$

of m-dimensional formal groups with

$$H^* \cong \mathfrak{O}_K[[\overline{x}]]_F/(\varphi(\overline{x}))$$

where the isogeny is defined as

$$\varphi(\overline{x}) = (\varphi_1(\overline{x}), \varphi_2(\overline{x}), \ldots, \varphi_m(\overline{x}))$$

with

$$\varphi_1(\overline{x}) \equiv x_1^q \bmod \mathrm{wt}\ q + 1,$$

$$\varphi_i(\overline{x}) = x_i, \quad 2 \le i \le m.$$

Here the equivalence is defined as follows: $\varphi_1(\overline{x}) \equiv x_1^q \bmod \mathrm{wt}\ q + 1$ if $\varphi_1 - x_1^q = \sum_\alpha c_\alpha \overline{x}^\alpha$ where $c_\alpha = 0$ for all α with $\mathrm{wt}(\overline{x}^\alpha) < q + 1$.

Let $c \in \mathfrak{p} \backslash \mathfrak{p}^2$, and let $A = \mathfrak{O}_K[[\overline{x}]]/J$, where J is the ideal

$$(\varphi_1(\overline{x}) - c, \varphi_2(\overline{x}), \ldots, \varphi_m(\overline{x})).$$

By [**Byo04c**, Lemma 3.7], A is a Galois H^*-object. Thus by Proposition 12.11, A is an H-Galois extension of $\mathfrak{O}_K$. We have

$$A \cong \mathfrak{O}_K[[x_1]]/(\psi(x_1) - c),$$

with $\psi(x_1) = \varphi_1(x_1, 0, 0, \ldots, 0)$.

The power series $\psi(x_1)$ satisfies $\psi(x_1) \equiv x_1^q \bmod \deg\ q + 1$. By the Weierstrass Preparation Theorem,

$$\psi(x_1) - c = f(x_1)u(x_1)$$

where $f(x_1)$ is an irreducible polynomial of degree q and $u(x_1)$ is an invertible power series. Consequently,

$$A \cong \mathfrak{O}_K[[x_1]]/(f(x_1)) \cong \mathfrak{O}_K[x_1]/(f(x_1)).$$

Let α be a zero of $f(x_1)$ in some algebraic closure K^{alg}. Then $A \cong \mathfrak{O}_K[\alpha]$ is the ring of integers of $L = K(\alpha)$.

We know that $\mathfrak{O}_K[\alpha]$ is an H-Galois extension of $\mathfrak{O}_K$. Thus (see discussion following Definition 7.1)

$$j : \mathfrak{O}_K[\alpha] \# H \to \mathrm{End}(\mathfrak{O}_K[\alpha])$$

is an isomorphism of algebras. Since $\mathfrak{O}_K[G] \subseteq H$, this says that L/K is Galois with group G and so H is realizable as a Galois group.

For the converse suppose there exists a Galois extension L/K with group G, $|G| = p^n$, for which $\mathfrak{O}_L$ is H-Galois. Let $k = \mathfrak{O}_K/\mathfrak{p}$ denote the residue class field of $\mathfrak{O}_K$. Since k is finite, $\mathfrak{O}_L$ is monogenic over $\mathfrak{O}_K$ [**CF67**, Proposition 1, p. 33]. Thus $\mathfrak{O}_L/\mathfrak{p}\mathfrak{O}_L$ is a monogenic k-algebra.

Now $k \otimes_{\mathfrak{O}_K} \mathfrak{O}_L = \mathfrak{O}_L/\mathfrak{p}\mathfrak{O}_L$ is a Galois $(k \otimes_{\mathfrak{O}_K} H)$-extension of k and so $k \otimes_{\mathfrak{O}_K} H^*$ is monogenic over k by [**Byo04c**, Corollary 4.4]. By Nakayama's lemma, H^* is then monogenic over $\mathfrak{O}_K$. □

REMARK 12.17. In the proof of Theorem 12.16 it is important that H^* is local. For then $k \otimes_{\mathfrak{O}_K} H^*$ is local with unique maximal ideal its augmentation ideal $(k \otimes_{\mathfrak{O}_K} H^*)^+$.

Now, the nilradical $\mathrm{nil}(k \otimes_{\mathfrak{O}_K} H^*)$ is the intersection of all prime ideals of $k \otimes_{\mathfrak{O}_K} H^*$. But every non-zero prime ideal of $k \otimes_{\mathfrak{O}_K} H^*$ is maximal. Thus, $\mathrm{nil}(k \otimes_{\mathfrak{O}_K} H^*) = (k \otimes_{\mathfrak{O}_K} H^*)^+$.

By [**Wat79**, 6.8, Lemma], the largest separable algebra of $k \otimes_{\mathfrak{O}_K} H^*$ is k and so by [**Wat79**, 6.6, Theorem], $k \otimes_{\mathfrak{O}_K} H^*$ represents a connected k-group scheme. We can then use [**Wat79**, 14.4, Theorem] to write

$$k \otimes_{\mathfrak{O}_K} H^* \cong k[x_1]/(x_1^q)$$

(since $k \otimes_{\mathfrak{O}_K} H^*$ is also monogenic). This is a key step in Byott's proof of his Embedding theorem.

Byott tells us precisely when a Hopf order is realizable in the sense of Definition 12.3.

THEOREM 12.18 (Byott). *Let K be a p-adic field, let G be an abelian group G of order p^n, $n \geq 1$. Let H be an $\mathfrak{O}_K$-Hopf order in $K[G]$ and suppose that both H and H^* are local rings. Then H is realizable if and only if H^* is a monogenic $\mathfrak{O}_K$-algebra. That is, there exists a Galois extension L/K with group G and $\mathfrak{A} = H$, if and only if H^* is monogenic.*

PROOF. If H^* is monogenic, then by Theorem 12.16, there exists a Galois extension L/K with group G for which $\mathfrak{O}_L$ is an H-Galois extension of $\mathfrak{O}_K$. Thus $H = \mathfrak{A}_{L/K}$, and so H is realizable.

Conversely, suppose H is realizable, i.e., there exists a Galois extension L/K with group G and $\mathfrak{A}_{L/K} = H$. Then by Childs' Theorem $\mathfrak{O}_L$ is is free rank one H-module and so $\mathfrak{O}_L$ is a tame H-extension.

Since H is local, [**Chi00**, (14.7) Theorem, (2) $\rightarrow$ (3)] applies to show that $\mathfrak{O}_L$ is an H-Galois extension of $\mathfrak{O}_K$. Thus by Theorem 12.16, H^* is monogenic. □

12.4. Group valuations and Larson orders

Let K be a p-adic field. For G arbitrary, the only general construction of Hopf orders in $K[G]$ is due to Larson [**Lar76**]. Larson's construction involves the notion of a group valuation. We follow the treatment given in [**Und11**, §5.2, §5.3].

12.4.1. Group valuations.

DEFINITION 12.19. Let G be a finite group. A *group valuation* is a function $\xi : G \rightarrow \mathbb{Z} \cup \{\infty\}$ which satisfies, for all $g, h \in G$,

(i) $\xi(g) \geq 0$;

(ii) $\xi(g) = \infty$ if and only if $g = 1$;

(iii) $\xi(gh) \geq \min\{\xi(g), \xi(h)\}$;

(iv) $\xi(ghg^{-1}h^{-1}) \geq \xi(g) + \xi(h)$.

We review some properties of the group valuation ξ on G.

PROPOSITION 12.20. *For all $g, h \in G$,*

(i) $\xi(g) = \xi(g^{-1})$;

(ii) $\xi(ghg^{-1}) = \xi(h)$.

Proof. For (i), let $g \in G$ and let $n = |\langle g \rangle|$, thus $g^{n-1} = g^{-1}$. Since $g = (g^{n-1})^{n-1}$,

$$\xi(g) = \xi((g^{n-1})^{n-1}) \geq \xi(g^{n-1}) = \xi(g^{-1})$$

by Definition 12.19(iii). On the other hand, $\xi(g^{-1}) = \xi(g^{n-1}) \geq \xi(g)$.

For (ii), let $g, h \in G$. Then

$$\begin{aligned} \xi(ghg^{-1}) &= \xi(ghg^{-1}h^{-1}h) \\ &\geq \min\{\xi(ghg^{-1}h^{-1}), \xi(h)\} \quad \text{by Definition 12.19(iii)} \\ &\geq \min\{\xi(g) + \xi(h), \xi(h)\} \quad \text{by Definition 12.19(iv)} \\ &\geq \xi(h). \end{aligned}$$

Now,

$$\xi(h) = \xi(g^{-1}(ghg^{-1})g) \geq \xi(ghg^{-1}) \geq \xi(h),$$

thus $\xi(h) = \xi(ghg^{-1})$. □

Let $\mathbb{Z}_{\geq 0} = \{r \in \mathbb{Z} | r \geq 0\}$ and let ξ be a group valuation on G. For each $r \in \mathbb{Z}_{\geq 0}$, put

$$G_r = \{g \in G \mid \xi(g) \geq r\}.$$

Proposition 12.21. *For each $r \in \mathbb{Z}_{\geq 0}$, G_r is a normal subgroup of G.*

Proof. By Definition 12.19(iii), G_r is closed under the binary operation of G. We have $1 \in G_r$ since $\xi(1) = \infty$ and $g^{-1} \in G_r$ whenever $g \in G_r$ by Proposition 12.20(i). Thus $G_r \leq G$.

Now let $gh \in gG_r$. Then $\xi(ghg^{-1}) = \xi(h) \geq r$ by Proposition 12.20(ii) and so, $ghg^{-1} \in G_r$. Thus $gG_r \subseteq G_r g$. Next, let $hg \in G_r g$. Then $\xi(g^{-1}hg) = \xi(h) \geq r$, and so $g^{-1}hg \in G_r$, thus, $G_r g \subseteq gG_r$. □

Proposition 12.22. *Let ξ be a group valuation on G, where G is a p-group of order p^n. Then ξ has at most n distinct finite values.*

Proof. Let $m = \max\{\xi(g) \mid g \in G, g \neq 1\}$. We have $G_0 = G$, $1 = G_{m+1}$, and $1 \subset G_m$ since there exists at least one $g \in G$ with $\infty > \xi(g) \geq m$. Now by Proposition 12.21, $G_j \lhd G$ for $0 \leq j \leq m$ and so there is a normal series

$$1 \subset G_m \subseteq G_{m-1} \subseteq \cdots \subseteq G_1 \subseteq G_0 = G$$

which can be refined to a composition series

$$1 \subset N_s \subset N_{s-1} \subset \cdots \subset N_1 \subset N_0 = G$$

with $N_i/N_{i+1} \cong C_p$, for $i = 0, 1, \ldots, s$ (here, $N_{s+1} = 1$). Note that $s = n - 1$.

For each i, $0 \leq i \leq s$, there is a subseries

$$G_{r+1} \subseteq N_{i+1} \subset N_i \subseteq G_r.$$

Let $g \in N_i \backslash N_{i+1}$. Since $g \in G_r$, $\xi(g) \geq r$. But since $g \notin N_{i+1}$, $g \notin G_{r+1}$ and thus, $\xi(g) \ngeq r + 1$. Thus $\xi(g) = r$. Let $r_i = r$ and let $\{r_i\}$ denote the collection of the r_i as i ranges from 0 to s.

Note that each non-trivial $g \in G$ is in $N_i \backslash N_{i+1}$ for exactly one i and so the collection $\{r_i\}_{i=0}^s$ determines the values of ξ. At most n of the values in $\{r_i\}_{i=0}^s$ are distinct since $s + 1 = n$. □

DEFINITION 12.23. Let G be a finite group. The group valuation $\xi : G \to \mathbb{Z} \cup \{\infty\}$ is *order bounded* with respect to the discrete valuation ν if

(i) $\xi(g) = 0$ for $|\langle g\rangle|$ not a power of p,

(ii) $\xi(g) \le e/(p^a - p^{a-1})$ for $|\langle g\rangle| = p^a$, $a \ge 1$ (where $e = \nu(p)$).

An order bounded group valuation (obgv) ξ on G is *p-adic* if $\xi(g^p) \ge p\xi(g)$ for all $g \in G$.

12.4.2. Hopf orders from group valuations (Larson orders). Let G be a finite group. Larson [**Lar76**, Proposition 3.2] has shown that order bounded group valuations on G give rise to Hopf orders in $K[G]$. We sketch Larson's proof, for complete details see [**Chi00**, (18.1)].

THEOREM 12.24 (Larson). *Let K be a p-adic field and let G be a finite group. Let ξ be a group valuation on G which is order bounded with respect to ν. Let $A(\xi)$ be the $\mathfrak{O}_K$-subalgebra of $K[G]$ generated by $\pi^{-\xi(g)}(g-1)$ for $g \ne 1$. Then $A(\xi)$ is an $\mathfrak{O}_K$-Hopf order in $K[G]$.*

PROOF. We consider first the case where $\xi(g) > 0$ for all $g \in G$. The elements of $G\backslash\{1\}$ can be ordered by the rule: $g \le h$ if and only if $\xi(g) \le \xi(h)$. We have the list of elements of $G\backslash\{1\}$:

$$g_1 \le g_2 \le g_3 \le \cdots \le g_l.$$

For each i, $1 \le i \le l$, let A_{g_i} be the $\mathfrak{O}_K$-subalgebra of $K\langle g_i\rangle$ of the form

$$A_{g_i} = \mathfrak{O}_K[\pi^{-\xi(g_i)}(g_i - 1)].$$

Since ξ is order bounded, each A_{g_i} is a finitely generated $\mathfrak{O}_K$-module and thus the tensor product

$$A_{g_1} \otimes A_{g_2} \otimes \cdots \otimes A_{g_l}$$

is finitely generated.

Larson next shows that there is a surjective map of $\mathfrak{O}_K$-modules

$$s : A_{g_1} \otimes A_{g_2} \otimes \cdots \otimes A_{g_l} \to A(\xi),$$

defined by $s(a_{g_1} \otimes a_{g_2} \otimes \cdots \otimes a_{g_l}) = a_{g_1}a_{g_2}\cdots a_{g_l}$. Thus $A(\xi)$ is finitely generated.

Now, let ξ be an arbitrary obgv on G and let $G_+ = \{g \in G \mid \xi(g) > 0\}$. By Proposition 12.21, G_+ is a normal subgroup of G. Now, $\xi|_{G_+}$ is an obgv on G_+ and as we have seen above, $A(\xi|_{G_+})$ is a finitely generated $\mathfrak{O}_K$-submodule of $K[G_+]$. It follows that

$$A(\xi) = A(\xi|_{G_+})\mathfrak{O}_K[G]$$

is finitely generated.

Since $K \otimes_{\mathfrak{O}_K} A(\xi) \cong K[G]$, $A(\xi)$ is an $\mathfrak{O}_K$-order in $K[G]$.

We have $\varepsilon_{K[G]}(A(\xi)) \subseteq \mathfrak{O}_K$. Also, since $\Delta_{K[G]}(g) = g \otimes g$ for all $g \in G$,

$$\Delta_{K[G]}\left(\frac{g-1}{\pi^{\xi(g)}}\right) = \frac{g-1}{\pi^{\xi(g)}} \otimes 1 + 1 \otimes \frac{g-1}{\pi^{\xi(g)}} + (g-1) \otimes \frac{g-1}{\pi^{\xi(g)}},$$

and so,

$$\Delta_{K[G]}(A(\xi)) \subseteq A(\xi) \otimes_{\mathfrak{O}_K} A(\xi).$$

Now, $S_{K[G]}(A(\xi)) \subseteq A(\xi)$ follows from Proposition 12.20(i), and so $A(\xi)$ is an $\mathfrak{O}_K$-Hopf algebra with structure induced from $K[G]$. Thus $A(\xi)$ is an $\mathfrak{O}_K$-Hopf order in $K[G]$. □

An $\mathfrak{O}_K$-Hopf order H in $K[G]$ is a *Larson order* in $K[G]$ if H is of the form $A(\xi)$ where ξ is a p-adic obgv on G. Note that the group ring $\mathfrak{O}_K[G]$ is a Larson order given by the trivial p-adic obgv: $\xi(g) = 0$, $\forall g \in G$.

We describe Larson orders in $K[G]$ where $G = C_{p^n}$ is the cyclic group of order p^n generated by g.

First, we consider the case $n = 1$. To characterize a Larson order in $K[C_p]$, we need only describe the possible p-adic obgvs on C_p. By Proposition 12.22, an obgv ξ on C_p has exactly one finite value: we have $\xi(g) = i$ for $g \neq 1$. Since ξ is order bounded, $0 \leq i \leq e'$, where $e' = e/(p-1)$. Since $\xi(g^p) = \xi(1) = \infty \geq \xi(g)$, ξ is p-adic. The Larson order $H = A(\xi)$ is written as

$$H = A(\xi) = \mathfrak{O}_K\left[\frac{g-1}{\pi^i}, \frac{g^2-1}{\pi^i}, \ldots, \frac{g^{p-1}-1}{\pi^i}\right].$$

PROPOSITION 12.25. $H = \mathfrak{O}_K\left[\frac{g-1}{\pi^i}\right]$.

PROOF. Clearly, $\mathfrak{O}_K\left[\frac{g-1}{\pi^i}\right] \subseteq A(\xi)$. So we prove the reverse containment. For $a = 2, 3, \ldots, p-1$, one has

$$\frac{g^a - 1}{\pi^i} = \sum_{m=0}^{a-1} \binom{a}{m} \pi^{i(a-m-1)} \left(\frac{g-1}{\pi^i}\right)^{a-m}$$

Thus $\frac{g^a-1}{\pi^i} \in \mathfrak{O}_K\left[\frac{g-1}{\pi^i}\right]$ which proves the proposition. □

We shall denote the Larson order of Proposition 12.25 by $H(i)$ since it depends on only one valuation parameter i, $0 \leq i \leq e'$.

We next consider Larson orders in $K[G]$ where $G = C_{p^2}$. By Proposition 12.22, a group valuation ξ on C_{p^2} has at most two finite values: $\xi(g^a) = i$, for $1 \leq a \leq p^2 - 1$, $a \equiv 0 \pmod{p}$, and $\xi(g^b) = j$, for $1 \leq b \leq p^2 - 1$, $b \not\equiv 0 \pmod{p}$. Since ξ is order bounded, $0 \leq i \leq e'$ and $0 \leq j \leq e'/p$, and since ξ is p-adic, $pj \leq i$. The Larson order $H = A(\xi)$ can be written as

$$H = A(\xi) = \mathfrak{O}_K\left[\left\{\frac{g^a-1}{\pi^i}, \frac{g^b-1}{\pi^j}\right\}_{a,b}\right],$$

where $1 \leq a \leq p^2 - 1$, $a \equiv 0 \pmod{p}$, $1 \leq b \leq p^2 - 1$, $b \not\equiv 0 \pmod{p}$.

PROPOSITION 12.26. $H = A(\xi) = \mathfrak{O}_K\left[\frac{g^p-1}{\pi^i}, \frac{g-1}{\pi^j}\right]$.

PROOF. The proof of Proposition 12.25 shows that

$$A(\xi) = \mathfrak{O}_K\left[\frac{g^p-1}{\pi^i}\right]\left[\left\{\frac{g^b-1}{\pi^j}\right\}_b\right],$$

$1 \leq b \leq p^2 - 1$, $b \not\equiv 0 \pmod{p}$. Certainly, $\mathfrak{O}_K\left[\frac{g^p-1}{\pi^i}, \frac{g-1}{\pi^j}\right] \subseteq A(\xi)$, so it remains to show that $A(\xi) \subseteq \mathfrak{O}_K\left[\frac{g^p-1}{\pi^i}, \frac{g-1}{\pi^j}\right]$. For $b = 2, 3, \ldots, p^2 - 1$, $b \not\equiv 0 \pmod{p}$, one has

$$\frac{g^b - 1}{\pi^j} = \sum_{m=0}^{b-1} \binom{b}{m} \pi^{j(b-m-1)} \left(\frac{g-1}{\pi^j}\right)^{b-m}.$$

Thus $\frac{g^b-1}{\pi^i} \in \mathfrak{O}_K\left[\frac{g^p-1}{\pi^i}, \frac{g-1}{\pi^j}\right]$ which proves the proposition. □

We denote the Larson order $H = A(\xi)$ in $K[C_{p^2}]$ by $H(i,j)$ since it depends on only two values of the group valuation ξ, namely, $\xi(g^p) = i$ and $\xi(g) = j$.

For the general case, Larson orders in $K[C_{p^n}]$ can be described as follows.

PROPOSITION 12.27. *Let $n \geq 1$, and let ξ be a p-adic obgv on C_{p^n}. Then the Larson order $A(\xi)$ in $K[C_{p^n}]$ has the form*

$$A(\xi) = \mathfrak{O}_K\left[\frac{g^{p^{n-1}}-1}{\pi^{i_1}}, \frac{g^{p^{n-2}}-1}{\pi^{i_2}}, \dots, \frac{g^p-1}{\pi^{i_{n-1}}}, \frac{g-1}{\pi^{i_n}}\right]$$

where i_r, $1 \leq r \leq n$, are integers satisfying $0 \leq pi_r \leq i_{r-1}$ for $2 \leq r \leq n$.

PROOF. By Proposition 12.22, a p-adic obgv on C_{p^n} has n distinct finite values $\xi(g^{p^{n-r}}) = i_r$, for $r = 1, \dots, n$, which satisfy $0 \leq pi_r \leq i_{r-1}$ for $2 \leq r \leq n$. We use induction on n. As we have shown, the proposition is true for the cases $n = 1, 2$, thus we assume that the $(n-1)$st case holds.

Let

$$B = \mathfrak{O}_K\left[\left\{\frac{g^a-1}{\pi^{\xi(g^a)}} \;\middle|\; 1 \leq a \leq p^n - 1, a \equiv 0 \pmod p\right\}\right].$$

Then

$$A(\xi) = B\left[\left\{\frac{g^b-1}{\pi^{i_n}} \;\middle|\; 1 \leq b \leq p^n - 1, b \not\equiv 0 \pmod p\right\}\right].$$

By the induction hypothesis, $B = C$ with

$$C = \mathfrak{O}_K\left[\frac{g^{p^{n-1}}-1}{\pi^{i_1}}, \frac{g^{p^{n-2}}-1}{\pi^{i_2}}, \dots, \frac{g^p-1}{\pi^{i_{n-1}}}\right],$$

and so,

$$A(\xi) = C\left[\left\{\frac{g^b-1}{\pi^{i_n}} \middle| 1 \leq b \leq p^n - 1, b \not\equiv 0 \pmod p\right\}\right].$$

Now,

$$\frac{g^b-1}{\pi^{i_n}} = \sum_{m=0}^{b-1} \binom{b}{m} \pi^{i_n(b-m-1)} \left(\frac{g-1}{\pi^{i_n}}\right)^{b-m}$$

and so,

$$A(\xi) = \mathfrak{O}_K\left[\frac{g^{p^{n-1}}-1}{\pi^{i_1}}, \frac{g^{p^{n-2}}-1}{\pi^{i_2}}, \dots, \frac{g^p-1}{\pi^{i_{n-1}}}\right]\left[\frac{g-1}{\pi^{i_n}}\right]. \qquad \square$$

Larson orders in $K[C_{p^n}]$ will be denoted as $H(i_1, i_2, \dots, i_n)$.

For c with $0 \leq c \leq e'/p^{n-1}$, $\xi(g) = c$ defines a p-adic order bounded group valuation on C_{p^n}; the corresponding Hopf order is the *one-parameter Larson order* in $K[C_{p^n}]$:

$$H(p^{n-1}c, \dots, pc, c) = \mathfrak{O}_K\left[\frac{g^{p^{n-1}}-1}{\pi^{p^{n-1}c}}, \dots, \frac{g^p-1}{\pi^{pc}}, \frac{g-1}{\pi^c}\right].$$

By construction one-parameter Larson orders are monogenic as $\mathfrak{O}_K$-algebras; we have

$$H(p^{n-1}c, \dots, pc, c) = \mathfrak{O}_K\left[\frac{g-1}{\pi^c}\right].$$

One-parameter Larson orders will be used to construct a class of Hopf orders in Section 12.8.

12.4.3. Group valuations from Hopf orders. Let G be a finite group. Then Larson shows that an $\mathfrak{O}_K$-Hopf order in $K[G]$ determines a p-adic obgv on G [**Lar76**, Section 3].

PROPOSITION 12.28. *Let G be a finite group and let H be an $\mathfrak{O}_K$-Hopf order in $K[G]$. Then H determines a p-adic obgv on G.*

PROOF. Define $\xi(g) = \infty$ if $g = 1$. For each $g \in G$, $g \neq 1$, let $|\langle g \rangle| = l$. If l is not a power of p, set $\xi(g) = 0$. If $|\langle g \rangle|$ is a power of p, then define

$$I_g = \{x \in K \mid x(g-1) \in H\}$$

and

$$I_g^{-1} = \{x \in K \mid xI_g \subseteq \mathfrak{O}_K\}.$$

By Proposition 12.10, $\mathfrak{O}_K[G] \subseteq H$ and so $\mathfrak{O}_K \subseteq I_g$. Thus I_g^{-1} is an ideal of $\mathfrak{O}_K$; I_g is a principal fractional ideal in K and we have $I_g^{-1} I_g = \mathfrak{O}_K$. Put $\xi(g) = \nu(I_g^{-1})$.

Thus $\xi(g) \geq 0$ for all $g \in G$ and so, the function $\xi : G \to \mathbb{Z} \cup \{\infty\}$ satisfies Definition 12.19(i),(ii).

For $g, h \in G$, we have

$$gh - 1 = (g-1)h + h - 1,$$

and so, $I_g \cap I_h \subseteq I_{gh}$. Thus, $I_{gh}^{-1} \subseteq (I_g \cap I_h)^{-1}$.

We can assume without loss of generality that $I_g = I_g \cap I_h$ and so, there exists $x \in I_g \cap I_h$ so that $xI_g^{-1} = \mathfrak{O}_K$. Now $x(I_g^{-1} + I_h^{-1}) = \mathfrak{O}_K + xI_h^{-1} = \mathfrak{O}_K$, hence $(I_g \cap I_h)(I_g^{-1} + I_h^{-1}) = \mathfrak{O}_K$. Thus

$$(I_g \cap I_h)^{-1} = I_g^{-1} + I_h^{-1}$$

and consequently, $I_{gh}^{-1} \subseteq I_g^{-1} + I_h^{-1}$. It follows that

$$\xi(gh) \geq \min\{\xi(g), \xi(h)\}.$$

By [**Chi00**, Theorem (17.1)(ii)], $\xi(ghg^{-1}h^{-1}) \geq \xi(g) + \xi(h)$, hence ξ is a group valuation.

To show that ξ is order bounded and p-adic, see [**Chi00**, Theorem (17.1)(iv), (v)]. □

Thus each $\mathfrak{O}_K$-Hopf order H in $K[G]$ gives rise to a p-adic obgv on G which we denote as $\Xi(H)$.

Given an $\mathfrak{O}_K$-Hopf order H in $K[G]$, $\Xi(H)$, in turn, yields a Larson order $A(\Xi(H))$ which is contained in H. For a given H, $A(\Xi(H))$ is the *largest Larson order contained in* H [**Und94**, §1.4], [**Und96**, §2.1], [**Chi00**, §19].

For valuations ξ and ξ' on G, define $\xi \leq \xi'$ if $\xi(g) \leq \xi'(g)$ for all $g \in G$. The relationship between A and Ξ is given as follows.

PROPOSITION 12.29. *Let H be an $\mathfrak{O}_K$-Hopf order in $K[G]$. Let ξ be an obgv on G. Then*

(i) $A(\Xi(H)) \subseteq H$, with equality holding if H is a Larson order.

(ii) $\xi \leq \Xi(A(\xi))$, with equality holding if ξ is p-adic.

PROOF. See [**Chi00**, §19]. □

Let G be any finite group and let H be an $\mathfrak{O}_K$-Hopf order in $K[G]$. Let $\Xi(H)$ be the p-adic obgv determined by H and let

$$G_+ = \{g \in G \mid \Xi(H)(g) > 0\}.$$

Then G_+ is a p-group, and by Proposition 12.21, G_+ is a normal subgroup of G.

Let $k = \mathfrak{O}_K/\mathfrak{p}$ denote the residue field of K. Following [**Chi00**, (20.1)], there is a homomorphism of k-Hopf algebras

$$j : k[G] \to k \otimes_{\mathfrak{O}_K} H,$$

which yields the group homomorphism

$$j : \Gamma(k[G]) \to \Gamma(k \otimes_{\mathfrak{O}_K} H),$$

where $\Gamma(-)$ denotes the set of grouplike elements. Now, $\Gamma(k[G]) = G$.

LEMMA 12.30. $\ker(j) = G_+$.

PROOF. Let $b_1, b_2, \ldots, b_n$ be an $\mathfrak{O}_K$-basis for H. Then their images form a k-basis for $k \otimes_{\mathfrak{O}_K} H$. Now if $j(g) = 1$, then $j(g-1) = 0$ in $k \otimes_{\mathfrak{O}_K} H$. Thus $g - 1 = \sum_{m=1}^n r_m b_m$, with $\nu(r_m) > 0$ for all m. Thus $g - 1 \in \pi H$, so that $\Xi(H)(g) > 0$, i.e., $g \in G_+$. Conversely, if $g \in G_+$, then $j(g-1) = 0$ in $k \otimes_{\mathfrak{O}_K} H$, hence $g \in \ker(j)$. □

PROPOSITION 12.31. *Suppose G is a group with no non-trivial normal subgroup which is a p-group. Then $\mathfrak{O}_K[G]$ is the only Hopf order in $K[G]$.*

PROOF. Let H be an $\mathfrak{O}_K$-Hopf order in $K[G]$ and let

$$j : \Gamma(k[G]) \to \Gamma(k \otimes_{\mathfrak{O}_K} H),$$

be the group homomorphism. By Lemma 12.30, $\ker(j) = G_+$.

Now, $\ker(j)$ is trivial since G admits no non-trivial normal subgroups that are p-groups. Consequently j is an injection, which yields $k[G] \cong k \otimes_{\mathfrak{O}_K} H$. By Nakayama's lemma, we conclude that $\mathfrak{O}_K[G] \cong H$. □

Here is another consequence of Lemma 12.30. Let G be a finite group with $|G| > 2$. Then $\mathbb{Z}_p[G]$ is the only $\mathbb{Z}_p$-Hopf order in $\mathbb{Q}_p[G]$. To see this, suppose that H is a $\mathbb{Z}_p$-Hopf order and let $\Xi(H)$ be the associated p-adic obgv. In this case, $\nu(p) = 1$, $k = \mathbb{Z}_p/\mathfrak{p} = \mathbb{F}_p$, and so by Definition 12.23(ii), G_+ is trivial. Thus

$$j : \Gamma(\mathbb{F}_p[G]) \to \Gamma(\mathbb{F}_p \otimes_{\mathfrak{O}_K} H)$$

is an injection. It follows that $H = \mathbb{Z}_p[G]$.

If the order of G is relatively prime to p then the only Hopf order in $K[G]$ is $\mathfrak{O}_K[G]$. So we look for Hopf orders in the cases where G is an abelian p-group, for instance $G = C_{p^n}$, where C_{p^n} is the cyclic group of order p^n or $G = C_p^n$, where C_p^n is the elementary abelian group of order p^n.

To give a complete accounting of Hopf orders in $K[C_{p^n}]$ or $K[C_p^n]$, we need to look beyond the Larson orders in $K[C_{p^n}]$ or $K[C_p^n]$; not every Hopf order in $K[C_{p^n}]$ is Larson. For an example, suppose $K = \mathbb{Q}_p(\zeta_{p^2})$, where ζ_{p^2} is a primitive p^2nd root of unity, and let $\widehat{C}_{p^2}$ denote the character group of C_{p^2}.

By Proposition 12.9, the dual $\mathfrak{O}_K[C_{p^2}]^*$ is an $\mathfrak{O}_K$-Hopf order in $K[C_{p^2}]^*$. But in view of Remark 12.8, $\mathfrak{O}_K[C_{p^2}]^*$ is an $\mathfrak{O}_K$-Hopf order in $K[\widehat{C}_{p^2}]$ and has the form

$$\mathfrak{O}_K[C_{p^2}]^* = \bigoplus_{i=0}^{p^2-1} \mathfrak{O}_K e_i,$$

where

$$e_i = \frac{1}{p^2} \sum_{j=0}^{p^2-1} \zeta_{p^2}^{-ij} \gamma^j.$$

We can take this one step further and identify e_i with

$$\frac{1}{p^2} \sum_{j=0}^{p^2-1} \zeta_{p^2}^{-ij} g^j,$$

and thus view $\mathfrak{O}_K[C_{p^2}]^*$ as an $\mathfrak{O}_K$-Hopf order in $K[C_{p^2}]$.

PROPOSITION 12.32. *The $\mathfrak{O}_K$-Hopf order $\mathfrak{O}_K[C_{p^2}]^*$ in $K[C_{p^2}]$ is not Larson, i.e. $\mathfrak{O}_K[C_{p^2}]^*$ is not of the form $A(\xi)$ for some p-adic obgv ξ on C_{p^2}.*

PROOF. By way of contradiction, we assume that $\mathfrak{O}_K[C_{p^2}]^* = A(\xi)$ for some p-adic obgv ξ on C_{p^2}. By Proposition 12.22, an obgv on C_{p^2} has at most two finite values: $\xi(g^{pa}) = i$ for $a = 1, 2, \ldots, p-1$, and $\xi(g^b) = j$ for $\gcd(b, p) = 1$. Since ξ is order bounded, $0 \le i \le e/(p-1) = p(p-1)/(p-1) = p$, and $0 \le j \le e/p(p-1) = 1$.

Let $I_{\mathfrak{O}_K[C_{p^2}]}, I_{\mathfrak{O}_K[C_{p^2}]^*}$ denote the ideal of left integrals of $\mathfrak{O}_K[C_{p^2}]$, $\mathfrak{O}_K[C_{p^2}]^*$, respectively. By [**Und11**, Proposition 4.4.11],

$$\nu(\varepsilon_{\mathfrak{O}_K[C_{p^2}]}(I_{\mathfrak{O}_K[C_{p^2}]})) + \nu(\varepsilon_{\mathfrak{O}_K[C_{p^2}]^*}(I_{\mathfrak{O}_K[C_{p^2}]^*})) = 2e = 2p(p-1),$$

and so, since $\nu(\varepsilon_{\mathfrak{O}_K[C_{p^2}]}(I_{\mathfrak{O}_K[C_{p^2}]})) = 2p(p-1)$, $\nu(\varepsilon_{\mathfrak{O}_K[C_{p^2}]^*}(I_{\mathfrak{O}_K[C_{p^2}]^*})) = 0$. Now, by [**Und11**, Proposition 5.3.6], $\nu(\varepsilon_{A(\xi)}(I_{A(\xi)})) = 2p(p-1) - (p-1)(i+j)$, and so, $2p(p-1) = (p-1)(i+j)$. Thus, $2p = i + j$. However, by the bounds on i, j, we have $1 + p \ge 2p$, which is impossible. Thus $A(\xi) \ne \mathfrak{O}_K[C_{p^2}]^*$. □

REMARK 12.33. As we will see in Section 12.6, the non-Larson Hopf order $\mathfrak{O}_K[C_{p^2}]^*$ in $K[C_{p^2}]$ is the dual of the Greither order $A(0,0,1) = H(0,0)$ and has the form $\mathfrak{O}_K[C_{p^2}]^* = A(e', e', \zeta_{p^2}^{-1})$. Thus $\mathfrak{O}_K[C_{p^2}]^*$ is given by two valuation parameters e', e', and one unit parameter $v = \zeta_{p^2}^{-1}$. We have

$$\mathfrak{O}_K[C_{p^2}]^* = \mathfrak{O}_K\left[\frac{g^p - 1}{\pi^{e'}}, \frac{g a_v - 1}{\pi^{e'}}\right],$$

where a_v is an element of $K[\langle g^p \rangle]$ of the form

$$a_v = \sum_{m=0}^{p-1} v^m f_m, \quad v = \zeta_{p^2}^{-1},$$

with $f_m = \frac{1}{p}\sum_{i=0}^{p-1} \zeta_p^{-mi} g^{pi}$.

12.5. Hopf orders in $K[G]$, $G = C_p$

In this section, and until Section 12.13, we assume that K is a p-adic field.

Let g be a generator for C_p. Then a Larson order in $K[C_p]$ can be written as

$$H(i) = \mathfrak{O}_K\left[\frac{g-1}{\pi^i}\right],$$

where i is an integer $0 \le i \le e'$ (Proposition 12.25).

In this section we show that every Hopf order in $K[C_p]$ is of the form $H(i)$ for some integer i, $0 \le i \le e'$.

To this end, we review the classification of rank p Hopf algebras given by Tate and Oort [**TO70**], following closely the treatment of Childs [**Chi00**, Chapter 4].

Let H be a rank p Hopf algebra over $\mathfrak{O}_K$. We assume that H is cocommutative, hence commutative by [**Chi00**, (15.1) Proposition]. For each $n \geq 1$, there is a Hopf algebra endomorphism $[n] : H \to H$ defined as $[1] = I$, the identity homomorphism, and

$$[n](h) = m^{(n-1)}(\Delta^{(n-1)}(h)),$$

where

$$m^{(n-1)} = m(I \otimes m)(I \otimes I \otimes m) \cdots (\underbrace{I \otimes I \otimes \cdots \otimes I}_{n-2} \otimes m),$$

$$\Delta^{(n-1)} = (\underbrace{I \otimes I \otimes \cdots \otimes I}_{n-2} \otimes \Delta) \cdots (I \otimes I \otimes \Delta)(I \otimes \Delta)\Delta,$$

for $n \geq 2$, $h \in H$ (see also [**Chi00**, §16]). The collection of endomorphisms $E = \{[n] \mid n \in \mathbb{F}_p^\times\}$ is a group under the operation $[n][n'] = [nn']$ with nn' taken modulo p. Clearly $E \cong \mathbb{F}_p^\times$ and henceforth, we identify $\mathbb{F}_p^\times$ with this collection of endomorphisms. In fact, since $\mathbb{F}_p^\times$ is a group, $\mathbb{F}_p^\times$ is a group of automorphisms of H. The group ring $\mathbb{Z}_p[\mathbb{F}_p^\times]$ is a collection of endomorphisms of H.

By Hensel's lemma, $\mathbb{Z}_p$ contains the $(p-1)$st roots of unity. For each $[n] \in \mathbb{F}_p^\times$, there exists a root of $x^{p-1} - 1$ in $\mathbb{Z}_p$ which is congruent to n modulo p. The *Teichmüller character* is the unique multiplicative group homomorphism $\chi : \mathbb{F}_p^\times \to \mathbb{Z}_p^\times$ where $\chi(n)$ is the root of $x^{p-1} - 1$ that satisfies $\chi(n) \equiv n \pmod{p}$.

For j, $1 \leq j \leq p-1$, put

$$\eta_j = \frac{1}{p-1} \sum_{n \in \mathbb{F}_p^\times} \chi^j(n)^{-1}[n].$$

Then the η_j form a set of pairwise mutually orthogonal minimal idempotents in $\mathbb{Z}_p[\mathbb{F}_p^\times]$, so that

$$\mathbb{Z}_p[\mathbb{F}_p^\times] = \mathbb{Z}_p\eta_1 \oplus \mathbb{Z}_p\eta_2 \oplus \cdots \oplus \mathbb{Z}_p\eta_{p-1}.$$

We let $\mathbb{Z}_p[\mathbb{F}_p^\times]$ act on H to obtain a decomposition of H.

Let H^+ denote the augmentation ideal of H. From the short exact sequence

$$0 \to H^+ \to H \xrightarrow{\varepsilon} \mathfrak{O}_K \to 0,$$

one obtains $H = \mathfrak{O}_K \oplus H^+$. Now, $\mathbb{Z}_p[\mathbb{F}_p^\times]$ acts on H yielding the decomposition

$$H = \mathfrak{O}_K \oplus \eta_1(H^+) \oplus \eta_2(H^+) \oplus \cdots \oplus \eta_{p-1}(H^+). \tag{12.1}$$

LEMMA 12.34. *For i, j, $1 \leq i, j \leq p-1$, we have $\eta_i(H^+)\eta_j(H^+) \subseteq \eta_{i+j}(H^+)$.*

PROOF. See [**Chi00**, (16.3) Lemma]. □

12.5.1. Decomposition of $\mathbb{Z}_p[C_p]$ and $\mathbb{Z}_p[C_p]^*$. We specialize to the case $\mathfrak{O}_K = \mathbb{Z}_p$ and $H = \mathbb{Z}_p[C_p]$. By (12.1) there is a decomposition

$$\mathbb{Z}_p[C_p] = \mathbb{Z}_p \oplus \eta_1(\mathbb{Z}_p[C_p]^+) \oplus \eta_2(\mathbb{Z}_p[C_p]^+) \oplus \cdots \oplus \eta_{p-1}(\mathbb{Z}_p[C_p]^+). \tag{12.2}$$

LEMMA 12.35. *Let*

$$x_i = -\sum_{n \in \mathbb{F}_p^\times} \chi(n)^{-i}(g^n - 1)$$

for $1 \leq i \leq p-1$. Then $\eta_i(\mathbb{Z}_p[C_p]^+) = \mathbb{Z}_p x_i$ for $1 \leq i \leq p-1$.

PROOF. See [**Chi00**, (16.5)]. □

LEMMA 12.36. *Let*

$$x = x_1 = -\sum_{n\in\mathbb{F}_p^\times} \chi(n)^{-1}(g^n - 1).$$

Then

$$\eta_i(\mathbb{Z}_p[C_p]^+) = \mathbb{Z}_p x_i = \mathbb{Z}_p x^i$$

for $1 \le i \le p-1$. *Moreover,* $x^p = w_p x$ *and* $x^i = w_i x_i$ *for elements* w_p, w_i, $1 \le i \le p-1$, *in* $\mathbb{Z}_p$ *satisfying*

$$w_i \equiv i! \pmod{p}, \quad 1 \le i \le p-1,$$

$$w_p = p w_{p-1}.$$

PROOF. From Lemma 12.35, $\eta_1(\mathbb{Z}_p[C_p]^+) = \mathbb{Z}_p x$, and from this and Lemma 12.34, one deduces that $\eta_i(\mathbb{Z}_p[C_p]^+) = \mathbb{Z}_p x^i$ for $1 \le i \le p-1$. As $x^i \in \mathbb{Z}_p x_i$, there exists an element $w_i \in \mathbb{Z}_p$ with $x^i = w_i x_i$, $1 \le i \le p-1$. Moreover, by Lemma 12.34, $x^p \in \eta_1(\mathbb{Z}_p[C_p]^+) = \mathbb{Z}_p x$ with $x^p = 0$ in $\mathbb{F}_p x$, and so there exists $w_p \in p\mathbb{Z}_p$ with $x^p = w_p x$.

To show that $w_i \equiv i! \pmod{p}$, $1 \le i \le p-1$ and $w_p = p w_{p-1}$, we refer the reader to [**Chi00**, (16.8)]. □

PROPOSITION 12.37. $\mathbb{Z}_p[C_p] = \mathbb{Z}_p[x]$ *where*

$$x = -\sum_{n\in\mathbb{F}_p^\times} \chi(n)^{-1}(g^n - 1)$$

with $x^p = w_p x$ *for some element* $w_p \in p\mathbb{Z}_p$. *The comultiplication is defined as*

$$\Delta(x) = x \otimes 1 + 1 \otimes x - \frac{1}{p-1}\sum_{j=1}^{p-1} \frac{x^j}{w_j} \otimes \frac{x^{p-j}}{w_{p-j}}.$$

PROOF. From the decomposition (12.2), Lemma 12.35 and Lemma 12.36, one has $\mathbb{Z}_p[C_p] = \mathbb{Z}_p[x]$ where $x^p = w_p x$. For the comultiplication, see [**Chi00**, (16.7)]. □

We give a similar characterization for the dual $\mathbb{Z}_p[C_p]^*$. First note that $\mathbb{Z}_p[C_p]^*$ decomposes as

$$\mathbb{Z}_p[C_p]^* = \mathbb{Z}_p \oplus \eta_1((\mathbb{Z}_p[C_p]^*)^+) \oplus \eta_2((\mathbb{Z}_p[C_p]^*)^+) \oplus \cdots \oplus \eta_{p-1}((\mathbb{Z}_p[C_p]^*)^+).$$

Let $\{e_i\}_{i=0}^{p-1}$ denote the basis dual to the basis $\{g^i\}_{i=0}^{p-1}$ of $\mathbb{Z}_p[C_p]$.

LEMMA 12.38. *Let*

$$\xi = \sum_{n\in\mathbb{F}_p^\times} \chi(n) e_n.$$

Then $\eta_i((\mathbb{Z}_p[C_p]^*)^+) = \mathbb{Z}_p \xi^i$ *for* $1 \le i \le p-1$ *and* $\mathbb{Z}_p[C_p]^* = \mathbb{Z}_p[\xi]$ *with* $\xi^p = \xi$.

PROOF. See [**Chi00**, (16.10)]. □

12.5.2. Tate-Oort classification. Now, let H be an $\mathfrak{O}_K$-Hopf order in $K[C_p]$.

PROPOSITION 12.39. *$H = \mathfrak{O}_K[z]$ with $z^p = bz$ for some $b \in \mathfrak{O}_K$, $z \in H$.*

PROOF. Recall the decomposition (12.1):

$$H \quad = \quad \mathfrak{O}_K \oplus \eta_1(H^+) \oplus \eta_2(H^+) \oplus \cdots \oplus \eta_{p-1}(H^+).$$

Let K^{alg} be an algebraic closure of K. Then $\eta_1(K^{alg}[C_p]^+)$ is one-dimensional over K^{alg} and so, $\eta_1(H^+)$ is free rank one over $\mathfrak{O}_K$, i.e., $\eta_1(H^+) = \mathfrak{O}_K z$ for some $z \in H$. By Lemma 12.34, $\mathfrak{O}_K z^i \subseteq \eta_i(H^+)$ for $1 \le i \le p-1$ and $z^p \in \eta_1(H^+)$.

We claim that $\mathfrak{O}_K z^i = \eta_i(H^+)$ for $1 \le i \le p-1$. To this end, let k be the residue class field of $\mathfrak{O}_K$ and let k^{alg} be an algebraic closure of k. By [**Wat79**, §6.8], $kH = k \otimes_{\mathfrak{O}_K} H$ is either separable as a k-algebra or a local ring. In the separable case, one has

$$k^{alg}H = k^{alg} \otimes_k kH \cong \underbrace{k^{alg} \oplus k^{alg} \oplus \cdots \oplus k^{alg}}_{p}$$

so that $k^{alg}H = k^{alg}[C_p]^*$. By Lemma 12.38, $\eta_i((\mathbb{Z}_p[C_p]^*)^+) = \mathbb{Z}_p\xi^i$ for $1 \le i \le p-1$. Reducing modulo $\mathfrak{p}$ yields $\eta_i((\mathbb{F}_p[C_p]^*)^+) = \mathbb{F}_p\xi^i$ for $1 \le i \le p-1$ with

$$\xi = \sum_{n \in \mathbb{F}_p^\times} n e_n.$$

Lifting to H then gives $\mathfrak{O}_K z^i = \eta_i(H^+)$ for $1 \le i \le p-1$. Consequently, $H = \mathfrak{O}_K[z]$, with $z^p = bz$ for some $b \in \mathfrak{O}_K$, $z \in H$.

If $k^{alg}H$ is local with separable dual, then $k^{alg}H = k^{alg}[C_p]$. By Lemma 12.36, $\eta_i(\mathbb{Z}_p[C_p]^+) = \mathbb{Z}_p x^i$, $1 \le i \le p-1$. Reducing modulo $\mathfrak{p}$ yields $\eta_i(\mathbb{F}_p[C_p]^+) = \mathbb{F}_p x^i$, $1 \le i \le p-1$, with

$$x = -\sum_{n \in \mathbb{F}_p^\times} n^{-1}(g^n - 1).$$

Lifting to H then gives $\mathfrak{O}_K z^i = \eta_i(H^+)$ for $1 \le i \le p-1$. Consequently, $H = \mathfrak{O}_K[z]$, with $z^p = bz$ for some $b \in \mathfrak{O}_K$, $z \in H$.

In the case that $k^{alg}H$ is local with local dual, we have $k^{alg}H \cong k^{alg}[t]$ with $t^p = 0$, t primitive, by [**Wat79**, §14.4]. As argued in [**Chi00**, (16.13)], one deduces that

$$\eta_i(k^{alg}H^+) \cong \eta_i(k^{alg}[t]^+) = k^{alg}t^i$$

for $1 \le i \le p-1$, thus $\eta_i(H^+) = \mathfrak{O}_K z^i$ for $1 \le i \le p-1$. Consequently, $H = \mathfrak{O}_K[z]$, with $z^p = bz$ for some $b \in \mathfrak{O}_K$, $z \in H$. □

THEOREM 12.40 (Tate-Oort). *Let H be an $\mathfrak{O}_K$-Hopf order in $K[C_p]$. Then $H = \mathfrak{O}_K[z]$ where $z^p = bz$ with $w_p = c^{p-1}b$ for $b, c \in \mathfrak{O}_K$, with w_p as in Lemma 12.36. The element z is of the form*

$$z = -\frac{1}{c}\sum_{n \in \mathbb{F}_p^\times} \chi(n)^{-1}(g^n - 1),$$

the comultiplication is given as

$$\Delta(z) = z \otimes 1 + 1 \otimes z - \frac{c^{p-1}}{p-1}\sum_{j=1}^{p-1} \frac{z^j}{w_j} \otimes \frac{z^{p-j}}{w_{p-j}},$$

where w_i, $1 \le i \le p-1$, as in Lemma 12.36.

PROOF. By Lemma 12.37, $\mathfrak{O}_K[C_p] = \mathfrak{O}_K[x]$ with

$$x = -\sum_{n\in\mathbb{F}_p^\times} \chi(n)^{-1}(g^n - 1)$$

and $x^p = w_p x$. By Proposition 12.39, $H = \mathfrak{O}_K[z]$ with $z^p = bz$ for some $b \in \mathfrak{O}_K$. By Proposition 12.10, $\mathfrak{O}_K[C_p] \subseteq H$, so there is a Hopf embedding $\varphi : \mathfrak{O}_K[C_p] \to H$.

Since φ is a Hopf map, we have, for $n \in \mathbb{F}_p^\times$,

$$\begin{aligned} \varphi([n]\mathfrak{O}_K[C_p]) &= \varphi(m^{(n-1)}\Delta^{(n-1)}\mathfrak{O}_K[C_p]) \\ &= m^{(n-1)}\Delta^{(n-1)}\varphi(\mathfrak{O}_K[C_p]) \\ &= [n]\varphi(\mathfrak{O}_K[C_p]) \\ &\subseteq [n]H. \end{aligned}$$

Thus $\varphi(\eta_1(\mathfrak{O}_K[C_p]^+)) \subseteq \eta_1(H^+)$ and so, $\varphi(\mathfrak{O}_K x) \subseteq \mathfrak{O}_K z$. Hence $\varphi(x) = cz$ for some $c \in \mathfrak{O}_K$. We make the identification $x = \varphi(x) = cz$. Thus $\varphi(x^p) = x^p = c^p z^p = c^p bz$ and $\varphi(w_p x) = w_p x = w_p cz$. As $\varphi(x^p) = \varphi(w_p x)$, we have $c^p bz = w_p cz$, hence $w_p = c^{p-1}b$.

Moreover, since $x = cz$,

$$z = -\frac{1}{c}\sum_{n\in\mathbb{F}_p^\times} \chi(n)^{-1}(g^n - 1).$$

By Proposition 12.37

$$\Delta(cz) = cz \otimes 1 + 1 \otimes cz - \frac{1}{p-1}\sum_{j=1}^{p-1} \frac{(cz)^j}{w_j} \otimes \frac{(cz)^{p-j}}{w_{p-j}},$$

thus

$$\Delta(z) = z \otimes 1 + 1 \otimes z - \frac{c^{p-1}}{p-1}\sum_{j=1}^{p-1} \frac{z^j}{w_j} \otimes \frac{z^{p-j}}{w_{p-j}}.$$

□

12.5.3. Every Hopf order in $K[C_p]$ is Larson. Using the Tate-Oort classification, we can show that every $\mathfrak{O}_K$-Hopf order H in $K[C_p]$ is a Larson order. By Theorem 12.40, $H = \mathfrak{O}_K[z]$, $z^p = bz$, with

$$z = -\frac{1}{c}\sum_{n\in\mathbb{F}_p^\times} \chi(n)^{-1}(g^n - 1).$$

We have

$$\begin{aligned} z &= -\frac{1}{c}\sum_{n\in\mathbb{F}_p^\times} \chi(n)^{-1}(g^n - 1) \\ &= -\frac{(g-1)}{c}\sum_{n\in\mathbb{F}_p^\times} \chi(n)^{-1}(1 + g + g^2 + \cdots + g^{n-1}) \\ &= -v\frac{(g-1)}{c} \end{aligned}$$

with

$$v = \sum_{n\in\mathbb{F}_p^\times} \chi(n)^{-1}(1 + g + g^2 + \cdots + g^{n-1}).$$

Observe that $v \in \mathbb{Z}_p[C_p] \subseteq \mathfrak{O}_K[C_p]$.

LEMMA 12.41. *v is a unit in $\mathbb{Z}_p[C_p]$.*

PROOF. Since $\mathbb{Z}_p[C_p] = \mathfrak{O}_K[x]/(x^p - 1)$, x indeterminate, there is an embedding

$$\mathbb{Z}_p[C_p] \to \mathbb{Z}_p \oplus \mathbb{Z}_p[\zeta_p]$$

defined by $g \mapsto (1, \zeta_p)$. The image of v under this embedding is $\widehat{v} = (\alpha, \beta)$ where $\alpha = \sum_{n \in \mathbb{F}_p^\times} \chi(n)^{-1} n$ and $\beta = \sum_{n \in \mathbb{F}_p^\times} \chi(n)^{-1}(1 + \zeta_p + \zeta_p^2 + \cdots + \zeta_p^{n-1})$.

By the method of [**Und11**, Lemma 7.1.5], α is a unit in $\mathbb{Z}_p$ and β is a unit in $\mathbb{Z}_p[\zeta_p]$. It follows that v is a unit in $\mathbb{Z}_p[C_p]$. □

THEOREM 12.42. *Let K be a p-adic field with uniformizing parameter π. Let $e' = \nu(p)/(p-1)$. Let C_p denote the cyclic group of order p generated by g and let H be an $\mathfrak{O}_K$-Hopf order in $K[C_p]$. Then*

$$H = \mathfrak{O}_K \left[\frac{g-1}{\pi^i} \right],$$

where i is an integer $0 \le i \le e'$.

PROOF. By Theorem 12.40, $H = \mathfrak{O}_K \left[\frac{-v(g-1)}{c} \right]$, where v is defined as above and c is some element of $\mathfrak{O}_K$. By Lemma 12.41, v is a unit of $\mathfrak{O}_K[C_p]$. Now by Proposition 12.10, $\mathfrak{O}_K[C_p] \subseteq H$, and so, v is a unit of H. It follows that $H = \mathfrak{O}_K \left[\frac{g-1}{c} \right]$.

Let $c = u\pi^i$ for some unit $u \in \mathfrak{O}_K$, integer $i \ge 0$. Then $H = \mathfrak{O}_K \left[\frac{g-1}{\pi^i} \right]$. It remains to show that $0 \le i \le e'$. Recall that $w_p = c^{p-1} b$ for some $b \in \mathfrak{O}_K$. Thus

$$e = \nu(w_p) = (p-1)\nu(c) + \nu(b) = (p-1)i + \nu(b).$$

Thus $(p-1)i \le e$ so that $i \le e'$. □

In our discussion of realizable Hopf orders in subsequent sections, we will need to know when $H(i)$ is a local ring.

PROPOSITION 12.43. *Let $H(i)$ be an $\mathfrak{O}_K$-Hopf order in $K[C_p]$, $0 \le i \le e'$. If $i < e'$, then $H(i)$ is a local ring.*

PROOF. We have $H(i) = \mathfrak{O}_K \left[\frac{g-1}{\pi^i} \right]$, $\langle g \rangle$. Now, $\mathrm{nil}(k \otimes_{\mathfrak{O}_K} H(i)) \subseteq (k \otimes_{\mathfrak{O}_K} H(i))^+$. Since $i < e'$, $\left(\frac{g-1}{\pi^i} \right)^p \equiv 0 \pmod{\pi \mathfrak{O}_K}$. But this says that $(k \otimes_{\mathfrak{O}_K} H(i))^+ \subseteq \mathrm{nil}(k \otimes_{\mathfrak{O}_K} H(i))$. Thus $(k \otimes_{\mathfrak{O}_K} H(i))^+ = \mathrm{nil}(k \otimes_{\mathfrak{O}_K} H(i))$, which implies that $k \otimes_{\mathfrak{O}_K} H(i)$ is local. Hence $H(i)$ is local. □

We can compute the linear dual of $H(i)$ in the case that K contains a primitive pth root of unity, ζ_p. Let $\widehat{C}_p$ denote the character group of C_p generated by γ with duality pairing $\langle \gamma^j, g^k \rangle = \zeta_p^{jk}$, $0 \le j, k \le p-1$. Let $H(i)$ be an $\mathfrak{O}_K$-Hopf order in $K[C_p]$, $0 \le i \le e'$. By Proposition 12.9, $H(i)^*$ is an $\mathfrak{O}_K$-Hopf order in $K[C_p]^* = K[\widehat{C}_p]$.

PROPOSITION 12.44. $H(i)^* = H(i') = \mathfrak{O}_K \left[\frac{\gamma - 1}{\pi^{i'}} \right]$, *where $i' = e' - i$.*

PROOF. Let $J = \mathfrak{O}_K\left[\frac{\gamma-1}{\pi^{i'}}\right]$. For $a = 1, \dots, p-1$,

$$\left\langle \frac{\gamma-1}{\pi^{i'}}, \left(\frac{g-1}{\pi^i}\right)^a \right\rangle = \frac{(\zeta_p-1)^a}{\pi^{i'+ai}}$$

which is in $\mathfrak{O}_K$ since $\nu(\zeta_p - 1) = e'$. Thus $J \subseteq H(i)^*$.

By [**Und11**, Proposition 5.3.3], $\varepsilon_{H(i)}(I_{H(i)}) = \frac{p}{\pi^{(p-1)i}}\mathfrak{O}_K$. Thus by [**Und11**, Proposition 4.4.11], $\varepsilon_{H(i)^*}(I_{H(i)^*}) = \pi^{(p-1)i}\mathfrak{O}_K$. Consequently, $\mathrm{disc}(H(i)^*) = \pi^{p(p-1)i}\mathfrak{O}_K$, by [**Und11**, Proposition 4.4.13].

On the other hand,

$$\varepsilon_{H(i')}(I_{H(i')}) = \frac{p}{\pi^{(p-1)i'}}\mathfrak{O}_K = \pi^{(p-1)i}\mathfrak{O}_K$$

by [**Und11**, Proposition 5.3.3] and so $\mathrm{disc}(H(i')) = \pi^{p(p-1)i}\mathfrak{O}_K$, by [**Und11**, Proposition 4.4.13]. It follows that $J = H(i)^*$. □

12.6. Hopf orders in $K[G]$, $G = C_p \times C_p$, $G = C_{p^2}$

Let $H(i)$, $H(j)$ be Hopf orders in $K[C_p]$ with $pj \le i$. We assume that K contains a primitive pth root of unity, ζ_p. Let $\mathbb{D}_i = \mathrm{Spec}\ H(i)$, $\mathbb{D}_j = \mathrm{Spec}\ H(j)$ denote the group schemes represented by $H(i)$ and $H(j)$.

For $q \ge 0$ in $\mathbb{Q}$, let $U_q = \{u \in \mathfrak{O}_K^\times \mid \nu(1-u) \ge q\}$. Greither [**Gre92**] has shown that the quotient group

$$(U_{i'+(j/p)} \cap U_{(i'/p)+j})/U_{i'+j}$$

is isomorphic to the group of generically trivial extensions $\mathrm{Ext}^1_{gt}(\mathbb{D}_i, \mathbb{D}_j)$. A class $[u]$ in the quotient group corresponds to an equivalence class of short exact sequences of group schemes

$$E_u: \ 1 \to \mathbb{D}_j \to \mathbb{D} \to \mathbb{D}_i \to 1,$$

which over K appears as

$$1 \to \boldsymbol{\mu}_{p,K} \to \boldsymbol{\mu}_{p,K} \times \boldsymbol{\mu}_{p,K} \to \boldsymbol{\mu}_{p,K} \to 1.$$

Here $\boldsymbol{\mu}_p = \mathrm{Spec}\ \mathfrak{O}_K[C_p]$ is the multiplicative group of the pth roots of unity.

The extension of group schemes E_u corresponds to a short exact sequence of Hopf orders

$$E_u: \ \mathfrak{O}_K \to H(i) \to H \to H(j) \to \mathfrak{O}_K$$

where H is an $\mathfrak{O}_K$-Hopf order in $K[C_p \times C_p]$ of the form

$$\mathfrak{O}_K\left[\frac{g-1}{\pi^i}, \frac{ha_u - 1}{\pi^j}\right],$$

where $\langle g, h\rangle = C_p \times C_p$ and $a_u = \sum_{m=0}^{p-1} u^m e_m$, where e_m are the minimal idempotents in $K\langle g\rangle$.

So, under the assumption $pj \le i$, Greither's construction yields all of the Hopf orders in $K[C_p \times C_p]$.

To obtain Hopf orders in $K[C_{p^2}]$, we need a way to multiply two extensions.

12.6.1. The Baer product. Quite generally, let H, H' be $\mathfrak{O}_K$-Hopf algebras and let $E(H', H)$ denote the collection of equivalence classes of short exact sequences of $\mathfrak{O}_K$-Hopf algebras,

$$E: \ \mathfrak{O}_K \to H \to J \to H' \to \mathfrak{O}_K.$$

On $E(H', H)$ we define a multiplication as follows. Let

$$E_1: \ \mathfrak{O}_K \to H \to H_1 \overset{s_1}{\to} H' \to \mathfrak{O}_K,$$

$$E_2: \ \mathfrak{O}_K \to H \to H_2 \overset{s_2}{\to} H' \to \mathfrak{O}_K,$$

be short exact sequences of Hopf algebras. Recalling that the tensor product of two Hopf algebras is again a Hopf algebra, we obtain a short exact sequence of $\mathfrak{O}_K$-Hopf algebras,

$$\mathfrak{O}_K \to H \otimes_{\mathfrak{O}_K} H \to H_1 \otimes_{\mathfrak{O}_K} H_2 \overset{s_1 \otimes s_2}{\to} H' \otimes_{\mathfrak{O}_K} H' \to \mathfrak{O}_K.$$

Let the pair of morphisms $\alpha : A \to H_1 \otimes_{\mathfrak{O}_K} H_2$, $\beta : A \to H'$ be the pull-back of $(s_1 \otimes s_2, \Delta_{H'})$. Here

$$A = \{x \otimes y \otimes z \in H_1 \otimes H_2 \otimes H' \mid (s_1 \otimes s_2)(x \otimes y) = \Delta_{H'}(z)\},$$

$\alpha(x \otimes y \otimes z) = x \otimes y$ and $\beta(x \otimes y \otimes z) = z$.

Then there is a commutative diagram with exact rows:

$$\begin{array}{ccccccccc} \mathfrak{O}_K & \to & H \otimes_{\mathfrak{O}_K} H & \to & H_1 \otimes_{\mathfrak{O}_K} H_2 & \overset{s_1 \otimes s_2}{\to} & H' \otimes_{\mathfrak{O}_K} H' & \to & \mathfrak{O}_K \\ \| & & \| & & \alpha \uparrow & & \Delta_{H'} \uparrow & & \| \\ \mathfrak{O}_K & \to & H \otimes_{\mathfrak{O}_K} H & \overset{j}{\to} & A & \overset{\beta}{\to} & H' & \to & \mathfrak{O}_K \end{array}$$

Next, let $m : H \otimes_{\mathfrak{O}_K} H \to H$ denote multiplication in H and let the pair of morphisms $\varrho : H \to B$, $\psi : A \to B$ be the push-out of (m, j). Here

$$B = (H \otimes A)/S$$

with

$$S = \{m(x \otimes y) \otimes 1 - 1 \otimes j(x \otimes y) \in H \otimes A \mid x \otimes y \in H \otimes H\},$$

$\varrho(h) = h \otimes 1 + S$ and $\psi(a) = 1 \otimes a + S$.

There is a commutative diagram with exact rows:

$$\begin{array}{cccccccccc} & \mathfrak{O}_K & \to & H \otimes_{\mathfrak{O}_K} H & \overset{j}{\to} & A & \to & H' & \to & \mathfrak{O}_K \\ & \| & & m \downarrow & & \psi \downarrow & & \| & & \| \\ E: & \mathfrak{O}_K & \to & H & \overset{\varrho}{\to} & B & \to & H' & \to & \mathfrak{O}_K \end{array}$$

The bottom row E is a short exact sequence of Hopf algebras. Let $[E]$ be the equivalence class of E, let $[E_1]$, $[E_2]$ be the classes of E_1, E_2, respectively. Then $[E]$ is the **Baer product** of classes of extensions

$$[E] = [E_1] * [E_2].$$

The Baer product endows $E(H', H)$ with the structure of a group.

REMARK 12.45. The Baer product (or sum) is defined for abelian groups (and hence for us on the group scheme side) in [**Rot17**, p. 302], [**Rot17**, Lemma C-3.87, Lemma C-3.88]. See also, [**HS71**, II, §6].

12.6.2. Greither orders. We take $H = H(i)$ and $H' = H(j)$ and apply the Baer product to the collection $E(H(j), H(i))$. Since $pj \leq i$, there exists a short exact sequence of Larson orders

$$E_0 : \mathfrak{O}_K \to H(i) \to \mathfrak{O}_K \left[\frac{g^p - 1}{\pi^i}, \frac{g-1}{\pi^j} \right] \to H(j) \to \mathfrak{O}_K, \quad \langle g \rangle = C_{p^2}.$$

Now the Baer product

$$[E_u] * [E_0]$$

yields a short exact sequence whose middle term is an $\mathfrak{O}_K$-Hopf order in $K[C_{p^2}]$ of the form

$$A(i,j,u) = \mathfrak{O}_K \left[\frac{g^p - 1}{\pi^i}, \frac{ga_u - 1}{\pi^j} \right],$$

known as a *Greither order.*

Note that the Greither order $A(i,j,u)$ is the Larson order $H(i,j)$ if and only if $\nu(1-u) \geq i' + j$, i.e., the class $[u]$ is trivial.

We can compute the linear dual of a Greither order assuming that K contains a primitive p^2nd root of unity.

PROPOSITION 12.46. *Suppose that $\zeta_{p^2} \in K$ and let $\widehat{C}_{p^2}$ denote the character group of C_{p^2}, generated by γ. Let $A(i,j,u)$ be a Greither order in $K[C_{p^2}]$. Then $A(i,j,u)^*$ is an $\mathfrak{O}_K$-Hopf order in $K\widehat{C}_{p^2}$ of the form*

$$A(j', i', \widetilde{u}) = \mathfrak{O}_K \left[\frac{\gamma^p - 1}{\pi^{j'}}, \frac{\gamma a_{\widetilde{u}} - 1}{\pi^{i'}} \right],$$

where $\widetilde{u} = \zeta_2^{-1} u^{-1}$, $i' = e' - i$, $j' = e' - j$ with $a_{\widetilde{u}} = \sum_{m=0}^{p-1} \widetilde{u}^m e_m$, where e_m are the minimal idempotents in $K\langle \gamma^p \rangle$.

PROOF. See [**Und11**, Proposition 8.3.9], [**UC06**, Theorem 1.2]. □

We note that Geither orders $A(i,j,u)$, and their duals $A(j', i', \widetilde{u})$ are determined by three parameters: two valuation parameters, i, j (or j',i' for duals) and one unit parameter, u ($\widetilde{u}$ for duals).

We may not have the p-adic condition $pi' \leq j'$, thus $A(j', i', \widetilde{u})$ may not be a Greither order. However, Byott, and later Underwood, have shown the following.

THEOREM 12.47. *Assume that $\zeta_{p^2} \in K$. Let H be an $\mathfrak{O}_K$-Hopf order in $K[C_{p^2}]$ and suppose that H induces the short exact sequence*

$$\mathfrak{O}_K \to H(i) \to H \to H(j) \to \mathfrak{O}_K,$$

where $H(i)$, $H(j)$ are Larson orders in $K[C_p]$. Then either $pj \leq i$ or $pi' \leq j'$.

PROOF. See [**Byo93**, Section 8, Theorem 5], [**Und94**, Theorem 2.6], [**Und11**, Section 8.1]. □

Thus, under the assumption $\zeta_{p^2} \in K$, an arbitrary $\mathfrak{O}_K$-Hopf order in $K[C_{p^2}]$ is either a Greither order $A(i,j,u)$ or the linear dual of a Greither order $A(i,j,u)^* = A(j', i', \widetilde{u})$. Consequently, $\mathfrak{O}_K$-Hopf orders in $K[C_{p^2}]$ can be completely determined by 3 parameters.

It has been conjectured [**CU03**, Introduction] that Hopf orders in $K[C_{p^n}]$ can be completely classified using $n(n+1)/2$ parameters.

We note the following additional facts.

Assume $\zeta_{p^2} \in K$ and let $A(i,j,u)$ be an $\mathfrak{O}_K$-Hopf order in $K[C_{p^2}]$.

PROPOSITION 12.48. *$\nu(1-u) \geq i' + (j/p)$.*

PROOF. Certainly if $A(i,j,u)$ is Greither, i.e., $pj \le i$, then $\nu(1-u) \ge i' + (j/p)$. So we assume that $A(j',i',\widetilde{u})$ is Greither, i.e., $pi' \le j'$. Now, if $\nu(1-u) \ge e'/p$, then $\nu(1-u) \ge i' + (j/p)$ since $pi' \le j'$ implies that $e'/p \ge i' + (j/p)$.

On the other hand, if $\nu(1-u) < e'/p$, then $\nu(1-u) = \nu(1-\widetilde{u}) \ge i' + (j/p)$. □

Without the assumption $\zeta_{p^2} \in K$, we still have the following.

PROPOSITION 12.49. *Let H be an $\mathfrak{O}_K$-Hopf order in $K[C_{p^2}]$ inducing the short exact sequence $\mathfrak{O}_K \to H(i) \to H \to H(j) \to \mathfrak{O}_K$. Then $i \ge j$.*

PROOF. The map $[p] : H \to H$, defined in Section 12.5, is an endomorphism of Hopf algebras, which over K yields the map $[p] : K[C_{p^2}] \to K[C_{p^2}]$, given as $g \mapsto g^p$. We have $[p](K[C_{p^2}]) = K[C_p]$ and hence $[p](H) \subseteq K[C_p]$. Thus $[p](H) \subseteq K[C_p] \cap H = H(i)$.

Moreover, $H(i)^+ H = \ker([p])$ and hence there is a Hopf algebra isomorphism $[p](H) \cong H(j)$, implying a Hopf algebra inclusion $H(j) \subseteq H(i)$ and thus $i \ge j$. □

12.7. Hopf orders in $K[G]$, $G = C_{p^3}$

We assume that K contains a primitive p^3rd root of unity, ζ_{p^3}. In contrast to the cases $n = 1, 2$, the classification of $\mathfrak{O}_K$-Hopf orders in $K[C_{p^3}]$ is not complete, though various authors have constructed large classes of Hopf orders in $K[C_{p^3}]$ (and $K[C_p^3]$). The method of Greither, successful for $n = 2$, does not carry forward in an obvious way to settle the $n = 3$ case. The obstruction lies in the computation of

$$\mathrm{Ext}^1(\mathbb{D}_{i,j,u}, \mathbb{E}_k) \cong \mathrm{H}_0^2(\mathbb{D}_{i,j,u}, \mathbb{E}_k),$$

where $\mathbb{D}_{i,j,u} = \mathrm{Spec}\ A(i,j,u)$, $\mathbb{E}_k = \mathrm{Spec}\ \mathfrak{O}_K[X, (1+\pi^k X)^{-1}]$, for X an indeterminate, $k \ge 0$, and $\mathrm{H}_0^2(\mathbb{D}_{i,j,u}, \mathbb{E}_k)$ is the 2nd cohomology group of the complex

$$\mathrm{Mor}(\mathbb{D}_{i,j,u}^{r-1}, \mathbb{E}_k) \xrightarrow{\partial} \mathrm{Mor}(\mathbb{D}_{i,j,u}^{r}, \mathbb{E}_k) \xrightarrow{\partial} \mathrm{Mor}(\mathbb{D}_{i,j,u}^{r+1}, \mathbb{E}_k).$$

The issue is that $\mathrm{Mor}(\mathbb{D}_{i,j,u}, \mathbb{E}_k)$ is difficult to compute: the $\mathfrak{O}_K$-algebra maps

$$\mathfrak{O}_K[X, (1+\pi^k X)^{-1}] \to A(i,j,u)$$

are not easily characterized (in contrast to $n = 2$, where it is just a matter of computing the maps $\mathfrak{O}_K[X, (1+\pi^j X)^{-1}] \to H(i)$).

One instance in which Greither's cohomological method works nicely is when

$$A(i,j,u) = H(d,c)^* = A(c', d', \zeta_{p^2}^{-1}),$$

i.e., $A(i,j,u)$ is the dual of a Larson order in $K[C_{p^2}]$, cf. Proposition 12.46.

Suppose that $pk \le l$, where $H(c', l) = \mathcal{A}(\Xi(H(d,c)^*))$ is the largest Larson order in $H(d,c)^*$. Then $\mathrm{Mor}(\mathrm{Spec}\ H(d,c)^*, \mathbb{E}_k)$ and (ultimately) $\mathrm{Ext}^1_{gt}(\mathrm{Spec}\ H(d,c)^*, \mathbb{D}_k)$ are computable.

Let (a,b) be a pair of units in $\mathfrak{O}_K$ and let u, v be elements of K^{alg} for which $a = u^p$, $b = v^{p^2}$. The classes $[(a,b)]$ in the quotient group M/N where

$$M = \{(a,b) \in U_{pc+k} \times U_{pd+k} \mid v^p \in U_{(d/p)+k}, uv \in U_{(c/p)+k}\},$$

$$N = \{(a,b) \in U_{pc+k} \times U_{pd+k} \mid v^p \in U_{d+k}, uv \in U_{c+k}\},$$

are in a one-to-one correspondence with elements in $\mathrm{Ext}^1(\mathrm{Spec}\ H(d,c)^*, \mathbb{D}_k)$, which over K appear as

$$1 \to \boldsymbol{\mu}_{p,K} \to \boldsymbol{\mu}_{p,K} \times \boldsymbol{\mu}_{p^2,K} \to \boldsymbol{\mu}_{p^2,K} \to 1.$$

The class $[(a,b)]$ corresponds to a short exact sequence of Hopf orders

$$E_{(a,b)}: \mathfrak{O}_K \to H(d,c)^* \to H \to H(k) \to \mathfrak{O}_K$$

where H is an $\mathfrak{O}_K$-Hopf order in $K[C_{p^2} \times C_p]$ of the form

$$\mathfrak{O}_K\left[\frac{g^p-1}{\pi^{c'}}, \frac{ga_{\zeta_{p^2}^{-1}}-1}{\pi^{d'}}, \frac{hb_{u,v}-1}{\pi^k}\right].$$

Here $\langle g,h\rangle = C_{p^2} \times C_p$ and

$$b_{u,v} = \sum_{m=0}^{p-1}\sum_{n=0}^{p-1} u^n v^{pm+n} e_{pm+n},$$

where e_{pm+n} are the minimal idempotents in $K\langle g\rangle$ [**CU03**, Theorem 4.0], [**UC06**, Theorem 1.7].

Now since $pk \le l$, there exists a short exact sequence

$$E_0: \mathfrak{O}_K \to H(d,c)^* \to \mathfrak{O}_K\left[\frac{g^{p^2}-1}{\pi^{c'}}, \frac{g^p a_{\zeta_{p^2}^{-1}}-1}{\pi^{d'}}, \frac{g-1}{\pi^k}\right] \to H(k) \to \mathfrak{O}_K.$$

Taking the Baer product

$$[E_{(a,b)}] * [E_0]$$

yields the $\mathfrak{O}_K$-Hopf order

$$\mathfrak{O}_K\left[\frac{g^{p^2}-1}{\pi^{c'}}, \frac{g^p a_{\zeta_{p^2}^{-1}}-1}{\pi^{d'}}, \frac{gb_{u,v}-1}{\pi^k}\right]$$

in $K[C_{p^3}]$ which is known as a *cohomological Hopf order.*

The special case $d = pc$ yields the subclass

$$\mathfrak{O}_K\left[\frac{g^{p^2}-1}{\pi^{c'}}, \frac{g^p a_{\zeta_{p^2}^{-1}}-1}{\pi^{(pc)'}}, \frac{gb_{1,w}-1}{\pi^k}\right],$$

where $[w] \in (U_{c+(k/p^2)} \cap U_{(c/p)+k})/U_{c+k}$. This special case can be extended to yield a collection of Hopf orders in $K[C_{p^n}]$ for $n \ge 1$ [**Und96**, §4.1]. (See Section 12.8.)

In the case that the kernel $A(i,j,u)$ is not a Larson dual, yet contains a largest Larson dual, we can use the construction as above to obtain a collection of $\mathfrak{O}_K$-Hopf orders $\mathfrak{O}_K\left[\frac{g^{p^2}-1}{\pi^i}, \frac{g^p a_u-1}{\pi^j}, \frac{ga_v b_w-1}{\pi^k}\right]$ in $K[C_{p^3}]$ known as *induced from Larson dual (ILD)* Hopf orders [**UC06**, Proposition 1.8]. Here the quantity $a_v b_w$ is the product of the Greither quantity a_v and quantity

$$b_w = \sum_{m=0}^{p-1}\sum_{n=0}^{p-1} w^m e_{pm+n}.$$

Observe that $b_{x,y} = a_{xy} b_{y^p}$.

In general, for Hopf orders H in $K[C_{p^3}]$ yielding the short exact sequences

$$\mathfrak{O}_K \to A(i,j,u) \to H \to H(k) \to \mathfrak{O}_K,$$

$$\mathfrak{O}_K \to H(i) \to H \to A(j,k,w) \to \mathfrak{O}_K,$$

the strategy has been to find an appropriate lift $\dfrac{gb-1}{\pi^k} \in H$, $b \in K\langle g^p\rangle$, which maps to the generator $\dfrac{\overline{g}a_w - 1}{\pi^k}$ in the Hopf order $\overline{H} = A(j,k,w)$. Necessarily, the Hopf orders are constructed in dual pairs, H, H^*.

In [**UC06**, §3], we choose a lift of the form $\dfrac{gb-1}{\pi^k} = \dfrac{ga_v b_w - 1}{\pi^k}$. We then find conditions under which the algebra

$$H = A(i,j,u)\left[\frac{ga_v b_w - 1}{\pi^k}\right] = \mathfrak{O}_K\left[\frac{g^{p^2}-1}{\pi^i}, \frac{g^p a_u - 1}{\pi^j}, \frac{ga_v b_w - 1}{\pi^k}\right]$$

is free of rank p over $A(i,j,u)$ [**UC06**, Proposition 3.5]. Similarly, we find conditions under which the algebra

$$J = A(k',j',z)\left[\frac{ga_x b_y - 1}{\pi^{i'}}\right] = \mathfrak{O}_K\left[\frac{g^{p^2}-1}{\pi^{k'}}, \frac{g^p a_z - 1}{\pi^{j'}}, \frac{ga_x b_y - 1}{\pi^{i'}}\right]$$

is free of rank p over $A(k',j',z)$ [**UC06**, Proposition 3.6].

Finally, we determine when $J \subseteq H^*$. Then both H and $J = H^*$ are $\mathfrak{O}_K$-Hopf orders in $K[C_{p^3}]$, called *duality Hopf orders* [**UC06**, Theorem 3.7].

12.8. General constructions in the cases $G = C_{p^n}$, $G = C_p^n$

12.8.1. The case $G = C_{p^n}$. There are a number of ways to construct Hopf orders in $K[C_{p^n}]$ for $n \geq 2$.

For instance, one can generalize Greither's construction in Section 12.6 to obtain a collection of $\mathfrak{O}_K$-Hopf orders in $K[C_{p^n}]$ [**Und96**, §4.1].

Assume that $\zeta_{p^n} \in K$. Let c be an integer, $0 \leq c \leq e'/p^{n-2}$, and let

$$H(p^{n-2}c, \ldots, pc, c)$$

be the one parameter Larson order in $K[C_{p^{n-1}}]$ with linear dual

$$H(p^{n-2}c, \ldots, pc, c)^*,$$

which is viewed as an $\mathfrak{O}_K$-Hopf order in $K[C_{p^{n-1}}]$. Assume that k is an integer with $pk \leq \Xi(H(p^{n-2}c, \ldots, pc, c)^*)(g^p)$, $\langle g\rangle = C_{p^n}$.

Given a class

$$[w] \in (U_{c+(k/p^{n-1})} \cap U_{(c/p)+k})/U_{c+k},$$

there exists an $\mathfrak{O}_K$-Hopf order in $K[C_{p^n}]$ of the form

$$H(p^{n-2}c, \ldots, p^2c, pc, c)^*\left[\frac{gb_{1,w}-1}{\pi^k}\right],$$

where $b_{1,w} = \sum_{m=0}^{p^{n-1}-1} w^m e_m$, with e_m the minimal idempotents in $K[C_{p^{n-1}}]$. These Hopf orders are called *generalized Greither orders.*

In [**CU03**, Theorem 2.1], we construct Hopf orders in $K[C_{p^n}]$ as the representing algebras of kernels of isogenies $f : F \to G$ of degree 2 dimension n polynomial formal groups. A certain lower triangular matrix $M \in \mathrm{Mat}_n(\mathfrak{O}_K)$ gives rise to an

$\mathfrak{O}_K$-Hopf order H_M of the form $H_M = \mathfrak{O}_K[z_1, z_2, \ldots, z_n]$, where

$$\begin{aligned} z_1 &= u_{1,1}(g^{p^{n-1}} - 1) \\ z_2 &= u_{2,1}(g^{p^{n-1}} - 1) + u_{2,2}(g^{p^{n-2}} - 1) \\ &\vdots \\ z_n &= u_{n,1}(g^{p^{n-1}} - 1)z_1 + \cdots + u_{n,n}(g-1), \end{aligned}$$

with $(u_{i,j}) = M^{-1}$. These are the *formal group Hopf orders.*

In [**CU04**, Theorem 2.2], we compute the duals of a certain collection of formal group Hopf orders to obtain a collection of *triangular Hopf orders* in $K[C_{p^n}]$. These are of the form $\mathfrak{O}_K[b_n, b_{n-1}, \ldots, b_1]$, where

$$\begin{aligned} b_n &= \pi^{-i'_n}(g^{p^{n-1}} - 1) \\ b_{n-1} &= \pi^{-i'_{n-1}}(g^{p^{n-2}} a^{(1)}_{x_{n-1,1}} - 1) \\ &\vdots \\ b_1 &= \pi^{-i'_1}(g a^{(1)}_{x_{1,1}} a^{(2)}_{x_{1,2}} \cdots a^{(n-1)}_{x_{1,n-1}} - 1), \end{aligned}$$

where $a^{(l)}_{x_{m,n}}$ are quantities that generalize Greither's a_u.

A duality Hopf order is a triangular Hopf order in $K[C_{p^3}]$.

12.8.2. The case $G = C_p^n$. We next discuss ways to construct collections of Hopf orders in $K[C_p^n]$, for $n \geq 1$.

In [**GC98**], Greither and Childs extend the construction of Raynaud orders [**Ray74**], and in [**CS05**] the authors have generalized the Hopf orders found in [**GC98**]. We describe (briefly) the Hopf orders in $K[C_p^n]$ constructed in [**CS05**].

Suppose g_j generates the jth component of C_p^n. Let $U = (u_{i,j})$ be a lower triangular matrix with $u_{i,i} = \zeta_p$. Let $a_{i,j}$ be the Greither quantity $a_{u_{i,j}}$ in $K\langle g_j \rangle$. Let H_n be the $\mathfrak{O}_K$-algebra

$$H_n = \mathfrak{O}_K\left[\frac{g_1 - 1}{\pi^{i_1}}, \frac{g_2 a_{2,1} - 1}{\pi^{i_2}}, \ldots, \frac{g_n a_{n,1} a_{n,2} \cdots a_{n,n-1} - 1}{\pi^{i_n}}\right].$$

Conditions are given for H_n to be an $\mathfrak{O}_K$-Hopf order in $K[C_p^n]$. Its dual H_n^* has an analogous form.

We note that all of the Hopf orders in $K[C_{p^n}]$ or $K[C_p^n]$ that we have constructed in this section satisfy the parameters conjecture (Section 12.6.2), i.e., they all can be described using $n(n+1)/2$ parameters.

However, for $n \geq 3$, these Hopf orders are almost never the associated orders of valuation rings of K; they are never realizable in the sense of Definition 12.3. By contrast, our next collection of Hopf orders is far more interesting in that regard.

12.9. Truncated exponential Hopf orders

Byott and Elder [**BE18**] have constructed a collection of Hopf orders in $K[C_p^n]$, $n \geq 1$, which employ "truncated exponentials." We assume that K is a p-adic field with ring of integers $\mathfrak{O}_K$ (though their construction also works in characteristic p).

For x, y, the *truncated exponential* is the quantity defined as

$$x^{[y]} = \sum_{m=0}^{p-1} \binom{y}{m} (x-1)^m,$$

where $\binom{y}{m}$ is the *generalized binomial coefficient*

$$\binom{y}{m} = y(y-1)(y-2)\cdots(y-m+1)/m!$$

Byott and Elder construct a totally ramified elementary abelian extension L/K with Galois group $C_p^n = \langle g_1, g_2, \dots g_n\rangle$ and a scaffold (see Chapter 15) for the action of $K[C_p^n]$ on L/K. It then follows from [**BCE18**] that $\mathfrak{O}_L$ is free of rank one over its associated order $\mathfrak{A}$. Furthermore, the associated order is of the form

$$\mathfrak{A} = \mathfrak{O}_K\left[\frac{\Theta_1 - 1}{\pi^{i_1}}, \frac{\Theta_2 - 1}{\pi^{i_2}}, \dots, \frac{\Theta_n - 1}{\pi^{i_n}}\right],$$

with

$$\Theta_1 = g_1,$$

and

$$\Theta_j = g_j\Theta_1^{[\mu_{j,1}]}\Theta_2^{[\mu_{j,2}]}\cdots\Theta_{j-1}^{[\mu_{j,j-1}]},$$

for $2 \le j \le n$, where

$$\mu_{j,1}, \mu_{j,2}, \dots, \mu_{j,j-1},$$

$2 \le j \le n$, are certain elements of K and i_j, $1 \le j \le n$, are certain non-negative integers.

A result of Bondarko [**Bon00**] then shows that $\mathfrak{A}$ is an $\mathfrak{O}_K$-Hopf order in $K[C_p^n]$. By construction, $\mathfrak{A}$ is realizable. The associated order $\mathfrak{A}$ is a type of *truncated exponential Hopf order*.

For $n = 2$ the truncated exponential Hopf order appears as

$$\mathfrak{O}_K\left[\frac{g_1 - 1}{\pi^{i_1}}, \frac{g_2 g_1^{[\mu_{2,1}]} - 1}{\pi^{i_2}}\right].$$

For $n = 3$ the truncated exponential Hopf order is

$$\mathfrak{O}_K\left[\frac{g_1 - 1}{\pi^{i_1}}, \frac{g_2 g_1^{[\mu_{2,1}]} - 1}{\pi^{i_2}}, \frac{g_3 g_1^{[\mu_{3,1}]}(g_2 g_1^{[\mu_{2,1}]})^{[\mu_{3,2}]} - 1}{\pi^{i_3}}\right].$$

In [**BE18**, §5.2] the authors provide a framework for the construction of a collection of (realizable) Hopf orders in $K[G]$ where G is any abelian p-group. Let L/K be a totally ramified Galois extension of degree p^n whose Galois group G is an abelian p-group. Let

$$\{e\} = G_m \subset G_{m-1} \subset \cdots \subset G_2 \subset G_1 \subset G_0 = G$$

be the ramification filtration of L/K and let

$$\{e\} = J_n \subset J_{n-1} \subset \cdots \subset J_2 \subset J_1 \subset J_0 = G$$

be a refinement with $J_{n-j}/J_{n-j+1} \cong C_p$. Let $g_j \in J_{n-j}\backslash J_{n-j+1}$ for $1 \le j \le n$. Then (in theory) there are elements

$$\mu_{j,1}, \mu_{j,2}, \dots, \mu_{j,j-1} \in K,$$

$2 \le j \le n$, and integers i_j, $1 \le j \le n$, so that the associated order

$$\mathfrak{A} = \mathfrak{O}_K\left[\frac{\Theta_1 - 1}{\pi^{i_1}}, \frac{\Theta_2 - 1}{\pi^{i_2}}, \dots, \frac{\Theta_n - 1}{\pi^{i_n}}\right]$$

is a truncated exponential Hopf order in $K[G]$, with Θ_j defined inductively as above.

12.10. Models of $\boldsymbol{\mu}_{p^n,K}$ for $n = 1, 2, 3$

Models are another way to study Hopf orders. A *model* of $\boldsymbol{\mu}_{p^n,K}$ is a finite, flat $\mathfrak{O}_K$-group scheme which is isomorphic to $\boldsymbol{\mu}_{p^n,K}$ on the generic fiber. Thus the representing Hopf algebra of a model is an $\mathfrak{O}_K$-Hopf order in $K[C_{p^n}]$.

Using Sekiguchi-Suwa theory, Tossici [**Tos10**] has classified models of $\boldsymbol{\mu}_{p^n,K}$ for $n = 1, 2$. By [**Tos10**, Proposition 1.3] a model of $\boldsymbol{\mu}_{p,K}$ is of the form

$$\mathrm{Spec}\ \mathfrak{O}_K[T]/(P(T)),$$

where $P(T) = \dfrac{(1+\pi^i T)^p - 1}{\pi^{pi}}$, $0 \le i \le e'$, T indeterminate. The quotient $\mathfrak{O}_K[T]/(P(T))$ is necessarily a Larson order: If $\langle g\rangle = C_p$, and we identify g with $1 + \pi^i\overline{T}$, then $\mathfrak{O}_K[T]/(P(T)) \cong H(i)$. Here $\overline{T}$ denotes the image of T under the canonical surjection $\mathfrak{O}_K[T] \to \mathfrak{O}_K[T]/(P(T))$.

By [**Tos10**, Lemma-Definition 3.9], a model of $\boldsymbol{\mu}_{p^2,K}$ or $\boldsymbol{\mu}_{p,K} \times \boldsymbol{\mu}_{p,K}$ is of the form

$$\mathrm{Spec}\ \mathfrak{O}_K[T_1, T_2]/(P(T_1), Q(T_1, T_2)),$$

where $P(T_1) = \dfrac{(1+\pi^i T_1)^p - 1}{\pi^{pi}}$, and

$$Q(T_1, T_2) = \frac{(D_1(T_1) + \pi^j T_2)^p (1 + \pi^i T_1)^{-r} - 1}{\pi^{pj}}$$

for some polynomial $D_1(T_1)$, $r \in \mathbb{Z}$. Following [**EU17**, Remark 3.5] (where the case C_p^2, characteristic p, is considered) we take $r = 0$,

$$D_1(T_1) = (1 + \pi^i T_1)^{[-\mu]} = \sum_{m=0}^{p-1} \binom{-\mu}{m} (\pi^i T_1)^m, \quad \mu \in K,$$

and we make the identifications $g_1 = 1 + \pi^i\overline{T}_1$, $g_2 = D_1(\overline{T}_1) + \pi^j\overline{T}_2$, with $\langle g_1, g_2\rangle = C_p^2$. Then $\mathfrak{O}_K[T_1, T_2]/(P(T_1), Q(T_1, T_2))$ is isomorphic to the truncated exponential Hopf order $\mathfrak{O}_K\left[\dfrac{g_1 - 1}{\pi^i}, \dfrac{g_2(g_1)^{[\mu]} - 1}{\pi^j}\right]$ in $K[C_p^2]$. In [**MRT13**] the authors conjecture that all models of $\boldsymbol{\mu}_{p^3,K}$ are of the form

$$\mathrm{Spec}\ \mathfrak{O}_K[T_1, T_2, T_3]/(P(T_1), Q(T_1, T_2), S(T_1, T_2, T_3)),$$

where $P(T_1) = \dfrac{(1+\pi^i T_1)^p - 1}{\pi^{pi}}$,

$Q(T_1, T_2) = \dfrac{(D_1(T_1) + \pi^j T_2)^p (1 + \pi^i T_1)^{-1} - 1}{\pi^{pj}}$, and

$$S(T_1, T_2, T_3) = \frac{(D_2(T_1, T_2) + \pi^k T_3)^p (D_1(T_1) + \pi^j T_2)^{-1} - 1}{\pi^{pk}},$$

for some polynomial $D_2(T_1, T_2)$. Constructing models via Sekiguchi-Suwa theory is comparable to the approach of Breuil-Kisin, see [**MRT13**, §8.2.6, §8.3.5].

Sekiguchi-Suwa theory can also be used to compute the abelian group of Galois extensions over Hopf orders in $K[C_{p^2}]$ [**Und97**].

12.11. Hopf orders and realizability

Let H be an $\mathfrak{O}_K$-Hopf order in $K[G]$ with $G = C_{p^n}$, or $G = C_p^n$.

Recall our definitions of realizable Hopf orders (Definition 12.3, Definition 12.4): H is *realizable* if there exists a Galois extension L/K with group G whose associated order $\mathfrak{A}$ is H.

H is *realizable as a Galois group* if there exists a Galois extension L/K with group G for which $\mathfrak{O}_L$ is a H-Galois extension of $\mathfrak{O}_K$.

We have seen that if H is realizable as a Galois group, then it is realizable. However, if H is realizable via L/K, then $\mathfrak{O}_L$ is not necessarily H-Galois (for that same L/K), see Example 12.6.

In this section we determine whether the Hopf orders constructed in the previous sections are realizable or not.

12.11.1. Hopf orders in $K[C_p]$ are realizable. Assume that $\zeta_p \in K$. We show that every Hopf order in $K[C_p]$ is realizable as a Galois group (and hence realizable).

By Theorem 12.42, an $\mathfrak{O}_K$-Hopf order H in $K[C_p]$ can be written in the form

$$H = H(i') = \mathfrak{O}_K\left[\frac{g-1}{\pi^{i'}}\right], \quad \langle g\rangle = C_p,$$

for some integer i with $i' = e' - i$, $0 \le i \le e'$.

In what follows, we construct an $H(i')$-Galois extension of $\mathfrak{O}_K$. Set $w = 1 + \pi^{pi+1}$. Then w cannot be a pth power in $\mathfrak{O}_K$ and the polynomial $x^p - w \in \mathfrak{O}_K[x]$ is irreducible. Set $L = K(z)$, with $z = w^{\frac{1}{p}}$. Then L/K is a Galois extension with group C_p.

Let λ denote the uniformizing parameter for L, thus $(\pi) = (\lambda)^{e_{L/K}}$, for some positive integer $e_{L/K}$. If $e_{L/K} = 1$ then L/K is unramified, and if $e_{L/K} = p$ then L/K is totally ramified.

THEOREM 12.50. *Let $x = (z-1)/\pi^i$. Then $\mathfrak{O}_K[x]$ is a $H(i')$-Galois extension of $\mathfrak{O}_K$.*

PROOF. We use Proposition 12.5, checking that $\mathrm{disc}_{H(i')}(\mathfrak{O}_K[x]/\mathfrak{O}_K) = \mathfrak{O}_K$.

By [**Und11**, Proposition 4.4.5] and [**Und11**, Proposition 5.3.3],

$$\Lambda = \pi^{-(p-1)i'} \sum_{g\in C_p} g = \mathrm{Tr}_{L/K}/\pi^{(p-1)i'}$$

is a generating integral for $H(i')$. Note that

$$\mathfrak{O}_K[\pi^i x] = \mathfrak{O}_K[z] \subseteq \mathfrak{O}_K\left[\frac{z-1}{\pi^i}\right].$$

With respect to the basis $\{1, z, z^2, \ldots, z^{p-1}\}$, we have

$$\mathrm{disc}(\mathfrak{O}_K[z]/\mathfrak{O}_K) = (-1)^{p(p-1)/2}\mathrm{N}_{L/K}(pz^{p-1})\mathfrak{O}_K = p^p\mathfrak{O}_K,$$

by [**Lan65**, p. 348].

For the convenience of the reader, we recall the argument of [**Und11**, Proposition 10.4.2]: The matrix which multiplies the basis $\{(z-1)^m/\pi^{mi}\}_{m=0}^{p-1}$ of $\mathfrak{O}_K[x]$

to give the basis $\{(z-1)^m\}$ of $\mathfrak{O}_K[z]$ is

$$\begin{pmatrix} 1 & 0 & 0 & \cdots & 0 \\ 0 & \pi^i & 0 & \cdots & 0 \\ 0 & 0 & \pi^{2i} & \cdots & 0 \\ \vdots & & & & \vdots \\ 0 & 0 & 0 & \cdots & \pi^{(p-1)i} \end{pmatrix},$$

and so, by a familiar property of discriminants,

$$p^p\mathfrak{O}_K = \pi^{p(p-1)i}\mathfrak{O}_K \cdot \mathrm{disc}(\mathfrak{O}_K[x]/\mathfrak{O}_K).$$

Thus, $\mathrm{disc}(\mathfrak{O}_K[x]/\mathfrak{O}_K) = \frac{p^p}{\pi^{p(p-1)i}}\mathfrak{O}_K$ and so,

$$\begin{aligned} \frac{p^p}{\pi^{p(p-1)i}}\mathfrak{O}_K &= \det(\mathrm{Tr}_{L/K}(x^a x^b))\mathfrak{O}_K \\ &= \pi^{p(p-1)i'}\det(\mathrm{Tr}_{L/K}(x^a x^b)/\pi^{(p-1)i'})\mathfrak{O}_K \\ &= \pi^{p(p-1)i'}\det(\Lambda(x^a x^b))\mathfrak{O}_K. \end{aligned}$$

Thus

$$\begin{aligned} \det(\Lambda(x^a x^b))\mathfrak{O}_K &= \pi^{pe-p(p-1)i-p(p-1)i'}\mathfrak{O}_K \\ &= \pi^{pe-p(p-1)(i+i')}\mathfrak{O}_K \\ &= \pi^{pe-p(p-1)e'}\mathfrak{O}_K \\ &= \mathfrak{O}_K. \end{aligned}$$

Hence $\mathfrak{O}_K[x]$ is a $H(i')$-Galois extension of $\mathfrak{O}_K$. □

We conclude that $H(i')$ is realizable if we can show that $\mathfrak{O}_K[x] = \mathfrak{O}_L$.

THEOREM 12.51. *Let $\mathfrak{O}_K[x]$ be defined as above with $0 \le i \le e'$. Then $\mathfrak{O}_K[x]$ is the ring of integers of $L = K(z)$, $z^p = w$, $w = 1 + \pi^{pi+1}$.*

PROOF. Clearly, $\mathfrak{O}_K[x] \subseteq \mathfrak{O}_L$. Suppose that $i = e'$. Then by Proposition 12.50, $\mathfrak{O}_K[x]$ is a $\mathfrak{O}_K[C_p]$-Galois extension of $\mathfrak{O}_K$ with

$$\mathrm{disc}_{\mathfrak{O}_K[C_p]}(\mathfrak{O}_K[x]/\mathfrak{O}_K) = \mathfrak{O}_K,$$

In this case, $\mathrm{disc}_{\mathfrak{O}_K[C_p]}(\mathfrak{O}_K[x]/\mathfrak{O}_K)$ is the classical discriminant of $\mathfrak{O}_K[x]$ over $\mathfrak{O}_K$. Thus $\mathrm{disc}(\mathfrak{O}_L/\mathfrak{O}_K) = \mathfrak{O}_K$ and so $\mathfrak{O}_K[x] = \mathfrak{O}_L$. We also conclude that L/K is unramified.

Next suppose that $0 \le i < e'$. Then [**Chi87**, Lemma 13.1] applies to show that $\nu_\lambda(x) = 1$. Thus $\mathfrak{O}_K[x]$ is the ring of integers of L. By [**Und11**, Proposition 10.4.1], L/K is totally ramified. □

We summarize the discussion above. For a given $\mathfrak{O}_K$-Hopf order $H(i')$ in $K[C_p]$, we have found a Galois extension L/K whose ring of integers $\mathfrak{O}_L$ is an $H(i')$-Galois extension, i.e., every Hopf order in $K[C_p]$ is realizable as a Galois group.

Hence, assuming $\zeta_p \in K$, every Hopf order in $K[C_p]$ is realizable.

12.11.2. Realizable Greither orders. Greither [**Gre92**, II, Theorem 3.2] has shown that Hopf orders of the form $A(i,j,u)$ are realizable.

THEOREM 12.52 (Greither). *Assume that* $\zeta_{p^2} \in K$. *Let* $A(i,j,u)$ *be a Greither order in* $K[C_{p^2}]$ *with* $j > 0$. *Then* $A(i,j,u)$ *is realizable as a Galois group if and only if* $\nu(1-u) = i' + (j/p)$.

PROOF. Suppose that $A(i,j,u)$ is realizable as a Galois group; let L/K be the Galois extension with group C_{p^2} that realizes $A(i,j,u)$. Let λ be the parameter for L. Then $\lambda^{p^2} = \pi$.

By Herbrand's Theorem, $pj - 1$ is the first ramification break of L/K.

If $\nu(1-u) \geq i'+j$, then $A(i,j,u)$ is the Larson order $H(i,j) = \mathfrak{O}_K\left[\frac{g^p-1}{\pi^i}, \frac{g-1}{\pi^j}\right]$. Since $\frac{g-1}{\pi^j}$ operates on $\mathfrak{O}_L$, we have

$$p^2 j \leq (pj-1)+1 = pj,$$

which is impossible since $j > 0$.

So we assume that $\nu(1-u) < i' + j$. Now, $H(i,l) = \mathfrak{O}_K\left[\frac{g^p-1}{\pi^i}, \frac{g-1}{\pi^l}\right]$, $l = \nu(1-u) - i'$, is the largest Larson order in $A(i,j,u)$. Since $\frac{g-1}{\pi^l}$ operates on $\mathfrak{O}_L$, we have

$$p^2 l \leq (pj-1)+1 = pj,$$

and so, $l \leq j/p$, thus

$$\nu(1-u) \leq i' + (j/p).$$

Since $\nu(1-u) \geq i' + (j/p)$, we have $\nu(1-u) = i' + (j/p)$.

For the converse we use Byott's theorem (Theorem 12.16). By Proposition 12.46, the linear dual of H is the $\mathfrak{O}_K$-Hopf order

$$A(i,j,u)^* = A(j',i',\widetilde{u}) = \mathfrak{O}_K\left[\frac{\gamma^p - 1}{\pi^{j'}}, \frac{\gamma a_{\widetilde{u}} - 1}{\pi^{i'}}\right].$$

We claim that $A(j',i',\widetilde{u})$ is a local ring. To this end, note that $j > 0$ implies that $i > 0$, hence the Larson orders $H(j')$ and $H(i')$ are local by Proposition 12.43. Consequently, $A(j',i',\widetilde{u})$ is local by Proposition 12.15. Thus Byott's theorem will apply.

For a proof that $A(j',i',\widetilde{u})$ is monogenic on the generator $\dfrac{\gamma a_{\widetilde{u}} - 1}{\pi^{i'}}$, i.e.,

$$A(j',i',\widetilde{u}) = \mathfrak{O}_K\left[\frac{\gamma a_{\widetilde{u}} - 1}{\pi^{i'}}\right],$$

see [**UC06**, Theorem 5.1]. □

REMARK 12.53. If $A(i,j,u)$ is a Greither order with $j = 0$, then $A(i,j)$ is the Larson order $H(i,j)$ in $K[C_{p^2}]$ and $H(i,j)$ is realizable (see the comments immediately before [**Gre92**, II, Lemma 3.2]).

If $j > 0$, and $H(i,j)$ is a Larson order, then $H(i,j) = A(i,j,u)$ for some $\nu(1-u) \geq i'+j$. Thus $\nu(1-u) \geq i'+j > i'+(j/p)$, and so $H(i,j)$ is not realizable as a Galois group by Theorem 12.52.

Suppose $\zeta_{p^2} \in K$ and let $A(i,j,u)$ be any Hopf order in $K[C_{p^2}]$, with $j > 0$, $\nu(1-u) < i' + j$. Then $A(i,j,u)$ is realizable only if $\nu(1-u) = i' + (j/p)$.

12.11.3. Realizable Hopf orders in $K[C_{p^3}]$. Assume that $\zeta_{p^3} \in K$. In Section 12.7, we constructed several collections of Hopf orders in $K[C_{p^3}]$, namely, *cohomological* Hopf orders, *ILD* Hopf orders and *duality* Hopf orders.

PROPOSITION 12.54. *An ILD Hopf order that is not cohomological is not realizable as Galois group.*

PROOF. Let

$$H = \mathfrak{O}_K\left[\frac{g^{p^2}-1}{\pi^i}, \frac{g^p a_u - 1}{\pi^j}, \frac{g a_v b_w - 1}{\pi^k}\right]$$

be an ILD Hopf order that is not cohomological, with $k > 0$. If H is realizable as a Galois group then so is $\overline{H} = A(j,k,w)$ by Proposition 12.14. Now by Remark 12.53 (essentially Theorem 12.52), $\nu(1-w) = j' + (k/p)$. But this contradicts [**UC06**, Proposition 1.8], cf. [**UC06**, Theorem 5.2]. □

PROPOSITION 12.55. *No duality Hopf order is realizable as a Galois group.*

PROOF. Suppose that $A(i,j,u)\left[\dfrac{g a_v b_w - 1}{\pi^k}\right]$, $k > 0$, is a duality Hopf order that is realizable as a Galois group. By Proposition 12.49, $j > 0$.

By Proposition 12.14, both $A(i,j,u)$ and $A(j,k,w)$ are realizable, thus by Remark 12.53, $\nu(1-u) = i' + (j/p)$ and $\nu(1-w) = j' + (k/p)$. As shown in [**UC06**, Theorem 5.3], this is impossible. □

So to find realizable Hopf orders we consider cohomological Hopf orders.
Let

$$H(d,c)^*\left[\frac{g b_{u,v} - 1}{\pi^k}\right] = \mathfrak{O}_K\left[\frac{g^{p^2}-1}{\pi^{c'}}, \frac{g^p a_{\zeta_{p^2}^{-1}} - 1}{\pi^{d'}}, \frac{g b_{u,v} - 1}{\pi^k}\right]$$

be a cohomological Hopf order in $K[C_{p^3}]$, $k > 0$.

PROPOSITION 12.56. *If the cohomological Hopf order $H(d,c)^*\left[\dfrac{g b_{u,v} - 1}{\pi^k}\right]$ is realizable as a Galois group, then $d = pc$.*

PROOF. By Proposition 12.49, $d' > 0$, and by Proposition 12.14, $A(c', d', \zeta_{p^2}^{-1})$ is realizable as a Galois group. Thus $\nu(1 - \zeta_{p^2}^{-1}) = e'/p = c + (d'/p)$ by Remark 12.53. Hence $d = pc$. □

In fact if $d = pc$, then the cohomological Hopf order reduces to the special form $H(pc,c)^*\left[\dfrac{g b_{1,w} - 1}{\pi^k}\right]$ with $[w] \in (U_{c+(k/p^2)} \cap U_{(c/p)+k})/U_{c+k}$ (Section 12.7).

We can use Byott's theorem to show realizability.

THEOREM 12.57. *Let $H = H(pc,c)^*\left[\dfrac{g b_{1,w} - 1}{\pi^k}\right]$ be a cohomological Hopf order in $K[C_{p^3}]$. Suppose that $k > 0$ and $\nu(1-w) = c + (k/p^2)$. Then $H(pc,c)^*\left[\dfrac{g b_{1,w} - 1}{\pi^k}\right]$ is realizable as a Galois group.*

PROOF. First note that H can be written as $H = H(pc, c)^* \left[\frac{ga_w b_{w^p} - 1}{\pi^k}\right]$. By [**UC06**, Theorem 5.6],

$$H^* = \mathfrak{O}_K \left[\frac{\gamma^{p^2} - 1}{\pi^{k'}}, \frac{\gamma^p a_{\widetilde{w^p}} - 1}{\pi^{pc}}, \frac{\gamma a_{\widetilde{\widetilde{w}}} - 1}{\pi^c}\right],$$

where $\widetilde{w^p} = w^{-p}\zeta_{p^2}^{-1}$ and $\widetilde{\widetilde{w}} = w^{-1}\zeta_{p^3}^{-1}$.

Now $k > 0$ implies that $e' > k'$, thus by Proposition 12.49, $e' > pc > c$. So by Proposition 12.15, H^* is local. The condition $\nu(1 - w) = c + (k/p^2)$ then yields H^* monogenic on the generator $\frac{\gamma a_{\widetilde{\widetilde{w}}} - 1}{\pi^c}$. For details see [**UC06**, Theorem 5.6]. □

We seek other Hopf orders in $K[C_{p^3}]$ that are realizable. We continue our assumption that $\zeta_{p^3} \in K$. In view of Byott's theorem (Theorem 12.16) we need to construct a local, monogenic Hopf order H in $K[C_{p^3}]$. Then H^* will be a realizable Hopf order in $K[\widehat{C}_{p^3}] \cong K[C_{p^3}]$.

Let $c \geq 0$ be an integer with $e' > p^2c$. Let

$$A(p^2c, c, u) = \mathfrak{O}_K \left[\frac{g^{p^2} - 1}{\pi^{p^2c}}, \frac{g^p a_u - 1}{\pi^c}\right]$$

be a Greither order in $K[\langle g^p \rangle] = K[C_{p^2}]$. Let

$$H = \mathfrak{O}_K \left[\frac{g^{p^2} - 1}{\pi^{p^2c}}, \frac{g^p - 1}{\pi^{pc}}, \frac{ga_u - 1}{\pi^c}\right].$$

PROPOSITION 12.58. *H is a local monogenic $\mathfrak{O}_K$-Hopf order in $K[C_{p^3}]$.*

PROOF. As shown in [**UC06**, Theorem 5.7], H is an $\mathfrak{O}_K$-subalgebra of $K[C_{p^3}]$, which is free of rank p^3 as an $\mathfrak{O}_K$-submodule. Moreover, $K \otimes_{\mathfrak{O}_K} H \cong K[C_{p^3}]$. Thus H is an $\mathfrak{O}_K$-order in $K[C_{p^3}]$. Moreover,

$$\Delta_{K[C_{p^3}]}(H) \subseteq H \otimes_{\mathfrak{O}_K} H,$$

thus by [**Und15**, Proposition 3.4.5], H is an $\mathfrak{O}_K$-Hopf order in $K[C_{p^3}]$.

Using a discriminant argument, one then shows that H is monogenic on the generator $\frac{ga_u - 1}{\pi^c}$.

Finally, the condition $e' > p^2c$ guarantees that H is a local ring. For details, see [**UC06**, Theorem 5.7]. □

THEOREM 12.59. *The linear dual of H is of the form*

$$H^* = \mathfrak{O}_K \left[\frac{\gamma^{p^2} - 1}{\pi^{c'}}, \frac{\gamma^p a_{\zeta_{p^2}^{-1}} - 1}{\pi^{(pc)'}}, \frac{\gamma a_{\widetilde{u}} b_{\zeta_{p^2}^{-1}} - 1}{\pi^{(p^2c)'}}\right],$$

and H^* is a realizable Hopf order in $K[\widehat{C}_{p^3}] \cong K[C_{p^3}]$.

PROOF. For the structure of the dual, see [**UC06**, Theorem 5.7]. Now, H^* is realizable by Proposition 12.58 and an application of Byott's theorem. □

REMARK 12.60. Note that the realizable Hopf order in Theorem 12.59 can be written as

$$H^* = H(pc, c)^* \left[\frac{\gamma a_{\tilde{\tilde{u}}} b_{\zeta_{p^2}^{-1}} - 1}{\pi^{(p^2 c)'}} \right],$$

and thus has a form similar to the realizable Hopf order in Theorem 12.57. Yet, as shown in [**UC06**, Example 5.8], these Hopf orders are distinct.

For another construction of realizable Hopf orders in $K[C_{p^3}]$, see [**Und08**].

12.12. Realizable Hopf orders in $K[C_{p^n}]$, $K[C_p^n]$

12.12.1. The case $G = C_{p^n}$. In Section 12.8, we constructed the collection of *formal group* Hopf orders in $K[C_{p^n}]$.

Unfortunately, most formal group Hopf orders are not realizable. For example, we have the following.

PROPOSITION 12.61. *Let $p > 2$ and let H_M be a formal group Hopf order in $K[C_{p^3}]$. Then H_M is not realizable as a Galois group.*

PROOF. See [**UC06**, Theorem 5.4]. □

However, on the strength of Byott's criterion (Theorem 12.18), there is an obvious way to construct a collection of realizable Hopf orders in $K[C_{p^n}]$.

Assume that $\zeta_{p^n} \in K$. Let c be an integer with $0 < c \leq (e'/p^{n-1})$. Let

$$H(p^{n-1}c, \ldots, pc, c) = \mathfrak{O}_K \left[\frac{g-1}{\pi^c} \right]$$

be a one-parameter Larson order in $K[C_{p^n}]$, $\langle g \rangle = C_{p^n}$. Then its dual $\mathfrak{O}_K \left[\frac{g-1}{\pi^c} \right]^*$ is an $\mathfrak{O}_K$-Hopf order in $K[\widehat{C}_{p^n}] \cong K[C_{p^n}]$.

THEOREM 12.62. *The $\mathfrak{O}_K$-Hopf order $\mathfrak{O}_K \left[\frac{g-1}{\pi^c} \right]^*$ in $K[C_{p^n}]$ is realizable.*

PROOF. Since $c > 0$, both $\mathfrak{O}_K \left[\frac{g-1}{\pi^c} \right]^*$ and its dual $(\mathfrak{O}_K \left[\frac{g-1}{\pi^c} \right]^*)^* = \mathfrak{O}_K \left[\frac{g-1}{\pi^c} \right]$ are local rings, with $\mathfrak{O}_K \left[\frac{g-1}{\pi^c} \right]$ monogenic. Thus by Theorem 12.18, $\mathfrak{O}_K \left[\frac{g-1}{\pi^c} \right]^*$ is realizable. □

12.12.2. The case $G = C_p^n$. In general, the Hopf orders H_n in $K[C_p^n]$ constructed by Childs and Smith in [**CS05**] are not realizable as Galois groups. For instance, we have the following.

THEOREM 12.63. *Assume $n = 3$ and $0 < i_1, i_2, i_3 < e'$. Then the Hopf order*

$$H_3 = \mathfrak{O}_K \left[\frac{g_1 - 1}{\pi^{i_1}}, \frac{g_2 a_{2,1} - 1}{\pi^{i_2}}, \frac{g_3 a_{3,1} a_{3,2} - 1}{\pi^{i_3}} \right]$$

is not realizable as a Galois group.

PROOF. See [**CS05**, Proposition 15]. □

So to find realizable Hopf orders in $K[C_p^n]$, we look to the construction of truncated exponential Hopf orders given by Byott and Elder in Section 12.9. As we recall, Byott and Elder construct a totally ramified elementary abelian extension L/K with Galois group $C_p^n = \langle g_1, g_2, \ldots g_n \rangle$ and a scaffold (see Chapter 15) for the action of $K[C_p^n]$ on L/K. The associated order $\mathfrak{A}$ is an $\mathfrak{O}_K$-Hopf order in $K[C_p^n]$ of the form

$$\mathfrak{A} = \mathfrak{O}_K\left[\frac{\Theta_1 - 1}{\pi^{i_1}}, \frac{\Theta_2 - 1}{\pi^{i_2}}, \ldots, \frac{\Theta_n - 1}{\pi^{i_n}}\right],$$

with

$$\Theta_1 = g_1,$$

and

$$\Theta_j = g_j \Theta_1^{[\mu_{j,1}]} \Theta_2^{[\mu_{j,2}]} \cdots \Theta_{j-1}^{[\mu_{j,j-1}]},$$

for $2 \leq j \leq n$.

By construction, $\mathfrak{A}$ is realizable. The associated order $\mathfrak{A}$ is a type of *truncated exponential Hopf order*.

For $n = 2$ the truncated exponential Hopf order appears as

$$\mathfrak{O}_K\left[\frac{g_1 - 1}{\pi^{i_1}}, \frac{g_2 g_1^{[\mu_{2,1}]} - 1}{\pi^{i_2}}\right].$$

For $n = 3$ the truncated exponential Hopf order is

$$\mathfrak{O}_K\left[\frac{g_1 - 1}{\pi^{i_1}}, \frac{g_2 g_1^{[\mu_{2,1}]} - 1}{\pi^{i_2}}, \frac{g_3 g_1^{[\mu_{3,1}]}(g_2 g_1^{[\mu_{2,1}]})^{[\mu_{3,2}]} - 1}{\pi^{i_3}}\right].$$

12.13. When K has characteristic p

For the remainder of this chapter, we assume that $\mathrm{char}(K) = p$ and $G = C_p^n$, $n \geq 1$. Now, K is the field of formal Laurent series $\mathbb{F}_q((T))$ over a finite field $\mathbb{F}_q$, where q is a power of p; $\mathfrak{O}_K$ is the ring of power series $\mathbb{F}_q[[T]]$. We consider Hopf orders in $K[C_p^n]$.

As in the characteristic 0 case, the problem of classifying $\mathfrak{O}_K$-Hopf orders in $K[C_p^n]$, $n \geq 1$, is difficult. The classification is complete for the cases $n = 1, 2$ [**TO70**], [**EU17**], but the case $n \geq 3$ remains open.

On the dual side the situation is much better. Koch [**Koc17**] has given a complete characterization of all $\mathfrak{O}_K$-Hopf orders in $K[C_p^n]^*$, $n \geq 1$. From Koch's work we conclude that Hopf orders in $K[C_p^n]^*$ are determined by $n(n+1)/2$ parameters, a result that recalls the parameters conjecture of Section 12.6.2.

One might hope to use Koch's result to classify Hopf orders in $K[C_p^n]$ by first computing the Hopf orders in $K[C_p^n]^*$ and then taking their duals to obtain all Hopf orders in $K[C_p^n]$ but difficulties arise in computing the algebraic structure of the duals in $K[C_p^n]$. At the very least we expect that Hopf orders in $K[C_p^n]$ can be classified using $n(n+1)/2$ parameters (since their duals in $K[C_p^n]^*$ require exactly $n(n+1)/2$ parameters).

12.13.1. Koch's classification of Hopf orders in $K[C_p^n]^*$. Let R be a commutative $\mathbb{F}_p$-algebra and let H be an R-Hopf algebra. An element $h \in H$ is *primitive* if $\Delta(h) = 1 \otimes h + h \otimes 1$. The Hopf algebra H is *primitively generated* if it is generated as an R-algebra by primitive elements.

If J is a primitively generated R-Hopf algebra of rank p^n, then there exists a matrix $A = (a_{i,j}) \in \mathrm{Mat}_n(R)$ for which

$$J = R[u_1, u_2, \ldots, u_n]/I$$

where I is the ideal in $R[u_1, u_2, \ldots, u_n]$ generated by $\{u_i^p - \sum_{j=1}^n a_{j,i} u_j\}_{1 \leq i \leq n}$, with u_i primitive for $1 \leq i \leq n$; A is the matrix *associated to* J.

In the case $R = K$, the K-Hopf algebra $K[C_p^n]^*$ is primitively generated of rank p^n. We have

$$K[C_p^n]^* = K[t_1, t_2, \ldots, t_n],$$

with $\Delta(t_i) = 1 \otimes t_i + t_i \otimes 1$ and $t_i^p = t_i$ for $1 \le i \le n$, see [**Koc17**, Example 2.6]. The matrix associated to $K[C_{p^n}]^*$ is the $n \times n$ identity matrix I_n.

Let F be an indeterminate in the non-commutative polynomial ring $\mathfrak{O}_K[F]$; F satisfies the condition $Fa = a^p F$ for all $a \in \mathfrak{O}_K$. In [**Koc17**] Koch has determined all $\mathfrak{O}_K$-Hopf orders in $K[C_p^n]^*$ using the equivalence between the category of primitively generated $\mathfrak{O}_K$-Hopf algebras and the category of finite $\mathfrak{O}_K[F]$-modules that are free as $\mathfrak{O}_K$-modules, i.e., the category of *Dieudonné modules* in the sense of [**dJ93**, §2].

We review Koch's result.

For a matrix $M = (m_{i,j}) \in \mathrm{Mat}_n(K)$, let $M^{(p)}$ be the matrix whose i,jth entry is $m_{i,j}^p$.

THEOREM 12.64 (Koch).

(i) Let $n \ge 1$, let $\Theta = (\theta_{i,j})$ be a lower triangular matrix in $\mathrm{GL}_n(K)$ and let $A = \Theta^{-1}\Theta^{(p)}$. Suppose that $A = (a_{i,j}) \in \mathrm{Mat}_n(\mathfrak{O}_K)$. Let

$$J = \mathfrak{O}_K[u_1, u_2, \ldots, u_n]/I$$

where I is the ideal in $\mathfrak{O}_K[u_1, u_2, \ldots, u_n]$ generated by $\{u_i^p - \sum_{j=1}^n a_{j,i}u_j\}_{1\le i\le n}$, with u_i primitive for $1 \le i \le n$. There is a Hopf embedding $J \to K[C_p^n]^*$ given as

$$u_i \mapsto \sum_{j=1}^n \theta_{j,i} t_j,$$

$t_j^p = t_j$, t_j primitive. The image of J under this embedding is

$$H_\Theta = \mathfrak{O}_K[\{\sum_{j=1}^n \theta_{j,i} t_j \mid 1 \le i \le n\}],$$

which is an $\mathfrak{O}_K$-Hopf order in $K[C_p^n]^*$.

(ii) Let J be an $\mathfrak{O}_K$-Hopf order in $K[C_p^n]^*$. Then J is a primitively generated $\mathfrak{O}_K$-Hopf algebra of rank p^n and there exists $A = (a_{i,j}) \in \mathrm{Mat}_n(\mathfrak{O}_K)$ with

$$J = \mathfrak{O}_K[u_1, u_2, \ldots, u_n]/I$$

where I is the ideal in $\mathfrak{O}_K[u_1, u_2, \ldots, u_n]$ generated by $\{u_i^p - \sum_{j=1}^n a_{j,i}u_j\}_{1\le i\le n}$, with u_i primitive for $1 \le i \le n$. Moreover, there exists a lower triangular matrix $\Theta = (\theta_{i,j}) \in \mathrm{GL}_n(K)$ so that

$$J \cong H_\Theta = \mathfrak{O}_K[\{\sum_{j=1}^n \theta_{j,i} t_j \mid 1 \le i \le n\}] \subseteq K[C_p^n]^*,$$

$t_j^p = t_j$, t_j primitive. Since Θ is lower triangular, J is determined by $n(n+1)/2$ parameters.

PROOF. (Sketch)

(i): Since $A = \Theta^{-1}\Theta^{(p)}$ is in $\mathrm{Mat}_n(\mathfrak{O}_K)$, [**Koc17**, Theorem 4.3] applies to show that H_Θ is an $\mathfrak{O}_K$-Hopf order in $K[C_p^n]^*$.

For (ii): Since $K[C_p^n]^*$ is primitively generated, J is a primitively generated $\mathfrak{O}_K$-Hopf algebra. Let $A = (a_{i,j}) \in \mathrm{Mat}_n(\mathfrak{O}_K)$ be the matrix associated to J, i.e.,

$$J = \mathfrak{O}_K[u_1, u_2, \ldots, u_n]/I$$

where I is the ideal in $\mathfrak{O}_K[u_1, u_2, \ldots, u_n]$ generated by $\{u_i^p - \sum_{j=1}^n a_{j,i}u_j\}_{1\le i\le n}$, with u_i primitive for $1 \le i \le n$. Note that J is generically isomorphic to $\mathfrak{O}_K[C_p^n]^*$, i.e.,

$$K \otimes_{\mathfrak{O}_K} J \cong K \otimes_{\mathfrak{O}_K} \mathfrak{O}_K[C_p^n]^* \cong K[C_p^n]^*$$

as K-Hopf algebras. Thus by [**Koc17**, Corollary 4.2], there exists a matrix $\Theta = (\theta_{i,j}) \in \mathrm{GL}_n(K)$ with $\Theta A = \Theta^{(p)}$, hence

$$A = \Theta^{-1}\Theta^{(p)}.$$

Let

$$H_\Theta = \mathfrak{O}_K[\{\sum_{j=1}^n \theta_{j,i}t_j \mid 1 \le i \le n\}] \subseteq K[C_p^n]^*,$$

$t_i^p = t_i$, t_i primitive. By [**Koc17**, Theorem 4.3], Θ determines an isomorphism $J \to H_\Theta$ (an embedding of J into $K[C_p^n]^*$) defined by

$$u_i \mapsto \sum_{j=1}^n \theta_{j,i}t_j$$

for $1 \le i \le n$.

From [**Koc17**, Theorem 6.1], we can assume that Θ is lower triangular, thus H_Θ is determined by $n(n+1)/2$ parameters. □

REMARK 12.65. Koch's classification applies more broadly to construct all of the $\mathfrak{O}_K$-Hopf orders in any primitively generated K-Hopf algebra, for instance $K[t]/(t^{p^n})$, which represents the group scheme $\boldsymbol{\alpha}_{p^n}$.

We illustrate how to use Theorem 12.64(i) to construct Hopf orders in the cases $n = 1, 2$.

EXAMPLE 12.66. $n = 1$. In this case we choose $\Theta = (\theta_{1,1}) \in \mathrm{GL}_1(K) \cong K^\times$, with $\theta_{1,1} = \pi^i$ for some integer i. To construct an $\mathfrak{O}_K$-Hopf order in $K[C_p]^*$, we require that

$$A = \Theta^{-1}\Theta^{(p)} = (\pi^{-i})(\pi^{pi}) = (\pi^{(p-1)i})$$

is in $\mathrm{Mat}_1(\mathfrak{O}_K)$. Thus we require $i \ge 0$. Let

$$J = \mathfrak{O}_K[u_1]/(u_1^p - \pi^{(p-1)i}u_1)$$

where u_1 is primitive. Then there is a Hopf embedding $J \to K[C_p]^* = K[t_1]$ defined as $u_1 \mapsto \pi^i t_1$. The image of J under this embedding is

$$H_i = \mathfrak{O}_K[\pi^i t_1] \subseteq K[C_p]^*;$$

H_i is a *Tate-Oort* Hopf order in $K[C_p]^*$ [**TO70**], [**EU17**, Theorem 2.3].

EXAMPLE 12.67. $n = 2$. In this case we choose $\Theta = \begin{pmatrix} \pi^i & 0 \\ \theta & \pi^j \end{pmatrix} \in \mathrm{GL}_2(K)$, where i, j are integers. To construct an $\mathfrak{O}_K$-Hopf order in $K[C_p]^*$, we require that

$$A = \Theta^{-1}\Theta^{(p)} = \begin{pmatrix} \pi^{(p-1)i} & 0 \\ \pi^{-j}\theta^p - \pi^{(p-1)i-j}\theta & \pi^{(p-1)j} \end{pmatrix} \in \mathrm{Mat}_2(\mathfrak{O}_K).$$

Thus we require $i, j \geq 0$ and

$$\nu(\pi^{-j}\theta^p - \pi^{(p-1)i-j}\theta) \geq 0.$$

(The condition $A \in \mathrm{Mat}_2(\mathfrak{O}_K)$ is equivalent to the conditions $i, j \geq 0$, $\nu(\pi^{-j}\theta^p - \pi^{(p-1)i-j}\theta) \geq 0$.)

Let

$$J = \mathfrak{O}_K[u_1, u_2]/I$$

with

$$I = (u_1^p - \pi^{(p-1)i}u_1 - (\pi^{-j}\theta^p - \pi^{(p-1)i-j}\theta)u_2, u_2^p - \pi^{(p-1)j}u_2),$$

u_1, u_2 primitive. Then there is a Hopf embedding $J \to K[C_p^2]^* = K[t_1, t_2]$ defined by

$$u_1 \mapsto \pi^i t_1 + \theta t_2, \quad u_2 \mapsto \pi^j t_2.$$

The image of J under this embedding is

$$H_{i,j,\theta} = \mathfrak{O}_K[\pi^i t_1 + \theta t_2, \pi^j t_2];$$

$H_{i,j,\theta}$ is an $\mathfrak{O}_K$-Hopf order in $K[C_p^2]^*$.

12.13.2. Classification of Hopf orders in $K[C_p^2]^*$ using cohomology. In [**EU17**], Elder and Underwood use Greither's cohomological method of [**Gre92**, I] to classify Hopf orders in $K[C_p^2]^*$, and hence recover case $n = 2$ of Theorem 12.64.

Let $C_p^2 = \langle g_1, g_2 \rangle$ and let $\xi_{1,0}$, $\xi_{0,1}$ be elements of $K[C_p^2]^*$ defined by

$$\langle \xi_{1,0}, (g_1 - 1)^j (g_2 - 1)^k \rangle = \delta_{1,j}\delta_{0,k},$$

$$\langle \xi_{0,1}, (g_1 - 1)^j (g_2 - 1)^k \rangle = \delta_{0,j}\delta_{1,k},$$

$0 \leq j, k \leq p-1$, where $\langle -, - \rangle : K[C_p^2]^* \times K[C_p^2] \to K$ is the duality map. Then $\xi_{1,0}$ and $\xi_{0,1}$ are primitive elements in $K[C_p^2]^*$ with $\xi_{1,0}^p = \xi_{1,0}$, $\xi_{0,1}^p = \xi_{0,1}$. The K-Hopf algebra $K[C_p^2]^*$ is primitively generated by $\xi_{1,0}$, $\xi_{0,1}$,

$$K[C_p^2]^* = K[\xi_{1,0}, \xi_{0,1}].$$

(The $\xi_{0,1}$, $\xi_{0,1}$ play the role of t_1, t_2 in Koch's construction given above.) The counit map of $K[C_p^2]^*$ is given as $\varepsilon(\xi_{1,0}) = \varepsilon(\xi_{0,1}) = 0$ and the coinverse is given as $S(\xi_{1,0}) = -\xi_{1,0}$, $S(\xi_{0,1}) = -\xi_{0,1}$.

For $x \in K$, let $\wp(x) = x^p - x$.

THEOREM 12.68 (Elder and Underwood). *Let $i_1, i_2 \geq 0$ be integers, let μ be an element of K satisfying $\nu(\wp(\mu)) \geq i_2 - pi_1$ and let*

$$J_{i_1,i_2,\mu} = \mathfrak{O}_K[\pi^{i_1}(\xi_{1,0} - \mu\xi_{0,1}), \pi^{i_2}\xi_{0,1}].$$

Then $J_{i_1,i_2,\mu}$ is an $\mathfrak{O}_K$-Hopf order in $K[C_p^2]^*$.

PROOF. We have

$$\begin{aligned} (\pi^{i_1}(\xi_{1,0} - \mu\xi_{0,1}))^p &= \pi^{pi_1}(\xi_{1,0} - \mu^p\xi_{0,1}) \\ &= \pi^{(p-1)i_1}\pi^{i_1}(\xi_{1,0} - \mu\xi_{0,1}) + \pi^{pi_1}\mu\xi_{0,1} - \pi^{pi_1}\mu^p\xi_{0,1} \\ &= \pi^{(p-1)i_1}\pi^{i_1}(\xi_{1,0} - \mu\xi_{0,1}) - \wp(\mu)\pi^{pi_1-i_2}\pi^{i_2}\xi_{0,1}. \end{aligned}$$

Since $i_1 \geq 0$ and $\nu(\wp(\mu)) \geq i_2 - pi_1$, $\pi^{i_1}(\xi_{1,0} - \mu\xi_{0,1})$ satisfies a monic polynomial with coefficients in $\mathfrak{O}_K[\pi^{i_2}\xi_{0,1}]$. Moreover,

$$(\pi^{i_2}\xi_{0,1})^p = \pi^{(p-1)i_2}\pi^{i_2}\xi_{0,1},$$

and so, $i_2 \geq 0$ guarantees that $\pi^{i_2}\xi_{0,1}$ satisfies a monic polynomial with coefficients in $\mathfrak{O}_K$.

Thus the $\mathfrak{O}_K$-algebra $J_{i_1,i_2,\mu}$ is an $\mathfrak{O}_K$-subalgebra of $K[C_p^2]^*$ and a free $\mathfrak{O}_K$-submodule of $K[C_p^2]^*$. Moreover,

$$K \otimes_{\mathfrak{O}_K} J_{i_1,i_2,\mu} \cong K[C_p^2]^*$$

as K-algebras and so, $J_{i_1,i_2,\mu}$ is an $\mathfrak{O}_K$-order in $K[C_p^2]^*$.

We have

$$\Delta_{K[C_p^2]^*}(\pi^{i_1}(\xi_{1,0} - \mu\xi_{0,1})) = 1 \otimes (\pi^{i_1}(\xi_{1,0} - \mu\xi_{0,1})) + (\pi^{i_1}(\xi_{1,0} - \mu\xi_{0,1})) \otimes 1,$$

$$\Delta_{K[C_p^2]^*}(\pi^{i_2}\xi_{0,1}) = 1 \otimes \pi^{i_2}\xi_{0,1} + \pi^{i_2}\xi_{0,1} \otimes 1,$$

$$\varepsilon_{K[C_p^2]^*}(\pi^{i_1}(\xi_{1,0} - \mu\xi_{0,1})) = 0, \quad \varepsilon_{K[C_p^2]^*}(\pi^{i_2}\xi_{0,1}) = 0,$$

and

$$S_{K[C_p^2]^*}(\pi^{i_1}(\xi_{1,0} - \mu\xi_{0,1})) = -\pi^{i_1}(\xi_{1,0} - \mu\xi_{0,1}), \quad S_{K[C_p^2]^*}(\pi^{i_2}\xi_{0,1}) = -\pi^{i_2}\xi_{0,1}.$$

Thus $J_{i_1,i_2,\mu}$ is an $\mathfrak{O}_K$-Hopf order in $K[C_p^2]^*$. □

We next show that an arbitrary $\mathfrak{O}_K$-Hopf order in $K[C_p^2]^*$ is of the form given in Theorem 12.68.

THEOREM 12.69. *Let J be an $\mathfrak{O}_K$-Hopf order in $K[C_p^2]^*$. Then*

$$J \cong J_{i_1,i_2,\mu} = \mathfrak{O}_K[\pi^{i_1}(\xi_{1,0} - \mu\xi_{0,1}), \pi^{i_2}\xi_{0,1}],$$

where $i_1, i_2 \geq 0$ are integers and μ is an element of K satisfying $\nu(\wp(\mu)) \geq i_2 - pi_1$.

PROOF. The $\mathfrak{O}_K$-Hopf order J induces a short exact sequence of $\mathfrak{O}_K$-Hopf algebras

$$\mathfrak{O}_K \to H \to J \to H' \to \mathfrak{O}_K \tag{12.3}$$

where H', H are Hopf orders in $K[C_p]^*$. By [**EU17**, Theorem 2.3], H' and H are Tate-Oort Hopf orders in $K[C_p]^*$, i.e., $H' \cong J_{i_1} = \mathfrak{O}_K[\pi^{i_1}\xi_{1,0}]$ and $H \cong J_{i_2} = \mathfrak{O}_K[\pi^{i_2}\xi_{0,1}]$ for integers $i_1, i_2 \geq 0$.

Let $\mathbb{D}_{i_1} = \text{Spec } J_{i_1}$, $\mathbb{D}_{i_2} = \text{Spec } J_{i_2}$, and $\mathbb{D} = \text{Spec } J$ denote the corresponding group schemes over $\mathfrak{O}_K$. From (12.3) we obtain the short exact sequence of group schemes

$$0 \longrightarrow \mathbb{D}_{i_1} \longrightarrow \mathbb{D} \longrightarrow \mathbb{D}_{i_2} \longrightarrow 0, \tag{12.4}$$

which over K appears as

$$0 \longrightarrow \text{Spec } K[C_p]^* \longrightarrow \text{Spec } K[C_p^2]^* \longrightarrow \text{Spec } K[C_p]^* \longrightarrow 0.$$

The short exact sequence (12.4) represents a class in $\text{Ext}^1_{gt}(\mathbb{D}_{i_2}, \mathbb{D}_{i_1})$, which is the group of generically trivial extensions of $\mathbb{D}_{i_2}$ by $\mathbb{D}_{i_1}$.

To determine the structure of J, we compute the representing algebras of the middle terms of $\text{Ext}^1_{gt}(\mathbb{D}_{i_2}, \mathbb{D}_{i_1})$.

In the remainder of the proof, we show how to compute J.

Let $\mathbb{G}_a$ denote the additive group scheme represented by the $\mathfrak{O}_K$-Hopf algebra $\mathfrak{O}_K[x]$, x indeterminate. There is a faithfully flat resolution of group schemes

$$0 \longrightarrow \mathbb{D}_{i_1} \longrightarrow \mathbb{G}_a \xrightarrow{\Psi} \mathbb{G}_a \longrightarrow 0, \tag{12.5}$$

with $\Psi(x) = x^p - \pi^{(p-1)i_1}x$.

From (12.5) we obtain the long exact sequence

$$\mathrm{Hom}(\mathbb{D}_{i_2}, \mathbb{G}_a) \xrightarrow{\Psi} \mathrm{Hom}(\mathbb{D}_{i_2}, \mathbb{G}_a) \xrightarrow{\omega} \mathrm{Ext}^1(\mathbb{D}_{i_2}, \mathbb{D}_{i_1}) \xrightarrow{\iota} \mathrm{Ext}^1(\mathbb{D}_{i_2}, \mathbb{G}_a),$$

with connecting homomorphism ω, which induces the map ρ in the exact sequence

$$0 \longrightarrow \mathrm{coker}(\Psi : \mathrm{Hom}(\mathbb{D}_{i_2}, \mathbb{G}_a)^{\circlearrowleft}) \xrightarrow{\rho} \mathrm{Ext}^1(\mathbb{D}_{i_2}, \mathbb{D}_{i_1}) \xrightarrow{\iota} \mathrm{Ext}^1(\mathbb{D}_{i_2}, \mathbb{G}_a).$$

Here we are denoting the cokernel of

$$\Psi : \mathrm{Hom}(\mathbb{D}_{i_2}, \mathbb{G}_a) \to \mathrm{Hom}(\mathbb{D}_{i_2}, \mathbb{G}_a)$$

as $\mathrm{coker}(\Psi : \mathrm{Hom}(\mathbb{D}_{i_2}, \mathbb{G}_a)^{\circlearrowleft})$.

Tensoring with K and considering kernels (indicated by gt) results in the following commutative diagram with exact rows and columns:

$$\begin{array}{ccccccc}
 & & 0 & & 0 & & 0 \\
 & & \downarrow & & \downarrow & & \downarrow \\
0 & \longrightarrow & \mathrm{coker}(\Psi : \mathrm{Hom}(\mathbb{D}_{i_2}, \mathbb{G}_a)^{\circlearrowleft})_{gt} & \xrightarrow{\rho} & \mathrm{Ext}^1_{gt}(\mathbb{D}_{i_2}, \mathbb{D}_{i_1}) & \xrightarrow{\iota} & \mathrm{Ext}^1_{gt}(\mathbb{D}_{i_2}, \mathbb{G}_a) \\
 & & \downarrow & & \downarrow & & \downarrow \\
0 & \longrightarrow & \mathrm{coker}(\Psi : \mathrm{Hom}(\mathbb{D}_{i_2}, \mathbb{G}_a)^{\circlearrowleft}) & \xrightarrow{\rho} & \mathrm{Ext}^1(\mathbb{D}_{i_2}, \mathbb{D}_{i_1}) & \xrightarrow{\iota} & \mathrm{Ext}^1(\mathbb{D}_{i_2}, \mathbb{G}_a) \\
 & & \downarrow & & \downarrow & & \downarrow \\
0 & \longrightarrow & \mathrm{coker}(\Psi : \mathrm{Hom}(\mathbb{D}_{i_2,K}, \mathbb{G}_{a,K})^{\circlearrowleft}) & \xrightarrow{\rho} & \mathrm{Ext}^1(\mathbb{D}_{i_2,K}, \mathbb{D}_{i_1,K}) & \xrightarrow{\iota} & \mathrm{Ext}^1(\mathbb{D}_{i_2,K}, \mathbb{G}_{a,K})
\end{array}$$

Now,

$$\mathrm{Ext}^1_{gt}(\mathbb{D}_{i_2}, \mathbb{G}_a) = 0$$

by [**EU17**, Proposition 4.3]. Thus, there is an isomorphism

$$\rho : \mathrm{coker}(\Psi : \mathrm{Hom}(\mathbb{D}_{i_2}, \mathbb{G}_a)^{\circlearrowleft})_{gt} \to \mathrm{Ext}^1_{gt}(\mathbb{D}_{i_2}, \mathbb{D}_{i_1}). \tag{12.6}$$

The isomorphism of (12.6) classifies all short exact sequences

$$0 \longrightarrow \mathbb{D}_{i_1} \longrightarrow \mathbb{D} \longrightarrow \mathbb{D}_{i_2} \longrightarrow 0,$$

which over K appear as

$$0 \longrightarrow \mathrm{Spec}\ K[C_p]^* \longrightarrow \mathrm{Spec}\ K[C_p^2]^* \longrightarrow \mathrm{Spec}\ K[C_p]^* \longrightarrow 0.$$

We seek to compute J, which is the representing algebra of $\mathbb{D}$.

Let $[E]$ be a class in $\mathrm{Ext}^1_{gt}(\mathbb{D}_{i_2}, \mathbb{D}_{i_1})$. In view of the isomorphism (12.6), we have $[h] = \rho^{-1}([E])$ for some homomorphism $h : \mathbb{D}_{i_2} \to \mathbb{G}_a$. Applying [**EU17**, Remark 4.1] to the resolution (12.5) we conclude that the image of h under the connecting homomorphism ω (inducing ρ) is the class $[\mathbb{D}_h]$ of the pull-back

$$\mathbb{D}_h = \{(w, x) \in \mathbb{D}_{i_2} \times \mathbb{G}_a \mid h(w) = \Psi(x)\}.$$

The homomorphism $h : \mathbb{D}_{i_2} \to \mathbb{G}_a$ is determined by a Hopf algebra map $f : \mathfrak{O}_K[x] \to J_{i_2}$ given as

$$x \mapsto -\wp(\mu)\pi^{pi_1}\xi_{0,1},$$

for some $\mu \in K$ with $\nu(\wp(\mu)) \geq i_2 - pi_1$ (use the isomorphism of [**EU17**, Proposition 4.5]).

The representing Hopf algebra $\mathcal{H}_h$ of $\mathbb{D}_h$ arises from the push-out diagram

$$\begin{array}{ccc}
\mathcal{H}_h & \leftarrow & \mathfrak{O}_K[x] \\
\uparrow & & \Psi \uparrow \\
J_{i_2} & \xleftarrow{f} & \mathfrak{O}_K[x].
\end{array}$$

Thus,

$$\begin{aligned}\mathcal{H}_h &= \frac{\mathfrak{O}_K[\pi^{i_2}\xi_{0,1}]\otimes_{\mathfrak{O}_K}\mathfrak{O}_K[x]}{(\Psi(\mu\pi^{i_1}\xi_{0,1})\otimes 1+1\otimes\Psi(x))}\\ &\cong \frac{\mathfrak{O}_K[\pi^{i_2}\xi_{0,1}][x]}{(\Psi(x)+\Psi(\mu\pi^{i_1}\xi_{0,1}))}\\ &\cong \frac{\mathfrak{O}_K[\pi^{i_2}\xi_{0,1}][x]}{(\Psi(x+\mu\pi^{i_1}\xi_{0,1}))}\\ &\cong \mathfrak{O}_K[\pi^{i_1}(\xi_{1,0}-\mu\xi_{0,1}),\pi^{i_2}\xi_{0,1}],\end{aligned}$$

since

$$J_{i_1}\cong\mathfrak{O}_K[x]/(\Psi(x))\cong\mathfrak{O}_K[\pi^{i_1}\xi_{1,0}].$$

It follows that

$$J\cong J_{i_1,i_2,\mu}=\mathfrak{O}_K[\pi^{i_1}(\xi_{1,0}-\mu\xi_{0,1}),\pi^{i_2}\xi_{0,1}],$$

where $i_1,i_2\geq 0$ are integers and μ is an element of K satisfying $\nu(\wp(\mu))\geq i_2-pi_1$. □

REMARK 12.70. Note that $J_{i_1,i_2,\mu}$ is Koch's $H_{i,j,\theta}$ of Example 12.67 with $\mu=-\pi^{-i}\theta$, $\xi_{1,0}=t_1$, $\xi_{0,1}=t_2$, $i_1=i$, and $i_2=j$. With this change of notation

$$J_{i_1,i_2,\mu}\cong\mathfrak{O}_K[u_1,u_2]/I$$

where I is generated by
$$\begin{aligned}&u_1^p-\pi^{(p-1)i}u_1-(\pi^{-j}\theta^p-\pi^{(p-1)i-j}\theta)u_2\\ =\ &u_1^p-\pi^{(p-1)i_1}u_1+(\pi^{-i_2}\pi^{pi_1}\mu^p-\pi^{(p-1)i_1-i_2}\pi^{i_1}\mu)u_2\\ =\ &u_1^p-\pi^{(p-1)i_1}u_1+\wp(\mu)\pi^{pi_1-i_2}u_2\end{aligned}$$

and

$$u_2^p-\pi^{(p-1)j}u_2=u_2^p-\pi^{(p-1)i_2}u_2.$$

Moreover, Koch's condition

$$\nu(\pi^{-j}\theta^p-\pi^{(p-1)i-j}\theta)\geq 0$$

becomes

$$\nu(\wp(\mu))\geq i_2-pi_1.$$

12.13.3. Hopf orders in $K[C_p^n]$, $n=1,2$. In (12.13.1) and (12.13.2), we constructed Hopf orders J in $K[C_p^n]^*$. To construct Hopf orders in $K[C_p^n]$, $n=1,2$, we need to compute the linear duals J^*.

12.13.3.1. *The case C_p.* Let $C_p=\langle g_1\rangle$, let $i\geq 0$ be an integer, and let

$$E(i)=\mathfrak{O}_K\left[\frac{g_1-1}{\pi^i}\right].$$

Then it is well-known that $E(i)$ is an $\mathfrak{O}_K$-Hopf order in $K[C_p]$ (see for instance [**EU17**, §2]).

PROPOSITION 12.71 (Elder and Underwood). *Let $i\geq 0$ be an integer and let $J_i=\mathfrak{O}_K[\pi^i\xi_1]$, ξ_1 primitive, $\xi_1^p=\xi_1$, be an $\mathfrak{O}_K$-Hopf order in $K[C_p]^*$. Then*

$$J_i^*=E(i).$$

PROOF. See [**EU17**, Proposition 2.2]. □

So every $\mathfrak{O}_K$-Hopf order in $K[C_p]$ can be written in the form

$$E(i) = \mathfrak{O}_K\left[\frac{g_1 - 1}{\pi^i}\right]$$

for some integer $i \geq 0$.

The case C_p^2. From Section 12.9 we have the truncated exponential, defined as

$$x^{[y]} = \sum_{m=0}^{p-1} \binom{y}{m}(x-1)^m,$$

where

$$\binom{y}{m} = y(y-1)(y-2)\cdots(y-m+1)/m!$$

PROPOSITION 12.72 (Elder and Underwood). *Let $C_p^2 = \langle g_1, g_2\rangle$, let $i_1, i_2 \geq 0$ be integers and let μ be an element of K that satisfies $\nu(\wp(\mu)) \geq i_2 - pi_1$. Let*

$$E(i_1, i_2, \mu) = \mathfrak{O}_K\left[\frac{g_1 - 1}{\pi^{i_1}}, \frac{g_2 g_1^{[\mu]} - 1}{\pi^{i_2}}\right].$$

Then $E(i_1, i_2, \mu)$ is an $\mathfrak{O}_K$-Hopf order in $K[C_p^2]$.

PROOF. Since $\mathrm{char}(K) = p$, we have $\left(\frac{g_1 - 1}{\pi^{i_1}}\right)^p = 0$ and $\left(\frac{g_2 g_1^{[\mu]} - 1}{\pi^{i_2}}\right)^p = 0.$ Thus the $\mathfrak{O}_K$-subalgebra $E(i_1, i_2, \mu)$ of $K[C_p^2]$ is free of finite rank as an $\mathfrak{O}_K$-submodule of $K[C_p^2]$. Also,

$$K \otimes_{\mathfrak{O}_K} E(i_1, i_2, \mu) \cong K[C_p^2]$$

and so, $E(i_1, i_2, \mu)$ is an $\mathfrak{O}_K$-order in $K[C_p^2]$.

In view of [**Und15**, Proposition 3.4.5], $E(i_1, i_2, \mu)$ is an $\mathfrak{O}_K$-Hopf order if

$$\Delta_{K[C_p^2]}(E(i_1, i_2, \mu)) \subseteq E(i_1, i_2, \mu) \otimes_{\mathfrak{O}_K} E(i_1, i_2, \mu). \tag{12.7}$$

In the remainder of the proof, we show that (12.7) holds. We will use [**Chi00**, (31.2) Proposition].

From [**EU17**, (2)], we have the polynomial identity

$$(1+x)^{[y]}(1+x)^{[z]} = (1+x)^{[y+z]} \in \mathcal{F}_p[x, y, z]/(x^p). \tag{12.8}$$

We claim that $g_1^{[\mu]}$ is a unit in $E(i_1)$. Since $x = g_1 - 1$ satisfies $x^p = 0$, (12.8) implies that

$$g_1^{[\mu]} g_1^{[-\mu]} = g_1^{[0]} = 1$$

and so, $g_1^{[\mu]}$ is a unit in $K\langle\sigma_1\rangle$. The quantity $g_1^{[\mu]}$ is a unit in $E(i_1)$ if

$$g_1^{[\pm\mu]} = \sum_{r=0}^{p-1} \binom{\pm\mu}{r} \pi^{ri_1} \left(\frac{g_1 - 1}{\pi^{i_1}}\right)^r \in E(i_1),$$

which is equivalent to $\binom{\pm\mu}{r}\pi^{ri_1} \in \mathfrak{O}_K$ for $0 \leq r \leq p-1$. For $\mu \in \mathfrak{O}_K$, this clearly holds. So we assume that $\nu(\mu) < 0$. Now, $p\nu(\mu) = \nu(\wp(\mu)) \geq i_2 - pi_1 \geq -pi_1$, and thus $\nu(\pm\mu) \geq -i_1$, which implies that $\nu(\binom{\pm\mu}{r}\pi^{ri_1}) \geq 0$.

Now by [**Chi00**, (31.2) Proposition], (12.7) holds if and only if

$$\Delta_{K[C_p^2]}(g_1^{[\mu]}) \equiv g_1^{[\mu]} \otimes g_1^{[\mu]} \pmod{\pi^{i_2}(E(i_1) \otimes E(i_1))}. \tag{12.9}$$

Let $X = (g_1-1)/\pi^{i_1}\otimes 1$ and $Y = 1\otimes(g_1-1)/\pi^{i_1}$, so that $E(i_1)\otimes E(i_1) = \mathfrak{O}_K[X, Y]$. Since $\Delta_{K[C_p^2]}$ is an algebra map and

$$\Delta_{K[C_p^2]}(g_1) = g_1 \otimes g_1 = (1 + \pi^{i_1}X)(1 + \pi^{i_2}Y),$$

condition (12.9) can be restated as

$$((1 + \pi^{i_1}X)(1 + \pi^{i_2}Y))^{[\mu]} \equiv (1 + \pi^{i_1}X)^{[\mu]}(1 + \pi^{i_2}Y)^{[\mu]} \pmod{\pi^{i_2}\mathfrak{O}_K[X, Y]}. \tag{12.10}$$

From [**BE05**, Lemma 2.2], [**EU17**, pp. 9-10], we have the formula

$$(1 + x + y + xy)^{[z]} = (1 + x)^{[z]}(1 + y)^{[z]}(1 + \wp(z)Q(x, y)) \tag{12.11}$$

in $\mathcal{F}_p[x, y, z]/(x^p, y^p)$, where

$$Q(x, y) = ((x + y + xy)^p - x^p - y^p - (xy)^p)/p \in (x, y)^p \subset \mathbb{Z}[x, y].$$

Note $X^p = Y^p = 0$. Using (12.11), we see that condition (12.10) is equivalent to

(12.12)
$$(1 + \pi^{i_1}X)^{[\mu]}(1 + \pi^{i_2}Y)^{[\mu]}(1 + \wp(\mu)Q(\pi^{i_1}X, \pi^{i_1}Y)) \equiv (1 + \pi^{i_1}X)^{[\mu]}(1 + \pi^{i_2}Y)^{[\mu]} \pmod{\pi^{i_2}\mathfrak{O}_K[X, Y]}.$$

Since $(1 + \pi^{i_1}X)^{[\mu]}$ and $(1 + \pi^{i_2}Y)^{[\mu]}$ are both units in $\mathfrak{O}_K[X, Y]$, condition (12.12) holds if and only if

$$\wp(\mu)Q(\pi^{i_1}X, \pi^{i_1}Y) \equiv 0 \pmod{\pi^{i_2}\mathfrak{O}_K[X, Y]}.$$

Since $Q(\pi^{i_1}X, \pi^{i_1}Y) \in \pi^{pi_1}\mathfrak{O}_K[X, Y]$, this congruence follows from $\pi^{pi_1}\wp(\mu) \in \pi^{i_2}\mathfrak{O}_K$, which is equivalent to $v_K(\wp(\mu)) \geq i_2 - pi_1$.

We conclude that

$$\Delta_{K[C_p^2]}(E(i_1, i_2, \mu)) \subseteq E(i_1, i_2, \mu) \otimes_{\mathfrak{O}_K} E(i_1, i_2, \mu)$$

and so $E(i_1, i_2, \mu)$ is an $\mathfrak{O}_K$-Hopf order in $K[C_p^2]$. □

PROPOSITION 12.73 (Elder and Underwood). *Let $i_1, i_2 \geq 0$ be integers and let μ be an element of K that satisfies $\nu(\wp(\mu)) \geq i_2 - pi_1$. Let*

$$J_{i_1,i_2,\mu} = \mathfrak{O}_K[\pi^{i_1}(\xi_{1,0} - \mu\xi_{0,1}), \pi^{i_2}\xi_{0,1}]$$

be an $\mathfrak{O}_K$-Hopf order in $K[C_p^2]^*$. Then

$$J^*_{i_1,i_2,\mu} = E(i_1, i_2, \mu).$$

PROOF. By Proposition 12.72, $E(i_1, i_2, \mu)$ is an $\mathfrak{O}_K$-Hopf order in $K[C_p^2]$. By [**EU17**, Proposition 3.2, Proposition 3.4],

$$\operatorname{disc}(J_{i_1,i_2,\mu}) = \operatorname{disc}(E(i_1, i_2, \mu)^*) = (\pi^{p^2(p-1)(i_1+i_2)}).$$

Therefore, we only need to prove that $J_{i_1,i_2,\mu} \subseteq E(i_1, i_2, \mu)^*$.

Let $\beta_1 = \pi^{i_1}(\xi_{1,0} - \mu\xi_{0,1})$, $\beta_2 = \pi^{i_2}\xi_{0,1}$. The set $\{\alpha_{a,b}\}_{a,b=0}^{p-1}$ with

$$\alpha_{a,b} = (g_1 - 1)^a(g_2g_1^{[\mu]} - 1)^b/\pi^{ai_1+bi_2},$$

is an $\mathfrak{O}_K$-basis for $E(i_1, i_2, \mu)$. So for each $r = 1, 2$, we need to show $\langle\beta_r, \alpha_{a,b}\rangle \in \mathfrak{O}_K$ for all $0 \leq a, b \leq p - 1$.

The values of $\langle\xi_{1,0}, \alpha_{a,b}\rangle$ and $\langle\xi_{0,1}, \alpha_{a,b}\rangle$ are determined modulo $((g_1 - 1), (g_2 - 1))^2$. Let $x = g_1 - 1$ and $y = g_2 - 1$ and note that $(\sigma_2\sigma_1^{[\mu]} - 1) \equiv y + \mu x \pmod{(x, y)^2}$.

Thus $\pi^{ai_1+bi_2}\alpha_{a,b} \equiv x^a(y+\mu x)^b \pmod{(x,y)^2}$. Using $\langle \xi_{1,0}, x^r y^s\rangle = \delta_{r,1}\delta_{s,0}$ and $\langle \xi_{0,1}, x^r y^s\rangle = \delta_{r,0}\delta_{s,1}$, we have

$$\langle \xi_{1,0}, \pi^{ai_1+bi_2}\alpha_{a,b}\rangle = \begin{cases} 1 & a=1, b=0,\\ \mu & a=0, b=1,\\ 0 & \text{else,}\end{cases} \quad \langle \xi_{0,1}, \pi^{ai_1+bi_2}\alpha_{a,b}\rangle = \begin{cases} 0 & a=1, b=0,\\ 1 & a=0, b=1,\\ 0 & \text{else.}\end{cases}$$

Therefore $\langle \beta_1, \alpha_{a,b}\rangle = \delta_{a,1}\delta_{b,0}$ and $\langle \beta_2, \alpha_{a,b}\rangle = \delta_{a,0}\delta_{b,1}$. □

From Theorem 12.69 and Proposition 12.73, we conclude that an arbitrary $\mathfrak{O}_K$-Hopf order in $K[C_p^2]$ is of the form

$$E(i_1, i_2, \mu) = \mathfrak{O}_K\left[\frac{g_1-1}{\pi^{i_1}}, \frac{g_2 g_1^{[\mu]}-1}{\pi^{i_2}}\right]$$

for integers $i_1, i_2 \geq 0$ and $\mu \in K$ with $\nu(\wp(\mu)) \geq i_2 - p i_1$.

12.13.4. A problem in the case $G = C_p^3$. Let $C_p^3 = \langle g_1, g_2, g_3\rangle$. Byott and Elder's construction of Hopf orders in $K[C_p^3]$ (Section 12.9) applies in the $\mathrm{char}(K) = p$ case to yield realizable Hopf orders in $K[C_p^3]$ of the form

$$E(j_1, j_2, j_3, \mu, \alpha, \beta) = \mathfrak{O}_K\left[\frac{g_1-1}{\pi^{j_1}}, \frac{g_2 g_1^{[\mu]}-1}{\pi^{j_2}}, \frac{g_3 g_1^{[\alpha]}(g_2 g_1^{[\mu]})^{[\beta]}-1}{\pi^{j_3}}\right].$$

Here $j_1, j_2, j_3 \geq 0$ are integers and μ, α, β are elements of K satisfying $\nu(\mu) = (j_2/p) - j_1$, $\nu(\alpha) = (j_3/p^2) - j_1$, and $\nu(\beta) = (j_3/p) - j_2$. See [**BE18**, §5.3].

We now consider the dual side.

Let $\xi_{1,0,0}$, $\xi_{0,1,0}$, $\xi_{0,0,1}$ be elements of $K[C_p^3]^*$ defined by

$$\langle \xi_{1,0,0}, (g_1-1)^j(g_2-1)^k(g_3-1)^l\rangle = \delta_{1,j}\delta_{0,k}\delta_{0,l},$$
$$\langle \xi_{0,1,0}, (g_1-1)^j(g_2-1)^k(g_3-1)^l\rangle = \delta_{0,j}\delta_{1,k}\delta_{0,l},$$
$$\langle \xi_{0,0,1}, (g_1-1)^j(g_2-1)^k(g_3-1)^l\rangle = \delta_{0,j}\delta_{0,k}\delta_{1,l},$$

$0 \leq j, k, l \leq p-1$, where $\langle -, -\rangle : K[C_p^3]^* \times K[C_p^3] \to K$ is the duality map. Then $\xi_{1,0,0}$, $\xi_{0,1,0}$ and $\xi_{0,0,1}$ are primitive elements in $K[C_p^3]^*$ with $\xi_{1,0,0}^p = \xi_{1,0,0}$, $\xi_{0,1,0}^p = \xi_{0,1,0}$ and $\xi_{0,0,1}^p = \xi_{0,0,1}$. The K-Hopf algebra $K[C_p^3]^*$ is primitively generated by $\xi_{1,0,0}, \xi_{0,1,0}, \xi_{0,0,1}$;

$$K[C_p^3]^* = K[\xi_{1,0,0}, \xi_{0,1,0}, \xi_{0,0,1}].$$

The dual $E(j_1, j_2, j_3, \mu, \alpha, \beta)^*$ is an $\mathfrak{O}_K$-Hopf order in $K[C_p^3]^*$.

By Theorem 12.64(ii), $E(j_1, j_2, j_3, \mu, \alpha, \beta)^*$ is determined by some matrix $\Theta = (\theta_{i,j})$ in $\mathrm{GL}_3(K)$ of the form

$$\Theta = \begin{pmatrix} \pi^{i_1} & 0 & 0\\ \theta_{2,1} & \pi^{i_2} & 0\\ \theta_{3,1} & \theta_{3,2} & \pi^{i_3}\end{pmatrix}.$$

We have

$$\begin{aligned} E(j_1, j_2, j_3, \mu, \alpha, \beta)^* &= H_\Theta\\ &= \mathfrak{O}_K[\pi^{i_1}\xi_{1,0,0} + \theta_{2,1}\xi_{0,1,0} + \theta_{3,1}\xi_{0,0,1}, \pi^{i_2}\xi_{0,1,0} + \theta_{3,2}\xi_{0,0,1}, \pi^{i_3}\xi_{0,0,1}]\\ &= \mathfrak{O}_K[\pi^{i_1}(\xi_{1,0,0} + \pi^{-i_1}\theta_{2,1}\xi_{0,1,0} + \pi^{-i_1}\theta_{3,1}\xi_{0,0,1}),\\ &\qquad \pi^{i_2}(\xi_{0,1,0} + \pi^{-i_2}\theta_{3,2}\xi_{0,0,1}), \pi^{i_3}\xi_{0,0,1}].\end{aligned}$$

Problem. It would be of interest to determine the relationship between the 6 parameters for $E(j_1, j_2, j_3, \mu, \alpha, \beta)$ and those for its dual H_Θ.

CHAPTER 13

Ramification theory for separable extensions of local fields

In this chapter we develop the theory of higher ramification for finite separable extensions L/K of local fields, without assuming that L/K is Galois. In Section 13.1 we define the lower ramification breaks of L/K. When L/K is not Galois, these are rational numbers, but not necessarily integers. We then construct the Hasse-Herbrand functions $\varphi_{L/K}$ and $\psi_{L/K}$, and use these to define the upper ramification breaks of L/K. In Section 13.2 we describe the behavior of ramification breaks and Hasse-Herbrand functions in towers of fields $K \subset L \subset M$. In Section 13.3 we show that if ℓ_i is a lower break of L/K then $v_p(\ell_i) \geq 0$, and in most cases $v_p(\ell_i) = 0$, where p is the residue characteristic of K. This information will be used in Chapter 14. The study of ramification theory for non-Galois extensions goes back at least to Krasner [**Kra37**]; more recent expositions are given in Deligne [**Del84**] and Helou [**Hel90, Hel91**]. Our presentation makes heavy use of power series, following Lubin [**Lub13**].

13.1. Basic theory

Let K be a local field with perfect residue field and let K^{sep} be a separable closure of K. All the fields that we consider in this chapter will be subextensions of K^{sep}/K. Let L/K be a finite subextension of K^{sep}/K. The normalized valuation v_L on L can be extended uniquely to a valuation on K^{sep} with values in $\mathbb{Q}$. We denote this extended valuation by v_L. Let

$$\mathfrak{O}_L = \{y \in L \mid v_L(y) \geq 0\}$$

be the ring of integers of L and let

$$\mathfrak{p}_L = \{y \in L \mid v_L(y) > 0\}$$

be the maximal ideal of $\mathfrak{O}_L$. Also let π_L be any element of L such that $v_L(\pi_L) = 1$; then $\mathfrak{p}_L = \mathfrak{O}_L\pi_L$. We say that π_L is a uniformizer for L. Let $\mathcal{E}_{L/K}$ be the set of K-embeddings of L into K^{sep}. Then $|\mathcal{E}_{L/K}| = [L : K]$. Let $\widetilde{L}/K$ be the Galois closure of L/K in K^{sep} and set $G = \mathrm{Gal}(\widetilde{L}/K)$, $G_L = \mathrm{Gal}(\widetilde{L}/L)$. Then there is a bijection from the coset space G/G_L to $\mathcal{E}_{L/K}$ given by $\gamma G_L \mapsto \gamma|_L$.

For $\sigma \in \mathcal{E}_{L/K}$ define

$$i_L(\sigma) = \min\{v_L(\sigma(y) - y) \mid y \in \mathfrak{O}_L\} - 1. \tag{13.1}$$

Then $i_L(\sigma) = \infty$ if and only if $\sigma = \mathrm{id}_L$. If $\sigma \neq \mathrm{id}_L$ then $i_L(\sigma) \in e_{\widetilde{L}/L}^{-1}\mathbb{Z}$, where $e_{\widetilde{L}/L}$ is the ramification index of $\widetilde{L}/L$. In particular, if L/K is Galois then $i_L(\sigma) \in \mathbb{Z}$. It is well-known that $\mathfrak{O}_L$ is monogenic over $\mathfrak{O}_K$, i.e., there is $\alpha \in \mathfrak{O}_L$ such that $\mathfrak{O}_L = \mathfrak{O}_K[\alpha]$ (see [**Ser62**, III, §6, Proposition 12]). Then $i_L(\sigma) = v_L(\sigma(\alpha) - \alpha) - 1$. This applies in particular when L/K is totally ramified, in which case $\mathfrak{O}_L = \mathfrak{O}_K[\pi_L]$

and $i_L(\sigma) = v_L(\sigma(\pi_L) - \pi_L) - 1$. For $t \in \mathbb{R}$ with $t \geq -1$ define the tth lower ramification set of $\mathcal{E}_{L/K}$ by

$$(\mathcal{E}_{L/K})_t = \{\sigma \in \mathcal{E}_{L/K} \mid i_L(\sigma) \geq t\}.$$

We have $(\mathcal{E}_{L/K})_t \subset (\mathcal{E}_{L/K})_s$ for $s \leq t$. Say that $\ell \geq -1$ is a lower ramification break for L/K if $(\mathcal{E}_{L/K})_\ell \neq (\mathcal{E}_{L/K})_{\ell+\epsilon}$ for all $\epsilon > 0$. This is equivalent to having $\ell = i_L(\sigma)$ for some $\sigma \in \mathcal{E}_{L/K}$.

EXAMPLE 13.1. Let K have characteristic 0 and let π_K be a uniformizer for K. Let $\pi_L \in K^{sep}$ be a root of the Eisenstein polynomial $f(X) = X^p - \pi_K$ and set $L = K(\pi_L)$. Thus L/K is a totally ramified extension of degree p, and $\mathfrak{O}_L = \mathfrak{O}_K[\pi_L]$. Let $\zeta_p \in K^{sep}$ be a primitive pth root of unity. Then the roots of $f(X)$ in K^{sep} are $\pi_L, \zeta_p\pi_L, \dots, \zeta_p^{p-1}\pi_L$. It follows that the set of K-embeddings of L into K^{sep} is $\mathcal{E}_{L/K} = \{\sigma_0, \sigma_1, \dots, \sigma_{p-1}\}$, where $\sigma_j(\pi_L) = \zeta_p^j\pi_L$. For $1 \leq j \leq p-1$ we have

$$i_L(\sigma_j) = v_L(\zeta_p^j\pi_L - \pi_L) - 1 = v_L(\zeta_p^j - 1) = \frac{v_L(p)}{p-1} = \frac{pe_K}{p-1},$$

where $e_K = v_K(p)$ is the absolute ramification index of K. Therefore $\ell_1 = pe_K/(p-1)$ is the unique lower ramification break of L/K. It follows that

$$(\mathcal{E}_{L/K})_t = \begin{cases} \mathcal{E}_{L/K} & \text{if } t \leq pe_K/(p-1), \\ \{\sigma_0\} & \text{if } t > pe_K/(p-1). \end{cases}$$

To help us prove the next proposition we recall the definition of the Teichmüller representatives of K. Let $\overline{K} = \mathfrak{O}_K/\mathfrak{p}_K$ be the residue field of K and let $\alpha \in \overline{K}$. Since $\overline{K}$ is assumed to be perfect, for each $n \geq 0$ there is a unique element $\alpha^{p^{-n}} \in \overline{K}$ whose p^nth power is α. For each $n \geq 0$ choose $a_n \in \mathfrak{O}_K$ such that $\alpha^{p^{-n}} = a_n + \mathfrak{p}_K$. The Teichmüller representative of α is defined to be $\tau(\alpha) = \lim_{n\to\infty} a_n^{p^n}$. It is shown in Proposition 7.1 of Chapter I in [**FV02**] that the limit exists and is independent of the choices of liftings a_n. Furthermore, we have $\alpha = \tau(\alpha) + \mathfrak{p}_K$, so the map $\tau : \overline{K} \to \mathfrak{O}_K$ is a canonical section of the reduction map $\rho : \mathfrak{O}_K \to \overline{K}$. The Teichmüller map τ can also be characterized as the unique section of ρ such that $\tau(\alpha\beta) = \tau(\alpha)\tau(\beta)$ for $\alpha, \beta \in \overline{K}$. Let $\mathcal{R}_K = \tau(\overline{K})$ denote the set of Teichmüller representatives of K and let L/K be a finite separable extension. Then L/K is totally ramified if and only if $\mathcal{R}_K$ is also the set of Teichmüller representatives of L.

The following facts are well-known in the case where the extension L/K is Galois:

PROPOSITION 13.2. *Let L/K be a finite separable extension.*

(a) *If L/K is totally ramified then the lower ramification breaks of L/K are all nonnegative.*
(b) *If L/K is not totally ramified then -1 is a lower ramification break of L/K.*

PROOF. Suppose L/K is totally ramified. Then every $y \in \mathfrak{O}_L$ can be written in the form $y = k + m$ with $k \in \mathcal{R}_K$ and $m \in \mathfrak{p}_L$. It follows that $v_L(\sigma(y) - y) = v_L(\sigma(m) - m) \geq 1$. Therefore the lower ramification breaks of L/K are nonnegative. Suppose L/K is not totally ramified. Let $t \in \mathcal{R}_L$ be such that $t \notin K$. Then there is $\sigma \in \mathcal{E}_{L/K}$ such that $\sigma(t) \neq t$. Hence $v_L(\sigma(t) - t) = 0$, so $i_L(\sigma) = -1$. □

We now define the Hasse-Herbrand function $\varphi_{L/K} : \mathbb{R}_{\geq -1} \to \mathbb{R}_{\geq -1}$ by

$$\varphi_{L/K}(x) = \int_0^x \frac{|(\mathcal{E}_{L/K})_t|}{|(\mathcal{E}_{L/K})_0|}\, dt. \tag{13.2}$$

Hence if $x > -1$ and x is not a lower ramification break of L/K then

$$\varphi'_{L/K}(x) = \frac{|(\mathcal{E}_{L/K})_x|}{|(\mathcal{E}_{L/K})_0|}.$$

Since $\varphi_{L/K} : \mathbb{R}_{\geq -1} \to \mathbb{R}_{\geq -1}$ is a bijection we may define $\psi_{L/K} : \mathbb{R}_{\geq -1} \to \mathbb{R}_{\geq -1}$ by $\psi_{L/K} = \varphi_{L/K}^{-1}$. This leads to the upper numbering on the ramification sets of L/K: For $t \geq -1$ set $\mathcal{E}_{L/K}^t = (\mathcal{E}_{L/K})_{\psi_{L/K}(t)}$. Say that $u \geq -1$ is an upper ramification break for L/K if $\mathcal{E}_{L/K}^u \neq \mathcal{E}_{L/K}^{u+\epsilon}$ for all $\epsilon > 0$. Thus ℓ is a lower ramification break if and only if $\varphi_{L/K}(\ell)$ is an upper ramification break. Since $\psi_{L/K}(\varphi_{L/K}(x)) = x$, if x is not a lower ramification break for L/K we get $\psi'_{L/K}(\varphi_{L/K}(x))\varphi'_{L/K}(x) = 1$, and hence

$$\psi'_{L/K}(\varphi_{L/K}(x)) = \frac{1}{\varphi'_{L/K}(x)} = \frac{|(\mathcal{E}_{L/K})_0|}{|(\mathcal{E}_{L/K})_x|},$$

$$\psi'_{L/K}(x) = \frac{|(\mathcal{E}_{L/K})_0|}{|(\mathcal{E}_{L/K})_{\psi_{L/K}(x)}|} = \frac{|\mathcal{E}_{L/K}^0|}{|\mathcal{E}_{L/K}^x|},$$

$$\psi_{L/K}(x) = \int_0^x \frac{|\mathcal{E}_{L/K}^0|}{|\mathcal{E}_{L/K}^t|}\, dt.$$

For the rest of this chapter we will restrict our attention to extensions L/K which are totally ramified. This assumption is justified by the following two propositions:

PROPOSITION 13.3. *Let L/K be a finite subextension of K^{sep}/K and let U/K be the maximum unramified subextension of L/K. Then $(\mathcal{E}_{L/K})_0 = \mathcal{E}_{L/U}$. Furthermore, for $x \geq 0$ we have $(\mathcal{E}_{L/K})_x = (\mathcal{E}_{L/U})_x$ and $\varphi_{L/K}(x) = \varphi_{L/U}(x)$.*

PROOF. We first show that $(\mathcal{E}_{L/K})_0 = \mathcal{E}_{L/U}$. Let $\sigma \in \mathcal{E}_{L/U}$ and let $y \in \mathfrak{O}_L$. Then there are $u \in \mathcal{R}_U$ and $m \in \mathfrak{p}_L$ such that $y = u + m$. We have $\sigma(y) - y = \sigma(m) - m$, and hence $v_L(\sigma(y) - y) \geq 1$. Therefore $i_L(\sigma) \geq 0$, so $\sigma \in (\mathcal{E}_{L/K})_0$. Let $\sigma \in (\mathcal{E}_{L/K})_0$ and let $u \in \mathcal{R}_U$. Then $v_L(\sigma(u) - u) \geq 1$. Let $\widetilde{L}/K$ be the Galois closure of L/K. Then u and $\sigma(u)$ are Teichmüller representatives for $\widetilde{L}$ such that $v_{\widetilde{L}}(\sigma(u) - u) > 0$. Hence $\sigma(u) = u$. Since U is unramified over K, it is generated over K by its Teichmüller representatives. Therefore $\sigma \in \mathcal{E}_{L/U}$, so we get $(\mathcal{E}_{L/K})_0 = \mathcal{E}_{L/U}$. It follows immediately that for $x \geq 0$ we have $(\mathcal{E}_{L/K})_x = (\mathcal{E}_{L/U})_x$ and $\varphi_{L/K}(x) = \varphi_{L/U}(x)$. □

PROPOSITION 13.4. *Let L/K be a finite subextension of K^{sep}/K and let U/K be the maximum unramified subextension of L/K. Let V/U be a finite unramified subextension of K^{sep}/U. Then the restriction map $\theta : (\mathcal{E}_{LV/V})_0 \to (\mathcal{E}_{L/K})_0$ defined by $\theta(\sigma) = \sigma|_L$ is a bijection. Furthermore, for $x \geq 0$ we have $\theta((\mathcal{E}_{LV/V})_x) = (\mathcal{E}_{L/K})_x$ and $\varphi_{L/K}(x) = \varphi_{LV/V}(x)$.*

PROOF. Let $\sigma, \tau \in (\mathcal{E}_{LV/V})_0$ be such that $\theta(\sigma) = \theta(\tau)$. Then since $\sigma|_L = \tau|_L$ and $\sigma|_V = \mathrm{id}_V = \tau|_V$ we get $\sigma = \tau$. Hence θ is one-to-one. Since L/U is totally ramified and V/U is unramified, LV/V is totally ramified of degree $[L:U]$. We have $(\mathcal{E}_{LV/V})_0 = \mathcal{E}_{LV/V}$ and $(\mathcal{E}_{L/K})_0 = \mathcal{E}_{L/U}$ by Proposition 13.3. Hence

$$|(\mathcal{E}_{LV/V})_0| = [LV:V] = [L:U] = |(\mathcal{E}_{L/K})_0|,$$

so θ is a bijection. Let $\sigma \in (\mathcal{E}_{LV/V})_0$ and let π_L be a uniformizer for L. Since LV/L is unramified, π_L is also a uniformizer for LV. Then since L/U and LV/V are totally ramified we get $\mathfrak{O}_L = \mathfrak{O}_U[\pi_L]$ and $\mathfrak{O}_{LV} = \mathfrak{O}_V[\pi_L]$. It follows that

$$i_L(\sigma|_L) = v_L(\sigma(\pi_L) - \pi_L) - 1 = v_{LV}(\sigma(\pi_L) - \pi_L) - 1 = i_{LV}(\sigma).$$

Hence for $x \geq 0$ we have $\theta((\mathcal{E}_{LV/V})_x) = (\mathcal{E}_{L/K})_x$. Since θ is a bijection it follows that $\varphi_{L/K}(x) = \varphi_{LV/V}(x)$. □

13.2. Power series and Herbrand's theorem

In this section we consider a tower $M/L/K$ of finite totally ramified subextensions of K^{sep}/K. Our goal is to determine the relations between the ramification breaks of these extensions, and also the relations between the Hasse-Herbrand functions $\varphi_{L/K}$, $\varphi_{M/L}$, and $\varphi_{M/K}$ associated to these extensions.

We will be using the Weierstrass preparation theorem, whose statement requires several definitions. Let R be a local integral domain with maximal ideal M. Assume that R is complete and Hausdorff with respect to the M-adic topology. Say that $g(X) \in R[X]$ is a distinguished polynomial if $g(X)$ is monic of degree $n \geq 0$, and for $0 \leq i < n$ the coefficient of X^i in $g(X)$ is an element of M. For $\varphi(X) \in R[[X]] \smallsetminus M[[X]]$ written $\varphi(X) = a_0 + a_1X + \cdots$, the Weierstrass degree of $\varphi(X)$ is defined to be the smallest $n \geq 0$ such that $a_n \notin M$.

A proof of the following result can be found in [**Abh64**, (10.2.4)].

THEOREM 13.5 (Weierstrass preparation theorem). *Let $\varphi(X) \in R[[X]] \smallsetminus M[[X]]$. Then there exist a unique distinguished polynomial $f(X) \in R[X]$ and unit power series $u(X) \in R[[X]]^\times$ such that $\varphi(X) = u(X)f(X)$. Furthermore, the degree of $f(X)$ is equal to the Weierstrass degree of $\varphi(X)$.*

Let L/K be a finite separable totally ramified extension of local fields, and set $n = [L:K]$. Let π_K, π_L be uniformizers for K, L and choose $F(X) \in \mathfrak{O}_K[[X]]$ such that $F(0) = 0$ and $F(\pi_L) = \pi_K$. Write $F(X) = a_1X + a_2X^2 + \cdots$. The assumption $F(\pi_L) = \pi_K$ implies that $a_n \in \mathfrak{O}_K^\times$ and $a_i \in \mathfrak{p}_K$ for $1 \leq i < n$. Hence the Weierstrass degree of $F(X) - \pi_K$ is equal to n. By the Weierstrass preparation theorem there exist $u(X) \in \mathfrak{O}_K[[X]]^\times$ and a distinguished polynomial $f(X)$ of degree n such that $F(X) - \pi_K = u(X)f(X)$ and $\deg(f) = n$. Since $f(\pi_L) = 0$ it follows that $f(X)$ is the minimum polynomial for π_L over K. In particular, $f(X)$ is independent of the choice of $F(X)$.

We define a new series associated to $F(X)$ by

$$F^*(X) = \pi_K^{-1}(F(\pi_L + \pi_L X) - \pi_K). \tag{13.3}$$

Since $F(X)$ has coefficients in $\mathfrak{O}_K$, the sum

$$F(\pi_L + \pi_L X) = \sum_{i=1}^{\infty} a_i(\pi_L + \pi_L X)^i$$

converges to a series in $\mathfrak{O}_L[[X]]$. Since $v_L(a_i) \geq n$ for $1 \leq i < n$, all coefficients in this series are divisible by π_K. Hence $F^*(X) \in \mathfrak{O}_L[[X]]$. Furthermore, since $F(\pi_L) = \pi_K$ we have $F^*(0) = 0$.

PROPOSITION 13.6. *Let L/K be a totally ramified extension of degree n and define $F^*(X)$ as in (13.3). Then $v_L((F^*)'(0)) = v_L(\mathfrak{D}_{L/K}) - n + 1$, where $\mathfrak{D}_{L/K}$ is the different of the extension L/K.*

PROOF. We have $F(X) = u(X)f(X) + \pi_K$, where $u(X) \in \mathfrak{O}_K[[X]]^\times$ and $f(X)$ is the minimum polynomial for π_L over K. It follows that

$$F'(X) = u(X)f'(X) + u'(X)f(X),$$

and hence that $F'(\pi_L) = u(\pi_L)f'(\pi_L)$. Using (13.3) we get

$$(F^*)'(X) = \pi_K^{-1}\pi_L F'(\pi_L + \pi_L X),$$

and hence $(F^*)'(0) = \pi_K^{-1}\pi_L u(\pi_L)f'(\pi_L)$. Since $u(\pi_L) \in \mathfrak{O}_L^\times$ this implies that $v_L((F^*)'(0)) = v_L(f'(\pi_L)) - n + 1$. The proposition now follows from the well-known formula $v_L(f'(\pi_L)) = v_L(\mathfrak{D}_{L/K})$ (see for instance [**Ser62**, III, §6, Corollary 2]). □

REMARK 13.7. The quantity $v_L(\mathfrak{D}_{L/K}) - n + 1$ is equal to the depth of ramification of L/K, as defined by Hyodo [**Hyo87**, Definition (1–3)]. It is also equal to the 0th index of inseparability i_0 of L/K, as defined by Heiermann [**Hei96**, Proposition 3.18].

Let $\eta(X)$ be a nonzero element of $X\mathfrak{O}_K[[X]]$. Write $\eta(X) = c_1X + c_2X^2 + \cdots$ and let L be a finite extension of K. We define the *valuation function* of η with respect to v_L by

$$\Psi_\eta^L(x) = \min\{v_L(c_i) + ix \mid i \geq 1\} \tag{13.4}$$

for $x \in [0, \infty)$. Since $\eta(X)$ has coefficients in $\mathfrak{O}_K$, Ψ_η^L is a well-defined continuous monotone increasing function on $[0, \infty)$. The graph of Ψ_η^L is known as the *Newton copolygon* of $\eta(X)$ with respect to v_L. Let $\alpha \in K^{sep}$ be such that $v_L(\alpha) > 0$ and set $a = v_L(\alpha)$. Then $v_L(\eta(\alpha)) \geq \Psi_\eta^L(a)$.

Let L be a finite extension of K. The following basic fact about valuation functions follows directly from (13.4):

LEMMA 13.8. *Let L/K be a finite extension with ramification index $e = e_{L/K}$ and let $\eta(X) \in X\mathfrak{O}_K[[X]]$ be nonzero. Then for $x \geq 0$ we have $\Psi_\eta^L(x) = e\Psi_\eta^K(x/e)$.*

We now show that for most positive rational numbers a, $\Psi_\eta^L(a)$ can indeed be interpreted in terms of valuations.

PROPOSITION 13.9. *Let $\eta(X)$ be a nonzero element of $X\mathfrak{O}_K[[X]]$ and let L be a finite extension of K. Then there is a finite subset $S \subset \mathbb{R}_{\geq 0}$ with the following property: Let $\alpha \in K^{sep}$ and set $a = v_L(\alpha)$. If $a > 0$ and $a \notin S$ then $v_L(\eta(\alpha)) = \Psi_\eta^L(a)$.*

PROOF. Define a partial order $\preceq$ on $\mathbb{N}$ by $h \preceq i$ if and only if $h \leq i$ and $v_L(c_h) \leq v_L(c_i)$. Let $\{m_1, \ldots, m_d\}$ be the (necessarily finite) set of minimal elements of $\mathbb{N}$ with respect to $\preceq$. If $i \in \mathbb{N} \smallsetminus \{m_1, \ldots, m_d\}$ then there is $1 \leq j \leq d$ such that $m_j \preceq i$ and $i \neq m_j$. Hence for $x > 0$ we get $v_L(c_i) + ix > v_L(c_{m_j}) + m_jx \geq \Psi_\eta^L(x)$. It follows that

$$\Psi_\eta^L(x) = \min\{v_L(c_{m_j}) + m_jx \mid 1 \leq j \leq d\}. \tag{13.5}$$

Let S be the set of $x > 0$ such that there are multiple values of j which achieve the minimum in (13.5). Then $|S| < d$, so S is finite. Let $\alpha \in K^{sep}$ satisfy $a = v_L(\alpha) > 0$. It follows from the above that if $a \notin S$ then there is a unique $i \in \mathbb{N}$ which minimizes

$v_L(c_i\alpha^i) = v_L(c_i) + ia$, and furthermore that $i = m_j$ for some $1 \leq j \leq d$. Hence $v_L(\eta(\alpha)) = v_L(c_{m_j}) + m_j a = \Psi^L_\eta(a)$. $\square$

LEMMA 13.10. *Let $\eta(X), \theta(X)$ be nonzero elements of $X\mathfrak{O}_K[[X]]$ and let L be a finite extension of K. Then $\Psi^L_{\eta\circ\theta} = \Psi^L_\eta \circ \Psi^L_\theta$.*

PROOF. Let $S_\eta, S_\theta, S_{\eta\circ\theta}$ be the finite sets associated to $\eta, \theta, \eta\circ\theta$ by Proposition 13.9. Then the set $T = S_\theta \cup S_{\eta\circ\theta} \cup (\Psi^L_\theta)^{-1}(S_\eta)$ is finite since Ψ^L_θ is one-to-one. Let $\alpha \in K^{sep}$ and set $a = v_L(\alpha)$. Suppose $a > 0$ and $a \notin T$. Then $v_L(\theta(\alpha)) = \Psi^L_\theta(a) \notin S_\eta$, so we get

$$\Psi^L_{\eta\circ\theta}(a) = v_L(\eta(\theta(\alpha)) = \Psi^L_\eta(v_L(\theta(\alpha))) = \Psi^L_\eta(\Psi^L_\theta(a)).$$

Since $\Psi^L_{\eta\circ\theta}$, Ψ^L_η, and Ψ^L_θ are all continuous functions it follows that $\Psi^L_{\eta\circ\theta}(x) = \Psi^L_\eta(\Psi^L_\theta(x))$ for all $x \in [0,\infty)$. $\square$

For our purposes, the significance of these constructions is that the Hasse-Herbrand functions $\varphi_{L/K}$ can be expressed in terms of valuation functions in a natural way. Gross [**Gro79**, Lemma 1.5] attributes the following observation to Tate:

PROPOSITION 13.11. *Let L/K be a totally ramified extension of degree n and let π_K, π_L be uniformizers for K, L. Choose $F(X) \in X\mathfrak{O}_K[[X]]$ so that $\pi_K = F(\pi_L)$. Then for $x \geq 0$ we have*

(a) $\Psi^L_{F^*}(x) = \sum_{\sigma\in\mathcal{E}_{L/K}} \min\{x, i_L(\sigma)\} = |(\mathcal{E}_{L/K})_x| \cdot x + \sum_{i_L(\sigma)<x} i_L(\sigma)$.

(b) $\varphi_{L/K}(x) = \Psi^K_{F^*}(x/n)$.

PROOF. We have $F(X) - \pi_K = u(X)f(X)$, with $u(X) \in \mathfrak{O}_K[[X]]^\times$ and $f(X)$ the minimum polynomial for π_L over K. Hence

$$F^*(X) = \pi_K^{-1}u(\pi_L + \pi_L X)f(\pi_L + \pi_L X). \tag{13.6}$$

We can express $f(X)$ in terms of K-embeddings of L into K^{sep} by

$$f(X) = \prod_{\sigma\in\mathcal{E}_{L/K}} (X - \sigma(\pi_L)).$$

It follows that

$$f(\pi_L + \pi_L X) = \prod_{\sigma\in\mathcal{E}_{L/K}} (\pi_L X + (\pi_L - \sigma(\pi_L))).$$

Let $\alpha \in K^{sep}$ and assume that $a = v_L(\alpha)$ satisfies $a > 0$ and $a \neq v_L(\pi_L - \sigma(\pi_L)) - v_L(\pi_L)$ for all $\sigma \in \mathcal{E}_{L/K}$. Then

$$\begin{aligned} v_L(f(\pi_L + \pi_L\alpha)) &= \sum_{\sigma\in\mathcal{E}_{L/K}} \min\{v_L(\pi_L) + a, v_L(\pi_L - \sigma(\pi_L))\} \\ &= n + \sum_{\sigma\in\mathcal{E}_{L/K}} \min\{a, i_L(\sigma)\}. \end{aligned}$$

Since $v_K(u(\pi_L + \pi_L\alpha)) = 0$ it follows from (13.6) that

$$\begin{aligned} v_L(F^*(\alpha)) &= \sum_{\sigma \in \mathcal{E}_{L/K}} \min\{a, i_L(\sigma)\} \\ &= \sum_{i_L(\sigma) \geq a} a + \sum_{i_L(\sigma)<a} i_L(\sigma) \\ &= |(\mathcal{E}_{L/K})_a| \cdot a + \sum_{i_L(\sigma)<a} i_L(\sigma). \end{aligned}$$

Since $\Psi^L_{F^*}(x)$ and $\min\{x, i_L(\sigma)\}$ are continuous on $[0,\infty)$ we get

$$\begin{aligned} \Psi^L_{F^*}(x) &= \sum_{\sigma \in \mathcal{E}_{L/K}} \min\{x, i_L(\sigma)\} \\ &= |(\mathcal{E}_{L/K})_x| x + \sum_{i_L(\sigma)<x} i_L(\sigma) \end{aligned}$$

for all $x \geq 0$. This proves (a).

Let $x > 0$ and $x \neq v_L(\pi_L - \sigma(\pi_L)) - v_L(\pi_L)$ for all $\sigma \in \mathcal{E}_{L/K}$. Then by part (a) we have $(\Psi^L_{F^*})'(x) = |(\mathcal{E}_{L/K})_x|$. By (13.2) we get

$$\varphi'_{L/K}(x) = \frac{|(\mathcal{E}_{L/K})_x|}{|(\mathcal{E}_{L/K})_0|} = \frac{|(\mathcal{E}_{L/K})_x|}{n}.$$

Since $\Psi^L_{F^*}$ and $\varphi_{L/K}$ are both continuous on $[0,\infty)$, and $\Psi^L_{F^*}(0) = \varphi_{L/K}(0) = 0$, it follows that $\Psi^L_{F^*}(x) = n\varphi_{L/K}(x)$ for all $x \geq 0$. By applying Lemma 13.8 we get $\varphi_{L/K}(x) = \Psi^K_{F^*}(x/n)$. □

EXAMPLE 13.12. Let $\mathrm{char}(K) = 0$ and $L = K(\pi_L)$ with $\pi_L^p = \pi_K$ as in Example 13.1. Choosing $F(X) = X^p$ we get

$$F^*(X) = \pi_K^{-1}((\pi_L + \pi_L X)^p - \pi_K) = (1+X)^p - 1.$$

Therefore $(F^*)'(0) = p$, so by Proposition 13.6 we have

$$v_L(\mathfrak{D}_{L/K}) = v_L(p) + p - 1 = pe_K + p - 1.$$

Let $\alpha \in K^{sep}$ with $v_K(p\alpha) \neq v_K(\alpha^p)$, that is, $v_K(\alpha) \neq pe_K/(p-1)$. Then

$$\begin{aligned} v_K(F^*(\alpha)) &= \min\{v_K(\alpha^p), v_K(p\alpha)\} \\ &= \min\{pv_K(\alpha), e_K + v_K(\alpha)\}. \end{aligned}$$

Hence by Proposition 13.9 and Proposition 13.11(b), for all $x \geq 0$ we have

$$\begin{aligned} \Psi^K_{F^*}(x) &= \min\{px, e_K + x\} \\ \varphi_{L/K}(x) &= \min\{x, e_K + p^{-1}x\}. \end{aligned}$$

Now let $M/L/K$ be finite totally ramified subextensions of K^{sep}/K. Let π_K, π_L, π_M be uniformizers for K, L, M and let $F(X), G(X) \in X\mathfrak{O}_K[[X]]$ satisfy $F(\pi_L) = \pi_K$ and $G(\pi_M) = \pi_L$. Set $n = [L:K]$, $m = [M:L]$, and $H(X) = F(G(X))$. Then $H(X) \in X\mathfrak{O}_K[[X]]$ satisfies $H(\pi_M) = \pi_K$. It follows that we can use the series

$$\begin{aligned} F^*(X) &= \pi_K^{-1}(F(\pi_L + \pi_L X) - \pi_K) \\ G^*(X) &= \pi_L^{-1}(G(\pi_M + \pi_M X) - \pi_L) \\ H^*(X) &= \pi_K^{-1}(H(\pi_M + \pi_M X) - \pi_K) \end{aligned}$$

to compute the Hasse-Herbrand functions for the extensions M/L and M/K. As Lubin points out in [**Lub13**, Theorem 1.6], by applying Proposition 13.11 to the relation $H^*(X) = F^*(G^*(X))$, we obtain the well-known composition formulas for the Hasse-Herbrand functions:

THEOREM 13.13. *Let $M/L/K$ be finite separable totally ramified extensions of local fields. Then $\varphi_{M/K} = \varphi_{L/K} \circ \varphi_{M/L}$ and $\psi_{M/K} = \psi_{M/L} \circ \psi_{L/K}$.*

PROOF. It follows from (13.3) that

$$F^*(G^*(X)) = \pi_K^{-1}(F(G(\pi_M + \pi_M X)) - \pi_K) = H^*(X).$$

Using Proposition 13.11(b), Lemma 13.10, and Lemma 13.8 we deduce that for all $x \geq 0$ we have

$$\begin{aligned}\varphi_{M/K}(x) &= \Psi_{H^*}^K(x/mn) \\ &= \Psi_{F^* \circ G^*}^K(x/mn) \\ &= \Psi_{F^*}^K(\Psi_{G^*}^K(x/mn)) \\ &= \Psi_{F^*}^K(\Psi_{G^*}^L(x/m)/n) \\ &= \varphi_{L/K}(\varphi_{M/L}(x)).\end{aligned}$$

Hence $\varphi_{M/K} = \varphi_{L/K} \circ \varphi_{M/L}$. It follows that $\psi_{M/K} = \psi_{M/L} \circ \psi_{L/K}$. □

In order to prove the functorial properties of the upper numbering of the ramification sets we need the following results.

PROPOSITION 13.14. *Let $M/L/K$ be finite totally ramified extensions and set $m = [M : L]$. Then for $\sigma \in \mathcal{E}_{L/K}$ we have*

(a) $i_L(\sigma) = \dfrac{1}{m} \displaystyle\sum_{\substack{\tau \in \mathcal{E}_{M/K} \\ \tau|_L = \sigma}} i_M(\tau)$.

(b) *Set* $c = \max\{i_M(\tau) \mid \tau \in \mathcal{E}_{M/K}, \tau|_L = \sigma\}$. *Then* $i_L(\sigma) = \varphi_{M/L}(c)$.

PROOF. Let $G(X) \in X\mathfrak{O}_K[[X]]$ satisfy $G(\pi_M) = \pi_L$. By the Weierstrass preparation theorem there is $u(X) \in \mathfrak{O}_L[[X]]^\times$ such that

$$G(X) - \pi_L = u(X) \cdot \prod_{\tau \in \mathcal{E}_{M/L}} (X - \tau(\pi_M)).$$

Let $\widetilde{M}/K$ be the Galois closure of M/K in K^{sep} and let $\tilde{\sigma}$ be a lifting of σ to $\mathrm{Gal}(\widetilde{M}/K)$. Then we get

$$G(X) - \sigma(\pi_L) = u^\sigma(X) \cdot \prod_{\gamma \in \mathcal{E}_{M/L}} (X - \tilde{\sigma}\gamma(\pi_M)), \tag{13.7}$$

where $u^\sigma(X)$ is obtained from $u(X)$ by applying σ to the coefficients. Since

$$\{\tilde{\sigma}\gamma \mid \gamma \in \mathcal{E}_{M/L}\} = \{\tau \in \mathcal{E}_{M/K} \mid \tau|_L = \sigma\},$$

by setting $X = \pi_M$ in (13.7) we get

$$\begin{aligned}\pi_L - \sigma(\pi_L) &= u^\sigma(\pi_M) \cdot \prod_{\gamma \in \mathcal{E}_{M/L}} (\pi_M - \tilde{\sigma}\gamma(\pi_M)) \\ &= u^\sigma(\pi_M) \cdot \prod_{\substack{\tau \in \mathcal{E}_{M/K} \\ \tau|_L = \sigma}} (\pi_M - \tau(\pi_M)).\end{aligned} \tag{13.8}$$

It follows that

$$\begin{aligned} i_L(\sigma)+1 &= v_L(\pi_L-\sigma(\pi_L)) \\ &= \sum_{\substack{\tau\in\mathcal{E}_{M/K}\\ \tau|_L=\sigma}} v_L(\pi_M-\tau(\pi_M)) \\ &= \frac{1}{m}\sum_{\substack{\tau\in\mathcal{E}_{M/K}\\ \tau|_L=\sigma}} v_M(\pi_M-\tau(\pi_M)) \\ &= \frac{1}{m}\sum_{\substack{\tau\in\mathcal{E}_{M/K}\\ \tau|_L=\sigma}} (i_M(\tau)+1) \\ &= 1+\frac{1}{m}\sum_{\substack{\tau\in\mathcal{E}_{M/K}\\ \tau|_L=\sigma}} i_M(\tau). \end{aligned}$$

This proves (a).

We now assume that the lifting $\tilde{\sigma}$ of σ to $\mathrm{Gal}(\widetilde{M}/K)$ is chosen to maximize $v_M(\tilde{\sigma}(\pi_M)-\pi_M)$; then $v_M(\tilde{\sigma}(\pi_M)-\pi_M)=c+1$. For $\gamma\in\mathcal{E}_{M/L}$ we get

$$\begin{aligned} v_M(\tilde{\sigma}\gamma(\pi_M)-\pi_M) &= v_M((\tilde{\sigma}\gamma(\pi_M)-\tilde{\sigma}(\pi_M))+(\tilde{\sigma}(\pi_M)-\pi_M)) \\ &\geq \min\{v_M(\tilde{\sigma}\gamma(\pi_M)-\tilde{\sigma}(\pi_M)), v_M(\tilde{\sigma}(\pi_M)-\pi_M)\} \\ &= \min\{v_M(\gamma(\pi_M)-\pi_M), c+1\}, \end{aligned}$$

with equality if $v_M(\tilde{\sigma}\gamma(\pi_M)-\tilde{\sigma}(\pi_M))\neq c+1$. However, if $v_M(\tilde{\sigma}\gamma(\pi_M)-\tilde{\sigma}(\pi_M))=c+1$ then by the maximality of c we get

$$c+1\geq v_M(\tilde{\sigma}\gamma(\pi_M)-\pi_M)\geq \min\{v_M(\gamma(\pi_M)-\pi_M), c+1\}=c+1.$$

It follows that

$$v_M(\tilde{\sigma}\gamma(\pi_M)-\pi_M)=\min\{v_M(\gamma(\pi_M)-\pi_M)), c+1\}$$

holds in all cases. Using (13.8) we get

$$\begin{aligned} i_L(\sigma)+1 &= \frac{1}{m}\sum_{\gamma\in\mathcal{E}_{M/L}} v_M(\tilde{\sigma}\gamma(\pi_M)-\pi_M) \\ &= \frac{1}{m}\sum_{\gamma\in\mathcal{E}_{M/L}} \min\{v_M(\gamma(\pi_M)-\pi_M), c+1\} \\ &= \frac{1}{m}\sum_{\gamma\in\mathcal{E}_{M/L}} \min\{i_M(\gamma), c\}+1. \end{aligned}$$

It follows from Proposition 13.11(a), Lemma 13.8, and Proposition 13.11(b) that

$$i_L(\sigma)=\frac{1}{m}\cdot\Psi^M_{G^*}(c)=\Psi^L_{G^*}(c/m)=\varphi_{M/L}(c).$$

This proves (b). □

Let $M/L/K$ be finite subextensions of K^{sep}/K. We may view the set $\mathcal{E}_{M/L}$ of L-embeddings of M into K^{sep} as a subset of $\mathcal{E}_{M/K}$. It follows directly from the definitions that for $x\geq 0$ we have $(\mathcal{E}_{M/L})_x=\mathcal{E}_{M/L}\cap(\mathcal{E}_{M/K})_x$. Let $\rho_{M/L}:\mathcal{E}_{M/K}\to\mathcal{E}_{L/K}$ be the restriction map.

THEOREM 13.15. *Let $x \geq 0$. Then $\rho_{M/L}(\mathcal{E}^x_{M/K}) = \mathcal{E}^x_{L/K}$.*

PROOF. Let $\tau \in \mathcal{E}^x_{M/K}$. Then $\tau \in (\mathcal{E}_{M/K})_{\psi_{M/K}(x)}$, so we have $i_M(\tau) \geq \psi_{M/K}(x)$. Hence by Proposition 13.14(b) and Theorem 13.13 we get

$$i_L(\tau|_L) \geq \varphi_{M/L}(\psi_{M/K}(x)) = \psi_{L/K}(x).$$

Therefore $\tau|_L \in (\mathcal{E}_{L/K})_{\psi_{L/K}(x)} = \mathcal{E}^x_{L/K}$. Suppose $\sigma \in \mathcal{E}^x_{L/K} = (\mathcal{E}_{L/K})_{\psi_{L/K}(x)}$. Then $i_L(\sigma) \geq \psi_{L/K}(x)$. Let $\tilde{\sigma}$ be a lifting of σ to $\mathcal{E}_{M/K}$ which maximizes $v_M(\tilde{\sigma}(\pi_M)-\pi_M)$ and set $u = i_M(\tilde{\sigma})$. Then by Proposition 13.14(b) we have $i_L(\sigma) = \varphi_{M/L}(u)$. Therefore by Theorem 13.13

$$u = \psi_{M/L}(i_L(\sigma)) \geq \psi_{M/L}(\psi_{L/K}(x)) = \psi_{M/K}(x),$$

and hence $\tilde{\sigma} \in (\mathcal{E}_{M/K})_{\psi_{M/K}(x)} = \mathcal{E}^x_{M/K}$. Since $\rho_{M/L}(\tilde{\sigma}) = \sigma$ this completes the proof. □

COROLLARY 13.16 (Herbrand's Theorem). *Let $x' \geq 0$ and set $x = \varphi_{M/L}(x')$. Then $\rho_{M/L}((\mathcal{E}_{M/K})_{x'}) = (\mathcal{E}_{L/K})_x$.*

PROOF. It follows from Theorem 13.13 that $\varphi_{M/K}(x') = \varphi_{L/K}(x)$. Hence by Theorem 13.15 we get

$$\begin{aligned}\rho_{M/L}((\mathcal{E}_{M/K})_{x'}) &= \rho_{M/L}(\mathcal{E}^{\varphi_{M/K}(x')}_{M/K}) \\ &= \rho_{M/L}(\mathcal{E}^{\varphi_{L/K}(x)}_{M/K}) \\ &= \mathcal{E}^{\varphi_{L/K}(x)}_{L/K} \\ &= (\mathcal{E}_{L/K})_x.\end{aligned}$$

□

PROPOSITION 13.17. *Let L/K be a finite separable totally ramified subextension of K^{sep}/K and let x, x' be positive real numbers related by the formulas $x = \varphi_{L/K}(x')$, $x' = \psi_{L/K}(x)$. Let M be the largest subfield of L which is fixed by every element of $\mathcal{E}^x_{L/K} = (\mathcal{E}_{L/K})_{x'}$. Then $\mathcal{E}^x_{L/K} = \mathcal{E}_{L/M}$. Furthermore, for $0 \leq t \leq x'$ we have $\varphi_{L/M}(t) = \psi_{L/M}(t) = t$.*

PROOF. Let $\widetilde{L}/K$ be the Galois closure of L/K in K^{sep} and set $G = \mathrm{Gal}(\widetilde{L}/K) = \mathcal{E}_{\widetilde{L}/K}$, $G_L = \mathrm{Gal}(\widetilde{L}/L) = \mathcal{E}_{\widetilde{L}/L}$. The restriction map $\rho_{\widetilde{L}/L} : G \to \mathcal{E}_{L/K}$ induces a bijection $\overline{\rho} : G/G_L \to \mathcal{E}_{L/K}$. It follows from Theorem 13.15 that $\rho_{\widetilde{L}/L}(G^x) = \mathcal{E}^x_{L/K}$. Therefore $\overline{\rho}(G^xG_L/G_L) = \mathcal{E}^x_{L/K}$. It follows that $M = \widetilde{L}^{G^xG_L}$, so we get

$$|\mathcal{E}_{L/M}| = [L : M] = |G^xG_L : G_L| = |\mathcal{E}^x_{L/K}|.$$

Since $\mathcal{E}^x_{L/K} \subset \mathcal{E}_{L/M}$ it follows that $\mathcal{E}^x_{L/K} = \mathcal{E}_{L/M}$. Let $0 \leq t \leq x'$. Then

$$(\mathcal{E}_{L/M})_t = \mathcal{E}_{L/M} \cap (\mathcal{E}_{L/K})_t = \mathcal{E}_{L/M} \cap \mathcal{E}^{\varphi_{L/K}(t)}_{L/K}.$$

Since $\varphi_{L/K}(t) \leq \varphi_{L/K}(x') = x$ we get $\mathcal{E}_{L/M} \subset \mathcal{E}^{\varphi_{L/K}(t)}_{L/K}$, and hence $(\mathcal{E}_{L/M})_t = \mathcal{E}_{L/M}$. Since this holds for all $0 \leq t \leq x'$, by (13.2) we get $\varphi_{L/M}(t) = t$ and $\psi_{L/M}(t) = \varphi^{-1}_{L/M}(t) = t$. □

PROPOSITION 13.18. *Let L/K be a finite separable totally ramified extension and let $b_1 < \cdots < b_r$ be the distinct lower ramification breaks of L/K. For $0 \leq i \leq r-1$ let L_i be the subfield of L which is fixed by every element of $(\mathcal{E}_{L/K})_{b_{i+1}}$. Also* *set $L_r = L$. Then for $1 \leq i \leq r$, b_i is the unique lower and upper ramification break of L_i/L_{i-1}.*

PROOF. It follows from Proposition 13.17 that for $1 \leq i \leq r$ we have $(\mathcal{E}_{L/K})_{b_i} = \mathcal{E}_{L/L_{i-1}}$ and $\psi_{L/L_{i-1}}(b_i) = b_i$. Therefore

$$\mathcal{E}^{b_i}_{L/L_{i-1}} = (\mathcal{E}_{L/L_{i-1}})_{b_i} = \mathcal{E}_{L/L_{i-1}} \cap (\mathcal{E}_{L/K})_{b_i} = \mathcal{E}_{L/L_{i-1}}.$$

By applying Theorem 13.15 we get

$$\mathcal{E}^{b_i}_{L_i/L_{i-1}} = \rho_{L/L_i}(\mathcal{E}^{b_i}_{L/L_{i-1}}) = \rho_{L/L_i}(\mathcal{E}_{L/L_{i-1}}) = \mathcal{E}_{L_i/L_{i-1}}.$$

On the other hand, if $x > b_i$ then $\psi_{L/L_{i-1}}(x) > \psi_{L/L_{i-1}}(b_i) = b_i$. Hence

$$\mathcal{E}^{x}_{L/L_{i-1}} = (\mathcal{E}_{L/L_{i-1}})_{\psi_{L/L_{i-1}}(x)} \subset (\mathcal{E}_{L/L_{i-1}})_{b_{i+1}} = \mathcal{E}_{L/L_i}.$$

It follows that $\mathcal{E}^x_{L_i/L_{i-1}} = \rho_{L/L_i}(\mathcal{E}^x_{L/L_{i-1}}) = \{\mathrm{id}_{L_i}\}$. Hence b_i is the unique upper ramification break of L_i/L_{i-1}. Therefore $\psi_{L_i/L_{i-1}}(b_i) = b_i$ is the unique lower ramification break of L_i/L_{i-1}. □

EXAMPLE 13.19. Let $\mathrm{char}(K) = 0$, let $L = K(\pi_L)$ with $\pi_L^p = \pi_K$, and let $M = L(\pi_M)$, with $\pi_M^p = \pi_L$. By Example 13.12 we have

$$\begin{aligned}\varphi_{L/K}(x) &= \min\{x, e_K + p^{-1}x\}\\ \varphi_{M/L}(x) &= \min\{x, e_L + p^{-1}x\}\\ &= \min\{x, pe_K + p^{-1}x\}.\end{aligned}$$

It follows from Theorem 13.13 that for $x \geq 0$ we have

$$\begin{aligned}\varphi_{M/K}(x) &= \varphi_{L/K}(\varphi_{M/L}(x))\\ &= \min\{x, pe_K + p^{-1}x, e_K + p^{-1}x, e_K + p^{-1}(pe_K + p^{-1}x)\}\\ &= \min\{x, e_K + p^{-1}x, 2e_K + p^{-2}x\}.\end{aligned}$$

Since the lines $y = x$ and $y = e_K + p^{-1}x$ meet at $(pe_K/(p-1), pe_K/(p-1))$, the first upper and lower ramification break of M/K is $\ell_1 = u_1 = pe_K/(p-1)$. Since the lines $y = e_K + p^{-1}x$ and $y = 2e_K + p^{-2}x$ meet at $(p^2e_K/(p-1), (2p-1)e_K/(p-1))$, the second lower ramification break of M/K is $\ell_2 = p^2e_K/(p-1)$ and the second upper ramification break of M/K is $u_2 = (2p-1)e_K/(p-1)$.

13.3. Properties of lower ramification breaks

Let L/K be a separable totally ramified extension of degree n and set $\nu = v_p(n)$. In this section we study the lower ramification breaks $\ell_1 \leq \ell_2 \leq \cdots \leq \ell_\nu$ of L/K. Since we are not assuming that L/K is Galois, the breaks ℓ_i are rational numbers, but not necessarily integers. We give a restriction on the denominator of ℓ_i which implies that $v_p(\ell_i) \geq 0$. We also determine upper bounds for ℓ_i and show that if ℓ_i does not achieve its upper bound then $v_p(\ell_i) = 0$. This implies that the condition $v_p(\ell_i) = 0$ in the definition of semistable extensions (Definition 14.26) is often satisfied.

Let π_K, π_L be uniformizers for K, L. Let $F(X) \in X\mathfrak{O}_K[[X]]$ satisfy $F(\pi_L) = \pi_K$, and write $F(X) = a_1X + a_2X^2 + \cdots$. For $m \geq 0$ set

$$(D^mF)(X) = \sum_{i=1}^{\infty} \binom{i}{m} a_i X^{i-m} \in \mathfrak{O}_K[[X]]. \tag{13.9}$$

As in the previous section we define

$$F^*(X) = \pi_K^{-1}(F(\pi_L + \pi_L X) - \pi_K) \in \mathfrak{O}_L[[X]].$$

Then we get the Taylor expansion

$$F^*(X) = \pi_K^{-1} \sum_{m=1}^{\infty} (D^m F)(\pi_L) \cdot \pi_L^m X^m.$$

It follows from Proposition 13.11(b) and (13.4) that for $x \geq 0$ we have

$$\varphi_{L/K}(x) = \frac{1}{n} \cdot \min\{m - n + v_L((D^m F)(\pi_L)) + mx \mid m \geq 1\}. \tag{13.10}$$

In order to obtain another formula for $\varphi_{L/K}(x)$ we will use the following elementary fact about binomial coefficients (see [**Hei96**, Lemma 5.6]).

LEMMA 13.20. *Let $a \geq b \geq 1$. Then*

$$v_p\left(\binom{a}{b}\right) \geq v_p(a) - v_p(b), \tag{13.11}$$

with equality if $v_p(a) \geq v_p(b)$ and b is a power of p.

PROOF. This could be deduced from Kummer's theorem, which states that $v_p\left(\binom{a}{b}\right)$ is equal to the number of carries that occur when $a - b$ is added to b in base p. Alternatively, note that since $v_p(a!) = \lfloor a/p \rfloor + \lfloor a/p^2 \rfloor + \cdots$ we have

$$v_p\left(\binom{a}{b}\right) = \sum_{i=1}^{\infty} \left(\left\lfloor \frac{a}{p^i} \right\rfloor - \left\lfloor \frac{b}{p^i} \right\rfloor - \left\lfloor \frac{a-b}{p^i} \right\rfloor \right). \tag{13.12}$$

Set $\alpha = v_p(a)$ and $\beta = v_p(b)$. We may assume that $\alpha \geq \beta$. For $1 \leq i \leq \beta$ the fractions on the right side of (13.12) are all integers, so the ith summand is 0. For $\beta < i \leq \alpha$, a/p^i is an integer but b/p^i is not, so the ith summand is 1 for these terms. Since the remaining summands are nonnegative we get (13.11). Suppose $b = p^\beta$ and write $a = a_0 p^\alpha$. Then for $i > \alpha$ we have

$$\begin{aligned} \left\lfloor \frac{a}{p^i} \right\rfloor - \left\lfloor \frac{b}{p^i} \right\rfloor - \left\lfloor \frac{a-b}{p^i} \right\rfloor &= \left\lfloor \frac{a_0 p^\alpha}{p^i} \right\rfloor - \left\lfloor \frac{p^\beta}{p^i} \right\rfloor - \left\lfloor \frac{a_0 p^\alpha - p^\beta}{p^i} \right\rfloor \\ &= \left\lfloor \frac{a_0 p^{\alpha-\beta}}{p^{i-\beta}} \right\rfloor - \left\lfloor \frac{a_0 p^{\alpha-\beta} - 1}{p^{i-\beta}} \right\rfloor. \end{aligned}$$

Since $p^{i-\beta} \nmid a_0 p^{\alpha-\beta}$ this last expression is equal to 0. Hence equality holds in (13.11) for this case. □

Let L/K be a totally ramified extension of degree n. In Section 13.2 we chose a series $F(X) \in X\mathfrak{O}_K[[X]]$ such that $F(\pi_L) = \pi_K$. In general there are many possibilities for $F(X)$, but there is a unique such series with coefficients in the set $\mathcal{R}_K$ of Teichmüller representatives of K.

PROPOSITION 13.21. *Let L/K be a totally ramified extension of degree n, and let $F(X)$ be the unique power series with coefficients in $\mathcal{R}_K$ such that $F(\pi_L) = \pi_K$. Set $\nu = v_p(n)$ and write $F(X) = a_n X^n + a_{n+1} X^{n+1} + \cdots$. Let*

$$S = \{(i, j) \mid i \geq n,\ a_i \neq 0,\ 0 \leq j \leq \nu\},$$

and for $x \geq 0$ set

$$T_x = \left\{ i - n + v_L\left(\binom{i}{p^j}\right) + p^j x \mid (i, j) \in S \right\}.$$

Then $\varphi_{L/K}(x) = \frac{1}{n} \cdot \min(T_x)$.

PROOF. Since $F(\pi_L) = \pi_K$ and $F(X)$ has coefficients in $\mathcal{R}_K$ we have $a_n \neq 0$. Hence $(n, j) \in S$ for $0 \leq j \leq \nu$. By Lemma 13.20 we have $v_p\left(\binom{n}{p^\nu}\right) = 0$. It follows that $\min(T_x)$ is finite for all $x \geq 0$. Furthermore, we get

$$\frac{1}{n} \cdot \min(T_0) = 0 = \varphi_{L/K}(0).$$

Let $x > 0$ and choose $(i_0, j_0) \in S$ so that

$$\min(T_x) = i_0 - n + v_L\left(\binom{i_0}{p^{j_0}}\right) + p^{j_0}x. \tag{13.13}$$

Set $h_0 = v_p(i_0)$. If $h_0 < j_0$ then

$$i_0 - n + v_L\left(\binom{i_0}{p^{h_0}}\right) + p^{h_0}x < i_0 - n + v_L\left(\binom{i_0}{p^{j_0}}\right) + p^{j_0}x$$

by Lemma 13.20. This contradicts the minimality of (13.13), so we must have $h_0 \geq j_0$. Suppose $h_0 > \nu$. Then $v_p\left(\binom{n}{p^{j_0}}\right) = \nu - j_0$ and $v_p\left(\binom{i_0}{p^{j_0}}\right) = h_0 - j_0$ by Lemma 13.20, so we get

$$n - n + v_L\left(\binom{n}{p^{j_0}}\right) + p^{j_0}x < i_0 - n + v_L\left(\binom{i_0}{p^{j_0}}\right) + p^{j_0}x.$$

Since $(n, j_0) \in S$, this contradicts the minimality of (13.13). Hence $h_0 \leq \nu$.

We claim that $i = i_0$ gives the unique term in the sum

$$(D^{p^{j_0}}F)(\pi_L) = \sum_{i=n}^{\infty} \binom{i}{p^{j_0}} a_i \pi_L^{i-p^{j_0}} \tag{13.14}$$

with minimum L-valuation. Suppose $i_1 \geq n$ satisfies

$$v_L\left(\binom{i_1}{p^{j_0}} a_{i_1} \pi_L^{i_1 - p^{j_0}}\right) \leq v_L\left(\binom{i_0}{p^{j_0}} a_{i_0} \pi_L^{i_0 - p^{j_0}}\right).$$

Since the nonzero Teichmüller representatives a_{i_1}, a_{i_0} are units we get

$$i_1 - p^{j_0} + v_L\left(\binom{i_1}{p^{j_0}}\right) \leq i_0 - p^{j_0} + v_L\left(\binom{i_0}{p^{j_0}}\right) \tag{13.15}$$

and hence

$$i_1 - n + v_L\left(\binom{i_1}{p^{j_0}}\right) + p^{j_0}x \leq i_0 - n + v_L\left(\binom{i_0}{p^{j_0}}\right) + p^{j_0}x. \tag{13.16}$$

By the minimality of (13.13) we have equality in (13.16), and hence also in (13.15). Since $n \mid v_L(p)$ this implies $i_1 \equiv i_0 \pmod{n}$. We showed above that $v_p(i_0) = h_0 \leq \nu$. Since

$$\min(T_x) = i_1 - n + v_L\left(\binom{i_1}{p^{j_0}}\right) + p^{j_0}x,$$

the same reasoning shows that $v_p(i_1) \leq \nu$. Therefore $v_p(i_1) = v_p(i_0) = h_0$. Hence by Lemma 13.20 we have

$$v_p\left(\binom{i_1}{p^{j_0}}\right) = v_p\left(\binom{i_0}{p^{j_0}}\right) = h_0 - j_0. \tag{13.17}$$

Since equality holds in (13.15) we deduce that $i_1 = i_0$, which verifies our claim.

It follows from this claim that

$$v_L((D^{p^{j_0}}F)(\pi_L)) = v_L\left(\binom{i_0}{p^{j_0}}a_{i_0}\pi_L^{i_0-p^{j_0}}\right) = i_0 - p^{j_0} + v_L\left(\binom{i_0}{p^{j_0}}\right). \tag{13.18}$$

Therefore by (13.13) we have

$$\min(T_x) = p^{j_0} - n + v_L((D^{p^{j_0}}F)(\pi_L)) + p^{j_0}x, \tag{13.19}$$

and hence $\varphi_{L/K}(x) \le \frac{1}{n}\cdot\min(T_x)$ by (13.10). We wish to show that equality holds. To this end we note that by (13.9) we have

$$v_L(D^mF(\pi_L)) \ge \min\left\{i - m + v_L\left(\binom{i}{m}\right) \,\middle|\, a_i \ne 0\right\}$$

for all $m \ge 1$. Hence by (13.10) we get

$$\varphi_{L/K}(x) \ge \frac{1}{n}\cdot\min\left\{i - n + v_L\left(\binom{i}{m}\right) + mx \,\middle|\, m \ge 1,\, a_i \ne 0\right\}.$$

Therefore it suffices to prove that

$$i - n + v_L\left(\binom{i}{m}\right) + mx \ge \min(T_x) \tag{13.20}$$

for all $m \ge 1$ and $i \ge 1$ with $a_i \ne 0$.

Let $m \ge 1$ and set $j = \min\{v_p(m), \nu\}$. Choose $i \ge n$ such that $a_i \ne 0$ and set $h = v_p(i)$. If $m > i$ then $\binom{i}{m} = 0$, and hence (13.20) holds. Suppose $j = \nu$. By Lemma 13.20 we have $v_p\left(\binom{n}{p^\nu}\right) = 0$. Since $m \ge p^\nu$ and $(n,\nu) \in S$ we get

$$\begin{aligned} i - n + v_L\left(\binom{i}{m}\right) + mx &\ge n - n + v_L\left(\binom{n}{p^\nu}\right) + p^\nu x \\ &\ge \min(T_x). \end{aligned}$$

We now consider the cases with $m \le i$ and $j < \nu$. In these cases we have $j = v_p(m)$. Suppose $j \le h$. Using Lemma 13.20 we get

$$v_p\left(\binom{i}{m}\right) \ge h - j = v_p\left(\binom{i}{p^j}\right).$$

Since $m \ge n > p^j$ and $(i,j) \in S$ it follows that

$$\begin{aligned} i - n + v_L\left(\binom{i}{m}\right) + mx &> i - n + v_L\left(\binom{i}{p^j}\right) + p^j x \\ &\ge \min(T_x). \end{aligned}$$

Suppose $h < j$. It follows from Lemma 13.20 that $v_L\left(\binom{i}{p^h}\right) = 0$. Since $m > p^h$ and $(i,h) \in S$ we get

$$\begin{aligned} i - n + v_L\left(\binom{i}{m}\right) + mx &> i - n + v_L\left(\binom{i}{p^h}\right) + p^h x \\ &\ge \min(T_x). \end{aligned}$$

Since (13.20) holds in all cases we conclude that $\varphi_{L/K}(x) = \frac{1}{n} \cdot \min(T_x)$. $\square$

COROLLARY 13.22. *Let L/K be a totally ramified extension of degree $n = up^\nu$, with $p \nmid u$. Then 0 is a lower ramification break for L/K if and only if $u > 1$.*

PROOF. Using Lemma 13.20 we see that $h - n + v_L\left(\binom{h}{p^j}\right) \geq 1$ for all $(h,j) \in S$ such that $(h,j) \neq (n,\nu)$. Let $0 < x < 1/(p^\nu - p^{\nu-1})$. Since $(n,\nu) \in S$ we get $\min(T_x) = p^\nu x$. Therefore $\varphi_{L/K}(x) = u^{-1}x$, and hence $|(\mathcal{E}_{L/K})_x| = p^\nu$. Since L/K is totally ramified we have $(\mathcal{E}_{L/K})_0 = \mathcal{E}_{L/K}$ by Proposition 13.2. If $u = 1$ then $(\mathcal{E}_{L/K})_x = (\mathcal{E}_{L/K})_0$, and hence 0 is not a lower ramification break, while if $u > 1$ then $(\mathcal{E}_{L/K})_x \neq (\mathcal{E}_{L/K})_0$, and hence 0 is a lower ramification break. $\square$

PROPOSITION 13.23. *Let L/K be a finite separable totally ramified extension of local fields of degree n and let $b_1 < \cdots < b_r$ be the distinct lower ramification breaks of L/K. Then for sufficiently small $\epsilon > 0$ the valuation of the different of L/K is given by*

$$v_L(\mathfrak{D}_{L/K}) = \sum_{j=1}^{r} (|(\mathcal{E}_{L/K})_{b_j}| - |(\mathcal{E}_{L/K})_{b_j+\epsilon}|)(b_j + 1).$$

PROOF. Let π_L be a uniformizer for L and let $f(X)$ be the minimum polynomial for π_L over K. Then

$$f(X) = \prod_{\sigma \in \mathcal{E}_{L/K}} (X - \sigma(\pi_L)).$$

It follows that

$$\begin{aligned} v_L(\mathfrak{D}_{L/K}) &= v_L(f'(\pi_L)) \\ &= \sum_{\substack{\sigma \in \mathcal{E}_{L/K} \\ \sigma \neq \mathrm{id}_L}} v_L(\pi_L - \sigma(\pi_L)) \\ &= \sum_{j=1}^{r} \sum_{\substack{\sigma \in \mathcal{E}_{L/K} \\ i_L(\sigma) = b_j}} (b_j + 1) \\ &= \sum_{j=1}^{r} (|(\mathcal{E}_{L/K})_{b_j}| - |(\mathcal{E}_{L/K})_{b_j+\epsilon}|)(b_j + 1). \end{aligned}$$ $\square$

REMARK 13.24. Let $\widetilde{L}/K$ be the Galois closure of L/K and let $e_{\widetilde{L}/L}$ be the ramification index of $\widetilde{L}/L$. It follows from (13.1) that $b_i \in e_{\widetilde{L}/L}^{-1}\mathbb{Z}$ for $1 \leq i \leq r$. Hence any ϵ satisfying $0 < \epsilon \leq e_{\widetilde{L}/L}^{-1}$ is sufficiently small for the purposes of Proposition 13.23.

COROLLARY 13.25. *Let L/K be a finite separable totally ramified extension of local fields of degree p^ν and let $b_1 < \cdots < b_r$ be the distinct lower ramification breaks of L/K. Set $L_r = L$, and for $0 \leq i \leq r-1$ let L_i be the subfield of L which is fixed by every element of $(\mathcal{E}_{L/K})_{b_{i+1}}$. Then $([L_i : L_{i-1}] - 1)b_i \in \mathbb{Z}$. In particular, $v_p(b_i) \geq 0$.*

PROOF. By Proposition 13.18, b_i is the unique lower ramification break of L_i/L_{i-1}. Hence by Proposition 13.23 we have

$$\begin{aligned} v_{L_i}(\mathfrak{D}_{L_i/L_{i-1}}) &= (|(\mathcal{E}_{L_i/L_{i-1}})_{b_i}| - |(\mathcal{E}_{L_i/L_{i-1}})_{b_i+\epsilon}|)(b_i+1) \\ &= ([L_i : L_{i-1}] - 1)(b_i+1). \end{aligned}$$

Since $v_{L_i}(\mathfrak{D}_{L_i/L_{i-1}}) \in \mathbb{Z}$ we get $([L_i : L_{i-1}] - 1)b_i \in \mathbb{Z}$. Since $[L_i : L_{i-1}] > 1$ and $[L_i : L_{i-1}]$ is a power of p this implies $v_p(b_i) \geq 0$. □

PROPOSITION 13.26. *Let L/K be a finite separable totally ramified extension of local fields of degree p^ν and let $b_1 < \cdots < b_r$ be the distinct lower ramification breaks of L/K. Set $L_r = L$, and for $0 \leq i \leq r-1$ let L_i be the subfield of L which is fixed by every element of $(\mathcal{E}_{L/K})_{b_{i+1}}$. Then for $1 \leq i \leq r$ we have*

(a) $b_i \leq pv_{L_{i-1}}(p)/(p-1)$.

(b) *If $v_p(b_i) \geq 1$ then $[L_i : L_{i-1}] = p$ and $b_i = pv_{L_{i-1}}(p)/(p-1)$.*

PROOF. By Proposition 13.18, b_i is the unique lower ramification break of L_i/L_{i-1}. Set $p^m = [L_i : L_{i-1}]$. Then

$$\varphi_{L_i/L_{i-1}}(x) = \begin{cases} x & \text{if } 0 \leq x \leq b_i, \\ b_i + p^{-m}(x - b_i) & \text{if } x \geq b_i. \end{cases} \tag{13.21}$$

It follows from Proposition 13.21 and Lemma 13.20 that

$$\begin{aligned} \varphi_{L_i/L_{i-1}}(b_i) &\leq \frac{1}{p^m}\left(p^m - p^m + v_{L_i}\left(\left(\binom{p^m}{p^{m-1}}\right)\right) + p^{m-1}b_i\right) \\ b_i &\leq p^{-m}(p^m v_{L_{i-1}}(p) + p^{m-1}b_i). \end{aligned}$$

Hence $b_i \leq pv_{L_{i-1}}(p)/(p-1)$. This proves (a).

Suppose $v_p(b_i) \geq 1$. By applying Proposition 13.21 to L_i/L_{i-1} we see that there is $h \geq p^m$ such that for all sufficiently large x we have

$$\begin{aligned} \varphi_{L_i/L_{i-1}}(x) &= \frac{1}{p^m}\left(h - p^m + v_{L_i}\left(\left(\binom{h}{p^0}\right)\right) + p^0 x\right) \\ &= p^{-m}(h - p^m + v_{L_i}(h) + x). \end{aligned} \tag{13.22}$$

More precisely, by (13.21) we see that (13.22) holds for all $x \geq b_i$. Since $\varphi_{L_i/L_{i-1}}(b_i) = b_i$ this implies

$$b_i = p^{-m}(h - p^m + v_{L_i}(h) + b_i)$$

$$(p^m - 1)b_i = h - p^m + v_{L_i}(h). \tag{13.23}$$

Since $v_p(b_i) \geq 1$ and $p^m \mid v_{L_i}(h)$ it follows from (13.23) that $v_p(h) \geq 1$. Therefore by Proposition 13.21 and Lemma 13.20 we get

$$\begin{aligned} \varphi_{L_i/L_{i-1}}(b_i) &\leq \frac{1}{p^m}\left(h - p^m + v_{L_i}\left(\left(\binom{h}{p^1}\right)\right) + p^1 b_i\right) \\ b_i &\leq p^{-m}(h - p^m + v_{L_i}(h) - v_{L_i}(p) + pb_i) \end{aligned}$$

$$(p^m - p)b_i \leq h - p^m + v_{L_i}(h) - v_{L_i}(p). \tag{13.24}$$

By subtracting (13.24) from (13.23) we get $(p-1)b_i \geq v_{L_i}(p)$. It follows that $b_i \geq p^m v_{L_{i-1}}(p)/(p-1)$. Using (a) we deduce that $m = 1$ and $b_i = pv_{L_{i-1}}(p)/(p-1)$. □

REMARK 13.27. In the case $\mathrm{char}(K) = p$, part (a) of Proposition 13.26 is trivial, while part (b) implies that $v_p(b_i) = 0$ for $1 \leq i \leq r$.

Let L/K be an extension satisfying the hypotheses of the preceding proposition. We say that the *multiplicity* of the lower ramification break b_i is equal to m_i, where $[L_i : L_{i-1}] = p^{m_i}$. The sequence of lower ramification breaks $\ell_1 \leq \cdots \leq \ell_\nu$, counted with multiplicity, consists of m_i copies of b_i for $1 \leq i \leq r$. Since L_i is the fixed field of $(\mathcal{E}_{L/K})_{b_{i+1}}$ we have $[L_i : K] = p^{m_1+\cdots+m_i}$.

COROLLARY 13.28. *Let L/K be a finite separable totally ramified extension of local fields of degree p^ν and let $\ell_1 \leq \cdots \leq \ell_\nu$ be the lower ramification breaks of L/K, counted with multiplicity. Then for $1 \leq j \leq \nu$ we have $\ell_j \leq p^j v_K(p)/(p-1)$.*

PROOF. In the notation of Proposition 13.26 we have $\ell_j = b_i$ for some $1 \leq i \leq r$. Hence the subfield of L fixed by the elements of $(\mathcal{E}_{L/K})_{\ell_j} = (\mathcal{E}_{L/K})_{b_i}$ is L_{i-1}. By Proposition 13.26(a) we have $\ell_j \leq p v_{L_{i-1}}(p)/(p-1)$. Since $j > m_1 + \cdots + m_{i-1}$ we get

$$[L_{i-1} : K] = p^{m_1+\cdots+m_{i-1}} \leq p^{j-1}.$$

Therefore $\ell_j \leq p^j v_K(p)/(p-1)$. □

Let M/K be the extension considered in Example 13.19. We saw that the lower ramification breaks of M/K are $\ell_1 = p v_K(p)/(p-1)$ and $\ell_2 = p^2 v_K(p)/(p-1)$. Thus the upper bounds given in Corollary 13.28 are achieved by M/K.

CHAPTER 14

Stable and semistable Hopf-Galois extensions

In the series of papers [**Bon00**, **Bon02**, **Bon06**], Bondarko introduced a method for studying the Galois module theory of extensions of local fields. Let L/K be a totally ramified G-Galois extension of local fields, with $[L:K]=p^n$. In [**Bon02**], Bondarko identified special classes of extensions which he called "semistable" and "stable". He showed that if L/K is a semistable extension then a certain fractional ideal $\mathfrak{p}_L^{-i_0}$ of L is free over its associated order $\mathfrak{A}(\mathfrak{p}_L^{-i_0})$ in $K[G]$. In [**Bon06**] Bondarko proved a partial converse to this result: He showed that if $\mathfrak{p}_L^h$ is free over $\mathfrak{A}(\mathfrak{p}_L^h)$ for some $h\in\mathbb{Z}$ and the ramification data of L/K satisfies certain conditions then L/K is semistable. Suppose L/K satisfies the stronger condition of being stable. In [**Bon02**] Bondarko showed how to compute the associated order $\mathfrak{A}(\mathfrak{p}_L^h)$ of $\mathfrak{p}_L^h$ for any $h\in\mathbb{Z}$. This leads to a numerical criterion for determining when $\mathfrak{p}_L^h$ is free over $\mathfrak{A}(\mathfrak{p}_L^h)$.

The conditions that L/K must satisfy to be semistable or stable depend on an isomorphism of K-vector spaces $\varphi: L\otimes_K L\to L[G]$ which is defined in terms of the trace map $\mathrm{Tr}_{L/K}: L\to K$. For L/K to be semistable there must exist $\beta\in L\otimes_K L$ such that β satisfies certain valuation conditions and $\varphi(\beta)\in K[G]$. The extension L/K is stable if β can be chosen to satisfy stronger valuation conditions. (See Definition 14.26 for precise statements.) Since there seems to be no simple description of $\varphi^{-1}(K[G])$, it can be difficult to determine whether a given extension L/K is semistable or stable. In the case of abelian p-extensions, however, Bondarko showed in [**Bon02**] that L/K is semistable if and only if there exists a one-dimensional formal group law $F(X,Y)$ over $\mathfrak{O}_K$ with certain properties.

In this chapter we present a part of Bondarko's theory in the more general setting of Hopf-Galois extensions. Let K be a local field with residue characteristic p and let L/K be a separable totally ramified extension of degree p^n. Assume that L/K is H-Galois for some Hopf algebra H over K. We first define what it means for L/K to be H-stable or H-semistable. We then show that H-stable and H-semistable extensions have properties analogous to those proved by Bondarko for stable and semistable Galois extensions in [**Bon00**, **Bon02**]. We do not include Bondarko's method for constructing abelian semistable extensions using formal group laws. Indeed, it is not clear that the formal group law approach can be extended to the Hopf-Galois setting.

14.1. Bondarko's map φ for Hopf-Galois extensions

Let K be a local field whose residue field $\overline{K}$ is perfect with characteristic p. Let L/K be a totally ramified G-Galois extension of degree p^n. In [**Bon00**] and [**Bon02**] Bondarko defines an isomorphism $\varphi: L\otimes_K L\to L[G]$ of vector spaces over K. For $\beta\in L\otimes_K L$ he derives a relation between the expression of β as a sum of simple tensors and the effect that $\varphi(\beta)$ has on the valuations of elements of L. This leads to methods for determining Galois module structure of ideals in

L. In this section we generalize the definition of φ to the setting of Hopf-Galois extensions, and prove some basic properties of this map.

Let H be a cocommutative Hopf algebra over K of dimension p^n with comultiplication $\Delta : H \to H \otimes_K H$ and counit $\varepsilon : H \to K$. Let L/K be a finite separable totally ramified H-Galois extension. Then L is free of rank 1 over H, so L/K has degree p^n. Let $\mathfrak{O}_K, \mathfrak{O}_L$ be the rings of integers of K, L, and let $\mathfrak{p}_K, \mathfrak{p}_L$ be the maximal ideals of these rings. For a fractional ideal $\mathfrak{p}_L^h$ of L define the associated order

$$\mathfrak{A}(\mathfrak{p}_L^h) = \{\gamma \in H \mid \gamma \cdot \mathfrak{p}_L^h \subset \mathfrak{p}_L^h\}$$

of $\mathfrak{p}_L^h$ in H. If $\mathfrak{p}_L^h$ is free over some order $\mathfrak{B}$ in H then $\mathfrak{B} = \mathfrak{A}(\mathfrak{p}_L^h)$, and $\mathfrak{p}_L^h$ is free of rank 1 over $\mathfrak{A}(\mathfrak{p}_L^h)$.

Let $\widetilde{L}/K$ be the Galois closure of L/K and set $G = \mathrm{Gal}(\widetilde{L}/K)$, $G_L = \mathrm{Gal}(\widetilde{L}/L)$, and $X = G/G_L$. Let $\mathrm{Map}(X, \widetilde{L})$ denote the $\widetilde{L}$-algebra of functions from X to $\widetilde{L}$. Extending the base field from K to $\widetilde{L}$ gives an $\widetilde{L}$-algebra isomorphism $\widetilde{L} \otimes_K L \cong \mathrm{Map}(X, \widetilde{L})$ which carries $a \otimes b \in \widetilde{L} \otimes_K L$ onto the function $f \in \mathrm{Map}(X, \widetilde{L})$ defined by $f(\overline{g}) = ag(b)$ for $\overline{g} = gG_L \in X$. We identify L with the K-subalgebra $K \otimes_K L$ of $\widetilde{L} \otimes_K L$. Thus $x \in L$ gets mapped onto $f_x \in \mathrm{Map}(X, \widetilde{L})$ defined by $f_x(\overline{g}) = g(x)$ for $\overline{g} \in X$, as in Subsection 11.1.1.

Let $\mathrm{Perm}(X)$ denote the group of permutations of X. For $\sigma \in G$ let $\lambda(\sigma) \in \mathrm{Perm}(X)$ denote the permutation of $X = G/G_L$ given by left multiplication. Since L/K is an H-Galois extension it follows from the Greither-Pareigis classification that there is a regular subgroup $N \leq \mathrm{Perm}(X)$ which is normalized by $\lambda(G)$ and an isomorphism of $\widetilde{L}$-Hopf algebras $\widetilde{L} \otimes_K H \cong \widetilde{L}[N]$. This induces an $\widetilde{L}$-Hopf algebra isomorphism $\widetilde{L} \otimes_K H^* \cong \mathrm{Map}(N, \widetilde{L})$, where H^* denotes the Hopf algebra dual of H. We identify H with $K \otimes_K H \subset \widetilde{L} \otimes_K H$ and H^* with $K \otimes_K H^* \subset \widetilde{L} \otimes_K H^*$. We get $\widetilde{L}$-algebra isomorphisms

$$\widetilde{L} \otimes_K L \otimes_K L \cong (\widetilde{L} \otimes_K L) \otimes_{\widetilde{L}} (\widetilde{L} \otimes_K L) \cong \mathrm{Map}(X, \widetilde{L}) \otimes_{\widetilde{L}} \mathrm{Map}(X, \widetilde{L}) \tag{14.1}$$

$$\widetilde{L} \otimes_K L \otimes_K H \cong (\widetilde{L} \otimes_K L) \otimes_{\widetilde{L}} (\widetilde{L} \otimes_K H) \cong \mathrm{Map}(X, \widetilde{L}) \otimes_{\widetilde{L}} \widetilde{L}[N] \tag{14.2}$$

$$\widetilde{L} \otimes_K H \otimes_K H \cong (\widetilde{L} \otimes_K H) \otimes_{\widetilde{L}} (\widetilde{L} \otimes_K H) \cong \widetilde{L}[N] \otimes_{\widetilde{L}} \widetilde{L}[N] \tag{14.3}$$

$$\widetilde{L} \otimes_K L \otimes_K H^* \cong (\widetilde{L} \otimes_K L) \otimes_{\widetilde{L}} (\widetilde{L} \otimes_K H^*) \cong \mathrm{Map}(X, \widetilde{L}) \otimes_{\widetilde{L}} \mathrm{Map}(N, \widetilde{L}). \tag{14.4}$$

We recall some basic facts about $\mathrm{Map}(X, \widetilde{L})$ (cf. Subsection 11.1.1). For $\overline{g} \in X$ define $u_{\overline{g}} : X \to \widetilde{L}$ by $u_{\overline{g}}(\overline{g}) = 1$ and $u_{\overline{g}}(\overline{h}) = 0$ for $\overline{h} \in X \smallsetminus \{\overline{g}\}$. Then $\{u_{\overline{g}} \mid \overline{g} \in X\}$ is an $\widetilde{L}$-basis for $\mathrm{Map}(X, \widetilde{L})$ consisting of primitive idempotents. In particular, for $x \in L$ we have $f_x = \sum_{\overline{g} \in X} g(x) u_{\overline{g}}$. For $\sigma \in \mathrm{Perm}(X)$ and $\overline{g}, \overline{h} \in X$ we have

$$(\sigma u_{\overline{g}})(\overline{h}) = u_{\overline{g}}(\sigma^{-1}[\overline{h}]) = \begin{cases} 1 & \text{if } \overline{h} = \sigma[\overline{g}], \\ 0 & \text{if } \overline{h} \neq \sigma[\overline{g}]. \end{cases}$$

Therefore $\sigma u_{\overline{g}} = u_{\sigma[\overline{g}]}$. This applies in particular to the cases where $\sigma \in N$ or $\sigma \in \lambda(G)$. For $\eta \in N$ define $u_\eta \in \mathrm{Map}(N, \widetilde{L})$ analogously to $u_{\overline{g}} \in \mathrm{Map}(X, \widetilde{L})$. By reasoning similar to the above we get $\eta \cdot u_\zeta = u_{\eta\zeta}$ for $\eta, \zeta \in N$.

Bondarko's work in [**Bon00, Bon02**] depends heavily on the trace element of $K[G]$. The following result, taken from [**Byo11**, Proposition 2.1], shows that H also has a trace element. Set

$$\theta = \sum_{\eta \in N} \eta \in \widetilde{L}[N] \cong \widetilde{L} \otimes_K H.$$

PROPOSITION 14.1. *Define θ as above. Then*

(a) $\theta \in K \otimes_K H \cong H$.
(b) *For every $h \in H$ we have $h\theta = \theta h = \varepsilon(h)\theta$.*
(c) *For every $x \in L$ we have $\theta(x) = \mathrm{Tr}_{L/K}(x)$.*

PROOF. Since $\lambda(G) \cong G$ normalizes N we get

$$\theta \in (\widetilde{L}[N])^G \cong (\widetilde{L} \otimes_K H)^G \cong \widetilde{L}^G \otimes_K H = K \otimes_K H,$$

which proves (a). For $\eta \in N$ we have $\eta\theta = \theta\eta = \theta = \varepsilon(\eta)\theta$, which implies that (b) holds for all $h \in \widetilde{L}[N]$. To prove (c), take $x \in L$ and let $f_x : X \to \widetilde{L}$ be the corresponding element of $\mathrm{Map}(X, \widetilde{L})$. Then

$$\theta \cdot f_x = \sum_{\eta \in N} \eta \cdot f_x = \sum_{\eta \in N} \sum_{\overline{g} \in X} g(x)\eta \cdot u_{\overline{g}} = \sum_{\overline{g} \in X} g(x) \sum_{\eta \in N} u_{\eta(\overline{g})}.$$

Since N acts simply transitively on set $X = G/G_L$ of K-embeddings of L into $\widetilde{L}$ we get

$$\theta \cdot f_x = \sum_{\overline{g} \in X} g(x) \sum_{\overline{h} \in X} u_{\overline{h}} = \sum_{\overline{h} \in X} \mathrm{Tr}_{L/K}(x) u_{\overline{h}} = f_{\mathrm{Tr}_{L/K}(x)}.$$

Since $x, \mathrm{Tr}_{L/K}(x)$ correspond to $f_x, f_{\mathrm{Tr}_{L/K}(x)}$ under the inclusion $L \hookrightarrow \mathrm{Map}(X, \widetilde{L})$, we get $\theta(x) = \mathrm{Tr}_{L/K}(x)$. □

REMARK 14.2. It follows from part (b) of the proposition that θ is both a left integral and a right integral for H (see Definition 3.1 in [**Chi00**]).

Write $\Delta(\theta) \in H \otimes_K H$ using Sweedler notation as $\Delta(\theta) = \sum_{(\theta)} \theta_{(1)} \otimes \theta_{(2)}$ and define

$$\varphi : L \otimes_K L \longrightarrow L \otimes_K H$$
$$\varphi(a \otimes b) = \sum_{(\theta)} a\theta_{(1)}(b) \otimes \theta_{(2)}.$$

Then for $y \in L$ we get

$$\varphi(a \otimes b)(y) = \sum_{(\theta)} a\theta_{(1)}(b)\theta_{(2)}(y) = a\theta(by). \tag{14.5}$$

Hence $\varphi(1 \otimes 1)(y) = \theta(y)$ for all $y \in L$, so we have $\varphi(1 \otimes 1) = 1 \otimes \theta$. Let

$$\Delta_{\widetilde{L}} : \widetilde{L}[N] \longrightarrow \widetilde{L}[N] \otimes_{\widetilde{L}} \widetilde{L}[N]$$
$$\varphi_{\widetilde{L}} : \mathrm{Map}(X, \widetilde{L}) \otimes_{\widetilde{L}} \mathrm{Map}(X, \widetilde{L}) \longrightarrow \mathrm{Map}(X, \widetilde{L}) \otimes_{\widetilde{L}} \widetilde{L}[N]$$

be the maps induced by $\mathrm{id}_{\widetilde{L}} \otimes \Delta$ and $\mathrm{id}_{\widetilde{L}} \otimes \varphi$ using the isomorphisms (14.3), (14.1), and (14.2). Then for $\eta \in N$ we have $\Delta_{\widetilde{L}}(\eta) = \eta \otimes \eta$. Hence for $a, b \in \mathrm{Map}(X, \widetilde{L})$ we get $\varphi_{\widetilde{L}}(a \otimes b) = \sum_{\eta \in N} a\eta(b) \otimes \eta$. In the classical case, with $\widetilde{L} = L$, $N = G = \mathrm{Gal}(L/K)$, and $H = K[G]$, we get $\mathrm{Map}(X, L) = L$ and

$$\varphi(a \otimes b) = \sum_{\sigma \in G} a\sigma(b)\sigma \in L[G]. \tag{14.6}$$

PROPOSITION 14.3. *Let L/K be a separable H-Galois extension and let $\alpha, \beta \in L \otimes_K L$ satisfy $\varphi(\alpha) \in H$ and $\varphi(\beta) \in H$. Then $\varphi(\alpha\beta) \in H$.*

PROOF. Recall that $\varepsilon^* : H^* \to K$ and $\Delta^* : H^* \to H^* \otimes_K H^*$ are the counit and comultiplication of H^*. We extend these maps by $\widetilde{L}$-linearity to give the counit and comultiplication of the Hopf algebra $\widetilde{L} \otimes_K H^* \cong \mathrm{Map}(N, \widetilde{L})$. Since $\varepsilon^*(u_1) = 1$ and $\varepsilon^*(u_\eta) = 0$ for $\eta \in N \smallsetminus \{1\}$, we have $u_1 f = f u_1 = \varepsilon^*(f) u_1$ for all $f \in H^*$. Hence $\theta^* = u_1$ is the unique integral in $\mathrm{Map}(N, \widetilde{L}) \cong \widetilde{L} \otimes_K H^*$ such that $\theta^*(\theta) = 1$. Since conjugation by $\lambda(G) \cong G$ fixes 1_N we get $\theta^* \in (\widetilde{L} \otimes_K H^*)^G \cong H^*$. We make H^* a left H-module as in [**Chi00**, p. 26]: For $h \in H$ and $f \in H^*$ write $\Delta^*(f) = \sum f_{(1)} \otimes f_{(2)}$ and set $h \cdot f = \sum f_{(2)}(h) f_{(1)}$. Define $\psi : H \to H^*$ by $\psi(h) = h \cdot \theta^*$. Then ψ is K-linear, and for $x \in H$ we have $\psi(h)(x) = \theta^*(xh)$. Furthermore, ψ induces a $\widetilde{L}$-linear map $\psi_{\widetilde{L}} : \widetilde{L}[N] \to \mathrm{Map}(N, \widetilde{L})$ such that $\psi_{\widetilde{L}}(\eta) = \eta \cdot u_1 = u_\eta$ for $\eta \in N$. It follows that $\psi_{\widetilde{L}}$ is an isomorphism of $\widetilde{L}$-vector spaces, and hence that ψ is an isomorphism of K-vector spaces.

Let $\overline{g}, \overline{h} \in X$, and let $u_{\overline{g}}, u_{\overline{h}} \in \mathrm{Map}(X, \widetilde{L})$ be the corresponding idempotents. Then

$$\begin{aligned} \varphi_{\widetilde{L}}(u_{\overline{g}} \otimes u_{\overline{h}}) &= \sum_{\eta \in N} u_{\overline{g}}(\eta \cdot u_{\overline{h}}) \otimes \eta \\ &= \sum_{\eta \in N} u_{\overline{g}} u_{\eta[\overline{h}]} \otimes \eta \\ &= u_{\overline{g}} \otimes \zeta^{-1}, \end{aligned} \tag{14.7}$$

where ζ is the unique element of N with $\zeta[\overline{g}] = \overline{h}$. It follows that

$$\begin{aligned} (\mathrm{id}_{\mathrm{Map}(X,\widetilde{L})} \otimes \psi_{\widetilde{L}})(\varphi_{\widetilde{L}}(u_{\overline{g}} \otimes u_{\overline{h}})) &= u_{\overline{g}} \otimes \psi_{\widetilde{L}}(\zeta^{-1}) \\ &= u_{\overline{g}} \otimes u_{\zeta^{-1}}. \end{aligned}$$

Since $\{u_{\overline{g}} \mid \overline{g} \in X\}$ and $\{u_\eta \mid \eta \in N\}$ are $\widetilde{L}$-bases for $\mathrm{Map}(X, \widetilde{L})$ and $\mathrm{Map}(N, \widetilde{L})$ consisting of orthogonal idempotents, it follows by $\widetilde{L}$-linearity that $(\mathrm{id}_{\mathrm{Map}(X,\widetilde{L})} \otimes \psi_{\widetilde{L}}) \circ \varphi_{\widetilde{L}}$ is an isomorphism of $\widetilde{L}$-algebras. Hence $(\mathrm{id}_L \otimes \psi) \circ \varphi$ is a K-algebra isomorphism.

Suppose $\alpha, \beta \in L \otimes_K L$ satisfy $\varphi(\alpha), \varphi(\beta) \in K \otimes_K H$. Then $(\mathrm{id}_L \otimes \psi)(\varphi(\alpha))$ and $(\mathrm{id}_L \otimes \psi)(\varphi(\beta))$ are in $K \otimes_K H^*$. It follows that

$$(\mathrm{id}_L \otimes \psi)(\varphi(\alpha\beta)) = (\mathrm{id}_L \otimes \psi)(\varphi(\alpha)) \cdot (\mathrm{id}_L \otimes \psi)(\varphi(\beta))$$

is in $K \otimes_K H^*$. Since the isomorphism

$$\mathrm{id}_L \otimes \psi : L \otimes_K H \longrightarrow L \otimes_K H^*$$

induces an isomorphism from $K \otimes_K H$ to $K \otimes_K H^*$, we get $\varphi(\alpha\beta) \in K \otimes_K H$. □

Define a K-bilinear pairing

$$\begin{aligned} \langle\, , \rangle_{L \otimes L} &: (L \otimes_K L) \times (L \otimes_K L) \longrightarrow K \\ \langle a \otimes b, c \otimes d \rangle_{L \otimes L} &= \theta(ac) \cdot \theta(bd). \end{aligned} \tag{14.8}$$

PROPOSITION 14.4. *$\langle\, , \rangle_{L \otimes L}$ is nondegenerate.*

PROOF. Let $\{x_1, x_2, \ldots, x_{p^n}\}$ be a basis for L over K, and let A be the matrix of the trace pairing $L \times L \to K$ with respect to this basis. Then the matrix B of $\langle\, , \rangle_{L \otimes L}$ with respect to the K-basis $\{x_i \otimes x_j \mid 1 \le i, j \le p^n\}$ for $L \otimes_K L$ is a Kronecker product of A with itself. Since A is invertible, so is B. □

Define a K-bilinear pairing

$$\langle\ ,\ \rangle_{L\otimes H} : (L \otimes_K H) \times (L \otimes_K H) \longrightarrow K$$

$$\langle a \otimes h, b \otimes k\rangle_{L\otimes H} = \theta(ab) \cdot \theta^*(hS(k)), \tag{14.9}$$

where $S : H \to H$ denotes the antipode of H.

PROPOSITION 14.5. *For $\alpha, \beta \in L \otimes_K L$ we have $\langle\varphi(\alpha), \varphi(\beta)\rangle_{L\otimes H} = \langle\alpha, \beta\rangle_{L\otimes L}$.*

PROOF. As in the proof of Proposition 14.3 we extend the base field from K to $\widetilde{L}$ and replace L, H with $\widetilde{L} \otimes_K L, \widetilde{L} \otimes_K H$. Let $\overline{g}, \overline{h}, \overline{g}', \overline{h}' \in X$ and $\eta, \eta' \in N$ satisfy $\eta[\overline{g}] = \overline{h}$ and $\eta'[\overline{g}'] = \overline{h}'$. By (14.8) we get

$$\langle u_{\overline{g}} \otimes u_{\overline{h}}, u_{\overline{g}'} \otimes u_{\overline{h}'}\rangle_{(\widetilde{L}\otimes L)\otimes(\widetilde{L}\otimes L)} = \theta(u_{\overline{g}}u_{\overline{g}'})\theta(u_{\overline{h}}u_{\overline{h}'}). \tag{14.10}$$

By (14.7) and (14.9) we get

$$\begin{aligned}\langle\varphi_{\widetilde{L}}(u_{\overline{g}} \otimes u_{\overline{h}}), \varphi_{\widetilde{L}}(u_{\overline{g}'} \otimes u_{\overline{h}'})\rangle_{(\widetilde{L}\otimes L)\otimes(\widetilde{L}\otimes H)} &= \langle u_{\overline{g}} \otimes \eta^{-1}, u_{\overline{g}'} \otimes (\eta')^{-1}\rangle_{(\widetilde{L}\otimes L)\otimes(\widetilde{L}\otimes H)} \\ &= \theta(u_{\overline{g}}u_{\overline{g}'})u_1(\eta^{-1}\eta'). \end{aligned} \tag{14.11}$$

Each of the expressions (14.10), (14.11) is equal to 1 if $(\overline{g}, \overline{h}) = (\overline{g}', \overline{h}')$, and 0 otherwise. Therefore by $\widetilde{L}$-linearity the formula holds over $\widetilde{L}$, and hence also over K. □

Using the nondegeneracy of $\langle\ ,\ \rangle_{L\otimes L}$ we deduce the following:

COROLLARY 14.6. *The map φ is an isomorphism of K-vector spaces.*

COROLLARY 14.7. *The pairing $\langle\ ,\ \rangle_{L\otimes H}$ is nondegenerate.*

Suppose M and N are $\mathfrak{O}_K$-lattices in L. Let M^* denote the dual of M with respect to the trace pairing, and let $(M \otimes_{\mathfrak{O}_K} N)^*$ denote the dual of $M \otimes_{\mathfrak{O}_K} N$ with respect to $\langle\ ,\ \rangle_{L\otimes L}$.

LEMMA 14.8. *Let M, N be $\mathfrak{O}_K$-lattices in L. Then $(M \otimes_{\mathfrak{O}_K} N)^* = M^* \otimes_{\mathfrak{O}_K} N^*$.*

PROOF. Let $\{x_1, \dots, x_{p^n}\}, \{y_1, \dots, y_{p^n}\}$ be $\mathfrak{O}_K$-bases for M, N, and let $\{x_1^*, \dots, x_{p^n}^*\}$, $\{y_1^*, \dots, y_{p^n}^*\}$ be the dual bases for M^*, N^* with respect to the trace pairing. Then

$$B = \{x_i \otimes y_j \mid 1 \le i, j \le p^n\}$$

is an $\mathfrak{O}_K$-basis for $M \otimes_{\mathfrak{O}_K} N$. Since θ induces $\mathrm{Tr}_{L/K}$ on L we get

$$\begin{aligned}\langle x_i \otimes y_j, x_k^* \otimes y_l^*\rangle_{L\otimes L} &= \theta(x_i x_k^*)\theta(y_j y_l^*) \\ &= \begin{cases} 1 & \text{if } (i,j) = (k,l) \\ 0 & \text{if } (i,j) \ne (k,l). \end{cases}\end{aligned}$$

for $1 \le i, j, k, l \le p^n$. Hence

$$B^* = \{x_i^* \otimes y_j^* \mid 1 \le i, j \le p^n\}$$

is the dual basis to B with respect to $\langle\ ,\ \rangle_{L\otimes L}$. Therefore

$$(M \otimes_{\mathfrak{O}_K} N)^* = \mathrm{Span}_{\mathfrak{O}_K}(B^*) = M^* \otimes_{\mathfrak{O}_K} N^*.$$

□

Let $\mathrm{End}_K(L)$ denote the vector space of K-linear maps from L to L. As in Section 7.1 we define $j : L \otimes_K H \to \mathrm{End}_K(L)$ by $j(y \otimes h)(x) = yh(x)$ for $x, y \in L$ and $h \in H$. Since L/K is an H-Galois extension, j is an isomorphism of vector spaces over K. If we replace $L \otimes_K H$ with the smash product $L\#H$ (see Definition 7.1) then j becomes an isomorphism of K-algebras. For fractional ideals $\mathfrak{J}_1, \mathfrak{J}_2$ of $\mathfrak{O}_L$ define

$$\mathfrak{C}(\mathfrak{J}_1, \mathfrak{J}_2) = \mathrm{Hom}_{\mathfrak{O}_K}(\mathfrak{J}_1, \mathfrak{J}_2) \subset \mathrm{End}_K(L).$$

Using the isomorphism $\mathrm{End}_K(L) \cong L \otimes_K H$ we may also define

$$\mathfrak{A}(\mathfrak{J}_1, \mathfrak{J}_2) = \mathfrak{C}(\mathfrak{J}_1, \mathfrak{J}_2) \cap (K \otimes_K H).$$

Then $\mathfrak{A}(\mathfrak{J}_1) := \mathfrak{A}(\mathfrak{J}_1, \mathfrak{J}_1)$ is the associated order of $\mathfrak{J}_1$ in H. Let $\mathfrak{D}_{L/K}$ denote the different of the extension L/K.

PROPOSITION 14.9. *Let $\mathfrak{J}_1, \mathfrak{J}_2$ be nonzero fractional ideals of $\mathfrak{O}_L$. Then*

$$\varphi(\mathfrak{J}_2 \otimes \mathfrak{D}_{L/K}^{-1}\mathfrak{J}_1^{-1}) = \mathfrak{C}(\mathfrak{J}_1, \mathfrak{J}_2).$$

PROOF. First, if $a \in \mathfrak{J}_2$, $b \in \mathfrak{D}_{L/K}^{-1}\mathfrak{J}_1^{-1}$, and $x \in \mathfrak{J}_1$ then $\varphi(a \otimes b)(x) = a\,\mathrm{Tr}_{L/K}(bx)$. Since $bx \in \mathfrak{D}_{L/K}^{-1}$ we get $\mathrm{Tr}_{L/K}(bx) \in \mathfrak{O}_K$, and hence $\varphi(a \otimes b)(x) \in \mathfrak{J}_2$. Thus $\varphi(\mathfrak{J}_2 \otimes \mathfrak{D}_{L/K}^{-1}\mathfrak{J}_1^{-1}) \subset \mathfrak{C}(\mathfrak{J}_1, \mathfrak{J}_2)$. Now suppose $f \in \mathfrak{C}(\mathfrak{J}_1, \mathfrak{J}_2)$. Then f extends uniquely to a K-linear map from L to L, which we also denote by f. Define $\omega_f : L \otimes_K L \to K$ by $\omega_f(a \otimes b) = \mathrm{Tr}_{L/K}(af(b))$. Then ω_f is K-linear. By the nondegeneracy of $\langle\ ,\ \rangle_{L\otimes L}$ there is $\alpha \in L \otimes_K L$ such that $\omega_f(\beta) = \langle \alpha, \beta \rangle_{L \otimes L}$ for all $\beta \in L \otimes_K L$.

Let $a \in \mathfrak{D}_{L/K}^{-1}\mathfrak{J}_2^{-1}$ and $b \in \mathfrak{J}_1$. Then

$$\langle \alpha, a \otimes b \rangle = \omega_f(a \otimes b) = \mathrm{Tr}_{L/K}(af(b)).$$

Since $f(b) \in \mathfrak{J}_2$ we get $af(b) \in \mathfrak{D}_{L/K}^{-1}$, and hence $\langle \alpha, a \otimes b \rangle \in \mathfrak{O}_K$. It follows that α is an element of

$$(\mathfrak{D}_{L/K}^{-1}\mathfrak{J}_2^{-1} \otimes_{\mathfrak{O}_K} \mathfrak{J}_1)^* = (\mathfrak{D}_{L/K}^{-1}\mathfrak{J}_2^{-1})^* \otimes_{\mathfrak{O}_K} \mathfrak{J}_1^* = \mathfrak{J}_2 \otimes \mathfrak{D}_{L/K}^{-1}\mathfrak{J}_1^{-1}.$$

We claim that for $\beta \in L \otimes_K L$ we have $\omega_{\varphi(\alpha)}(\beta) = \langle \alpha, \beta \rangle_{L \otimes L}$. Assume for now that this claim holds. Then $\omega_{\varphi(\alpha)} = \omega_f$, so for $a, b \in L$ we have $\mathrm{Tr}_{L/K}(a\varphi(\alpha)(b)) = \mathrm{Tr}_{L/K}(af(b))$. By the nondegeneracy of the trace we get $\varphi(\alpha)(b) = f(b)$ for all $b \in L$. Thus $\varphi(\alpha) = f$, which completes the proof.

To prove the claim we extend the base field from K to $\widetilde{L}$. The trace map $\mathrm{Tr}_{L/K} : L \to K$ extends to the $\widetilde{L}$-linear map $\mathrm{Tr}_{\mathrm{Map}(X,\widetilde{L})/\widetilde{L}} : \mathrm{Map}(X, \widetilde{L}) \to \widetilde{L}$ defined by $\mathrm{Tr}_{\mathrm{Map}(X,\widetilde{L})/\widetilde{L}}(u_{\overline{h}}) = 1$ for $\overline{h} \in X$. Using linearity we may assume that $\alpha = u_{\overline{g}} \otimes u_{\overline{h}}$ and $\beta = u_{\overline{g}'} \otimes u_{\overline{h}'}$ with $\overline{g}, \overline{h}, \overline{g}', \overline{h}' \in X$. Let $\eta, \eta' \in N$ satisfy $\eta[\overline{g}] = \overline{h}$ and $\eta'[\overline{g}'] = \overline{h}'$. By (14.7) we have $\varphi_{\widetilde{L}}(\alpha) = u_{\overline{g}} \otimes \eta^{-1}$, and hence

$$\begin{aligned}
\omega_{\varphi_{\widetilde{L}}(\alpha)}(\beta) &= \mathrm{Tr}_{\mathrm{Map}(X,\widetilde{L})/\widetilde{L}}(u_{\overline{g}'}(u_{\overline{g}} \otimes \eta^{-1})(u_{\overline{h}'})) \\
&= \mathrm{Tr}_{\mathrm{Map}(X,\widetilde{L})/\widetilde{L}}(u_{\overline{g}'}u_{\overline{g}}u_{\eta^{-1}[\overline{h}']}).
\end{aligned}$$

It follows that $\omega_{\varphi_{\widetilde{L}}(\alpha)}(\beta) = 1$ if $\overline{g}' = \overline{g} = \eta^{-1}[\overline{h}']$, and $\omega_{\varphi_{\widetilde{L}}(\alpha)}(\beta) = 0$ otherwise. Since $\overline{g} = \eta^{-1}[\overline{h}]$ we get $\omega_{\varphi_{\widetilde{L}}(\alpha)}(\beta) = 1$ if and only if $\overline{g} = \overline{g}'$ and $\overline{h} = \overline{h}'$. On the other hand,

$$\langle \alpha, \beta \rangle_{(\widetilde{L}\otimes L)\otimes(\widetilde{L}\otimes L)} = \mathrm{Tr}_{\mathrm{Map}(X,\widetilde{L})/\widetilde{L}}(u_{\overline{g}}u_{\overline{g}'})\,\mathrm{Tr}_{\mathrm{Map}(X,\widetilde{L})/\widetilde{L}}(u_{\overline{h}}u_{\overline{h}'})$$

is equal to 1 if $\overline{g} = \overline{g}'$ and $\overline{h} = \overline{h}'$, and equal to 0 otherwise. Therefore

$$\omega_{\varphi_{\widetilde{L}}(\alpha)}(\beta) = \langle \alpha, \beta \rangle_{(\widetilde{L}\otimes L)\otimes(\widetilde{L}\otimes L)} \tag{14.12}$$

holds with $\alpha = u_{\overline{g}} \otimes u_{\overline{h}}$ and $\beta = u_{\overline{g}'} \otimes u_{\overline{h}'}$ for all $\overline{g}, \overline{h}, \overline{g}', \overline{h}' \in X$. It follows that (14.12) holds for all $\alpha, \beta \in (\widetilde{L} \otimes_K L) \otimes_{\widetilde{L}} (\widetilde{L} \otimes_K L)$. We conclude that $\omega_{\varphi(\alpha)}(\beta) = \langle \alpha, \beta \rangle_{L\otimes L}$ for all $\alpha, \beta \in L \otimes_K L$. □

14.2. Some technical lemmas

In this section we prove some lemmas that will be needed later in this chapter. Some of these statements are easily verified for classical Galois extensions, but are not obvious for Hopf-Galois extensions. We assume throughout the section that K is a local field and L/K is a separable totally ramified H-Galois extension of degree p^n. We let $\widetilde{L}/K$ be the Galois closure of L/K, $G = \mathrm{Gal}(\widetilde{L}/K)$, $G_L = \mathrm{Gal}(\widetilde{L}/L)$, and $X = G/G_L$. We also let N denote the regular subgroup of $\mathrm{Perm}(X)$ which is associated to our H-Galois extension by the Greither-Pareigis classification. Thus N is normalized by $\lambda(G) \leq \mathrm{Perm}(X)$. The action of G on N is given by ${}^{\gamma}\eta = \lambda(\gamma)\eta\lambda(\gamma^{-1})$. This extends to an action of G on $\widetilde{L}[N]$ using the Galois action of G on $\widetilde{L}$, with ${}^{\gamma}(c\eta) = \gamma(c){}^{\gamma}\eta$ for $c \in \widetilde{L}$. Then H is the K-subalgebra of $\widetilde{L}[N]$ fixed by G under this action.

LEMMA 14.10. *Let $\eta \in N$, let G_η be the stabilizer of η in G, and let $T = \widetilde{L}^{G_\eta}$ be the subfield of $\widetilde{L}$ fixed by G_η. For $\alpha \in T$ set*

$$\chi_\eta(\alpha) = \sum_{\gamma G_\eta \in G/G_\eta} \gamma(\alpha)\, {}^{\gamma}\eta.$$

Then $\chi_\eta(\alpha) \in H \subset \widetilde{L}[N]$. Furthermore, for $y \in L$ we have

$$\chi_\eta(\alpha)(y) = \sum_{\gamma G_\eta \in G/G_\eta} \gamma(\alpha) g_\gamma(y),$$

where g_γ is any element of G such that $\overline{g}_\gamma = ({}^{\gamma}\eta^{-1})[\overline{1}_G] = \gamma\eta^{-1}[\overline{\gamma^{-1}}]$.

PROOF. By construction we have ${}^{\delta}\chi_\eta(\alpha) = \chi_\eta(\alpha)$ for all $\delta \in G$, and hence $\chi_\eta(\alpha) \in (\widetilde{L}[N])^G = H$. Let

$$i(y) = \sum_{\overline{g} \in G/G_L} g(y) u_{\overline{g}}$$

denote the image of y in $\mathrm{Map}(X, \widetilde{L}) \cong \widetilde{L} \otimes_K L$. Then $\chi_\eta(\alpha)(y) \in L$ is equal to the coefficient of $u_{\overline{1}_G}$ in

$$\begin{aligned}
\chi_\eta(\alpha)(i(y)) &= \left(\sum_{\gamma G_\eta \in G/G_\eta} \gamma(\alpha)\, {}^{\gamma}\eta \right) \left(\sum_{\overline{g} \in G/G_L} g(y) u_{\overline{g}} \right) \\
&= \sum_{\gamma G_\eta \in G/G_\eta} \sum_{\overline{g} \in G/G_L} \gamma(\alpha) g(y)\, {}^{\gamma}\eta(u_{\overline{g}}) \\
&= \sum_{\gamma G_\eta \in G/G_\eta} \sum_{\overline{g} \in G/G_L} \gamma(\alpha) g(y) \gamma\eta(u_{\gamma^{-1}\overline{g}}).
\end{aligned}$$

The coefficient of $u_{\overline{1}_G}$ in the last expression is equal to $\sum \gamma(\alpha)g(y)$, where the sum is over pairs $(\gamma G_\eta, \overline{g})$ such that $\gamma\eta[\gamma^{-1}\overline{g}] = \overline{1}_G$. Since this condition is equivalent to $\overline{g} = \gamma\eta^{-1}\overline{[\gamma^{-1}]}$, the lemma follows. □

PROPOSITION 14.11. *Let T/K be a subextension of $\widetilde{L}/K$ which is at most tamely ramified. Let $\mathcal{E}_T$ denote the set of K-embeddings of T into $\widetilde{L}$, and for each $\sigma \in \mathcal{E}_T$ choose $c_\sigma \in \mathfrak{O}_{\widetilde{L}}$. Assume that for every $\alpha \in T^\times$ we have*

$$\sum_{\sigma\in\mathcal{E}_T} c_\sigma \frac{\sigma(\alpha)}{\alpha} \equiv 0 \pmod{\mathfrak{p}_{\widetilde{L}}}. \tag{14.13}$$

Then $c_\sigma \in \mathfrak{p}_{\widetilde{L}}$ for all $\sigma \in \mathcal{E}_T$.

PROOF. Let U/K be the maximum unramified subextension of T/K. Then (14.13) can be rewritten

$$\sum_{\tau\in\mathcal{E}_U} \sum_{\sigma|_U=\tau} c_\sigma \frac{\sigma(\alpha)}{\alpha} \equiv 0 \pmod{\mathfrak{p}_{\widetilde{L}}}. \tag{14.14}$$

Let $t \in \mathcal{R}_U$ be a nonzero Teichmüller representative for U. By replacing α with $t\alpha$ and multiplying by t we get

$$\sum_{\tau\in\mathcal{E}_U} d_\tau(\alpha)\tau(t) \equiv 0 \pmod{\mathfrak{p}_{\widetilde{L}}},$$

where $d_\tau(\alpha) = \sum_{\sigma|_U=\tau} c_\sigma\sigma(\alpha)/\alpha$. By the linear independence of characters (with values in $\mathfrak{O}_{\widetilde{L}}/\mathfrak{p}_{\widetilde{L}}$) we get $d_\tau(\alpha) \equiv 0 \pmod{\mathfrak{p}_{\widetilde{L}}}$ for all $\tau \in \mathcal{E}_U$ and $\alpha \in T^\times$.

Let V/K be the maximum unramified subextension of $\widetilde{L}/K$ and let S/K be the maximum subextension of $\widetilde{L}/K$ which is at most tamely ramified. Then V/K, S/K, and S/V are all Galois, and $\text{Gal}(S/V)$ is cyclic. Since $V \subset TV \subset S$, it follows that TV/K and TV/V are Galois. Since V/U is unramified and T/U is totally ramified we have $[T:U] = [TV:V]$.

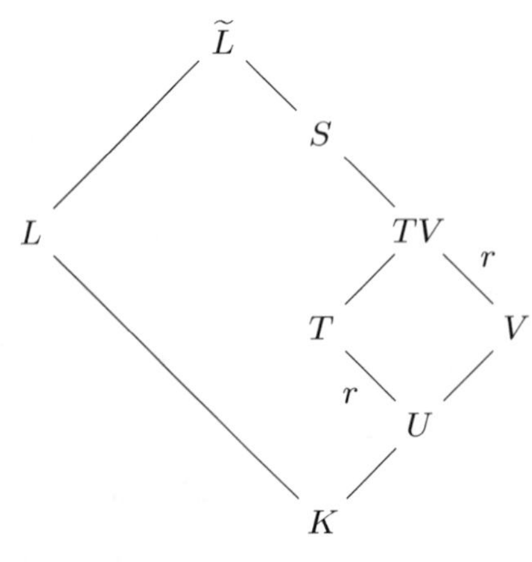

FIGURE 1. Field diagram for Proposition 14.11

Set $r = [TV : V]$ and let π_{TV} be a uniformizer of TV. For $1 \le i \le r$ define a character

$$\psi_i : \mathrm{Gal}(TV/V) \longrightarrow (\mathfrak{O}_{\widetilde{L}}/\mathfrak{p}_{\widetilde{L}})^\times$$
$$\psi_i(\sigma) = \frac{\sigma(\pi_{TV}^i)}{\pi_{TV}^i} + \mathfrak{p}_{\widetilde{L}}.$$

Then ψ_i does not depend on the choice of uniformizer π_{TV}. Since TV/V is totally ramified, $\psi_1, \dots, \psi_r$ are distinct. Hence $\psi_1, \dots, \psi_r$ are linearly independent over $\mathfrak{O}_{\widetilde{L}}/\mathfrak{p}_{\widetilde{L}}$. Write $\mathrm{Gal}(TV/V) = \{\sigma_1, \dots, \sigma_r\}$ and let A be the $r \times r$ matrix with entries in $\mathfrak{O}_{\widetilde{L}}/\mathfrak{p}_{\widetilde{L}}$ whose (i,j) entry is $\psi_i(\sigma_j)$. Then A is invertible.

Let $\tau \in \mathcal{E}_U$ and set $\mathcal{E}_{T,\tau} = \{\sigma \in \mathcal{E}_T \mid \sigma|_U = \tau\}$. Then $|\mathcal{E}_{T,\tau}| = [T : U] = r$. Since TV/K is Galois there is an embedding $\tilde{\tau} : T \to TV$ such that $\tilde{\tau}|_U = \tau$. Since $U \subset V$, for each $\sigma_i \in \mathrm{Gal}(TV/V)$ we have $\sigma_i\tilde{\tau} \in \mathcal{E}_{T,\tau}$. Let $\sigma_i, \sigma_j \in \mathrm{Gal}(TV/V)$ be such that $\sigma_i\tilde{\tau} = \sigma_j\tilde{\tau}$. Then $\sigma_j^{-1}\sigma_i$ is an element of $\mathrm{Gal}(TV/V)$ which fixes $\tilde{\tau}(T)$. Hence $\sigma_j^{-1}\sigma_i$ fixes $\tilde{\tau}(T)V$. Since TV/K and V/K are Galois, we get $\tilde{\tau}(T)V = TV$, and hence $\sigma_i = \sigma_j$. It follows that the set $\{\sigma_1\tilde{\tau}, \dots, \sigma_r\tilde{\tau}\}$ contains r distinct elements. Hence

$$\{\sigma \in \mathcal{E}_T \mid \sigma|_U = \tau\} = \{\sigma_1\tilde{\tau}, \dots, \sigma_r\tilde{\tau}\}. \tag{14.15}$$

Let $\alpha \in T$. It follows from (14.15) that

$$d_\tau(\alpha) = \sum_{j=1}^r c_{\sigma_j\tilde{\tau}} \frac{\sigma_j(\tilde{\tau}(\alpha))}{\alpha} = \frac{\tilde{\tau}(\alpha)}{\alpha} \sum_{j=1}^r c_{\sigma_j\tilde{\tau}} \frac{\sigma_j(\tilde{\tau}(\alpha))}{\tilde{\tau}(\alpha)}.$$

Let $1 \le i \le r$ and set $\alpha = \pi_T^i$. Since TV/T is unramified, so is $TV/\tilde{\tau}(T)$. Hence $\tilde{\tau}(\pi_T)$ is a uniformizer for TV, so we get

$$\begin{aligned}
d_\tau(\pi_T^i) &= \frac{\tilde{\tau}(\pi_T)^i}{\pi_T^i} \sum_{j=1}^r c_{\sigma_j\tilde{\tau}} \frac{\sigma_j(\tilde{\tau}(\pi_T)^i)}{\tilde{\tau}(\pi_T)^i} \\
d_\tau(\pi_T^i) + \mathfrak{p}_{\widetilde{L}} &= \frac{\tilde{\tau}(\pi_T)^i}{\pi_T^i} \sum_{j=1}^r c_{\sigma_j\tilde{\tau}} \psi_i(\sigma_j) \\
0 + \mathfrak{p}_{\widetilde{L}} &= \sum_{j=1}^r c_{\sigma_j\tilde{\tau}} \psi_i(\sigma_j).
\end{aligned}$$

Let $\vec{c} \in (\mathfrak{O}_{\widetilde{L}}/\mathfrak{p}_{\widetilde{L}})^r$ be the column vector whose jth entry is $c_{\sigma_j\tilde{\tau}} + \mathfrak{p}_{\widetilde{L}}$. It follows from the above that $A\vec{c} = \vec{0}$. Hence $\vec{c} = \vec{0}$, as required. $\square$

Note that since we are not assuming that L/K is Galois, the lower ramification breaks ℓ_i of L/K need not be integers. We do, however, have $\ell_i \in \mathbb{Z}_{(p)}$ by Corollary 13.25.

PROPOSITION 14.12. *There exists $\nu \neq 0$ in the center of H and $h \in \mathbb{N}$ such that*

(a) $h \equiv \ell \pmod{p^n\mathbb{Z}_{(p)}}$ *for some lower ramification break ℓ of L/K.*
(b) *For all $y \in L^\times$ we have $v_L(\nu \cdot y) \ge v_L(y) + h$, with equality if and only if $p \nmid v_L(y)$.*

PROOF. The center $Z(N)$ of N is stabilized by G. Since $p \mid |Z(N)|$ and $\{1_G\}$ is a G-orbit of $Z(N)$, there is a G-orbit $S \subset Z(N)$ such that $p \nmid |S|$ and $S \neq \{1_G\}$. Let $\eta \in S$ and let $G_\eta \le G$ be the stabilizer of η in G. Let $T = \widetilde{L}^{G_\eta}$ be the subfield of $\widetilde{L}$ fixed by G_η. Then $[T:K]=|G:G_\eta|=|S|$ is prime to p, so T/K is at most tamely ramified. We may identify G/G_η with the set $\mathcal{E}_T$ of K-embeddings of T into $\widetilde{L}$ using the bijection $\gamma G_\eta \mapsto \gamma|_T$. Let $\gamma G_\eta \in G/G_\eta$; then $\gamma\eta^{-1}[\overline{\gamma^{-1}}]$ does not depend on the choice of coset representative. For each $\gamma G_\eta \in G/G_\eta$ choose $g_\gamma \in G$ such that $g_\gamma G_L = \gamma\eta^{-1}[\overline{\gamma^{-1}}]$. Set

$$\ell = \min\{v_L(g_\gamma(\pi_L)-\pi_L)-1 \mid \gamma G_\eta \in G/G_\eta\}.$$

Since L/K is totally ramified, ℓ is a lower ramification break of L/K (cf. (13.1)). Let $e_{\widetilde{L}/L}$ denote the ramification index of $\widetilde{L}/L$ and set $m=e_{\widetilde{L}/L}\ell$. Then $m\in\mathbb{Z}$ and $\ell = v_L(\pi_{\widetilde{L}}^m)$.

For each $\gamma G_\eta \in G/G_\eta$ set $c_\gamma = (g_\gamma(\pi_L)-\pi_L)/(\pi_L\pi_{\widetilde{L}}^m)$. Then $c_\gamma \in \mathfrak{O}_{\widetilde{L}}$ for all $\gamma G_\eta \in G/G_\eta$, and by the definition of ℓ there exists $\tau G_\eta \in G/G_\eta$ such that $c_\tau \in \mathfrak{O}_{\widetilde{L}}^\times$. Hence by Proposition 14.11 there is $\alpha \in T^\times$ such that

$$\sum_{\gamma G_\eta \in G/G_\eta} c_\gamma \frac{\gamma(\alpha)}{\alpha} \not\equiv 0 \pmod{\mathfrak{p}_{\widetilde{L}}}. \tag{14.16}$$

We may assume that $\alpha\in\mathfrak{O}_T$. Define

$$\begin{aligned}\nu &= \chi_\eta(\alpha)-\mathrm{Tr}_{T/K}(\alpha)\\ &= \sum_{\gamma G_\eta\in G/G_\eta}\gamma(\alpha)\,{}^\gamma\eta - \sum_{\gamma G_\eta\in G/G_\eta}\gamma(\alpha)\\ &= \sum_{\gamma G_\eta\in G/G_\eta}\gamma(\alpha)({}^\gamma\eta-1).\end{aligned}$$

It follows from Lemma 14.10 that $\nu\in H$. In fact, since ${}^\gamma\eta\in Z(N)$ for $\gamma\in G$, we see that ν lies in the center of H. Furthermore, for $y\in L$ we have

$$\nu\cdot y = \sum_{\gamma G_\eta\in G/G_\eta}\gamma(\alpha)(g_\gamma(y)-y). \tag{14.17}$$

In particular, we get

$$\nu\cdot\pi_L = \sum_{\gamma G_\eta\in G/G_\eta} c_\gamma\pi_L\pi_{\widetilde{L}}^m\gamma(\alpha).$$

Hence $(\nu\cdot\pi_L)/\pi_L \in \alpha\mathfrak{p}_{\widetilde{L}}^m$. In addition, by (14.16) we have $(\nu\cdot\pi_L)/\pi_L\notin\alpha\mathfrak{p}_{\widetilde{L}}^{m+1}$. Therefore $v_L((\nu\cdot\pi_L)/\pi_L) = v_L(\alpha\pi_{\widetilde{L}}^m)$.

Let $\gamma G_\eta\in G/G_\eta$. Then for $i\ge 1$ we have

$$\frac{g_\gamma(\pi_L^i)-\pi_L^i}{\pi_L^i} \equiv i\left(\frac{g_\gamma(\pi_L)-\pi_L}{\pi_L}\right) \pmod{\mathfrak{p}_{\widetilde{L}}^{m+1}}.$$

Hence by (14.17) we get

$$\frac{\nu\cdot(\pi_L^i)}{\pi_L^i} \equiv i\left(\frac{\nu\cdot\pi_L}{\pi_L}\right) \pmod{\alpha\mathfrak{p}_{\widetilde{L}}^{m+1}}. \tag{14.18}$$

Set $h=v_L(\alpha\pi_{\widetilde{L}}^m)$. It follows from (14.18) that $v_L(\nu\cdot(\pi_L^i))\ge i+h$, with equality if and only if $p\nmid i$. Hence for $y\in L^\times$ we get $v_L(\nu\cdot y)\ge v_L(y)+h$, with equality if and only if $p\nmid v_L(y)$. Since $\nu\cdot y\in L$ for $y\in L$ it follows that $h\in\mathbb{Z}$. Since $m>0$

and $\alpha \in \mathfrak{O}_T$ we get $h \geq 1$. Let $e_{T/K}$ be the ramification index of T/K. Since L/K is totally ramified of degree p^n we get

$$v_L(\alpha) = p^n v_K(\alpha) = \frac{p^n}{e_{T/K}} v_T(\alpha).$$

Since $p \nmid e_{T/K}$ we get $v_p(v_L(\alpha)) \geq n$, and hence $h \equiv v_L(\pi_{\widetilde{L}}^m) \pmod{p^n\mathbb{Z}_{(p)}}$. Therefore $h \equiv \ell \pmod{p^n\mathbb{Z}_{(p)}}$. □

REMARK 14.13. Suppose we are in the classical setting, with L/K Galois, $\mathrm{Gal}(L/K) = G$, and $H = K[G]$. Let $\sigma \in Z(G)$ with $\sigma \neq 1_G$. Then $\nu = \sigma - 1_G \in K[G]$ satisfies the conditions of Proposition 14.12.

Let $s_L : L \otimes_K L \to L \otimes_K L$ be the map which switches the two factors in the tensor product.

LEMMA 14.14. *Let $\beta \in L \otimes_K L$. Then $\varphi(\beta) \in H$ if and only if $\varphi(s_L(\beta)) \in H$.*

PROOF. Let $\alpha \in (\widetilde{L} \otimes_K L) \otimes_{\widetilde{L}} (\widetilde{L} \otimes_K L)$. It suffices to show that if $\varphi_{\widetilde{L}}(\alpha)$ lies in

$$\begin{aligned}(\widetilde{L} \otimes_K K) \otimes_{\widetilde{L}} (\widetilde{L} \otimes_K H) &\cong \widetilde{L} \otimes_K K \otimes_K H \\ &\cong \widetilde{L}[N] \subset (\widetilde{L} \otimes_K L)[N]\end{aligned}$$

then so does $\varphi_{\widetilde{L}}(s_{\widetilde{L}}(\alpha))$. Write $\alpha = \sum_{i=1}^r a_i \otimes b_i$ with $a_i, b_i \in \widetilde{L} \otimes_K L$. Then

$$\varphi_{\widetilde{L}}(\alpha) = \sum_{i=1}^r \sum_{\eta \in N} a_i \eta(b_i)\eta = \sum_{\eta \in N} \psi_\eta(\alpha)\eta,$$

with $\psi_\eta(\alpha) = \sum_{i=1}^r a_i\eta(b_i) \in \widetilde{L}$. Hence

$$\psi_\eta(s_{\widetilde{L}}(\alpha)) = \sum_{i=1}^r b_i\eta(a_i) = \eta\left(\sum_{i=1}^r a_i\eta^{-1}(b_i)\right) = \eta(\psi_{\eta^{-1}}(\alpha)).$$

Since $\psi_{\eta^{-1}}(\alpha) \in \widetilde{L}$ we get $\psi_\eta(s_{\widetilde{L}}(\alpha)) = \psi_{\eta^{-1}}(\alpha)$. Therefore $\varphi_{\widetilde{L}}(s_{\widetilde{L}}(\alpha)) \in \widetilde{L}[N]$. □

LEMMA 14.15. *Let $\beta \in L \otimes_K L$ and let $h \in H$. Then $\varphi((1 \otimes h)(\beta)) = \varphi(\beta)S(h)$, where $S : H \to H$ is the antipode map. Hence if $\varphi(\beta) \in H$ then $\varphi((1 \otimes h)(\beta)) \in H$.*

PROOF. As in the previous lemma, we work over $\widetilde{L}$ rather than K. Let $a, b \in \widetilde{L} \otimes_K L$ and $\kappa \in N$. It suffices to prove the claim with $\beta = a \otimes b$ and $h = \kappa$. In fact we have

$$\begin{aligned}\varphi((1 \otimes \kappa)(a \otimes b)) &= \varphi(a \otimes \kappa(b)) \\ &= \sum_{\eta \in N} a\eta(\kappa(b))\eta \\ &= \sum_{\eta \in N} a\eta(b)\eta\kappa^{-1} \\ &= \varphi(a \otimes b)S(\kappa).\end{aligned}$$

□

14.3. Diagrams of elements of $L \otimes_K L$

Let L/K be a separable totally ramified H-Galois extension of degree p^n. An element $\beta \in L \otimes_K L$ can be expressed as a finite sum of simple tensors. By considering the valuations of the factors of these simple tensors we get a collection of discrete invariants of β which together form what we call the diagram of β. In this section we give the definition of the diagram of β, and show how this data gives information about the effect that $\varphi(\beta) \in L \otimes_K H$ has on valuations of elements of L.

We wish to use the valuation v_L on L to associate invariants to the nonzero elements of $L \otimes_K L$. One problem that arises, even with simple tensors, is that each element of $L \otimes_K L$ can be represented in multiple ways, for instance $\pi_K \otimes 1 = 1 \otimes \pi_K$. To account for this, our invariants will not lie in $\mathbb{Z} \times \mathbb{Z}$, but rather in a quotient of this group. Let $A = \langle (p^n, -p^n) \rangle$ be the subgroup of $\mathbb{Z} \times \mathbb{Z}$ generated by the element $(p^n, -p^n)$, and for $(a, b) \in \mathbb{Z} \times \mathbb{Z}$ write $[a, b]$ for the coset $(a, b) + A$. We define a partial order on $(\mathbb{Z} \times \mathbb{Z})/A$ by $[a, b] \leq [c, d]$ if there is $(c', d') \in [c, d]$ with $a \leq c'$ and $b \leq d'$.

LEMMA 14.16. *Let $[a, b], [c, d] \in (\mathbb{Z} \times \mathbb{Z})/A$.*

(a) *Let (a', b') be the unique element of $[a, b]$ such that $c - p^n < a' \leq c$. Then $[a, b] \leq [c, d]$ if and only if $b' \leq d$.*
(b) *Let (a'', b'') be the unique element of $[a, b]$ such that $d - p^n < b'' \leq d$. Then $[a, b] \leq [c, d]$ if and only if $a'' \leq c$.*
(c) *$[a, b] \not\leq [c, d]$ if and only if $[c + 1, d - p^n + 1] \leq [a, b]$.*
(d) *If $a + b + p^n - 1 \leq c + d$ then $[a, b] \leq [c, d]$.*

PROOF. Parts (a) and (b) are clear. Suppose $[a, b] \not\leq [c, d]$ and let (c', d') be the unique element of $[c, d]$ such that $b \leq d' < b + p^n$. Then $c' + 1 \leq a$, so we get

$$[c + 1, d - p^n + 1] = [c' + 1, d' - p^n + 1] \leq [a, b].$$

To prove the converse, suppose $[c + 1, d - p^n + 1] \leq [a, b]$. If $[a, b] \leq [c, d]$ then $[c + 1, d - p^n + 1] \leq [c, d]$ by transitivity. This contradicts (b), so we must have $[a, b] \not\leq [c, d]$. If $a + b + p^n - 1 \leq c + d$ then $(c + 1) + (d - p^n + 1) > a + b$, so we can't have $[c + 1, d - p^n + 1] \leq [a, b]$. Hence by (c) we get $[a, b] \leq [c, d]$. $\square$

Fix a uniformizer π_L for L and let $\mathcal{R}_K$ be the set of Teichmüller representatives for K. Note that

$$\mathcal{F} = \{(a, b) \in \mathbb{Z} \times \mathbb{Z} \mid 0 \leq b < p^n\}$$

is a set of coset representatives for $(\mathbb{Z} \times \mathbb{Z})/A$. Let $\beta \in L \otimes_K L$. Then there are uniquely determined $c_j \in L$ and $d_{ij} \in \mathcal{R}_K$ such that

$$\beta = \sum_{j=0}^{p^n - 1} c_j \otimes \pi_L^j = \sum_{(i,j) \in \mathcal{F}} d_{ij} \pi_L^i \otimes \pi_L^j. \tag{14.19}$$

Set

$$R(\beta) = \{[i, j] \in (\mathbb{Z} \times \mathbb{Z})/A \mid (i, j) \in \mathcal{F},\ d_{ij} \neq 0\}.$$

DEFINITION 14.17 (**[Bon02]**, Definition 2.4.1). Define the diagram of β to be

$$D(\beta) = \{[a, b] \in (\mathbb{Z} \times \mathbb{Z})/A \mid [i, j] \leq [a, b] \text{ for some } [i, j] \in R(\beta)\}.$$

Let $G(\beta)$ denote the set of minimal elements of $D(\beta)$ with respect to $\leq$. Then $G(\beta)$ is finite and

$$D(\beta) = \{[a, b] \in (\mathbb{Z} \times \mathbb{Z})/A \mid [i, j] \leq [a, b] \text{ for some } [i, j] \in G(\beta)\}.$$

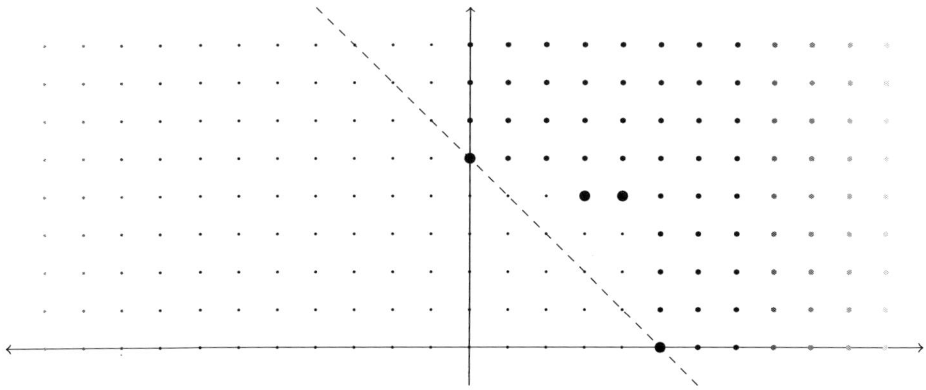

FIGURE 2. Diagram for Example 14.18

It follows from Proposition 14.20 below that while $R(\beta)$ can depend on the choice of uniformizer π_L for L, $D(\beta)$ does not depend on this choice. Hence $G(\beta)$ is also independent of the choice of π_L.

EXAMPLE 14.18. Let $p^n = 9$ and set

$$\beta = a_{50}\pi_L^5 \otimes \pi_L^0 + a_{44}\pi_L^4 \otimes \pi_L^4 + a_{34}\pi_L^3 \otimes \pi_L^4 + a_{05}\pi_L^0 \otimes \pi_L^5.$$

with $a_{ij} \in \mathcal{R}_K \smallsetminus \{0\}$. We get

$$R(\beta) = \{[5,0],[4,4],[3,4],[0,5]\}$$
$$G(\beta) = \{[5,0],[3,4],[0,5]\}.$$

The subset of $\mathcal{F}$ corresponding to $D(\beta)$ is given in Figure 2. Of course, this example is atypical in that $R(\beta)$ is a finite set. For most choices of β and π_L, $R(\beta)$ will be infinite.

For $\xi \in H \smallsetminus \{0\}$ define $f_\xi : \mathbb{Z} \to \mathbb{Z}$ by

$$f_\xi(b) = \min\{v_L(\xi \cdot y) \mid y \in \mathfrak{p}_L^b\}. \tag{14.20}$$

Then $f_\xi(b+1) \geq f_\xi(b)$. Furthermore, for every $b \in \mathbb{Z}$ there is $z \in L$ with $v_L(z) = b$ and $v_L(\xi \cdot z) = f_\xi(b)$. Hence

$$f_\xi(b) = \min\{v_L(\xi \cdot y) \mid v_L(y) = b\}. \tag{14.21}$$

PROPOSITION 14.19. *Let $\xi \in H \smallsetminus \{0\}$ and let $a, b \in \mathbb{Z}$. Then the following are equivalent:*

(a) *$f_\xi(b) = a$ and $f_\xi(b+1) > a$.*
(b) *For all $y \in L$ with $v_L(y) = b$ we have $v_L(\xi \cdot y) = a$.*

PROOF. Suppose (a) holds. Then there is $y_0 \in L$ such that $v_L(y_0) = b$ and $v_L(\xi \cdot y_0) = a$. Let $y \in L$ satisfy $v_L(y) = b$. Then there are $u \in \mathfrak{O}_K^\times$ and $z \in \mathfrak{p}_L^{b+1}$ such that $y = uy_0 + z$. Since $v_L(\xi \cdot z) > a$ we get $v_L(\xi \cdot y) = v_L(u\xi \cdot y_0 + \xi \cdot z) = a$. Suppose (b) holds. Then $f_\xi(b) = a$ by (14.21). Suppose $f_\xi(b+1) = a$. Then there is $z \in L$ with $v_L(z) = b+1$ and $v_L(\xi \cdot z) = a$. Let $y \in L$ satisfy $v_L(y) = b$. Then there is $u \in \mathfrak{O}_K^\times$ such that

$$v_L(\xi \cdot (y + uz)) = v_L(\xi \cdot y + u\xi \cdot z) > a.$$

Since $v_L(y + uz) = b$, this is a contradiction. Hence $f_\xi(b+1) > a$. □

Recall from Proposition 14.1 that there is $\theta \in H$ such that $\theta(y) = \mathrm{Tr}_{L/K}(y)$ for all $y \in L$. We wish to determine the function $f_\theta : \mathbb{Z} \to \mathbb{Z}$. For this it suffices to determine the function $r : \mathbb{Z} \to \mathbb{Z}$ such that $\mathrm{Tr}_{L/K}(\mathfrak{p}_L^a) = \mathfrak{p}_K^{r(a)}$ for $a \in \mathbb{Z}$. Let $\mathfrak{D}_{L/K}$ denote the different of L/K. Then for $a, b \in \mathbb{Z}$ we have

$$\begin{aligned}\mathrm{Tr}_{L/K}(\mathfrak{p}_L^a) \subset \mathfrak{p}_K^b &\Leftrightarrow \mathfrak{p}_K^{-b}\,\mathrm{Tr}_{L/K}(\mathfrak{p}_L^a) \subset \mathfrak{O}_K \\ &\Leftrightarrow \mathrm{Tr}_{L/K}(\mathfrak{p}_K^{-b}\mathfrak{p}_L^a) \subset \mathfrak{O}_K \\ &\Leftrightarrow \mathrm{Tr}_{L/K}(\mathfrak{p}_L^{a-p^n b}) \subset \mathfrak{O}_K \\ &\Leftrightarrow a - p^n b \geq -v_L(\mathfrak{D}_{L/K}) \\ &\Leftrightarrow (a + v_L(\mathfrak{D}_{L/K}))/p^n \geq b.\end{aligned}$$

Hence $\mathrm{Tr}_{L/K}(\mathfrak{p}_L^a) = \mathfrak{p}_K^{r(a)}$, with

$$r(a) = \left\lfloor \frac{a + v_L(\mathfrak{D}_{L/K})}{p^n} \right\rfloor. \tag{14.22}$$

Set

$$i_0 = v_L(\mathfrak{D}_{L/K}) - p^n + 1. \tag{14.23}$$

Then i_0 is the 0th index of inseparability of L/K (see Remark 13.7), and $r(a) = \lceil (a + i_0)/p^n \rceil$. Thus

$$f_\theta(a) = p^n r(a) = p^n \left\lceil \frac{a + i_0}{p^n} \right\rceil. \tag{14.24}$$

The next proposition gives information about the effect that ξ has on valuations of elements of L.

PROPOSITION 14.20. *Let $\xi \in H \smallsetminus \{0\}$, let $\beta \in L \otimes_K L$ satisfy $\xi = \varphi(\beta)$, and let $a, b \in \mathbb{Z}$. Then the following are equivalent:*

(a) *$[a, b] \in D(\beta)$.*

(b) *$f_\xi(-b - i_0) \leq a$.*

PROOF. Suppose $[a, b] \in D(\beta)$. Then there is $[a', b'] \in G(\beta)$ such that $[a', b'] \leq [a, b]$. We may assume that $(a', b') \in \mathcal{F}$, $a' \leq a$, and $b' \leq b$. Let $y \in L$ satisfy $v_L(y) = -b' - i_0$. It follows from (14.19) and (14.5) that

$$\xi \cdot y = \sum_{(i,j) \in \mathcal{F}} d_{ij} \pi_L^i \theta(\pi_L^j y) \tag{14.25}$$

with $d_{ij} \in \mathcal{R}_K$. Since $(a', b') \in G(\beta)$ we have $d_{a'b'} \neq 0$, and hence $v_L(d_{a'b'}) = 0$. In addition, $v_L(\pi_L^{b'} y) = -i_0$, so by (14.24) and Proposition 14.19 we get $v_L(\theta(\pi_L^{b'} y)) = 0$. Hence $v_L(d_{a'b'} \pi_L^{a'} \theta(\pi_L^{b'} y)) = a'$. Let $(i, j) \in \mathcal{F}$ be such that $(i, j) \neq (a', b')$ and $d_{ij} \neq 0$. Since $[i, j] \in D(\beta)$ and $[a', b'] \in G(\beta)$ we must have $[i, j] \not\leq [a', b']$. Hence by Lemma 14.16(c) we get

$$[a' - p^n + 1, b' + 1] = [a' + 1, b' - p^n + 1] \leq [i, j].$$

It follows from Lemma 14.16(b) that if $0 \leq j \leq b'$ then $i \geq a' + 1$, while if $b' < j < p^n$ then $i \geq a' - p^n + 1$. In the first case we get $v_L(\pi_L^j y) \geq -b' - i_0 > -p^n - i_0$, and hence $v_K(\theta(\pi_L^j y)) \geq 0$ by (14.24). It follows that $v_L(d_{ij} \pi_L^i \theta(\pi_L^j y)) \geq i > a'$. In the second case we have $v_L(\pi_L^j y) > -i_0$, and hence $v_K(\theta(\pi_L^j y)) \geq 1$ by (14.24). It follows that $v_L(d_{ij} \pi_L^i \theta(\pi_L^j y)) \geq i + p^n > a'$. Hence by (14.25) we get $v_L(\xi \cdot y) = a' \leq a$.

Since y is an arbitrary element of $\mathfrak{p}_L^{-b'-i_0} \setminus \mathfrak{p}_L^{-b'-i_0+1}$ it follows from (14.21) that $f_\xi(-b-i_0) \leq a$.

Suppose $[a,b] \notin D(\beta)$ and let $(i,j) \in \mathcal{F}$ satisfy $d_{ij} \neq 0$. Then $[i,j] \not\leq [a,b]$, so by Lemma 14.16(c) we get $[a+1, b-p^n+1] \leq [i,j]$. Hence there is $(a',b') \in [a,b]$ with $a'+1 \leq i$ and $b'-p^n+1 \leq j$. Since $[a,b] = [a',b']$ there is $\ell \in \mathbb{Z}$ such that $a'-a = b-b' = \ell p^n$. Let $y \in L$ with $v_L(y) \geq -b-i_0$. Then

$$v_L(\pi_K^\ell \pi_L^j y) = \ell p^n + j + v_L(y) \geq \ell p^n + j - b - i_0 = j - b' - i_0 > -p^n - i_0.$$

Hence by (14.24) we get $v_L(\theta(\pi_K^\ell \pi_L^j y)) \geq 0$. It follows that

$$v_L(\pi_L^i \theta(\pi_L^j y)) + \ell p^n = v_L(\pi_L^i \theta(\pi_K^\ell \pi_L^j y)) \geq i > a' = a + \ell p^n.$$

Therefore $v_L(\pi_L^i \theta(\pi_L^j y)) > a$. Since this holds for all $(i,j) \in \mathcal{F}$ such that $d_{ij} \neq 0$ we get $f_\xi(-b-i_0) > a$. □

COROLLARY 14.21. *Let $\xi \in H \setminus \{0\}$, let $\beta \in L \otimes_K L$ satisfy $\xi = \varphi(\beta)$, and let $h \in \mathbb{Z}$. Then the following are equivalent:*

(a) $[h-1, -h-i_0] \notin D(\beta)$.

(b) $\xi \in \mathfrak{A}(\mathfrak{p}_L^h)$.

PROOF. We have $\xi \in \mathfrak{A}(\mathfrak{p}_L^h)$ if and only if $f_\xi(h) \geq h$, or equivalently, $f_\xi(h) \not\leq h-1$. By Proposition 14.20 this is equivalent to $[h-1, -h-i_0] \notin D(\beta)$. □

The following is a version of Proposition 2.4.2 of [**Bon02**]:

PROPOSITION 14.22. *Let $\xi \in H \setminus \{0\}$, let $\beta \in L \otimes_K L$ satisfy $\xi = \varphi(\beta)$, and let $a, b \in \mathbb{Z}$. Then the following are equivalent:*

(a) $f_\xi(-b-i_0) = a$ *and* $f_\xi(-b-i_0+1) > a$.

(b) *For all $y \in L$ with $v_L(y) = -b-i_0$ we have $v_L(\xi \cdot y) = a$.*

(c) $[a,b] \in G(\beta)$.

PROOF. It follows from Proposition 14.19 that (a) and (b) are equivalent. We have $[a,b] \in G(\beta)$ if and only if $[a,b] \in D(\beta)$, $[a-1,b] \notin D(\beta)$, and $[a,b-1] \notin D(\beta)$. By Proposition 14.20 these conditions are equivalent to $f_\xi(-b-i_0) \leq a$, $f_\xi(-b-i_0) \not\leq a-1$, and $f_\xi(-b-i_0+1) \not\leq a$. There are in turn equivalent to $f_\xi(-b-i_0) = a$ and $f_\xi(-b-i_0+1) > a$. Hence (a) and (c) are equivalent. □

For later use we prove some basic facts about the relation between addition in $L \otimes_K L$ and diagrams.

PROPOSITION 14.23. *Let β, β' be nonzero elements of $L \otimes_K L$ such that $\beta + \beta' \neq 0$. Then*

(a) $D(\beta + \beta') \subset D(\beta) \cup D(\beta')$.

(b) *If $G(\beta) \cap G(\beta') = \emptyset$ then $D(\beta+\beta') = D(\beta) \cup D(\beta')$.*

PROOF. Set $\xi = \varphi(\beta)$ and $\xi' = \varphi(\beta')$. Let $[a,b] \in (\mathbb{Z} \times \mathbb{Z})/A$ be such that $[a,b] \notin D(\beta)$ and $[a,b] \notin D(\beta')$. Then by Proposition 14.20 we have $f_\xi(-b-i_0) > a$ and $f_{\xi'}(-b-i_0) > a$. Let $y \in \mathfrak{p}_L^{-b-i_0}$. Then $v_L(\xi \cdot y) > a$ and $v_L(\xi' \cdot y) > a$, which implies $v_L((\xi + \xi') \cdot y) > a$. Since $\xi + \xi' = \varphi(\beta + \beta')$, by Proposition 14.20 we get $[a,b] \notin D(\beta + \beta')$. This proves (a).

In order to prove (b), we first show that $G(\beta) \subset D(\beta+\beta')$. Let $[a,b] \in G(\beta)$. If $[a,b] \notin D(\beta')$ then $f_{\xi'}(-b-i_0) > a$ by Proposition 14.20. Let $y \in L$ with $v_L(y) = -b-i_0$. Then $v_L(\xi \cdot y) = a$ by Proposition 14.22 and $v_L(\xi' \cdot y) > a$. Hence $v_L((\xi+\xi') \cdot y) = a$, so we get $[a,b] \in G(\beta+\beta')$ by Proposition 14.22. Thus $[a,b] \in D(\beta+\beta')$. Suppose $[a,b] \in D(\beta')$. Then $[a,b] \notin G(\beta')$ by assumption.

Hence by Proposition 14.20 and Proposition 14.22 we get $f_{\xi'}(-b-i_0) < a$. Let $y \in \mathfrak{p}_L^{-b-i_0}$ be such that $v_L(\xi' \cdot y) < a$. Since $v_L(\xi \cdot y) \geq a$ we get $v_L((\xi+\xi') \cdot y) < a$. Therefore $f_{\xi+\xi'}(-b-i_0) < a$, so $[a,b] \in D(\beta+\beta')$ by Proposition 14.20. Thus $G(\beta) \subset D(\beta+\beta')$. It follows that $D(\beta) \subset D(\beta+\beta')$. By the same reasoning we get $D(\beta') \subset D(\beta+\beta')$. Using (a) we conclude that $D(\beta+\beta') = D(\beta) \cup D(\beta')$. □

Let $\gamma \in \mathrm{End}_K(L) \cong L \otimes_K H$. Motivated by the definition of the norm of a linear operator we define

$$\hat{v}_L(\gamma) = \min\{v_L(\gamma(y)) - v_L(y) \mid y \in L^\times\} \tag{14.26}$$

for $\gamma \neq 0$. Thus $\hat{v}_L(0) = \infty$ and $\hat{v}_L(\mathrm{id}_L) = 0$. Furthermore, for $\gamma, \delta \in \mathrm{End}_K(L)$ we have

$$\begin{aligned} \hat{v}_L(\gamma+\delta) &\geq \min\{\hat{v}_L(\gamma), \hat{v}_L(\delta)\} \\ \hat{v}_L(\gamma \circ \delta) &\geq \hat{v}_L(\gamma) + \hat{v}_L(\delta). \end{aligned} \tag{14.27}$$

Therefore $\hat{v}_L$ is a pseudovaluation on $\mathrm{End}_K(L)$ (see [**Ree55**, p. 108]). It follows that if $\hat{v}_L(\gamma) \neq \hat{v}_L(\delta)$ then equality holds in (14.27). For $r \in \mathbb{Z}$ set

$$\mathfrak{C}_r = \{\gamma \in \mathrm{End}_K(L) \mid \hat{v}_L(\gamma) \geq r\}.$$

Then $\mathfrak{C}_0$ is a subring of $\mathrm{End}_K(L)$, and $\mathfrak{C}_r$ is an ideal in $\mathfrak{C}_0$ for $r \geq 0$. Let X_r be the $\mathfrak{O}_K$-submodule of $L \otimes_K L$ generated by all elements of the form $c \otimes d$ with $v_L(c) + v_L(d) \geq r$.

PROPOSITION 14.24. *Let $r \in \mathbb{Z}$. Then $\varphi(X_{r-i_0}) = \mathfrak{C}_r$.*

PROOF. Let $y, z \in L^\times$ and set $h = v_L(y)$, $k = v_L(z)$. Assume that $h + k \geq r - i_0$ and set $\beta = y \otimes z$, $\xi = \varphi(\beta)$. We have $[h,k] \not\leq [h-1, k+p^n-1]$ by Lemma 14.16(b). It follows that $[h-1, k+p^n-1] \notin D(\beta)$, so by Proposition 14.20 we get $f_\xi(-k-p^n+1-i_0) > h-1$. Let $x \in L^\times$ satisfy $v_L(x) = -i_0 - u$ with $k \leq u \leq k+p^n-1$. Then $v_L(x) \geq -k-p^n+1-i_0$, so we get

$$\begin{aligned} v_L(\xi \cdot x) - v_L(x) &\geq h - (-i_0 - u) \\ &\geq h + i_0 + k \\ &\geq r. \end{aligned}$$

Since ξ is K-linear it follows that $v_L(\xi \cdot x) - v_L(x) \geq r$ for all $x \in L^\times$. Therefore $\xi \in \mathfrak{C}_r$, so we get $\varphi(X_{r-i_0}) \subset \mathfrak{C}_r$.

On the other hand, suppose $\xi \in \mathfrak{C}_r$. Let $\beta \in L \otimes_K L$ be such that $\varphi(\beta) = \xi$ and choose $[a,b] \in D(\beta)$. Then by Proposition 14.20 we have $f_\xi(-b-i_0) \leq a$. Since $\xi \in \mathfrak{C}_r$ we get $a - (-i_0 - b) \geq r$, and hence $a + b \geq r - i_0$. Therefore $\beta \in X_{r-i_0}$, so we have $\mathfrak{C}_r \subset \varphi(X_{r-i_0})$. We conclude that $\mathfrak{C}_r = \varphi(X_{r-i_0})$. □

For $\beta \in L \otimes_K L$ with $\beta \neq 0$ set

$$d(\beta) = \min\{i + j \mid [i,j] \in D(\beta)\}.$$

Define the diagonal of β to be

$$N(\beta) = \{[i,j] \in D(\beta) \mid i + j = d(\beta)\}.$$

Then $N(\beta) \subset G(\beta)$.

COROLLARY 14.25. *Let $\beta \in L \otimes_K L$ and set $\xi = \varphi(\beta)$. Then $\hat{v}_L(\xi) = d(\beta) + i_0$.*

PROOF. By Proposition 14.24 we have $\hat{v}_L(\xi) \geq r$ if and only if $d(\beta) \geq r - i_0$. □

14.4. H-stable and H-semistable extensions

In this section we give the definitions of H-stable and H-semistable extensions. We then prove some facts about the numerical data associated to these types of extensions. In Section 14.5 we will prove theorems about the Hopf-Galois module structure of ideals in such extensions.

DEFINITION 14.26. Let L/K be a totally ramified H-Galois extension of degree p^n. Say that L/K is an H-semistable extension if the following conditions hold:

(i) The lower ramification breaks $\ell_1 \leq \cdots \leq \ell_n$ of L/K satisfy $v_p(\ell_i) = 0$ for $1 \leq i \leq n$.

(ii) There is $\beta \in L \otimes_K L$ such that $\varphi(\beta) \in H$ and $N(\beta) = \{[a_1, b_1], [a_2, b_2]\}$ with $p \nmid a_1 + b_1$ and $p \nmid a_1 - a_2$.

Say that L/K is an H-stable extension if β can be chosen to satisfy the additional condition $G(\beta) = N(\beta)$.

REMARK 14.27. We will prove in Proposition 14.32 that if L/K is H-semistable then $N(\beta) = \{[0, d(\beta)], [d(\beta), 0]\}$.

REMARK 14.28. The definition of a semistable Galois extension given in Definition 3.1.1 of [**Bon02**] does not include the conditions $v_p(\ell_i) = 0$ and $p \nmid a_1 - a_2$. Instead, they are proved as consequences of the definition (see [**Bon02**, Proposition 3.2]). It seems to be necessary to include these conditions as assumptions in the setting of Hopf-Galois extensions. It follows from Proposition 13.26 that the assumption $v_p(\ell_i) = 0$ is quite weak, and holds automatically if $\mathrm{char}(K) = p$.

We give an example which illustrates Definition 14.26 in the classical Galois setting:

EXAMPLE 14.29. Let p be a prime and let $\zeta_{p^2} \in \mathbb{Q}_p^{alg}$ be a primitive p^2 root of unity. Set $\zeta_p = \zeta_{p^2}^p$ and define $K = \mathbb{Q}_p(\zeta_p)$, $L = \mathbb{Q}_p(\zeta_{p^2})$. Then L/K is a totally ramified Galois extension whose Galois group is isomorphic to C_p. Let σ be the generator for $G = \mathrm{Gal}(L/K)$ which satisfies $\sigma(\zeta_{p^2}) = \zeta_{p^2}^{1+p} = \zeta_p\zeta_{p^2}$. Define $\beta \in L \otimes_K L$ by $\beta = -1 \otimes 1 + \zeta_{p^2}^{-1} \otimes \zeta_{p^2}$. Then by (14.6) we get

$$\begin{aligned}\varphi(\beta) &= -\sum_{i=0}^{p-1} 1 \cdot \sigma^i(1)\sigma^i + \sum_{i=0}^{p-1} \zeta_{p^2}^{-1} \cdot \sigma^i(\zeta_{p^2})\sigma^i \\ &= \sum_{i=0}^{p-1} (\zeta_p^i - 1)\sigma^i.\end{aligned}$$

Hence $\varphi(\beta) \in K[G]$. Set $\pi_L = 1 - \zeta_{p^2}$. Then π_L is a uniformizer for L and

$$\begin{aligned}\beta &= -1 \otimes 1 + (1 - \pi_L)^{-1} \otimes (1 - \pi_L) \\ &= (\pi_L \otimes 1 + \pi_L^2 \otimes 1 + \cdots) - (1 \otimes \pi_L + \pi_L \otimes \pi_L + \pi_L^2 \otimes \pi_L + \cdots).\end{aligned}$$

It follows that $N(\beta) = G(\beta) = \{[1, 0], [0, 1]\}$. Therefore L/K is a stable extension.

The significance of the element $\beta \in L \otimes_K L$ in the definition of semistable extension is that it can be used to construct a K-basis for H with some useful properties:

PROPOSITION 14.30. *Let L/K be a separable totally ramified H-semistable extension of degree p^n and let $\beta \in L \otimes_K L$ satisfy the conditions of Definition 14.26. Then*

(a) *For* $0 \le s < p^n$ *we have* $d(\beta^s) = sd(\beta)$ *and* $\hat{v}_L(\varphi(\beta^s)) = sd(\beta) + i_0$.
(b) $\{\varphi(\beta^s) \mid 0 \le s < p^n\}$ *is a* K*-basis for* $H \cong K \otimes_K H$.
(c) *Let* $c_0, c_1, \dots, c_{p^n-1} \in K$ *and set* $\xi = \sum_{s=0}^{p^n-1} c_s\varphi(\beta^s)$. *Then*

$$\hat{v}_L(\xi) = \min\{\hat{v}_L(c_s\varphi(\beta^s)) \mid 0 \le s < p^n\}.$$

PROOF. Write $N(\beta) = \{[a_1, b_1], [a_2, b_2]\}$ and set $c = d(\beta) = a_1 + b_1 = a_2 + b_2$. Then there are $t, u \in \mathcal{R}_K \smallsetminus \{0\}$ and $E \in X_{c+1}$ such that

$$\beta = t\pi_L^{a_1} \otimes \pi_L^{b_1} + u\pi_L^{a_2} \otimes \pi_L^{b_2} + E. \tag{14.28}$$

It follows that for $0 \le s < p^n$ there are $E_s \in X_{sc+1}$ such that

$$\beta^s = \sum_{i=0}^{s} \binom{s}{i} t^i u^{s-i} \pi_L^{ia_1+(s-i)a_2} \otimes \pi_L^{ib_1+(s-i)b_2} + E_s. \tag{14.29}$$

Since $p \nmid a_1 - a_2$ the integers $ia_1 + (s-i)a_2$ for $0 \le i \le s$ represent distinct congruence classes modulo p^n. It follows that $[sa_2, sb_2] \in N(\beta^s)$, and hence that $d(\beta^s) = sc = sd(\beta)$. Using Corollary 14.25 we get $\hat{v}_L(\varphi(\beta^s)) = sd(\beta) + i_0$, which completes the proof of (a). Since $p \nmid c$ statement (a) implies that $\{d(\beta^s) \mid 0 \le s < p^n\}$ is a complete set of residue classes modulo p^n. Hence by Corollary 14.25 we deduce that $\{\hat{v}_L(\varphi(\beta^s)) \mid 0 \le s < p^n\}$ is also a complete set of residue classes modulo p^n. Therefore the integers $\hat{v}_L(c_s\varphi(\beta^s))$ for $0 \le s < p^n$ with $c_s \ne 0$ are distinct. Statement (c) now follows from the fact that $\hat{v}_L$ is a pseudovaluation on H. By Proposition 14.3 we have $\varphi(\beta^s) \in K \otimes_K H$ for $0 \le s < p^n$. Using (c) we deduce that $\{\varphi(\beta^s) \mid 0 \le s < p^n\}$ is a K-basis for $K \otimes_K H$, which gives (b). □

COROLLARY 14.31. *Let* L/K *be an* H*-semistable extension of degree* p^n *and let* $\beta \in L \otimes_K L$ *satisfy the conditions of Definition* 14.26. *Let* $c_0, c_1, \dots, c_{p^n-1} \in K$ *and set* $\alpha = \sum_{s=0}^{p^n-1} c_s\beta^s$. *Then*

$$d(\alpha) = \min\{d(c_s\beta^s) \mid 0 \le s < p^n\}.$$

Furthermore, if $0 \le t < p^n$ *satisfies* $d(\alpha) = d(c_t\beta^t)$ *then* $N(\alpha) = N(c_t\beta^t)$.

PROOF. The first statement follows from Proposition 14.30(c) and Corollary 14.25. If $d(\alpha) = d(c_t\beta^t)$ and $s \ne t$ then $d(\beta^s) \not\equiv d(\beta^t) \pmod{p^n}$ by Proposition 14.30(a). It follows that $d(c_s\beta^s) > d(c_t\beta^t)$. Therefore $N(\alpha) = N(c_t\beta^t)$. □

PROPOSITION 14.32. *Let* L/K *be an* H*-semistable extension of degree* p^n *and let* $\beta \in L \otimes_K L$ *satisfy the conditions of Definition* 14.26. *Then* $N(\beta) = \{[0, d(\beta)], [d(\beta), 0]\}$.

PROOF. Let $\nu \in H$ and $h \in \mathbb{N}$ satisfy the conditions of Proposition 14.12. Then $\nu \ne 0$ and ν is in the center of H. We have $v_L(\nu \cdot y) \ge v_L(y) + h$ for all $y \in L^\times$, with equality if and only if $p \nmid v_L(y)$. Furthermore, there is a lower ramification break ℓ of L/K such that $h \equiv \ell \pmod{p\mathbb{Z}_{(p)}}$. Since L/K is H-semistable, by Definition 14.26 we get $p \nmid h$. Write $N(\beta) = \{[a_1, b_1], [a_2, b_2]\}$ and let $t \in \mathbb{Z}$ be such that $th \equiv -1 \pmod{p}$. For $i = 1, 2$ let r_i denote the least nonnegative residue of tb_i modulo p. It follows from Definition 14.26 that $p \nmid b_1 - b_2$. Hence $r_1 \ne r_2$, so we may assume without loss of generality that $r_1 > r_2$. An inductive argument based on (14.28) and Proposition 14.12(b) shows that for $0 \le i \le r_2$ we have

$$N((1 \otimes \nu^i)\beta) = \{[a_1, b_1 + ih], [a_2, b_2 + ih]\}.$$

Applying Proposition 14.12(b) to the above with $i = r_2$ gives

$$N((1 \otimes \nu^{r_2+1})\beta) = \{[a_1, b_1 + (r_2 + 1)h]\}. \tag{14.30}$$

By Lemma 14.15 we have $\varphi((1\otimes\nu^{r_2+1})\beta)\in H$. Using (14.29) we see that for $1\le s<p^n$ we have $[sa_1,sb_1],[sa_2,sb_2]\in N(\beta^s)$. On the other hand, $N(\beta^0)=N(1\otimes 1)=\{[0,0]\}$. Hence the only $0\le s<p^n$ such that $N(\beta^s)$ consists of a single element is $s=0$. It follows from (14.30) and Corollary 14.31 that there is $c_0\in K^\times$ such that $d((1\otimes\nu^{r_2+1})\beta)=d(c_0\beta^0)$ and

$$N((1\otimes\nu^{r_2+1})\beta)=N(c_0\beta^0)=\{[p^nv_K(c_0),0]\}.$$

Therefore $p^n\mid a_1$, so we get $[a_1,b_1]=[0,a_1+b_1]=[0,d(\beta)]$. By Lemma 14.14 we have $\varphi(s_L(\beta))\in H$, where $s_L:L\otimes_K L\to L\otimes_K L$ is defined by $s_L(y\otimes z)=z\otimes y$. Since $N(s_L(\beta))=\{[b_1,a_1],[b_2,a_2]\}$ we can apply the same arguments to $s_L(\beta)$ and deduce that either $p^n\mid b_1$ or $p^n\mid b_2$. If $p^n\mid b_1$ then we contradict the assumption $p\nmid a_1+b_1$. It follows that $p^n\mid b_2$, and hence that $N(\beta)=\{[0,d(\beta)],[d(\beta),0]\}$. □

For $r\in\mathbb{Z}$ define $\mathfrak{A}_r=\{\xi\in H\mid \hat{v}_L(\xi)\ge r\}$. Thus $\mathfrak{A}_r=\mathfrak{C}_r\cap H$.

COROLLARY 14.33. *Let L/K be an H-semistable extension of degree p^n and let $\beta\in L\otimes_K L$ satisfy the conditions of Definition 14.26. Let $r\in\mathbb{Z}$, and for $0\le s<p^n$ set*

$$c(s,r)=\left\lceil\frac{r-i_0-sd(\beta)}{p^n}\right\rceil.$$

Then $\{\pi_K^{c(s,r)}\varphi(\beta^r)\mid 0\le s<p^n\}$ is an $\mathfrak{O}_K$-basis for $\mathfrak{A}_r$.

PROOF. By Proposition 14.30(a) we have $\hat{v}_L(\varphi(\beta^s))=sd(\beta)+i_0$. Let $c_s\in K$ for $0\le s<p^n$. It follows by Proposition 14.30(c) that $\sum_{s=0}^{p^n-1}c_s\varphi(\beta^s)\in\mathfrak{A}_r$ if and only if $p^nv_K(c_s)+sd(\beta)+i_0\ge r$ for $0\le s<p^n$. Since this condition is equivalent to $v_K(c_s)\ge c(s,r)$, the claim follows. □

14.5. Hopf-Galois module structure

Let L/K be a separable totally ramified H-Galois extension of degree p^n. In this section we prove theorems about the H-Galois module structure of $\mathfrak{O}_L$-ideals in the cases where L/K is either H-stable or H-semistable. We also give a characterization of H-semistable extensions in terms of the pseudovaluation $\hat{v}_L$ on H.

Let $f\in\mathfrak{A}_h$ and $g\in\mathfrak{A}_k$. Then $f\circ g\in\mathfrak{A}_{h+k}$. It follows that $\mathfrak{A}_0$ is a $\mathfrak{O}_K$-algebra, and that $\mathfrak{A}_h$ is a left and right $\mathfrak{A}_0$-module for all $h\in\mathbb{Z}$.

THEOREM 14.34. *Let L/K be an H-semistable extension and let $h\in\mathbb{Z}$. Then for every $\rho\in L$ such that $v_L(\rho)=-i_0$ we have $\mathfrak{A}_{h+i_0}\cdot\rho=\mathfrak{p}_L^h$. Hence there is an isomorphism of left $\mathfrak{A}_0$-modules $\mathfrak{A}_{h+i_0}\cong\mathfrak{p}_L^h$.*

PROOF. By Proposition 14.32 there is $\beta\in L\otimes_K L$ such that $\varphi(\beta)\in H$, $p\nmid d(\beta)$, and $N(\beta)=\{[0,d(\beta)],[d(\beta),0]\}$. By Corollary 14.33 the set

$$\{\pi_K^{c(s,h+i_0)}\varphi(\beta^s)\mid 0\le s<p^n\}$$

is an $\mathfrak{O}_K$-basis for $\mathfrak{A}_{h+i_0}$, with $c(s,h+i_0)=\lceil(h-sd(\beta))/p^n\rceil$. Let $\rho\in L$ with $v_L(\rho)=-i_0$ and let $0\le s<p^n$. It follows from (14.29) (with $a_1=b_2=0$ and $a_2=b_1=d(\beta)$) that $[sd(\beta),0]\in N(\beta^s)$. Hence by Proposition 14.22 we get

$$v_L(\pi_K^{c(s,h+i_0)}\varphi(\beta^s)\cdot\rho)=p^n\left\lceil\frac{h-sd(\beta)}{p^n}\right\rceil+sd(\beta). \tag{14.31}$$

Therefore we have

$$h\le v_L(\pi_K^{c(s,h+i_0)}\varphi(\beta^s)\cdot\rho)<h+p^n.$$

Since $p \nmid d(\beta)$ it follows from (14.31) that

$$\{v_L(\pi_K^{c(s,h+i_0)}\varphi(\beta^s)\cdot\rho) \mid 0 \le s < p^n\}$$

is a complete set of residue classes modulo p^n. Hence

$$\{\pi_K^{c(s,h+i_0)}\varphi(\beta^s)\cdot\rho \mid 0 \le s < p^n\}$$

is an $\mathfrak{O}_K$-basis for $\mathfrak{p}_L^h$. Therefore the map which carries $\xi \in \mathfrak{A}_{h+i_0}$ onto $\xi \cdot \rho$ gives an $\mathfrak{A}_0$-module isomorphism from $\mathfrak{A}_{h+i_0}$ onto $\mathfrak{p}_L^h$. □

COROLLARY 14.35. *Let L/K be an H-semistable extension. Then $\mathfrak{p}_L^{-i_0}$ is free over its associated H-order $\mathfrak{A}(\mathfrak{p}_L^{-i_0})$.*

PROOF. We have $\mathfrak{A}_0 \subset \mathfrak{A}(\mathfrak{p}_L^{-i_0})$ by definition. Let $\rho \in L$ with $v_L(\rho) = -i_0$. By the theorem $\mathfrak{p}_L^{-i_0}$ is free over $\mathfrak{A}_0$, with generator ρ. Let $\xi \in \mathfrak{A}(\mathfrak{p}_L^{-i_0})$. Then there is $r \ge 0$ such that $\pi_K^r\xi \in \mathfrak{A}_0$. In addition, there is $\nu \in \mathfrak{A}_0$ such that $\xi\cdot\rho = \nu\cdot\rho$. Hence $\pi^r(\xi-\nu)(\rho) = 0$ and $\pi^r(\xi-\nu) \in \mathfrak{A}_0$. Therefore $\pi^r(\xi-\nu) = 0$, so we get $\xi = \nu \in \mathfrak{A}_0$. It follows that $\mathfrak{A}_0 = \mathfrak{A}(\mathfrak{p}_L^{-i_0})$, and hence that $\mathfrak{p}_L^{-i_0}$ is free over $\mathfrak{A}(\mathfrak{p}_L^{-i_0})$. □

In order to prove Theorem 14.43, our main result on Hopf-Galois module theory for H-stable extensions, we need several lemmas about tensor products and diagrams.

LEMMA 14.36. *Let $\alpha \in \mathfrak{O}_L \otimes_{\mathfrak{O}_K} \mathfrak{O}_L$, and write $\alpha = \sum_{i=0}^{\infty}\sum_{j=0}^{\infty} t_{ij}\pi_L^i \otimes \pi_L^j$ with $t_{ij} \in \mathcal{R}_K$. Then $\alpha \in (\mathfrak{O}_L \otimes_{\mathfrak{O}_K} \mathfrak{O}_L)^\times$ if and only if $t_{00} \ne 0$.*

PROOF. Since $p\mathfrak{O}_L \subset \pi_K\mathfrak{O}_L = \pi_L^{p^n}\mathfrak{O}_L$ we get

$$\alpha^{p^n} \equiv \sum_{i=0}^{\infty}\sum_{j=0}^{\infty} t_{ij}^{p^n}\pi_L^{ip^n} \otimes \pi_L^{jp^n} \pmod{p\mathfrak{O}_L \otimes_{\mathfrak{O}_K} \mathfrak{O}_L}$$

$$\alpha^{p^n} \equiv t_{00}^{p^n} \pmod{\pi_K\mathfrak{O}_L \otimes_{\mathfrak{O}_K} \mathfrak{O}_L}.$$

Suppose $t_{00} \ne 0$. Then $t_{00} \in \mathfrak{O}_K^\times$, and $(t_{00}^{-1}\alpha)^{p^n} = 1 + \pi_K\beta$ for some $\beta \in \mathfrak{O}_L \otimes_{\mathfrak{O}_K} \mathfrak{O}_L$. Since $\mathfrak{O}_L \otimes_{\mathfrak{O}_K} \mathfrak{O}_L$ is a π_K-adically complete $\mathfrak{O}_K$-algebra it follows that $(t_{00}^{-1}\alpha)^{p^n}$ is a unit. Hence α is a unit. Suppose $t_{00} = 0$. Then $\alpha^{p^n} \in \pi_K\mathfrak{O}_L \otimes_{\mathfrak{O}_K} \mathfrak{O}_L$, so α^{p^n} is not a unit. Hence α is not a unit. □

REMARK 14.37. The set of $\alpha \in \mathfrak{O}_L \otimes_{\mathfrak{O}_K} \mathfrak{O}_L$ such that $t_{00} = 0$ is an ideal in $\mathfrak{O}_L \otimes_{\mathfrak{O}_K} \mathfrak{O}_L$. This implies that $\mathfrak{O}_L \otimes_{\mathfrak{O}_K} \mathfrak{O}_L$ is a local ring.

PROPOSITION 14.38. *Let $\beta \in L \otimes_K L$ with $\beta \ne 0$ and let $S = \{[a_1, b_1], \ldots, [a_r, b_r]\}$ be an antichain in $(\mathbb{Z}\times\mathbb{Z})/A$. (In other words, $[a_h, b_h] \not\le [a_i, b_i]$ for all $h \ne i$.) Then the following are equivalent:*

(a) *$G(\beta) = S$.*

(b) *There are $u_h \in (\mathfrak{O}_L \otimes_{\mathfrak{O}_K} \mathfrak{O}_L)^\times$ such that $\beta = \sum_{h=1}^{r} u_h(\pi_L^{a_h} \otimes \pi_L^{b_h})$.*

PROOF. Suppose (a) holds. Then there are $t_{hij} \in \mathcal{R}_K$ with $t_{h00} \neq 0$ such that

$$\begin{aligned}\beta &= \sum_{h=1}^{r} \sum_{i=0}^{\infty} \sum_{j=0}^{\infty} t_{hij} \pi_L^{a_h+i} \otimes \pi_L^{b_h+j} \\ &= \sum_{h=1}^{r} (\pi_L^{a_h} \otimes \pi_L^{b_h}) \sum_{i=0}^{\infty} \sum_{j=0}^{\infty} t_{hij} \pi_L^{i} \otimes \pi_L^{j}.\end{aligned}$$

It follows from Lemma 14.36 that

$$u_h = \sum_{i=0}^{\infty} \sum_{j=0}^{\infty} t_{hij} \pi_L^{i} \otimes \pi_L^{j} \tag{14.32}$$

is a unit in $\mathfrak{O}_L \otimes_{\mathfrak{O}_K} \mathfrak{O}_L$.

Suppose (b) holds. For $1 \leq h \leq r$ let $\beta_h = u_h(\pi_L^{a_h} \otimes \pi_L^{b_h})$. By Lemma 14.36 we can write u_h in the form (14.32), with $t_{hij} \in \mathcal{R}_K$ and $t_{h00} \neq 0$. It follows that $G(\beta_h) = \{[a_h, b_h]\}$. Using Proposition 14.23 and the fact that S is an antichain one proves by induction that for $1 \leq k \leq r$ we have

$$\begin{aligned}D(\beta_1 + \cdots + \beta_k) &= D(\beta_1) \cup \cdots \cup D(\beta_k) \\ G(\beta_1 + \cdots + \beta_k) &= \{[a_1, b_1], \ldots, [a_k, b_k]\}.\end{aligned}$$

The case $k = r$ gives (a). □

Define $\mathbb{S}_{p^n} = \{0, 1, \ldots, p^n - 1\}$. Write $s \in \mathbb{S}_{p^n}$ in base p as

$$s = s_{(0)}p^0 + s_{(1)}p^1 + \cdots + s_{(n-1)}p^{n-1}$$

with $0 \leq s_{(i)} < p$. Define a partial order on $\mathbb{S}_{p^n}$ by $s \preceq t$ if $s_{(i)} \leq t_{(i)}$ for $0 \leq i \leq n-1$. For the next four lemmas we assume that L/K is an H-stable extension of degree p^n. By Proposition 14.32 there is $\beta \in L \otimes_K L$ such that $\varphi(\beta) \in K \otimes_K H$, $p \nmid d(\beta)$, and $G(\beta) = \{[0, d(\beta)], [d(\beta), 0]\}$.

LEMMA 14.39. *Let $s \in \mathbb{S}_{p^n}$. Then $d(\beta^s) = sd(\beta)$ and*

$$G(\beta^s) = \{[jd(\beta), (s-j)d(\beta)] \mid 0 \leq j \leq s,\ j \preceq s\}.$$

PROOF. Set $c = d(\beta)$. Since $G(\beta) = \{[0, c], [c, 0]\}$, by Proposition 14.38 there are $u, v \in (\mathfrak{O}_L \otimes_{\mathfrak{O}_K} \mathfrak{O}_L)^\times$ such that

$$\beta = u(1 \otimes \pi_L^c) + v(\pi_L^c \otimes 1).$$

It follows that

$$\beta^s = \sum_{j=0}^{s} \binom{s}{j} u^{s-j} v^j (\pi_L^{jc} \otimes \pi_L^{(s-j)c}).$$

By Lucas' theorem we have $p \nmid \binom{s}{j}$ if and only if $j \preceq s$. Suppose $p \mid \binom{s}{j}$. By Lemma 14.16(d) we have $[sc, 1] \leq [jc + p^n, (s-j)c]$. Since $p \in \pi_L^{p^n} \mathfrak{O}_L$, there is $\delta_j \in \mathfrak{O}_L \otimes_{\mathfrak{O}_K} \mathfrak{O}_L$ such that

$$\binom{s}{j} u^{s-j} v^j (\pi_L^{jc} \otimes \pi_L^{(s-j)c}) = \delta_j(\pi_L^{sc} \otimes \pi_L) = \delta_j(1 \otimes \pi_L)(\pi_L^{sc} \otimes 1).$$

It follows that

$$\beta^s = \sum_{0 \leq j \leq s,\ j \preceq s} w_j(\pi_L^{jc} \otimes \pi_L^{(s-j)c})$$

with $w_j = \binom{s}{j} u^{s-j}v^j$ for all j such that $0 \le j < s$ and $j \preceq s$, and

$$w_s = v^s + \sum_{0 \le j < s,\, j \not\preceq s} \delta_j(1 \otimes \pi_L).$$

Since $u, v \in (\mathfrak{O}_L \otimes_{\mathfrak{O}_K} \mathfrak{O}_L)^\times$ we have $w_j \in (\mathfrak{O}_L \otimes_{\mathfrak{O}_K} \mathfrak{O}_L)^\times$ for all $0 \le j < s$ such that $j \preceq s$. Since $\delta_j(1 \otimes \pi_L)$ is in the maximal ideal of the local ring $\mathfrak{O}_L \otimes_{\mathfrak{O}_K} \mathfrak{O}_L$, we also have $w_s \in (\mathfrak{O}_L \otimes_{\mathfrak{O}_K} \mathfrak{O}_L)^\times$. The lemma now follows by Proposition 14.38. □

LEMMA 14.40. *Let $c_s \in K$ for $0 \le s < p^n$. Then*

$$D\left(\sum_{s=0}^{p^n-1} c_s\beta^s\right) = \bigcup_{s=0}^{p^n-1} D(c_s\beta^s).$$

PROOF. Set $\alpha = \sum_{s=0}^{p^n-1} c_s\beta^s$. By Proposition 14.23(a) we have

$$D(\alpha) \subset \bigcup_{s=0}^{p^n-1} D(c_s\beta^s).$$

Let $[a, b]$ be a minimal element of $\bigcup_{s=0}^{p^n-1} D(c_s\beta^s)$. Then $[a, b] \in G(c_s\beta^s)$ for some $0 \le s < p^n$. By Lemma 14.39 we have $a + b \equiv sd(\beta) \pmod{p^n}$. Let $t \in \mathbb{S}_{p^n}$ with $t \neq s$. Since $p \nmid d(\beta)$ we have $a + b \not\equiv td(\beta) \pmod{p^n}$, and hence $[a, b] \notin G(c_t\beta^t)$ by Lemma 14.39. It follows from the minimality of $[a, b]$ that $[a, b] \notin D(c_t\beta^t)$. Therefore $[a, b] \in D(\alpha)$, so we get $D(\alpha) = \bigcup_{s=0}^{p^n-1} D(c_s\beta^s)$. □

For $0 \le s < p^n$ set $\xi^{(s)} = \varphi(\beta^s)$ and define $f_{\xi^{(s)}} : \mathbb{Z} \to \mathbb{Z}$ as in (14.20).

LEMMA 14.41. *For $h \in \mathbb{Z}$ we have*

$$f_{\xi^{(s)}}(h) = \min\left\{(s-j)d(\beta) + p^n \left\lceil \frac{jd(\beta) + i_0 + h}{p^n} \right\rceil \,\middle|\, j \preceq s\right\}.$$

PROOF. Using Proposition 14.20 we get

$$\begin{aligned} f_{\xi^{(s)}}(h) &= \min\{a \in \mathbb{Z} \mid [a, -h - i_0] \in D(\beta^s)\} \\ &= \min\{a \in \mathbb{Z} \mid \exists b \in \mathbb{Z} \text{ such that } [a, b] \in D(\beta^s) \text{ and } -b - i_0 \ge h\} \\ &= \min\{a \in \mathbb{Z} \mid \exists b \in \mathbb{Z} \text{ such that } [a, b] \in G(\beta^s) \text{ and } -b - i_0 \ge h\}. \qquad (14.33) \end{aligned}$$

By Lemma 14.39 the elements of $G(\beta^s)$ have the form $[jd(\beta), (s-j)d(\beta)]$ with $j \preceq s$. For every $e \in \mathbb{Z}$ we have

$$[jd(\beta), (s-j)d(\beta)] = [jd(\beta) + ep^n, (s-j)d(\beta) - ep^n].$$

For $0 \le j \le s$ such that $j \preceq s$ set

$$\begin{aligned} e_j &= \min\{e \in \mathbb{Z} \mid -((s-j)d(\beta) - ep^n) - i_0 \ge h\} \\ &= \left\lceil \frac{(s-j)d(\beta) + i_0 + h}{p^n} \right\rceil. \end{aligned}$$

It follows from (14.33) that

$$\begin{aligned} f_{\xi^{(s)}}(h) &= \min\{jd(\beta) + e_jp^n \mid j \preceq s\} \\ &= \min\left\{jd(\beta) + p^n \left\lceil \frac{(s-j)d(\beta) + i_0 + h}{p^n} \right\rceil \,\middle|\, j \preceq s\right\}. \end{aligned}$$

Since $j \preceq s$ if and only if $s - j \preceq s$ the lemma follows. □

LEMMA 14.42. *Let $h \in \mathbb{Z}$ and for $h \le i < h+p^n$ let $y_i \in L$ with $v_L(y_i) = i$. Then $T = \{y_h, y_{h+1}, \dots, y_{h+p^n-1}\}$ is an $\mathfrak{O}_K$-basis for $\mathfrak{p}_L^h$.*

PROOF. Let $c_i \in K$ for $h \le i < h+p^n$ and set

$$z = c_h y_h + c_{h+1} y_{h+1} + \cdots + c_{h+p^n-1} y_{h+p^n-1}. \tag{14.34}$$

Since $v_L(y_i) = i$ are distinct modulo p^n we get

$$v_L(z) = \min\{v_L(c_i y_i) : h \le i < h+p^n\}.$$

Hence $z = 0$ if and only if $c_i = 0$ for $h \le i < h+p^n$. Therefore T is a K-basis for L. It follows that every $z \in L$ can be written uniquely in the form (14.34). Furthermore, $z \in \mathfrak{p}_L^h$ if and only if $c_i \in \mathfrak{O}_K$ for $h \le i < h+p^n$. □

For $g, h, s \in \mathbb{Z}$ such that $0 \le s < p^n$ define

$$c(g,h) = \left\lfloor \frac{gd(\beta) - h}{p^n} \right\rfloor \tag{14.35}$$

$$w(s,h) = \min\{c(s-j,h) - c(-j,h+i_0) \mid j \preceq s\}. \tag{14.36}$$

The following is a version of Theorem 3.6 of [**BCE18**] which applies to H-stable extensions, rather than to extensions with an H-scaffold. (The theory of H-scaffolds will be developed in Chapter 15.)

THEOREM 14.43. *Let L/K be an H-stable extension of degree p^n, let $h \in \mathbb{Z}$, and let $\beta \in L \otimes_K L$ satisfy $\varphi(\beta) \in K \otimes_K H$, $p \nmid d(\beta)$, and $G(\beta) = \{[0,d(\beta)], [d(\beta),0]\}$. Then*

(a) *An $\mathfrak{O}_K$-basis for the associated order $\mathfrak{A}(\mathfrak{p}_L^h)$ of $\mathfrak{p}_L^h$ is given by*
$$B = \{\pi_K^{-w(s,h)} \varphi(\beta^s) \mid s \in \mathbb{S}_{p^n}\}.$$
(b) *If $w(s,h) = c(s,h) - c(0,h+i_0)$ for all $s \in \mathbb{S}_{p^n}$ then $\mathfrak{p}_L^h$ is free over $\mathfrak{A}(\mathfrak{p}_L^h)$.*
(c) *If $\mathfrak{p}_L^h$ is free over $\mathfrak{A}(\mathfrak{p}_L^h)$ and $\mathfrak{A}(\mathfrak{p}_L^h)$ has a unique maximal two-sided ideal then $w(s,h) = c(s,h) - c(0,h+i_0)$ for all $s \in \mathbb{S}_{p^n}$.*

PROOF. (a) Using Lemma 14.41 we get

$$\begin{aligned}
\left\lfloor \frac{f_{\xi^{(s)}}(h) - h}{p^n} \right\rfloor &= \min\left\{ \left\lfloor \frac{(s-j)d(\beta) - h}{p^n} \right\rfloor + \left\lceil \frac{jd(\beta) + i_0 + h}{p^n} \right\rceil \,\middle|\, j \preceq s \right\} \\
&= \min\left\{ \left\lfloor \frac{(s-j)d(\beta) - h}{p^n} \right\rfloor - \left\lfloor \frac{-jd(\beta) - i_0 - h}{p^n} \right\rfloor \,\middle|\, j \preceq s \right\} \\
&= \min\{c(s-j,h) - c(-j,h+i_0) \mid j \preceq s\} \\
&= w(s,h).
\end{aligned}$$

It follows that

$$f_{\xi^{(s)}}(h) - h - p^n < p^n w(s,h) \le f_{\xi^{(s)}}(h) - h,$$

and hence that $f_{\pi_K^{-w(s,h)}\xi^{(s)}}(h) \ge h$ and $f_{\pi_K^{-w(s,h)-1}\xi^{(s)}}(h) < h$. Thus for $0 \le s < p^n$ we have $\pi_K^{-w(s,h)} \varphi(\beta^s) \in \mathfrak{A}(\mathfrak{p}_L^h)$ and $\pi_K^{-w(s,h)-1} \varphi(\beta^s) \notin \mathfrak{A}(\mathfrak{p}_L^h)$. In particular, B is contained in $\mathfrak{A}(\mathfrak{p}_L^h)$.

Suppose $\mu = \varphi(\alpha) \in \mathfrak{A}(\mathfrak{p}_L^h)$. By Proposition 14.30(b) there are $a_s \in K$ such that $\alpha = \sum_{s=0}^{p^n-1} a_s \beta^s$. By Corollary 14.21 we have $[h-1, -h-i_0] \notin D(\alpha)$, and by Lemma 14.40 we have $D(\alpha) = \bigcup_{s=0}^{p^n-1} D(a_s \beta^s)$. Hence $[h-1, -h-i_0] \notin D(a_s \beta^s)$ for all $0 \le s < p^n$. Applying Corollary 14.21 again we get $a_s \varphi(\beta^s) \in \mathfrak{A}(\mathfrak{p}_L^h)$. It follows from the preceding paragraph that $a_s \in \pi_K^{-w(s,h)} \mathfrak{O}_K$. Hence $\mu \in \mathrm{Span}_{\mathfrak{O}_K}(B)$. We conclude that B is an $\mathfrak{O}_K$-basis for $\mathfrak{A}(\mathfrak{p}_L^h)$.

(b) Let $\rho \in L$ satisfy $v_L(\rho) = -i_0 - p^n c(0, h+i_0)$; then $v_L(\rho) \equiv -i_0 \pmod{p^n}$ and $h \le v_L(\rho) < h + p^n$. It follows from Lemma 14.39 that

$$[sd(\beta), 0] = [sd(\beta) - p^n c(0, h+i_0), p^n c(0, h+i_0)]$$

is in $G(\beta^s)$. Hence by Proposition 14.22 and the assumption on $w(s,h)$ we get

$$\begin{aligned} v_L(\pi_K^{-w(s,h)}\xi^{(s)} \cdot \rho) &= -p^n w(s,h) + sd(\beta) - p^n c(0, h+i_0) \\ &= -p^n(w(s,h) + c(0,h+i_0)) + sd(\beta) \\ &= -p^n c(s,h) + sd(\beta). \end{aligned} \tag{14.37}$$

By (14.35) we have

$$\frac{sd(\beta) - h}{p^n} - 1 < c(s,h) \le \frac{sd(\beta) - h}{p^n}. \tag{14.38}$$

Therefore by (14.37) we get

$$h \le v_L(\pi_K^{-w(s,h)}\xi^{(s)} \cdot \rho) < h + p^n. \tag{14.39}$$

Set $T = \{\pi_K^{-w(s,h)}\xi^{(s)} \cdot \rho \mid s \in \mathbb{S}_{p^n}\}$ and $A = \{v_L(t) \mid t \in T\}$. Since $p \nmid d(\beta)$, it follows from (14.37) that A is a complete set of residues modulo p^n. Using (14.39) we deduce that

$$A = \{h, h+1, \dots, h+p^n-1\}. \tag{14.40}$$

It follows from Lemma 14.42 that $\mathfrak{p}_L^h$ is a free $\mathfrak{O}_K$-module with basis T. Using (a) we conclude that $\mathfrak{p}_L^h = \mathfrak{A}(\mathfrak{p}_L^h) \cdot \rho$ is a free $\mathfrak{A}(\mathfrak{p}_L^h)$-module of rank 1.

(c) Recall from Proposition 14.1 that $\theta \in H$ induces the trace map $\mathrm{Tr}_{L/K} : L \to K$. We have $\theta = \varphi(1 \otimes 1)$ and $G(1 \otimes 1) = \{[0,0]\}$. It follows from Proposition 14.22 that for all $\rho \in L$ with $v_L(\rho) \equiv -i_0 \pmod{p^n}$ we have $v_L(\theta \cdot \rho) = v_L(\rho) + i_0$. In particular, $\theta \cdot \rho \ne 0$. Since $\mathrm{Tr}_{L/K}$ is an onto K-linear map, its kernel has K-dimension $p^n - 1$. Since $\ker(\mathrm{Tr}_{L/K}) \cap K\rho = \{0\}$ we get a K-vector space decomposition $L = \ker(\mathrm{Tr}_{L/K}) \oplus K\rho$. For each $j \in \mathbb{Z}$ with $j \not\equiv i_0 \pmod{p^n}$ choose $y_j' \in L$ with $v_L(y_j') = j$. Write $y_j' = y_j + z_j$ with $y_j \in \ker(\mathrm{Tr}_{L/K})$ and $z_j \in K\rho$. Then $v_L(y_j) \not\equiv -i_0 \pmod{p^n}$, so $v_L(y_j) \ne v_L(z_j)$. Hence $v_L(y_j) = v_L(y_j') = j$. For $j \in \mathbb{Z}$ with $j \equiv -i_0 \pmod{p^n}$ let y_j be any element of L with $v_L(y_j) = j$. By Lemma 14.42 we see that $\{y_j \mid h \le j < h+p^n\}$ is an $\mathfrak{O}_K$-basis for $\mathfrak{p}_L^h$.

Since $\mathfrak{p}_L^h$ is a free $\mathfrak{A}(\mathfrak{p}_L^h)$-module of rank 1, there is an $\mathfrak{A}(\mathfrak{p}_L)$-module isomorphism $\psi : \mathfrak{p}_L^h \to \mathfrak{A}(\mathfrak{p}_L^h)$. Hence there are $c_j \in \mathfrak{O}_K$ such that

$$\psi(c_h y_h + \cdots + c_{h+p^n-1} y_{h+p^n-1}) = 1. \tag{14.41}$$

For $h \le j < h + p^n$ such that $j \not\equiv -i_0 \pmod{p^n}$ we have

$$\theta\psi(c_j y_j) = \psi(\theta \cdot (c_j y_j)) = \psi(0) = 0.$$

Therefore $\psi(c_j y_j)$ is not a unit in $\mathfrak{A}(\mathfrak{p}_L^h)$. It follows that $\psi(c_j y_j)$ is in the unique maximal ideal of $\mathfrak{A}(\mathfrak{p}_L^h)$. Set $b = -i_0 - p^n c(0, h+i_0)$. Then $h \le b < h + p^n$ and $b \equiv -i_0 \pmod{p^n}$. Using (14.41) we deduce that $c_b\psi(y_b) = \psi(c_b y_b) \in \mathfrak{A}(\mathfrak{p}_L^h)^\times$. Therefore $\psi(y_b) \in \mathfrak{A}(\mathfrak{p}_L^h)^\times$ and $\mathfrak{p}_L^h = \mathfrak{A}(\mathfrak{p}_L^h) \cdot y_b$.

Let $\mu \in H$ satisfy $\mu \cdot y_b \in \mathfrak{p}_L^h$. Then

$$\mu\psi(y_b) = \psi(\mu \cdot y_b) \in \psi(\mathfrak{p}_L^h) = \mathfrak{A}(\mathfrak{p}_L^h).$$

Since $\psi(y_b) \in \mathfrak{A}(\mathfrak{p}_L^h)^\times$ it follows that $\mu \in \mathfrak{A}(\mathfrak{p}_L^h)$. On the other hand, if $\mu \in \mathfrak{A}(\mathfrak{p}_L^h)$ then $\mu \cdot y_b \in \mathfrak{p}_L^h$ since $y_b \in \mathfrak{p}_L^h$. Hence

$$\mathfrak{A}(\mathfrak{p}_L^h) = \{\mu \in H \mid \mu \cdot y_b \in \mathfrak{p}_L^h\}. \tag{14.42}$$

Suppose $w(s,h) \neq c(s,h) - c(0,h+i_0)$ for some $s \in \mathbb{S}_{p^n}$. Then

$$w(s,h) + 1 \leq c(s,h) - c(0,h+i_0).$$

By Lemma 14.39, Proposition 14.22, and (14.38) we get

$$\begin{aligned} v_L(\pi_K^{-w(s,h)-1}\xi^{(s)} \cdot y_b) &= -p^n w(s,h) - p^n - i_0 - p^n c(0,h+i_0) + sd(\beta) + i_0 \\ &\geq -p^n c(s,h) + sd(\beta) \\ &\geq h. \end{aligned}$$

Hence (14.42) implies $\pi_K^{-w(s,h)-1}\varphi(\beta^s) \in \mathfrak{A}(\mathfrak{p}_L^h)$. This contradicts (a), so we must have $w(s,h) = c(s,h) - c(0,h+i_0)$ for all $s \in \mathbb{S}_{p^n}$. □

We now apply the results of this section to the example from Section 14.4.

EXAMPLE 14.44. Let $K = \mathbb{Q}_p(\zeta_p)$ and $L = \mathbb{Q}_p(\zeta_{p^2})$. We saw in Example 14.29 that L/K is a totally ramified G-Galois extension with $G \cong C_p$, and that L/K is a $K[G]$-stable extension with respect to

$$\begin{aligned} \beta &= -1 \otimes 1 + \zeta_{p^2}^{-1} \otimes \zeta_{p^2} \\ &= \pi_L \otimes 1 - 1 \otimes \pi_L + \cdots . \end{aligned}$$

Since the ramification break of L/K is $b = p-1$, we have

$$\begin{aligned} v_L(\mathfrak{D}_{L/K}) &= (p-1)(b+1) = p^2 - p \\ i_0 &= v_L(\mathfrak{D}_{L/K}) - p + 1 = (p-1)^2. \end{aligned}$$

It follows from Corollary 14.35 that $\mathfrak{p}_L^{-(p-1)^2}$ is free over its associated order. Since $-(p-1)^2 \equiv p-1 \pmod{p}$ this implies that $\mathfrak{p}_L^{p-1}$ is free over $\mathfrak{A}(\mathfrak{p}_L^{p-1})$.

We claim that $\mathfrak{O}_L$ is also free over its associated order. Since $d(\beta) = 1$ we get $c(x,y) = \lfloor (x-y)/p \rfloor$ by (14.35). Hence for $0 \leq j \leq s < p$ we have $c(s-j,0) = 0$ and

$$\begin{aligned} c(-j,i_0) &= \left\lfloor \frac{-j-(p-1)^2}{p} \right\rfloor \\ &= -p+2+\left\lfloor \frac{-j-1}{p} \right\rfloor \\ &= -p+1. \end{aligned}$$

Therefore by (14.36) for $0 \leq s < p$ we have

$$w(s,0) = 0 - (-p+1) = p-1 = c(s,0) - c(0,i_0).$$

It follows by Theorem 14.43(b) that $\mathfrak{O}_L$ is free over $\mathfrak{A}(\mathfrak{O}_L)$.

14.6. A non-classical example

In this section we give an example of an extension L/K of local fields with a non-classical H-Galois structure. We show that L/K is H-stable, and use this fact to obtain information about the H-Galois module structure of L/K.

Let K be a local field of characteristic 2, let $b > 0$ with $2 \nmid b$, and let $\omega \in K$ with $v_K(\omega) = -b$. Using Artin-Schreier-Witt theory we construct a cyclic extension L/K of degree 4: Let $x_1 \in K^{sep}$ be a root of the Artin-Schreier polynomial $X^2 - X - \omega$, and set $K_1 = K(x_1)$. Since $v_K(\omega) < 0$ and $v_K(\omega)$ is odd, K_1/K is a cyclic extension of degree 2. Now let $x_2 \in K^{sep}$ be a root of $X^2 - X - \omega x_1$, and set $L = K_1(x_2)$. The first 2-adic Witt addition polynomial is $S_1((X_0, X_1); (Y_0, Y_1)) = X_1 + Y_1 - X_0Y_0$ (see Section 10.3). Hence by [**Wit37**], L/K is a Galois extension with Galois group isomorphic to C_4. The lower ramification breaks of L/K are $-v_K(\omega) = b$ and $-v_{K_1}(\omega x_1) = 3b$. Therefore the L-valuation of the different $\mathfrak{D}_{L/K}$ of L/K is

$$v_L(\mathfrak{D}_{L/K}) = 2(b+1) + (3b+1) = 5b+3,$$

and $i_0 = v_L(\mathfrak{D}_{L/K}) - 4 + 1 = 5b$.

Set $G = \mathrm{Gal}(L/K)$ and let σ be a generator of G. Then $\sigma(x_1) = x_1 + 1$ and

$$\begin{aligned}(\sigma(x_2) - x_2)^2 - (\sigma(x_2) - x_2) &= \sigma(x_2^2 - x_2) - (x_2^2 - x_2) \\ &= \sigma(\omega x_1) - \omega x_1 = \omega.\end{aligned}$$

Thus $\sigma(x_2) - x_2 \in \{x_1, x_1 + 1\}$. Therefore by replacing σ by σ^3 if necessary, we may assume that $\sigma(x_2) = x_2 + x_1$. Let N be the Klein 4-subgroup of $\mathrm{Perm}(G)$. Then N is a regular subgroup of $\mathrm{Perm}(G)$ which is normalized by $\lambda(G)$. Let H be the K-subalgebra of $L[N]$ fixed by the action of G. Then L/K is an H-Galois extension.

We label the elements of N as follows:

$$\begin{aligned}\eta_0 &= \mathrm{id}_G \\ \eta_1 &= (1_G\ \sigma)(\sigma^2\ \sigma^3) \\ \eta_2 &= (1_G\ \sigma^2)(\sigma\ \sigma^3) \\ \eta_3 &= (1_G\ \sigma^3)(\sigma\ \sigma^2).\end{aligned}$$

The stabilizers of these elements in G are $G_{\eta_0} = G_{\eta_2} = G$ and $G_{\eta_1} = G_{\eta_3} = \langle \sigma^2 \rangle$. Hence $L^{G_{\eta_0}} = L^{G_{\eta_2}} = K$ and $L^{G_{\eta_1}} = L^{G_{\eta_3}} = K_1$. Let $\alpha \in K$ and $\delta \in K_1$. Using Lemma 14.10 we get the following elements of H:

$$\begin{aligned}\chi_{\eta_0}(\alpha) &= \alpha\eta_0 \\ \chi_{\eta_1}(\delta) &= \delta\eta_1 + \sigma(\delta)\eta_3 \\ \chi_{\eta_2}(\alpha) &= \alpha\eta_2 \\ \chi_{\eta_3}(\delta) &= \delta\eta_3 + \sigma(\delta)\eta_1.\end{aligned}$$

For $y \in L$ we have

$$\begin{aligned}\chi_{\eta_0}(\alpha) \cdot y &= \alpha y \\ \chi_{\eta_1}(\delta) \cdot y &= \delta\sigma(y) + \sigma(\delta)\sigma^3(y) \\ \chi_{\eta_2}(\alpha) \cdot y &= \alpha\sigma^2(y) \\ \chi_{\eta_3}(\delta) \cdot y &= \delta\sigma^3(y) + \sigma(\delta)\sigma(y).\end{aligned}$$

Let $\xi = \chi_{\eta_0}(1) + \chi_{\eta_1}(x_1) \in H$. Using the formulas above we find that for $y \in L$ we have

$$\xi = \eta_0 + x_1\eta_1 + (x_1+1)\eta_3$$
$$\xi \cdot y = y + x_1\sigma(y) + (x_1+1)\sigma^3(y).$$

In particular, we find that

$$\begin{aligned} \xi \cdot 1 &= 1 + x_1 + (x_1+1) = 0 \\ \xi \cdot x_1 &= x_1 + x_1(x_1+1) + (x_1+1)(x_1+1) = 1 \\ \xi \cdot x_2 &= x_2 + x_1(x_2+x_1) + (x_1+1)(x_2+x_1+1) = 1 \\ \xi \cdot (x_1x_2) &= x_1x_2 + x_1(x_1+1)(x_2+x_1) + (x_1+1)^2(x_2+x_1+1) \\ &= x_2 + x_1 + 1. \end{aligned}$$

Since $v_L(1) = 0$, $v_L(x_1) = -2b$, $v_L(x_2) = -3b$, and $v_L(x_1x_2) = -5b$ form a complete set of congruence classes modulo 4, $\{1, x_1, x_2, x_1x_2\}$ is a K-basis for L. Let $y \in L^\times$ and write

$$y = c_0 + c_1x_1 + c_2x_2 + c_3x_1x_2 \tag{14.43}$$

with $c_i \in K$. Then

$$\begin{aligned} v_L(y) &= \min\{v_L(c_0), v_L(c_1x_1), v_L(c_2x_2), v_L(c_3x_1x_2)\} \\ &= \min\{v_L(c_0), v_L(c_1) - 2b, v_L(c_2) - 3b, v_L(c_3) - 5b\}. \end{aligned}$$

We also get

$$\xi \cdot y = c_1 + c_2 + c_3(x_2 + x_1 + 1)$$

and hence

$$\begin{aligned} v_L(\xi \cdot y) &\geq \min\{v_L(c_1), v_L(c_2), v_L(c_3(x_2+x_1+1))\} \\ &= \min\{v_L(c_1), v_L(c_2), v_L(c_3) - 3b\}, \end{aligned}$$

with equality if the minimum is achieved only once.

PROPOSITION 14.45. *The extension L/K constructed above is H-stable.*

PROOF. We first show that L/K is H-semistable. Let $y \in L^\times$. We claim that

$$v_L(\xi \cdot y) > v_L(y) + 2b \text{ if } v_L(y) \equiv 0, b \pmod 4$$
$$v_L(\xi \cdot y) = v_L(y) + 2b \text{ if } v_L(y) \equiv 2b, 3b \pmod 4.$$

It suffices to prove these claims in the cases where $v_L(y) \in \{0, -3b, -2b, -5b\}$. Suppose $v_L(y) = 0$. Then $v_L(c_0) = 0$, $v_L(c_1) > 2b$, $v_L(c_2) > 3b$, and $v_L(c_3) > 5b$. Hence

$$v_L(\xi \cdot y) > \min\{2b, 3b, 2b\} = 2b = 0 + 2b.$$

Suppose $v_L(y) = -3b$. Then $v_L(c_0) > -3b$, $v_L(c_1) > -b$, $v_L(c_2) = 0$, and $v_L(c_3) > 2b$. Hence

$$v_L(\xi \cdot y) > \min\{-b, 0, -b\} = -b = -3b + 2b.$$

Suppose $v_L(y) = -2b$. Then $v_L(c_0) > -2b$, $v_L(c_1) = 0$, $v_L(c_2) > b$, and $v_L(c_3) > 3b$. Hence $v_L(c_3(x_2+x_1+1)) > 0$, so we get $v_L(\xi \cdot y) = 0 = -2b + 2b$. Finally, suppose $v_L(y) = -5b$. Then $v_L(c_0) > -5b$, $v_L(c_1) > -3b$, $v_L(c_2) > -2b$, and $v_L(c_3) = 0$. Hence $v_L(c_3(x_2+x_1+1)) = -3b$, so we get $v_L(\xi \cdot y) = -3b = -5b + 2b$.

It follows from the above that for $y \in L^\times$ we have $v_L(\xi \cdot y) - v_L(y) \geq 2b$, with equality if and only if either $v_L(y) \equiv 2b \pmod 4$ or $v_L(y) \equiv 3b \pmod 4$. In terms of the function $f_\xi : \mathbb{Z} \to \mathbb{Z}$ defined in (14.20), we get $f_\xi(a) \geq a + 2b$, with equality if

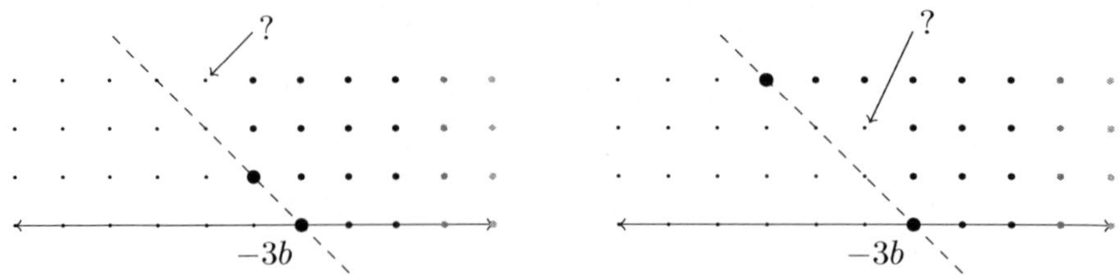

Case 1: $b \equiv 1 \pmod 4$ Case 2: $b \equiv 3 \pmod 4$

FIGURE 3. Diagrams for Proposition 14.45

and only if either $a \equiv 2b \pmod 4$ or $a \equiv 3b \pmod 4$. Let β be the unique element of $L \otimes_K L$ such that $\varphi(\beta) = \xi$. Since $i_0 = 5b$, it follows from Proposition 14.20 that $[-3b, 0]$ and $[0, -3b]$ are elements of $D(\beta)$, but $[-b, -2b]$ and $[-2b, -b]$ are not. Therefore the diagonal of β is $N(\beta) = \{[-3b, 0], [0, -3b]\}$. Since b is odd it follows that the extension L/K is H-semistable in the sense of Definition 14.26.

We now prove that the H-semistable extension L/K is actually H-stable. For this we need to show that $G(\beta) = N(\beta)$. Suppose $b \equiv 1 \pmod 4$. Then

$$N(\beta) = \{[-3b, 0], [0, -3b]\} = \{[-3b, 0], [-3b - 1, 1]\}.$$

To verify that $G(\beta) = N(\beta)$ it suffices to show that $[-3b - 2, 3] \notin D(\beta)$ (see Figure 3). Let $y \in L$ with $v_L(y) = -3 - i_0 = -3 - 5b$. Express y as in (14.43). Then $v_L(c_0) = -3 - 5b$, $v_L(c_1) \geq -1 - 3b$, $v_L(c_2) \geq -2 - 2b$, and $v_L(c_3) \geq 0$, so we get

$$v_L(\xi \cdot y) \geq \min\{-1 - 3b, -2 - 2b, -3b\} = -1 - 3b.$$

Hence $f_\xi(-3 - i_0) \geq -1 - 3b > -3b - 2$. By Proposition 14.20 we deduce that $[-3b - 2, 3] \notin D(\beta)$, and hence that $G(\beta) = N(\beta)$.

Suppose $b \equiv 3 \pmod 4$. Then $N(\beta) = \{[-3b, 0], [-3b - 3, 3]\}$, so we need to show that $[-3b - 1, 2] \notin D(\beta)$. Let $y \in L$ with $v_L(y) = -2 - i_0 = -2 - 5b$. Express y as in (14.43). Then $v_L(c_0) \geq -1 - 5b$, $v_L(c_1) \geq 1 - 3b$, $v_L(c_2) = -2 - 2b$, and $v_L(c_3) \geq 0$, so we get

$$v_L(\xi \cdot y) \geq \min\{1 - 3b, -2 - 2b, -3b\} = -3b.$$

Hence $f_\xi(-2 - i_0) \geq -3b > -3b - 1$. By Proposition 14.20 we get $[-3b - 1, 2] \notin D(\beta)$, and hence $G(\beta) = N(\beta)$. □

We now use the proposition to get information about the H-Galois module structure of L/K. Since L/K is H-semistable, it follows from Corollary 14.35 that the ideal $\mathfrak{p}_L^{-i_0} = \mathfrak{p}_L^{-5b}$ is free over its associated order $\mathfrak{A}(\mathfrak{p}_L^{-5b})$ in H. Since L/K is H-stable, Theorem 14.43 gives sufficient conditions for an ideal $\mathfrak{p}_L^h$ to be free over its associated order $\mathfrak{A}(\mathfrak{p}_L^h)$. We will use these conditions to show that $\mathfrak{O}_L$ is free over $\mathfrak{A}(\mathfrak{O}_L)$.

Using (14.36) we compute $w(s,0)$ for $0 \le s < 4$:

$$w(0,0) = c(0,0) - c(0,5b) = -\left\lfloor \frac{-5b}{4} \right\rfloor$$

$$\begin{aligned} w(1,0) &= \min\{c(1,0) - c(0,5b),\, c(0,0) - c(-1,5b)\} \\ &= \min\left\{\left\lfloor \frac{-3b}{4} \right\rfloor - \left\lfloor \frac{-5b}{4} \right\rfloor, -\left\lfloor \frac{-2b}{4} \right\rfloor\right\} \\ &= \left\lfloor \frac{-3b}{4} \right\rfloor - \left\lfloor \frac{-5b}{4} \right\rfloor \end{aligned}$$

$$\begin{aligned} w(2,0) &= \min\{c(2,0) - c(0,5b),\, c(0,0) - c(-2,5b)\} \\ &= \min\left\{\left\lfloor \frac{-6b}{4} \right\rfloor - \left\lfloor \frac{-5b}{4} \right\rfloor, -\left\lfloor \frac{b}{4} \right\rfloor\right\} \\ &= \left\lfloor \frac{-6b}{4} \right\rfloor - \left\lfloor \frac{-5b}{4} \right\rfloor \end{aligned}$$

$$\begin{aligned} w(3,0) &= \min\{c(3-j,0) - c(-j,5b) : 0 \le j \le 3\} \\ &= \min\{-b,-b,-b,-b\} = -b. \end{aligned}$$

For $0 \le s < 4$ we get

$$w(s,0) = c(s,0) - c(0,5b) = c(s,0) - c(0,i_0).$$

It follows by Theorem 14.43(b) that $\mathfrak{O}_L$ is free over $\mathfrak{A}(\mathfrak{O}_L)$.

CHAPTER 15

Hopf-Galois scaffolds

The theory of scaffolds for totally ramified p-extensions of local fields was developed mainly by G. G. Elder and N. Byott. It provides an alternative to Bondarko's theory for answering questions about the Hopf-Galois module theory of an extension L/K of local fields. In some cases, a scaffold allows one to obtain more detailed information about Hopf-Galois module structure than is provided by Bondarko's theory, which is focused largely on determining whether an ideal $\mathfrak{p}_L^h$ in L is free over its associated order. For instance, in Theorem 3.6 of [**BCE18**], a method is given for computing the minimum number of generators for an ideal $\mathfrak{p}_L^h$ in L as a module over its associated order $\mathfrak{A}(\mathfrak{p}_L^h)$, assuming that L/K has a scaffold with sufficiently large precision (see Definition 15.1). Unfortunately, extensions that admit scaffolds are comparatively rare, perhaps even less common than semistable extensions. In fact Theorem 15.9 shows that if L/K is a Hopf-Galois extension which admits a scaffold then L/K is semistable. This chapter is not intended as a general overview of the theory of H-scaffolds. Instead we focus on the connection between H-scaffolds and H-semistable extensions given by Theorem 15.9. For a more complete exposition of the theory of H-scaffolds see [**BCE18**].

15.1. H-scaffolds

Let L/K be a totally ramified H-Galois extension of local fields of degree p^n. The notion of an H-scaffold for L/K was introduced in [**BCE18**]. An H-scaffold, if it exists, allows one to answer various questions about the H-Galois module structure of L/K which would otherwise be difficult to address. Among these questions are some which, as far as we know, have not been answered for H-semistable or H-stable extensions (see for instance Theorems 3.6 and 4.2 of [**BCE18**]).

In order to give the definition of an H-scaffold we recall some notation from Section 14.5. Let $\mathbb{S}_{p^n} = \{0, 1, \ldots, p^n - 1\}$ and write $s \in \mathbb{S}_{p^n}$ in base p as

$$s = s_{(0)}p^0 + s_{(1)}p^1 + \cdots + s_{(n-1)}p^{n-1}$$

with $0 \leq s_{(i)} < p$. Define a partial order on $\mathbb{S}_{p^n}$ by $s \preceq t$ if $s_{(i)} \leq t_{(i)}$ for $0 \leq i \leq n-1$. Let $b_1, b_2, \ldots, b_n$ be integers which are relatively prime to p and define $\mathfrak{b} : \mathbb{S}_{p^n} \to \mathbb{Z}$ by

$$\mathfrak{b}(s) = s_{(0)}p^0 b_n + s_{(1)}p^1 b_{n-1} + \cdots + s_{(n-1)}p^{n-1} b_1.$$

Let $r : \mathbb{Z} \to \mathbb{S}_{p^n}$ be the function which maps $t \in \mathbb{Z}$ onto its least nonnegative residue modulo p^n. The function $r \circ (-\mathfrak{b}) : \mathbb{S}_{p^n} \to \mathbb{S}_{p^n}$ is a bijection since $p \nmid b_i$. Therefore we may define $\mathfrak{a} : \mathbb{S}_{p^n} \to \mathbb{S}_{p^n}$ to be the inverse of $r \circ (-\mathfrak{b})$. We extend $\mathfrak{a}$ to a function from $\mathbb{Z}$ to $\mathbb{S}_{p^n}$ by setting $\mathfrak{a}(t) = \mathfrak{a}(r(t))$ for $t \in \mathbb{Z}$. We easily see that $\mathfrak{a}(-\mathfrak{b}(s)) = s$ for $s \in \mathbb{S}_{p^n}$ and $\mathfrak{b}(\mathfrak{a}(t)) \equiv -t \pmod{p^n}$ for $t \in \mathbb{Z}$. Recall that $v_L : L \to \mathbb{Z} \cup \{\infty\}$ is the normalized valuation on L.

DEFINITION 15.1 ([**BCE18**], Def. 2.6). Let $\mathfrak{c} \geq 1$. An H-scaffold $(\{\Psi_i\}, \lambda_t\})$ for L/K with precision $\mathfrak{c}$ and shift parameters $b_1, b_2, \ldots, b_n$ consists of elements $\Psi_i \in H$ for $1 \leq i \leq n$ and $\lambda_t \in L$ for all $t \in \mathbb{Z}$ such that the following hold:

(i) $v_L(\lambda_t) = t$ for all $t \in \mathbb{Z}$.
(ii) $\lambda_{t_1}\lambda_{t_2}^{-1} \in K$ whenever $t_1 \equiv t_2 \pmod{p^n}$.
(iii) $\Psi_i(1) = 0$ for $1 \leq i \leq n$.
(iv) For $1 \leq i \leq n$ and $t \in \mathbb{Z}$ there exists $u_{it} \in \mathfrak{O}_K^\times$ such that the following congruence modulo $\lambda_{t+p^{n-i}b_i}\mathfrak{p}_L^{\mathfrak{c}}$ holds:

$$\Psi_i(\lambda_t) \equiv \begin{cases} u_{it}\lambda_{t+p^{n-i}b_i} & \text{if } \mathfrak{a}(t)_{(n-i)} \geq 1, \\ 0 & \text{if } \mathfrak{a}(t)_{(n-i)} = 0. \end{cases}$$

An H-scaffold for L/K with infinite precision consists of the above data with the congruence in (iv) replaced by equality.

Suppose we are in the classical setting, where L/K is G-Galois and $H = K[G]$. We say that the scaffold $(\{\Psi_i\}, \{\lambda_t\})$ is a Galois scaffold for L/K if the shift parameters $b_1 \leq b_2 \leq \cdots \leq b_n$ are the same as the lower ramification breaks of L/K, counted with multiplicity. This is the case encountered most often in the literature.

The idea behind Definition 15.1 is that if we can find p^n elements of L that take on all valuations between 0 and $p^n - 1$, then they form an $\mathfrak{O}_K$-basis for $\mathfrak{O}_L$ because L/K is totally ramified. If the Ψ_i's all map each λ_t to either zero or a K-multiple of an element of the set $\{\lambda_s \mid 0 \leq s < p^n\}$, then we might be able to form a basis of the associated order $\mathfrak{A}(\mathfrak{O}_L)$ of $\mathfrak{O}_L$ using products of the elements Ψ_i. When that occurs, we can determine whether $\mathfrak{O}_L$ is free over $\mathfrak{A}(\mathfrak{O}_L)$.

15.2. Examples of H-scaffolds

To help motivate the concept of scaffold we consider two examples.

15.2.1. A scaffold for a purely inseparable extension of local fields. Our first example is a special case of a family of examples of scaffolds for purely inseparable extensions given in Section 5.2 of [**BCE18**]. See Example 10.4 for another example of a scaffold for a purely inseparable extension. In the example given in this section, the action of $\Psi_1, \ldots, \Psi_n$ on L can be understood in terms of derivations.

Let K be a field of characteristic p and let L be a primitive purely inseparable extension of K. Thus $L \cong K[x]/(x^{p^n} - c)$ for some $n \geq 0$ and $c \in K$ such that $c \notin K^p$. For $0 \leq i < p^n$ define $D_i : L \to L$ to be the K-linear map such that

$$D_i(x^j) = \binom{j}{i} x^{j-i} \tag{15.1}$$

for $0 \leq j < p^n$, with the usual convention that $\binom{j}{i} = 0$ if $i > j$. We may view D_i as a characteristic-p analog of the operator $\dfrac{1}{i!}\dfrac{d^i}{dx^i}$ which is defined even for $i \geq p$. Set

$$D = (D_0, D_1, \ldots, D_{p^n-1}).$$

In Example 10.18 we showed that D is a higher K-derivation on L. Thus $D_0 = \mathrm{id}_L$, $D_i(c) = 0$ for $1 \leq i < p^n$ and $c \in K$, and for $0 \leq m < p^n$ and $a, b \in L$ we have

$$D_m(ab) = \sum_{i=0}^{m} D_i(a)D_{m-i}(b).$$

For $m = 1$ this gives

$$D_1(ab) = D_0(a)D_1(b) + D_1(a)D_0(b) = aD_1(b) + D_1(a)b.$$

Therefore $D_1 : L \to L$ is a K-derivation in the usual sense.

Set $\mathcal{A}(n) = K[D_0, D_1, \ldots, D_{p^n-1}]$. Then $\mathcal{A}(n)$ is a K-Hopf algebra, with comultiplication

$$\Delta(D_r) = \sum_{j=0}^{r} D_j \otimes D_{r-j},$$

counit $\varepsilon(D_r) = \delta_{0,r}$, and antipode $S(D_r) = (-1)^r D_r$. We say that $\mathcal{A}(n)$ is the divided power Hopf algebra of dimension p^n over K. It follows from (15.1) that

$$D_i D_j = \binom{i+j}{j} D_{i+j}$$

for $i+j < p^n$, and $D_i D_j = 0$ for $i+j \geq p^n$. It follows that there is an isomorphism of K-algebras

$$\mathcal{A}(n) \cong K[X_0, X_1, \ldots X_{n-1}]/(X_0^p, \ldots, X_{n-1}^p)$$

which maps D_{p^i} onto X_i. In addition, $\mathcal{A}(n)$ acts on $L = K[x]$ via

$$D_{p^r}(x^a) = \binom{a}{p^r} x^{a-p^r} = a_{(r)} x^{a-p^r},$$

where $a = a_{(0)} + a_{(1)}p + \ldots + a_{(n-1)}p^{n-1}$ is the base-p expansion of a. This action makes L/K an $\mathcal{A}(n)$-Galois extension. (See Proposition 5.4 of [**BCE18**] for details.)

Now assume that K is a local field of characteristic p and let L/K be a purely inseparable extension of degree p^n. Since the residue field of K is assumed to be perfect, L/K is totally ramified. Hence if π_L is a uniformizer for L then $L = K(\pi_L)$. Therefore L/K is a primitive extension, and L/K can be given the structure of an $\mathcal{A}(n)$-Galois extension in multiple ways, depending on a choice of $x \in L$ such that $L = K(x)$. We now construct an action of $\mathcal{A}(n)$ on L such that there exists an $\mathcal{A}(n)$-scaffold on L/K with infinite precision and all shift parameters equal to $p^n - 1$.

Let $x = \pi_L^{-p^n+1}$. Then $L = K[x]$, so we can define an action of $\mathcal{A}(n)$ on L by

$$D_{p^r}(x^a) = \binom{a}{p^r} x^{a-p^r} = a_{(r)} x^{a-p^r}$$

for $0 \leq r < n$ and $0 \leq a < p^n$. Let $t \in \mathbb{Z}$. Since we are assuming that all shift parameters are equal to $p^n - 1$ we have $\mathfrak{b}(s) = (p^n - 1)s$ for $s \in \mathbb{S}_{p^n}$. Hence for $t \in \mathbb{Z}$ we have $\mathfrak{a}(t) \equiv t \pmod{p^n}$, so $\mathfrak{a}(t) = r(t)$ is the least nonnegative residue of t modulo p^n. Write $\mathfrak{a}(t)$ in base p:

$$\mathfrak{a}(t) = \mathfrak{a}(t)_{(0)} + \mathfrak{a}(t)_{(1)}p + \ldots + \mathfrak{a}(t)_{(n-1)}p^{n-1}$$

and let $f_t \in \mathbb{Z}$ be such that $t = (1 - p^n)\mathfrak{a}(t) + p^n f_t$. Define

$$\lambda_t = \frac{\pi_K^{f_t} x^{\mathfrak{a}(t)}}{(\mathfrak{a}(t)_{(0)})! \cdot (\mathfrak{a}(t)_{(1)})! \cdot \ldots \cdot (\mathfrak{a}(t)_{(n-1)})!}.$$

Then by construction we have $v_K(\lambda_t) = t$ and $\lambda_{t_1}\lambda_{t_2}^{-1} = \pi_K^{f_{t_1} - f_{t_2}} \in K$ if $t_1 \equiv t_2 \pmod{p^n}$. For $1 \leq i \leq n$ set $\Psi_i = D_{p^{n-i}}$. Then $\Psi_i(1) = 0$ and

$$\begin{aligned}\Psi_i(\lambda_t) &= D_{p^{n-i}}\lambda_t \\ &= \begin{cases} \lambda_{t+p^{n-i}(p^n-1)} & \text{if } \mathfrak{a}(t)_{(n-i)} > 0, \\ 0 & \text{if } \mathfrak{a}(t)_{(n-i)} = 0. \end{cases}\end{aligned}$$

It follows that $(\{\Psi_i\}, \{\lambda_t\})$ is an $\mathcal{A}(n)$-scaffold for L/K with infinite precision.

More generally, let $0 < b < p^n$ be relatively prime to p. A similar construction gives an $\mathcal{A}(n)$-module structure on L and an $\mathcal{A}(n)$-scaffold with shift parameter b, as explained in Section 5.2 of [**BCE18**].

15.2.2. A Galois scaffold. Let K be a finite extension of $\mathbb{Q}_p$ and let L/K be a totally ramified Galois extension. Assume that $G = \mathrm{Gal}(L/K)$ is cyclic of order p, and let σ be a generator for G. Assume further that the ramification break of L/K is $b = 1$; thus L/K is weakly ramified.

Let $\rho \in L$ satisfy $v_L(\rho) = b = 1$; then $L = K(\rho)$. Set $\Psi_1 = \sigma - 1 \in K[G]$. Define $\lambda_1, \ldots, \lambda_p \in L$ recursively by $\lambda_1 = \rho$ and $\lambda_{r+1} = \Psi_1(\lambda_r)$ for $1 \le r < p$. Then for $1 \le t \le p$ we have $\lambda_t = \Psi_1^{t-1}\rho$ and $v_L(\lambda_t) = t$. Since $(\Psi_1 + 1)^p = \sigma^p = 1$ we get

$$\Psi_1^p = -\sum_{r=1}^{p-1} \binom{p}{r} \Psi_1^r.$$

Since $v_L(\Psi_1^r(\rho)) = r + 1$ for $1 \le r \le p - 1$ it follows that

$$v_L(\Psi_1^p(\rho)) = v_L(-p\Psi_1(\rho)) = v_L(p) + v_L(\Psi_1(\rho)) = pe_K + 2.$$

Thus $v_L(\Psi_1^p(\rho)) \ge p + 1 + \mathfrak{c}$, with $\mathfrak{c} = pe_K - (p-1)$.

For $t \in \mathbb{Z}$ let t_0 satisfy $t_0 \equiv t \pmod{p}$ and $1 \le t_0 \le p$, and set $\lambda_t = \pi_K^{(t-t_0)/p}\lambda_{t_0}$. Then $v_L(\lambda_t) = t$, and $\lambda_{t_1}\lambda_{t_2}^{-1} = \pi_K^{(t_1-t_2)/p} \in K$ for $t_1 \equiv t_2 \pmod{p}$. Furthermore, we have $\Psi_1(1) = 0$ and

$$\begin{aligned}
&\Psi_1(\lambda_t) = \lambda_{t+1} && \text{if } t \not\equiv 0 \pmod{p},\\
&\Psi_1(\lambda_t) \equiv 0 \pmod{\mathfrak{p}_L^{t+1+\mathfrak{c}}} && \text{if } t \equiv 0 \pmod{p}.
\end{aligned}$$

It follows that $(\{\Psi_1\}, \{\lambda_t\})$ is a Galois scaffold for L/K with precision $\mathfrak{c} \ge 1$.

We now show how the Galois scaffold constructed above gives us information about the Galois module structure of L/K. Proposition 2.12 of [**BCE18**] implies that $\{\Psi_1^s \mid 0 \le s < p\}$ is a K-basis for $K[G]$,

$$\{\Psi_1^s(\rho) \mid 0 \le s < p\} = \{\lambda_t \mid 1 \le t \le p\}$$

is a K-basis for L, and $K[G] \cdot \rho = L$. Theorem 3.1 of [**BCE18**] gives a method for determining the associated order

$$\mathfrak{A}(\mathfrak{O}_L) = \{\alpha \in K[G] \mid \alpha \cdot \mathfrak{O}_L \subset \mathfrak{O}_L\}$$

of $\mathfrak{O}_L$ in $K[G]$. For $0 \le s < p$, let

$$\begin{aligned}
d(s) &= \left\lfloor \frac{s+1}{p} \right\rfloor,\\
w(s) &= \min\{d(n) - d(n-s) \mid n \in \mathbb{S}_p, n \ge s\}.
\end{aligned}$$

Then $d(s) = 0$ for $0 \le s < p - 1$ and $d(p-1) = 1$. Hence $w(s) = d(s)$ for all $s \in \mathbb{S}_p$. Applying Theorem 3.1(1) of [**BCE18**] we find that $\{\pi_K^{-w(s)}\Psi_1^s \mid s \in \mathbb{S}_p\}$ is an $\mathfrak{O}_K$-basis for $\mathfrak{A}(\mathfrak{O}_L)$. Furthermore, since $w(s) = d(s)$ for $s \in \mathbb{S}_p$, $\mathfrak{O}_L$ is a free $\mathfrak{A}(\mathfrak{O}_L)$-module of rank 1. (This fact also follows from work of Bertrandias and Ferton [**BF72**].) We will carry out a more general investigation of separable totally ramified extensions of degree p in Section 15.5.

15.3. Some basic properties of H-scaffolds

In this section we prove some fundamental properties of H-scaffolds $(\{\Psi_i\},\{\lambda_t\})$. In particular, we study elements $\xi \in H$ which are formed by composing elements of $\{\Psi_i\}_{1\le i\le n}$. We also consider how ξ acts on λ_t for $t\in\mathbb{Z}$.

Let $(\{\Psi_i\},\{\lambda_t\})$ be an H-scaffold for L/K. For $s\in\mathbb{S}_{p^n}$ let M_s be the multiset which consists of $s_{(n-i)}$ copies of Ψ_i for $1\le i\le n$. Let $\Upsilon^{(s)}$ denote the set of products of all the elements of M_s in every possible order. For instance, if $p\ge 3$ and $n\ge 2$ then

$$\begin{aligned} M_{p+2} &= \{\Psi_{n-1},\Psi_n,\Psi_n\} \\ \Upsilon^{(p+2)} &= \{\Psi_n^2\Psi_{n-1},\Psi_n\Psi_{n-1}\Psi_n,\Psi_{n-1}\Psi_n^2\}. \end{aligned}$$

Also let

$$\Psi^{(s)} = \Psi_n^{s_{(0)}}\Psi_{n-1}^{s_{(1)}}\dots\Psi_2^{s_{(n-2)}}\Psi_1^{s_{(n-1)}}.$$

Then $\Psi^{(s)}\in\Upsilon^{(s)}$, and $\Upsilon^{(s)}=\{\Psi^{(s)}\}$ if H is commutative.

PROPOSITION 15.2. *Let $(\{\Psi_i\},\{\lambda_t\})$ be an H-scaffold for L/K with precision $\mathfrak{c}\ge 1$ and shift parameters $b_1,\dots,b_n$. Then for every $t\in\mathbb{Z}$, $s\in\mathbb{S}_{p^n}$, and $\xi\in\Upsilon^{(s)}$, there is $U_{\xi,t}\in\mathfrak{O}_K^\times$ such that the following congruence holds modulo $\lambda_{t+\mathfrak{b}(s)}\mathfrak{p}_L^{\mathfrak{c}}$:*

$$\xi\cdot\lambda_t \equiv \begin{cases} U_{\xi,t}\lambda_{t+\mathfrak{b}(s)} & \text{if } s\preceq\mathfrak{a}(t), \\ 0 & \text{if } s\not\preceq\mathfrak{a}(t). \end{cases}$$

If the H-scaffold has infinite precision then the congruence is replaced by equality.

PROOF. We use induction on the poset $(\mathbb{S}_{p^n},\preceq)$. The base case $s=0$ is clear. Let $s\in\mathbb{S}_{p^n}$ with $0\prec s$ and assume the claim holds for all $s'\in\mathbb{S}_{p^n}$ with $s'\prec s$. If $\xi\in\Upsilon^{(s)}$ then $\xi=\Psi_i\xi'$ for some $1\le i\le n$ such that $s_{(n-i)}\ge 1$ and some $\xi'\in\Upsilon^{(s-p^{n-i})}$. Set $s'=s-p^{n-i}$. Then $s'\prec s$, $s'_{(n-i)}=s_{(n-i)}-1$, and $s'_{(j)}=s_{(j)}$ for $j\ne n-i$. Since $\Psi_i(\mathfrak{p}_L^{t+\mathfrak{b}(s')+\mathfrak{c}})\subset\mathfrak{p}_L^{t+\mathfrak{b}(s')+\mathfrak{c}+p^{n-i}b_i}$ and $\mathfrak{b}(s')+p^{n-i}b_i=\mathfrak{b}(s)$, the following congruences hold modulo $\lambda_{t+\mathfrak{b}(s)}\mathfrak{p}_L^{\mathfrak{c}}$:

$$\begin{aligned} \xi\cdot\lambda_t &\equiv \Psi_i(\xi'(\lambda_t)) \\ &\equiv \begin{cases} U_{\xi',t}\Psi_i(\lambda_{t+\mathfrak{b}(s')}) & \text{if } s'\preceq\mathfrak{a}(t) \\ 0 & \text{if } s'\not\preceq\mathfrak{a}(t) \end{cases} \\ &\equiv \begin{cases} U_{\xi',t}u_{i,t+\mathfrak{b}(s')}\lambda_{t+\mathfrak{b}(s)} & \text{if } s'\preceq\mathfrak{a}(t) \text{ and } \mathfrak{a}(t+\mathfrak{b}(s'))_{(n-i)}\ge 1 \\ 0 & \text{otherwise.} \end{cases} \end{aligned}$$

Since $U_{\xi',t}u_{i,t+\mathfrak{b}(s')}\in\mathfrak{O}_K^\times$, to complete the induction we need to show that $s\preceq\mathfrak{a}(t)$ if and only if $s'\preceq\mathfrak{a}(t)$ and $\mathfrak{a}(t+\mathfrak{b}(s'))_{(n-i)}\ge 1$.

Set $w=\mathfrak{a}(t)\in\mathbb{S}_{p^n}$. Since $-\mathfrak{b}(w)=-\mathfrak{b}(\mathfrak{a}(t))\equiv t \pmod{p^n}$ it suffices to prove the claim with t replaced by $-\mathfrak{b}(w)$. Suppose $s'\preceq w$ and $\mathfrak{a}(-\mathfrak{b}(w)+\mathfrak{b}(s'))_{(n-i)}\ge 1$. Then $s'_{(j)}\le w_{(j)}$ for $0\le j<n$, so $w-s'\in\mathbb{S}_{p^n}$, $(w-s')_{(j)}=w_{(j)}-s'_{(j)}$ for $0\le j<n$, and $\mathfrak{b}(w-s')=\mathfrak{b}(w)-\mathfrak{b}(s')$. Hence

$$w_{(n-i)}-s'_{(n-i)}=(w-s')_{(n-i)}=\mathfrak{a}(-\mathfrak{b}(w-s'))_{(n-i)}=\mathfrak{a}(-\mathfrak{b}(w)+\mathfrak{b}(s'))_{(n-i)}. \tag{15.2}$$

It follows that $w_{(n-i)}\ge s'_{(n-i)}+1$, and hence that $s=s'+p^{n-i}\preceq w$. Suppose $s\preceq w$. Then $w_{(n-i)}\ge s_{(n-i)}=s'_{(n-i)}+1$. Since $s'\preceq s$ we get $s'\preceq w$. Hence

(15.2) holds, so we have

$$\mathfrak{a}(-\mathfrak{b}(w)+\mathfrak{b}(s'))_{(n-i)} = w_{(n-i)} - s'_{(n-i)} \geq 1. \qquad \square$$

Recall that in (14.26) we defined a pseudovaluation $\hat{v}_L$ on $\mathrm{End}_K(L)$.

COROLLARY 15.3. *Let $(\{\Psi_i\},\{\lambda_t\})$ be an H-scaffold for L/K and let $s \in \mathbb{S}_{p^n}$. Then for $\xi \in \Upsilon^{(s)}$ we have $\hat{v}_L(\xi) = \mathfrak{b}(s)$.*

In (14.23) we defined the 0th index of inseparability of L/K to be

$$i_0 = v_L(\mathfrak{D}_{L/K}) - p^n + 1,$$

where $\mathfrak{D}_{L/K}$ is the different of L/K. The following gives a connection between H-scaffolds and Bondarko's diagrams:

PROPOSITION 15.4. *Let $(\{\Psi_i\},\{\lambda_t\})$ be an H-scaffold for L/K with precision $\mathfrak{c} \geq 1$ and shift parameters $b_1,\dots,b_n$. Let $s \in \mathbb{S}_{p^n}$ and $\xi \in \Upsilon^{(s)}$. Let $\beta \in L \otimes_K L$ be such that $\xi = \varphi(\beta)$. Then $d(\beta) = \mathfrak{b}(s) - i_0$ and*

$$N(\beta) = \{[t+\mathfrak{b}(s), -t-i_0] \mid t \text{ satisfies } s \preceq \mathfrak{a}(t)\}. \tag{15.3}$$

PROOF. Since $\hat{v}_L(\xi) = \mathfrak{b}(s)$, it follows from Corollary 14.25 that $d(\beta) = \mathfrak{b}(s) - i_0$. Let $t \in \mathbb{Z}$. By Proposition 15.2 we have $v_L(\xi \cdot \lambda_t) \geq t + \mathfrak{b}(s)$, with equality if and only if $s \preceq \mathfrak{a}(t)$. It follows that for $y \in L^\times$ with $v_L(y) = t$ we have $v_L(\xi \cdot y) \geq t + \mathfrak{b}(s)$, again with equality if and only if $s \preceq \mathfrak{a}(t)$. Therefore $f_\xi(t) \geq t + \mathfrak{b}(s)$, with equality if and only if $s \preceq \mathfrak{a}(t)$. Using Proposition 14.20 we deduce that

$$[t+\mathfrak{b}(s), -t-i_0] \in D(\beta)$$

if and only if $s \preceq \mathfrak{a}(t)$. Since $d(\beta) = \mathfrak{b}(s) - i_0$ this gives (15.3). $\square$

One reason that an H-scaffold is useful is that it gives a K-basis for H which is compatible with the pseudovaluation $\hat{v}_L$ on H:

PROPOSITION 15.5. *Let L/K be a totally ramified H-Galois extension of degree p^n which has an H-scaffold $(\{\Psi_i\},\{\lambda_t\})$ with precision $\mathfrak{c} \geq 1$. Then*

(a) *$B = \{\Psi^{(s)} \mid s \in \mathbb{S}_{p^n}\}$ is a K-basis for H.*
(b) *Let $c_s \in K$ and set $\xi = \sum_{s=0}^{p^n-1} c_s \Psi^{(s)} \in H$. Then there is $s_0 \in \mathbb{S}_{p^n}$ such that $\hat{v}_L(\xi) = \hat{v}_L(c_{s_0}\Psi^{(s_0)})$ and $\hat{v}_L(\xi - c_{s_0}\Psi^{(s_0)}) > \hat{v}_L(\xi)$. Hence*

$$\hat{v}_L(\xi) = \min\{\hat{v}_L(c_s\Psi^{(s)}) \mid s \in \mathbb{S}_{p^n}\}.$$

PROOF. If $c_s \neq 0$ then by Corollary 15.3 we get $\hat{v}_L(c_s\Psi^{(s)}) \equiv \mathfrak{b}(s) \pmod{p^n}$. Since $\mathfrak{b}(0), \mathfrak{b}(1), \dots, \mathfrak{b}(p^n-1)$ are distinct modulo p^n, the integers $\hat{v}_L(c_s\Psi^{(s)})$ are distinct for $s \in \mathbb{S}_{p^n}$ such that $c_s \neq 0$. Hence (b) follows from the additive property (14.27) of $\hat{v}_L$. Therefore B is linearly independent over K, so B is a K-basis for H. $\square$

In Section 15.4 we will need the following fact, which relates the shift parameters of an H-scaffold for L/K to the ramification data of L/K.

COROLLARY 15.6. *Let L/K be a totally ramified H-Galois extension of degree p^n which has as H-scaffold $(\{\Psi_i\},\{\lambda_t\})$ with precision $\mathfrak{c} \geq 1$ and shift parameters $b_1,\dots,b_n$. Then $\mathfrak{b}(p^n-1) \equiv i_0 \pmod{p^n}$.*

PROOF. We showed in Proposition 14.1 that H contains an element θ which induces the trace map from L to K. By Proposition 15.5 there are $c_s \in K$ such that $\theta = \sum_{s=0}^{p^n-1} c_s \Psi^{(s)}$. Furthermore, there is $s_0 \in \mathbb{S}_{p^n}$ such that $\hat{v}_L(\theta) = \hat{v}_L(c_{s_0}\Psi^{(s_0)})$ and $\hat{v}_L(\theta - c_{s_0}\Psi^{(s_0)}) > \hat{v}_L(\theta)$. As we saw in (14.22), for $a \in \mathbb{Z}$ we have $\mathrm{Tr}_{L/K}(\mathfrak{p}_L^a) = \mathfrak{p}_K^r$, with

$$r = \left\lfloor \frac{a + v_L(\mathfrak{D}_{L/K})}{p^n} \right\rfloor = \left\lceil \frac{a + i_0}{p^n} \right\rceil.$$

Hence for $y \in L^\times$ we have

$$v_L(\theta(y)) = v_L(\mathrm{Tr}_{L/K}(y)) \geq v_L(y) + i_0, \tag{15.4}$$

with equality if and only if $v_L(y) \equiv -i_0 \pmod{p^n}$. Using (14.5) we get $\varphi(1\otimes 1) = \theta$. Let $\gamma \in L \otimes_K L$ satisfy $\varphi(\gamma) = \Psi^{(s_0)}$. Since $\hat{v}_L(\theta - c_{s_0}\Psi^{(s_0)}) > \hat{v}_L(\theta)$, it follows from Corollary 14.25 that $d(1 \otimes 1 - c_{s_0}\gamma) > d(1 \otimes 1) = 0$. Therefore $N(c_{s_0}\gamma) = N(1 \otimes 1) = \{[0,0]\}$. It follows that $|N(\gamma)| = 1$, so by Proposition 15.4 we have $s_0 = p^n - 1$. Using (15.4) and Corollary 15.3 we get

$$\begin{aligned} i_0 &= \hat{v}_L(\theta) \\ &= \hat{v}_L(c_{p^n-1}\Psi^{(p^n-1)}) \\ &= p^n v_K(c_{p^n-1}) + \hat{v}_L(\Psi^{(p^n-1)}) \\ &\equiv \mathfrak{b}(p^n - 1) \pmod{p^n}. \end{aligned}$$

□

15.4. H-semistable extensions and H-scaffolds

Let L/K be a totally ramified extension of degree p^n. In this section we show that if L/K has an H-scaffold then L/K is H-semistable. We also show that if L/K has an H-scaffold with sufficiently high precision then L/K is H-stable. These results generalize theorems from [**Kea20**]. We start by giving a definition of precision for H-semistable extensions.

DEFINITION 15.7. Let L/K be a totally ramified H-semistable extension of degree p^n and let $\mathfrak{c} \geq 1$. We say that L/K is H-semistable with precision $\mathfrak{c}$ if we can choose $\beta \in L \otimes_K L$ satisfying the conditions of Definition 14.26 and the additional condition that $a + b \geq d(\beta) + \mathfrak{c}$ for all $[a,b] \in G(\beta) \smallsetminus N(\beta)$.

If L/K is H-stable then we may choose β so that $G(\beta) = N(\beta)$. Hence L/K is H-semistable with infinite precision in this case. On the other hand, if L/K is H-semistable with sufficiently high precision then L/K is H-stable:

PROPOSITION 15.8. *Let L/K be a totally ramified H-Galois extension of degree p^n which is H-semistable with precision $\mathfrak{c}$. Let $\beta \in L \otimes_K L$ satisfy the conditions of Definition 15.7 and let h be the unique element of $\mathbb{S}_{p^n}$ such that $h \equiv d(\beta) \pmod{p^n}$. Set*

$$m_\beta = \max\{h - 1, p^n - h - 1\}$$

and assume $\mathfrak{c} \geq m_\beta$. Then L/K is H-stable.

PROOF. By Proposition 14.32 we have $N(\beta) = \{[0, d(\beta)], [d(\beta), 0]\}$. Set $\ell = (d(\beta) - h)/p^n$. Then $\ell \in \mathbb{Z}$, so we may define $\gamma = \pi_K^{-\ell}\beta$. We get $\varphi(\gamma) = \pi_K^{-\ell}\varphi(\beta) \in H$ and $D(\gamma) = \{[-\ell p^n, 0]\} + D(\beta)$. It follows that $N(\gamma) = \{[h,0],[0,h]\}$. Suppose there exists $[a,b] \in \mathcal{F}$ with $[a,b] \in G(\gamma) \smallsetminus N(\gamma)$. Then $a+b \geq h+\mathfrak{c}$. Furthermore, we have $[h,0] \not\leq [a,b]$ and $[0,h] \not\leq [a,b]$. It follows by Lemma 14.16(c) that

$$[a+1, b-p^n+1] \leq [h,0] \tag{15.5}$$

$$[a+1, b-p^n+1] \leq [0,h]. \tag{15.6}$$

By (15.5) we have $a+1 \leq h$. Hence if $0 \leq b \leq h-1$ then

$$a+b \leq 2h-2 < 2h-1.$$

If $h \leq b \leq p^n - 1$ then by (15.6) we have $a+1 \leq 0$, and hence

$$a+b \leq -1+p^n-1 < p^n-1.$$

In either case we get $a+b < h+m_\beta \leq h+\mathfrak{c}$, a contradiction. Hence $G(\gamma) \smallsetminus N(\gamma) = \emptyset$, so L/K is H-stable. □

THEOREM 15.9. *Let L/K be a totally ramified H-Galois extension of degree p^n which has an H-scaffold $(\{\Psi_i\}, \{\lambda_t\})$ with precision $\mathfrak{c} \geq 1$ and shift parameters $b_1, \ldots, b_n$. Assume further that the lower ramification breaks $\ell_1 \leq \cdots \leq \ell_n$ of L/K satisfy $v_p(\ell_i) = 0$ for $1 \leq i \leq n$. Then L/K is H-semistable with precision $\mathfrak{c}$.*

PROOF. Let $\beta \in L \otimes_K L$ be such that $\Psi^{(p^n-2)} = \varphi(\beta)$. Then by Proposition 15.4 we have $d(\beta) = \mathfrak{b}(p^n-2) - i_0$ and

$$N(\beta) = \{[t+\mathfrak{b}(p^n-2), -t-i_0] \mid p^n-2 \preceq \mathfrak{a}(t)\}.$$

The condition $p^n - 2 \preceq \mathfrak{a}(t)$ is equivalent to $\mathfrak{a}(t) \in \{p^n-1, p^n-2\}$, which is valid if and only if one of the following congruences holds:

$$t \equiv -\mathfrak{b}(p^n-1) \pmod{p^n}$$
$$t \equiv -\mathfrak{b}(p^n-2) \pmod{p^n}.$$

It follows that

$$\begin{aligned} N(\beta) &= \{[-\mathfrak{b}(p^n-1)+\mathfrak{b}(p^n-2), \mathfrak{b}(p^n-1)-i_0], [0, \mathfrak{b}(p^n-2)-i_0]\} \\ &= \{[-b_n, \mathfrak{b}(p^n-1)-i_0], [0, \mathfrak{b}(p^n-2)-i_0]\}. \end{aligned}$$

By Corollary 15.6 we have $i_0 \equiv \mathfrak{b}(p^n-1) \pmod{p^n}$, and hence

$$\begin{aligned} d(\beta) &= \mathfrak{b}(p^n-2) - i_0 \\ &\equiv \mathfrak{b}(p^n-2) - \mathfrak{b}(p^n-1) \pmod{p^n} \\ &= -b_n \end{aligned} \tag{15.7}$$

Since b_n is a shift parameter for an H-scaffold we have $p \nmid b_n$ by assumption. Therefore L/K satisfies the conditions of Definition 14.26, so L/K is H-semistable.

Suppose $[e,f] \in G(\beta)$ and $[e,f] \notin N(\beta)$. Then

$$f \not\equiv \mathfrak{b}(p^n-1) - i_0 \pmod{p^n}$$
$$f \not\equiv \mathfrak{b}(p^n-2) - i_0 \pmod{p^n},$$

and hence $\mathfrak{a}(-f-i_0) \notin \{p^n-1, p^n-2\}$. Therefore by Proposition 14.22 and Proposition 15.2 we get

$$\begin{aligned} e &= v_L(\Psi^{(p^n-2)}(\lambda_{-f-i_0})) \\ &\geq -f - i_0 + \mathfrak{b}(p^n-2) + \mathfrak{c} \\ &= -f + d(\beta) + \mathfrak{c}. \end{aligned}$$

Hence $e+f \geq d(\beta)+\mathfrak{c}$, so L/K is H-semistable with precision $\mathfrak{c}$. □

COROLLARY 15.10. *Let L/K be a totally ramified H-Galois extension of degree p^n which has an H-scaffold $(\{\Psi_i\},\{\lambda_t\})$ with precision $\mathfrak{c} \geq 1$ and shift parameters $b_1,\dots,b_n$. Assume that the lower ramification breaks $\ell_1 \leq \cdots \leq \ell_n$ of L/K satisfy $v_p(\ell_i) = 0$ for $1 \leq i \leq n$. Then $\mathfrak{p}_L^{-i_0}$ is free over its associated H-order $\mathfrak{A}(\mathfrak{p}_L^{-i_0})$.*

PROOF. It follows from the theorem that L/K is H-semistable. Hence $\mathfrak{p}_L^{-i_0}$ is free over $\mathfrak{A}(\mathfrak{p}_L^{-i_0})$ by Corollary 14.35. □

COROLLARY 15.11. *Let L/K be a totally ramified H-Galois extension of degree p^n which has an H-scaffold $(\{\Psi_i\},\{\lambda_t\})$ with precision $\mathfrak{c} \geq 1$ and shift parameters $b_1,\dots,b_n$. Let h_0 be the unique element of $\mathbb{S}_{p^n}$ such that $h_0 \equiv -b_n \pmod{p^n}$ and set*

$$m = \max\{h_0 - 1, p^n - h_0 - 1\}.$$

If $\mathfrak{c} \geq m$ then L/K is an H-stable extension.

PROOF. Let $\beta \in L \otimes_K L$ be such that $\varphi(\beta) = \Psi^{(p^n-2)}$. Then by the proof of Theorem 15.9, β satisfies the conditions of Definition 15.7. Hence L/K is H-semistable with precision $\mathfrak{c}$. By (15.7) we have $d(\beta) \equiv -b_n \pmod{p^n}$, and hence $m_\beta = m$. It follows by Proposition 15.8 that L/K is H-stable. □

If L/K has an H-scaffold which satisfies the conditions of Corollary 15.11 then we can apply Theorem 14.43 to fractional ideals $\mathfrak{p}_L^h$ of L. Thus we get an $\mathfrak{O}_K$-basis for the associated order $\mathfrak{A}(\mathfrak{p}_L^h)$ and sufficient conditions for $\mathfrak{p}_L^h$ to be free over $\mathfrak{A}(\mathfrak{p}_L^h)$. Note that since $p \nmid b_n$, Corollary 15.11 applies to any extension which has an H-scaffold with precision $\mathfrak{c} \geq p^n - 2$. Also compare with Theorem 3.1 of [**BCE18**], which gives similar results, but with different lower bounds for $\mathfrak{c}$.

15.5. Ramified extensions of degree p

Let L/K be a separable totally ramified extension of degree p. In [**Eld18**] it is shown that L/K has a unique Hopf-Galois structure with respect to some K-Hopf algebra H. Furthermore, sufficient conditions are given for L/K to admit an H-scaffold. In this section we use scaffolds to give sufficient conditions for L/K to be H-semistable, and sufficient conditions for L/K to be H-stable.

In general, a separable field extension L/K of prime degree has at most one Hopf-Galois structure [**Chi89**, §2]. In our setting, there is exactly one Hopf-Galois structure:

PROPOSITION 15.12. *Let K be a local field with residue characteristic p and let L/K be a separable totally ramified extension of degree p. Then L/K admits a unique Hopf-Galois structure.*

PROOF. Let $\widetilde{L}/K$ be the Galois closure of L/K and set $G = \mathrm{Gal}(\widetilde{L}/K)$, $G_L = \mathrm{Gal}(\widetilde{L}/L)$. Let G_1 be the wild ramification subgroup of G. Since L/K is a ramified extension of degree p we have $|G_1| \geq p$. Since G is isomorphic to a subgroup of S_p we have $|G_1| \leq p$. Hence $|G_1| = p$. Recall that $X = G/G_L$ and $\lambda : G \to \mathrm{Perm}(X)$ is the left translation map. Set $N = \lambda(G_1)$. Since $G_1 \trianglelefteq G$, we see that N is normalized by $\lambda(G)$. Because $\widetilde{L}/K$ is the Galois closure of L/K, the action of N on X is faithful. Since $|N| = |G_1| = p$, it follows that N acts simply transitively on X.

Suppose N' is another regular subgroup of $\mathrm{Perm}(X)$ which is normalized by $\lambda(G)$. Let A be the normalizer of N' in $\mathrm{Perm}(X)$. We have $N' \trianglelefteq A$ and $A \leq \mathrm{Perm}(X)$ with $|X| = p$. Therefore N' is the unique subgroup of A with order p.

Since $\lambda(G)$ normalizes N' we have $\lambda(G) \leq A$. Hence $N \leq A$, so $N' = N$. Therefore $N = \lambda(G_1)$ determines the unique Hopf-Galois structure on L/K. $\square$

In [**Bon00, Bon02**] Bondarko uses formal group laws to show that certain abelian extensions L/K are $K[G]$-semistable. Since we haven't developed formal group methods for H-Galois extensions, to show that L/K is semistable we first construct an H-scaffold for L/K and then apply Theorem 15.9. Recall that $i_0 = v_L(\mathfrak{D}_{L/K}) - p + 1$, where $\mathfrak{D}_{L/K}$ is the different of the extension L/K.

PROPOSITION 15.13. *Let L/K be a separable totally ramified H-Galois extension of degree p whose unique (upper and lower) ramification break b satisfies $v_p(b) = 0$. Set $\mathfrak{c} = v_L(p) - i_0$. Then $\mathfrak{c} \geq 1$, L/K has an H-scaffold with precision $\mathfrak{c}$, and L/K is H-semistable with precision $\mathfrak{c}$.*

PROOF. As in the proof of Proposition 15.12 let $\widetilde{L}/K$ be the Galois closure of L/K and set $G = \mathrm{Gal}(\widetilde{L}/K)$, $G_L = \mathrm{Gal}(\widetilde{L}/L)$. Then $\widetilde{L}/L$ is a cyclic extension of degree d, with $d \mid p-1$. Let G_1 be the wild ramification subgroup of G. Then $G_1 \trianglelefteq G$ and G_1 is cyclic of order p. Let η be a generator for G_1 and let γ be a generator of $\mathrm{Gal}(\widetilde{L}/L)$. Then $\gamma\eta\gamma^{-1} = \eta^r$ for some $r \in \mathbb{Z}$ such that $r+p\mathbb{Z}$ has order d in $(\mathbb{Z}/p\mathbb{Z})^\times$. Therefore there is a primitive dth root of unity $\zeta_d \in K$ such that $r \equiv \zeta_d \pmod{\mathfrak{p}_K}$. Let $M = \widetilde{L}^{G_1}$ be the fixed field of G_1. Then $\mathrm{Gal}(\widetilde{L}/L) \cong \mathrm{Gal}(M/K)$, so there is $\alpha \in M^\times$ such that $\gamma(\alpha)/\alpha = \zeta_d^{-1}$. Set $N = \lambda(G_1)$ and identify G_1 with N. Define

$$\Psi_1 = \sum_{i=0}^{d-1} \alpha\zeta_d^{-i}(\eta^{r^i} - 1) \in \widetilde{L}[N].$$

Then $\Psi_1(1) = 0$. Since G acts on $\widetilde{L}$ through Galois and on N by ${}^g\eta = g\eta g^{-1}$, we get an action of G on $\widetilde{L}[N]$. By construction we have ${}^\eta\Psi_1 = \Psi_1$ and ${}^\gamma\Psi_1 = \Psi_1$. Hence $\Psi_1 \in (\widetilde{L}[N])^G = H$. (This fact could also be deduced from Lemma 14.10.)

Let $e = v_{\widetilde{L}}(\pi_L)$ be the ramification index of $\widetilde{L}/L$. Then

$$eb = v_{\widetilde{L}}(\eta(\pi_L) - \pi_L) - v_{\widetilde{L}}(\pi_L)$$

is a positive integer, and there is $\kappa \in \mathfrak{O}_{\widetilde{L}}^\times$ such that

$$\eta(\pi_L) \equiv \pi_L(1 + \kappa\pi_{\widetilde{L}}^{eb}) \pmod{\pi_L\mathfrak{p}_{\widetilde{L}}^{eb+1}}.$$

Since $r \equiv \zeta_d \pmod{\mathfrak{p}_K}$ it follows that

$$\begin{aligned}
\eta^{r^i}(\pi_L^j) &\equiv \pi_L^j(1 + \kappa j r^i \pi_{\widetilde{L}}^{eb}) \pmod{\pi_L^j\mathfrak{p}_{\widetilde{L}}^{eb+1}} \\
&\equiv \pi_L^j(1 + \kappa j \zeta_d^i \pi_{\widetilde{L}}^{eb}) \pmod{\pi_L^j\mathfrak{p}_{\widetilde{L}}^{eb+1}}.
\end{aligned}$$

Therefore we have

$$\begin{aligned}
\Psi_1(\pi_L^j) &\equiv \sum_{i=0}^{d-1} \alpha\zeta_d^{-i}\pi_L^j\kappa j\zeta_d^i\pi_{\widetilde{L}}^{eb} \pmod{\alpha\pi_L^j\mathfrak{p}_{\widetilde{L}}^{eb+1}} \\
&\equiv dj\kappa\alpha\pi_L^j\pi_{\widetilde{L}}^{eb} \pmod{\alpha\pi_L^j\mathfrak{p}_{\widetilde{L}}^{eb+1}}. \qquad (15.8)
\end{aligned}$$

Set $b' = b + v_L(\alpha)$. Since $\alpha \in M$ we have $v_p(v_L(\alpha)) \geq 1$, and hence $b' \equiv b \pmod{p\mathbb{Z}_{(p)}}$. It follows from (15.8) that $v_L(\Psi_1(\pi_L^j)) \geq j + b'$, with equality if and only if $p \nmid j$. Since $\Psi_1(\pi_L) \in L$ this implies $b' \in \mathbb{Z}$. Let $\lambda \in L^\times$. Then $v_L(\Psi_1(\lambda)) \geq v_L(\lambda) + b'$, again with equality if and only if $p \nmid v_L(\lambda)$. Let $\rho \in L$ satisfy $v_L(\rho) = b'$. Then for $0 \leq s < p$ we have $v_L(\Psi^s(\rho)) = (s+1)b'$. For $t \in \mathbb{Z}$

let $c_t \in \mathbb{S}_p$ be such that $b'c_t \equiv t - b' \pmod{p}$. Set $f_t = (t - b' - b'c_t)/p$ and $\lambda_t = \pi_K^{f_t}\Psi_1^{c_t}(\rho)$. Then $v_L(\lambda_t) = pf_t + (c_t + 1)b' = t$. Furthermore, $\Psi_1(\lambda_t) = \lambda_{t+b'}$ for all $t \in \mathbb{Z}$ such that $p \nmid t$. Let I denote the augmentation ideal of $\mathfrak{O}_{\tilde{L}}[N]$. Then $\alpha^{-1}\Psi_1 \in I$. Since $\eta^p = \mathrm{id}_{\tilde{L}}$ this implies $(\alpha^{-1}\Psi_1)^p \equiv 0 \pmod{p\mathfrak{O}_{\tilde{L}}[N]}$, and hence $\alpha^{-p}\Psi_1^p \in (p\mathfrak{O}_{\tilde{L}}[N]) \cap I = pI$. Since $\lambda_{pb'} = \Psi_1^{p-1}(\rho)$ we get

$$v_L(\Psi_1(\lambda_{pb'})) = v_L(\Psi_1^p(\rho)) \geq v_L(\rho) + b + pv_L(\alpha) + v_L(p).$$

In general, if $p \mid t$ then $\lambda_t = \pi_K^{(t/p)-b'}\Psi_1^{p-1}(\rho)$, so we get

$$\begin{aligned} v_L(\Psi_1(\lambda_t)) &= p((t/p) - b') + v_L(\Psi_1^p(\rho)) \\ &\geq t - pb' + v_L(\rho) + b + pv_L(\alpha) + v_L(p) \\ &= t + b' + (v_L(p) - (p-1)b). \end{aligned}$$

Setting $\mathfrak{c} = v_L(p) - (p-1)b$ we get $v_L(\Psi_1(\lambda_t)) \geq t + b' + \mathfrak{c}$ for all $t \in p\mathbb{Z}$. By Proposition 13.23 we have $\mathfrak{c} = v_L(p) - i_0$, and by Proposition 13.26(a) we have $\mathfrak{c} > 0$. Hence $\mathfrak{c} \in \mathbb{Z}$ and $\mathfrak{c} \geq 1$. Therefore $(\{\Psi_1\}, \{\lambda_t\}_{t\in\mathbb{Z}})$ is an H-scaffold for L/K with precision $\mathfrak{c}$.

It follows from Theorem 15.9 that L/K is H-semistable with precision $\mathfrak{c}$. Indeed, if we set $\xi = \Psi_1^{p-2}$ and let $\beta \in L \otimes_K L$ be such that $\varphi(\beta) = \xi$, then for $\lambda \in L^\times$ we have

$$\begin{aligned} v_L(\xi \cdot \lambda) &= v_L(\lambda) + (p-2)b' && \text{if } v_L(\lambda) \equiv b' \pmod{p}, \\ v_L(\xi \cdot \lambda) &= v_L(\lambda) + (p-2)b' && \text{if } v_L(\lambda) \equiv 2b' \pmod{p}, \\ v_L(\xi \cdot \lambda) &\geq v_L(\lambda) + (p-2)b' + \mathfrak{c} && \text{otherwise.} \end{aligned}$$

Hence $d(\beta) = (p-2)b' - i_0$, and for all $[a, b] \in G(\beta) \smallsetminus N(\beta)$ we have $a + b \geq d(\beta) + \mathfrak{c}$. We conclude that L/K is an H-semistable extension with precision $\mathfrak{c}$. □

REMARK 15.14. It follows from Proposition 13.26 that for a ramified extension of degree p with ramification break b the conditions $v_p(b) = 0$, $p \nmid i_0$, and $i_0 < v_L(p)$ are all equivalent.

COROLLARY 15.15. *Let L/K be a separable totally ramified extension of degree p. Let H denote the K-Hopf algebra which acts on L and let $i_0 = v_L(\mathfrak{D}_{L/K}) - p + 1$. If $i_0 < v_L(p)$ then $\mathfrak{p}_L^{-i_0}$ is free over its associated H-order $\mathfrak{A}(\mathfrak{p}_L^{-i_0})$.*

PROOF. It follows from Proposition 15.13 that L/K is H-semistable with precision $\mathfrak{c} = v_L(p) - i_0$. By applying Corollary 14.35 we deduce that $\mathfrak{p}_L^{-i_0}$ is free over $\mathfrak{A}(\mathfrak{p}_L^{-i_0})$. □

COROLLARY 15.16. *Let $p > 2$. If $i_0 \leq v_L(p) - p + 2$ then L/K is H-stable.*

PROOF. Since $i_0 < pv_K(p)$ we have $p \nmid i_0$. Therefore in the notation of Corollary 15.11 we have $1 \leq h_0 \leq p - 1$, and hence $m \leq p - 2$. We get $\mathfrak{c} \geq p - 2$ by Proposition 15.13, so Corollary 15.11 implies that L/K is H-stable. □

Bibliography

[AB18] Ali A. Alabdali and Nigel P. Byott, *Counting Hopf-Galois structures on cyclic field extensions of squarefree degree*, J. Algebra **493** (2018), 1–19, DOI 10.1016/j.jalgebra.2017.09.009. MR3715201 ↑35, 53, 66, 92

[AB20a] E. Acri and M. Bonatto, *Skew braces of size pq*, Comm. Algebra **48** (2020), no. 5, 1872–1881, DOI 10.1080/00927872.2019.1709480. MR4085764 ↑110

[AB20b] Ali A. Alabdali and Nigel P. Byott, *Hopf-Galois structures of squarefree degree*, J. Algebra **559** (2020), 58–86, DOI 10.1016/j.jalgebra.2020.04.019. MR4093704 ↑35, 53, 66

[AB20c] A. A. Alabdali and N. P. Byott, *Skew braces of squarefree order*, J. Algebra Appl. (2020), 2150128. ↑53

[Abh64] Shreeram Shankar Abhyankar, *Local analytic geometry*, Pure and Applied Mathematics, Vol. XIV, Academic Press, New York-London, 1964. MR0175897 ↑240

[Art42] E. Artin, *Galois Theory - Notre Dame mathematical lectures no. 2*, Field theory: Lectures delivered at the university of Notre Dame, 1942. ↑1

[AW73] J. C. Ault and J. F. Watters, *Circle groups of nilpotent rings*, Amer. Math. Monthly **80** (1973), 48–52, DOI 10.2307/2319260. MR316493 ↑38

[Bac15] David Bachiller, *Classification of braces of order p^3*, J. Pure Appl. Algebra **219** (2015), no. 8, 3568–3603, DOI 10.1016/j.jpaa.2014.12.013. MR3320237 ↑29, 30, 67

[Bac16] David Bachiller, *Counterexample to a conjecture about braces*, J. Algebra **453** (2016), 160–176, DOI 10.1016/j.jalgebra.2016.01.011. MR3465351 ↑14, 28, 29, 30, 38

[Bat18] Giulia Battiston, *A theory of Galois descent for finite inseparable extensions*, Proc. Amer. Math. Soc. **146** (2018), no. 1, 69–83, DOI 10.1090/proc/13713. MR3723121 ↑157

[BB99] Werner Bley and Robert Boltje, *Lubin-Tate formal groups and module structure over Hopf orders* (English, with English and French summaries), J. Théor. Nombres Bordeaux **11** (1999), no. 2, 269–305. MR1745880 ↑181

[BC12] Nigel P. Byott and Lindsay N. Childs, *Fixed-point free pairs of homomorphisms and nonabelian Hopf-Galois structures*, New York J. Math. **18** (2012), 707–731. MR2991421 ↑34, 42, 50, 52, 53, 113

[BCE18] Nigel P. Byott, Lindsay N. Childs, and G. Griffith Elder, *Scaffolds and generalized integral Galois module structure* (English, with English and French summaries), Ann. Inst. Fourier (Grenoble) **68** (2018), no. 3, 965–1010. MR3805766 ↑8, 9, 217, 277, 285, 286, 287, 288, 293

[BE05] Nigel P. Byott and G. Griffith Elder, *New ramification breaks and additive Galois structure* (English, with English and French summaries), J. Théor. Nombres Bordeaux **17** (2005), no. 1, 87–107. MR2152213 ↑233

[BE07] Nigel P. Byott and G. Griffith Elder, *A valuation criterion for normal bases in elementary abelian extensions*, Bull. Lond. Math. Soc. **39** (2007), no. 5, 705–708, DOI 10.1112/blms/bdm036. MR2365217 ↑168

[BE13] Nigel P. Byott and G. Griffith Elder, *Galois scaffolds and Galois module structure in extensions of characteristic p local fields of degree p^2*, J. Number Theory **133** (2013), no. 11, 3598–3610, DOI 10.1016/j.jnt.2013.04.021. MR3084290 ↑174

[BE14] Nigel P. Byott and G. Griffith Elder, *Integral Galois module structure for elementary abelian extensions with a Galois scaffold*, Proc. Amer. Math. Soc. **142** (2014), no. 11, 3705–3712, DOI 10.1090/S0002-9939-2014-12126-5. MR3251712 ↑174

[BE18] Nigel P. Byott and G. Griffith Elder, *Sufficient conditions for large Galois scaffolds*, J. Number Theory **182** (2018), 95–130, DOI 10.1016/j.jnt.2017.06.004. MR3703934 ↑8, 9, 216, 217, 234

[Ber79] Françoise Bertrandias, *Sur les extensions cycliques de degré p^n d'un corps local* (French), Acta Arith. **34** (1979), no. 4, 361–377, DOI 10.4064/aa-34-4-361-377. MR543208 ↑173

[BF72] Françoise Bertrandias and Marie-Josée Ferton, *Sur l'anneau des entiers d'une extension cyclique de degré premier d'un corps local* (French), C. R. Acad. Sci. Paris Sér. A-B **274** (1972), A1330–A1333. MR296047 ↑288

[BH36] Garrett Birkhoff and Philip Hall, *On the order of groups of automorphisms*, Trans. Amer. Math. Soc. **39** (1936), no. 3, 496–499, DOI 10.2307/1989764. MR1501860 ↑78

[BJ08] Werner Bley and Henri Johnston, *Computing generators of free modules over orders in group algebras*, J. Algebra **320** (2008), no. 2, 836–852, DOI 10.1016/j.jalgebra.2008.01.042. MR2422318 ↑186

[BL96] Nigel P. Byott and Günter Lettl, *Relative Galois module structure of integers of abelian fields* (English, with English and French summaries), J. Théor. Nombres Bordeaux **8** (1996), no. 1, 125–141. MR1399950 ↑176, 178

[Bon00] M. V. Bondarko, *Local Leopoldt's problem for rings of integers in abelian p-extensions of complete discrete valuation fields*, Doc. Math. **5** (2000), 657–693. MR1808921 ↑7, 9, 217, 255, 256, 294

[Bon02] M. V. Bondarko, *Local Leopoldt's problem for ideals in totally ramified p-extensions of complete discrete valuation fields*, Algebraic number theory and algebraic geometry, Contemp. Math., vol. 300, Amer. Math. Soc., Providence, RI, 2002, pp. 27–57, DOI 10.1090/conm/300/05142. MR1936366 ↑8, 9, 255, 256, 266, 269, 271, 294

[Bon06] M. V. Bondarko, *The Leopoldt problem for totally ramified abelian extensions of complete discrete valuation fields* (Russian, with Russian summary), Algebra i Analiz **18** (2006), no. 5, 99–129, DOI 10.1090/S1061-0022-07-00972-7; English transl., St. Petersburg Math. J. **18** (2007), no. 5, 757–778. MR2301042 ↑8, 9, 255

[Bur11] W. Burnside, *Theory of groups of finite order*, Cambridge University Press, 1911. ↑80

[Byo00] N. P. Byott, *Galois module theory and Kummer theory for Lubin-Tate formal groups*, Algebraic number theory and diophantine analysis (proceedings of conference in graz, 1998), 2000, pp. 55–67. ↑4, 165, 171, 173

[Byo02] Nigel P. Byott, *Integral Hopf-Galois structures on degree p^2 extensions of p-adic fields*, J. Algebra **248** (2002), no. 1, 334–365, DOI 10.1006/jabr.2001.9053. MR1879021 ↑4, 6, 7, 66, 108, 172, 173, 184

[Byo04a] Nigel P. Byott, *Hopf-Galois structures on field extensions with simple Galois groups*, Bull. London Math. Soc. **36** (2004), no. 1, 23–29, DOI 10.1112/S0024609303002595. MR2011974 ↑16, 33, 45

[Byo04b] Nigel P. Byott, *Hopf-Galois structures on Galois field extensions of degree pq*, J. Pure Appl. Algebra **188** (2004), no. 1-3, 45–57, DOI 10.1016/j.jpaa.2003.10.010. MR2030805 ↑35, 36, 59, 101, 107, 113, 140

[Byo04c] Nigel P. Byott, *Monogenic Hopf orders and associated orders of valuation rings*, J. Algebra **275** (2004), no. 2, 575–599, DOI 10.1016/j.jalgebra.2003.07.003. MR2052627 ↑76, 195, 196

[Byo07] Nigel P. Byott, *Hopf-Galois structures on almost cyclic field extensions of 2-power degree*, J. Algebra **318** (2007), no. 1, 351–371, DOI 10.1016/j.jalgebra.2007.04.010. MR2363137 ↑28, 30, 32, 59

[Byo08] Nigel P. Byott, *On the integral Galois module structure of cyclic extensions of p-adic fields*, Q. J. Math. **59** (2008), no. 2, 149–162, DOI 10.1093/qmath/ham037. MR2428073 ↑173

[Byo11] Nigel P. Byott, *A valuation criterion for normal basis generators of Hopf-Galois extensions in characteristic p* (English, with English and French summaries), J. Théor. Nombres Bordeaux **23** (2011), no. 1, 59–70. MR2780619 ↑169, 256

[Byo13] Nigel P. Byott, *Nilpotent and abelian Hopf-Galois structures on field extensions*, J. Algebra **381** (2013), 131–139, DOI 10.1016/j.jalgebra.2013.02.008. MR3030514 ↑14, 31, 34, 52

[Byo15] Nigel P. Byott, *Solubility criteria for Hopf-Galois structures*, New York J. Math. **21** (2015), 883–903. MR3425626 ↑31, 36, 37, 53, 55, 94

[Byo93] N. P. Byott, *Cleft extensions of Hopf algebras. II*, Proc. London Math. Soc. (3) **67** (1993), no. 2, 277–304, DOI 10.1112/plms/s3-67.2.277. MR1226603 ↑172, 212

[Byo96] N. P. Byott, *Corrigendum: "Uniqueness of Hopf Galois structure for separable field extensions"*, Comm. Algebra **24** (1996), no. 11, 3705, DOI 10.1080/00927879608825779. MR1405283 ↑4, 13, 16, 27, 33, 52, 66, 69, 113

[Byo97] Nigel P. Byott, *Galois structure of ideals in wildly ramified abelian p-extensions of a p-adic field, and some applications* (English, with English and French summaries), J. Théor. Nombres Bordeaux **9** (1997), no. 1, 201–219. MR1469668 ↑7

[Byo99] Nigel P. Byott, *Integral Galois module structure of some Lubin-Tate extensions*, J. Number Theory **77** (1999), no. 2, 252–273, DOI 10.1006/jnth.1999.2385. MR1702149 ↑6, 7

[Car13] A. Caranti, *Quasi-inverse endomorphisms*, J. Group Theory **16** (2013), no. 5, 779–792, DOI 10.1515/jgt-2013-0012. MR3101012 ↑47, 48, 49, 50

[Car20] A. Caranti, *Bi-skew braces and regular subgroups of the holomorph*, 2020. ↑56

[CC07] Lindsay N. Childs and Jesse Corradino, *Cayley's theorem and Hopf Galois structures for semidirect products of cyclic groups*, J. Algebra **308** (2007), no. 1, 236–251, DOI 10.1016/j.jalgebra.2006.09.016. MR2290920 ↑44, 49

[CC99] Scott Carnahan and Lindsay Childs, *Counting Hopf Galois structures on non-abelian Galois field extensions*, J. Algebra **218** (1999), no. 1, 81–92, DOI 10.1006/jabr.1999.7861. MR1704676 ↑34, 35, 45, 48

[CCDC20] E. Campedel, A. Caranti, and I. Del Corso, *Hopf-Galois structures on extensions of degree p^2q and skew braces of order p^2q: the cyclic Sylow p-subgroup case*, J. Algebra **556** (2020), 1165–1210, DOI 10.1016/j.jalgebra.2020.04.009. MR4089566 ↑110

[CDVS06] A. Caranti, F. Dalla Volta, and M. Sala, *Abelian regular subgroups of the affine group and radical rings*, Publ. Math. Debrecen **69** (2006), no. 3, 297–308. MR2273982 ↑4, 13, 22, 25, 28, 61, 64, 97

[CF67] J. W. S. Cassels and A. Fröhlich (eds.), *Algebraic number theory*, Academic Press, London, 1967. ↑196

[CG18] Lindsay N. Childs and Cornelius Greither, *Bounds on the number of ideals in finite commutative nilpotent $\mathbb{F}_p$-algebras*, Publ. Math. Debrecen **92** (2018), no. 3-4, 495–516, DOI 10.5486/pmd.2018.8081. MR3789700 ↑92, 98, 99

[CH86] Lindsay N. Childs and Susan Hurley, *Tameness and local normal bases for objects of finite Hopf algebras*, Trans. Amer. Math. Soc. **298** (1986), no. 2, 763–778, DOI 10.2307/2000648. MR860392 ↑3, 6, 170

[Cha72] Stephen U. Chase, *On the automorphism scheme of a purely inseparable field extension*, Ring theory (Proc. Conf., Park City, Utah, 1971), Academic Press, New York, 1972, pp. 75–106. MR0354629 ↑157

[Cha76] Stephen U. Chase, *Infinitesimal group scheme actions on finite field extensions*, Amer. J. Math. **98** (1976), no. 2, 441–480, DOI 10.2307/2373897. MR424773 ↑157, 158, 159, 160

[Chi00] Lindsay N. Childs, *Taming wild extensions: Hopf algebras and local Galois module theory*, Mathematical Surveys and Monographs, vol. 80, American Mathematical Society, Providence, RI, 2000, DOI 10.1090/surv/080. MR1767499 ↑4, 6, 7, 9, 14, 16, 28, 45, 48, 59, 84, 86, 87, 99, 105, 145, 165, 166, 168, 169, 170, 171, 183, 189, 190, 194, 197, 199, 202, 203, 205, 206, 207, 232, 257, 258

[Chi03] Lindsay N. Childs, *On Hopf Galois structures and complete groups*, New York J. Math. **9** (2003), 99–115. MR2016184 ↑35, 50, 53, 77

[Chi05] Lindsay N. Childs, *Elementary abelian Hopf Galois structures and polynomial formal groups*, J. Algebra **283** (2005), no. 1, 292–316, DOI 10.1016/j.jalgebra.2004.07.009. MR2102084 ↑20, 21, 22, 61, 62

[Chi07] Lindsay N. Childs, *Some Hopf Galois structures arising from elementary abelian p-groups*, Proc. Amer. Math. Soc. **135** (2007), no. 11, 3453–3460, DOI 10.1090/S0002-9939-07-08888-0. MR2336557 ↑29, 62, 65, 98

[Chi11] Lindsay N. Childs, *Hopf Galois structures on Kummer extensions of prime power degree*, New York J. Math. **17** (2011), 51–74. MR2781908 ↑49

[Chi13] Lindsay N. Childs, *Fixed-point free endomorphisms and Hopf Galois structures*, Proc. Amer. Math. Soc. **141** (2013), no. 4, 1255–1265, DOI 10.1090/S0002-9939-2012-11418-2. MR3008873 ↑49, 160

[Chi15] Lindsay N. Childs, *On abelian Hopf Galois structures and finite commutative nilpotent rings*, New York J. Math. **21** (2015), 205–229. MR3336553 ↑65, 68, 98

[Chi17] Lindsay N. Childs, *On the Galois correspondence for Hopf Galois structures*, New York J. Math. **23** (2017), 1–10. MR3611070 ↑90, 92, 97, 98

[Chi18] Lindsay N. Childs, *Skew braces and the Galois correspondence for Hopf Galois structures*, J. Algebra **511** (2018), 270–291, DOI 10.1016/j.jalgebra.2018.06.023. MR3834774 ↑91, 92, 93, 94, 96

[Chi19a] Lindsay N. Childs, *Abelian Hopf Galois structures from almost trivial commutative nilpotent algebras*, New York J. Math. **25** (2019), 1421–1437. MR4044375 ↑62, 65

[Chi19b] Lindsay N. Childs, *Bi-skew braces and Hopf Galois structures*, New York J. Math. **25** (2019), 574–588. MR3982254 ↑54, 55, 56

[Chi21] Lindsay N. Childs, *On the Galois correspondence for Hopf Galois structures arising from finite radical algebras and Zappa-Szép products*, Publ. Mat. **65** (2021), no. 1, 141–163, DOI 10.5565/PUBLMAT6512105. MR4185830 ↑92, 94, 95, 96

[Chi87] Lindsay N. Childs, *Taming wild extensions with Hopf algebras*, Trans. Amer. Math. Soc. **304** (1987), no. 1, 111–140, DOI 10.2307/2000707. MR906809 ↑6, 171, 220

[Chi89] Lindsay N. Childs, *On the Hopf Galois theory for separable field extensions*, Comm. Algebra **17** (1989), no. 4, 809–825, DOI 10.1080/00927878908823760. MR990979 ↑13, 15, 100, 108, 113, 186, 293

[Chi96] Lindsay N. Childs, *Hopf Galois structures on degree p^2 cyclic extensions of local fields*, New York J. Math. **2** (1996), 86–102. MR1420597 ↑6, 172

[CHR65] S. U. Chase, D. K. Harrison, and Alex Rosenberg, *Galois theory and Galois cohomology of commutative rings*, Mem. Amer. Math. Soc. **52** (1965), 15–33. MR195922 ↑122, 171, 180

[Cla58] L. E. Clarke, *On Cayley's formula for counting trees*, J. London Math. Soc. **33** (1958), 471–474, DOI 10.1112/jlms/s1-33.4.471. MR100854 ↑45

[CM94] L. N. Childs and D. J. Moss, *Hopf algebras and local Galois module theory*, Advances in Hopf algebras, 1994, pp. 1–24. ↑3

[CR90] Charles W. Curtis and Irving Reiner, *Methods of representation theory. Vol. I*, Wiley Classics Library, John Wiley & Sons, Inc., New York, 1990. With applications to finite groups and orders; Reprint of the 1981 original; A Wiley-Interscience Publication. MR1038525 ↑181

[CRV15a] T. Crespo, A. Rio, and M. Vela, *From Galois to Hopf Galois: theory and practice*, Trends in number theory, Contemp. Math., vol. 649, Amer. Math. Soc., Providence, RI, 2015, pp. 29–46, DOI 10.1090/conm/649/13018. MR3415265 ↑97, 101

[CRV15b] Teresa Crespo, Anna Rio, and Montserrat Vela, *Non-isomorphic Hopf Galois structures with isomorphic underlying Hopf algebras*, J. Algebra **422** (2015), 270–276, DOI 10.1016/j.jalgebra.2014.07.038. MR3272077 ↑160

[CRV16a] T. Crespo, A. Rio, and M. Vela, *Induced Hopf Galois structures*, J. Algebra **457** (2016), 312–322. ↑55

[CRV16b] Teresa Crespo, Anna Rio, and Montserrat Vela, *On the Galois correspondence theorem in separable Hopf Galois theory*, Publ. Mat. **60** (2016), no. 1, 221–234. MR3447739 ↑85, 100, 101, 111, 113

[CRV18] T. Crespo, A. Rio, and M. Vela, *Hopf Galois structures on symmetric and alternating extensions*, New York J. Math. **24** (2018), 451–457. ↑35, 108, 113

[CS05] Lindsay N. Childs and Harold H. Smith III, *Dual Hopf orders in group rings of elementary abelian p-groups*, J. Algebra **294** (2005), no. 2, 489–518, DOI 10.1016/j.jalgebra.2005.06.015. MR2183362 ↑216, 224

[CS19] Teresa Crespo and Marta Salguero, *Hopf Galois structures on separable field extensions of odd prime power degree*, J. Algebra **519** (2019), 424–439, DOI 10.1016/j.jalgebra.2018.11.004. MR3880638 ↑28, 97

[CS69] Stephen U. Chase and Moss E. Sweedler, *Hopf algebras and Galois theory*, Lecture Notes in Mathematics, Vol. 97, Springer-Verlag, Berlin-New York, 1969. MR0260724 ↑1, 2, 5, 9, 83, 84, 87, 103

[CU03] Lindsay N. Childs and Robert G. Underwood, *Cyclic Hopf orders defined by isogenies of formal groups*, Amer. J. Math. **125** (2003), no. 6, 1295–1334. MR2018662 ↑212, 214, 215

[CU04] Lindsay N. Childs and Robert G. Underwood, *Duals of formal group Hopf orders in cyclic groups*, Illinois J. Math. **48** (2004), no. 3, 923–940. MR2114259 ↑216

[DG18] Willem A. De Graaf, *Classification of nilpotent associative algebras of small dimension*, Internat. J. Algebra Comput. **28** (2018), no. 1, 133–161, DOI 10.1142/S0218196718500078. MR3768261 ↑67, 96

[Del84] P. Deligne, *Les corps locaux de caractéristique p, limites de corps locaux de caractéristique* 0 (French), Representations of reductive groups over a local field, Travaux en Cours, Hermann, Paris, 1984, pp. 119–157. MR771673 ↑237

[Dic05] Leonard Eugene Dickson, *Definitions of a group and a field by independent postulates*, Trans. Amer. Math. Soc. **6** (1905), no. 2, 198–204, DOI 10.2307/1986298. MR1500706 ↑36, 52

[dJ93] A. J. de Jong, *Finite locally free group schemes in characteristic p and Dieudonné modules*, Invent. Math. **114** (1993), no. 1, 89–137, DOI 10.1007/BF01232664. MR1235021 ↑226

[DM96] J. Dixon and B. Mortimer, *Permutation groups*, GTM, vol. 163, Springer, New York, 1996. ↑79

[DCR13] Ilaria Del Corso and Lorenzo Paolo Rossi, *Normal integral bases and tameness conditions for Kummer extensions*, Acta Arith. **160** (2013), no. 1, 1–23, DOI 10.4064/aa160-1-1. MR3085149 ↑183

[dSFT12] B. de Smit, M. Florence, and L. Thomas, *The valuation criterion for normal basis generators*, Bull. Lond. Math. Soc. **44** (2012), no. 4, 729–737, DOI 10.1112/blms/bds005. MR2967240 ↑168, 169

[Eld09] G. Griffith Elder, *Galois scaffolding in one-dimensional elementary abelian extensions*, Proc. Amer. Math. Soc. **137** (2009), no. 4, 1193–1203, DOI 10.1090/S0002-9939-08-09710-4. MR2465640 ↑174

[Eld10] G. Griffith Elder, *A valuation criterion for normal basis generators in local fields of characteristic p*, Arch. Math. (Basel) **94** (2010), no. 1, 43–47, DOI 10.1007/s00013-009-0076-6. MR2581332 ↑168, 169

[Eld18] G. Griffith Elder, *Ramified extensions of degree p and their Hopf-Galois module structure* (English, with English and French summaries), J. Théor. Nombres Bordeaux **30** (2018), no. 1, 19–40. MR3809707 ↑293

[ESS99] Pavel Etingof, Travis Schedler, and Alexandre Soloviev, *Set-theoretical solutions to the quantum Yang-Baxter equation*, Duke Math. J. **100** (1999), no. 2, 169–209, DOI 10.1215/S0012-7094-99-10007-X. MR1722951 ↑31

[EU17] G. Griffith Elder and Robert G. Underwood, *Finite group scheme extensions, and Hopf orders in KC_p^2 over a characteristic p discrete valuation ring*, New York J. Math. **23** (2017), 11–39. MR3611071 ↑9, 218, 225, 227, 228, 229, 230, 231, 232, 233

[FCC12] S. C. Featherstonhaugh, A. Caranti, and L. N. Childs, *Abelian Hopf Galois structures on prime-power Galois field extensions*, Trans. Amer. Math. Soc. **364** (2012), no. 7, 3675–3684, DOI 10.1090/S0002-9947-2012-05503-6. MR2901229 ↑4, 13, 14, 15, 28, 29

[Fea03] S. C. Featherstonhaugh, *Hopf algebra structures on abelian Galois extensions of fields*, Ph.D. Thesis, 2003. ↑15, 28

[Fer74] Marie-José Ferton, *Sur l'anneau des entiers d'extensions cycliques d'un corps local* (French), Journées Arithmétiques (Grenoble, 1973), Soc. Math. France, Paris, 1974, pp. 69–74. Bull. Soc. Math. France Mém., No. 37, DOI 10.24033/msmf.130. MR0374104 ↑171

[Frö83] A. Fröhlich, *Galois module structure of algebraic integers*, Springer Verlag, 1983. ↑165, 169, 183

[FT93] A. Fröhlich and M. J. Taylor, *Algebraic number theory*, Cambridge Studies in Advanced Mathematics, vol. 27, Cambridge University Press, Cambridge, 1993. MR1215934 ↑178

[Fuj90] Genjiro Fujisaki, *An elementary construction of Galois quaternion extension*, Proc. Japan Acad. Ser. A Math. Sci. **66** (1990), no. 3, 80–83. MR1051598 ↑185

[FV02] I. B. Fesenko and S. V. Vostokov, *Local fields and their extensions*, 2nd ed., Translations of Mathematical Monographs, vol. 121, American Mathematical Society, Providence, RI, 2002. With a foreword by I. R. Shafarevich, DOI 10.1090/mmono/121. MR1915966 ↑238

[GA94] E. J. Gómez Ayala, *Bases normales d'entiers dans les extensions de Kummer de degré premier* (French, with English and French summaries), J. Théor. Nombres Bordeaux **6** (1994), no. 1, 95–116. MR1305289 ↑183, 186, 187

[GC98] C. Greither and L. N. Childs, *p-elementary group schemes–constructions, and Raynaud's theory*, Hopf algebras, polynomial formal groups, and raynaud orders, 1998. ↑216

[Ger68] Murray Gerstenhaber, *On modular field extensions*, J. Algebra **10** (1968), 478–484, DOI 10.1016/0021-8693(68)90073-2. MR231814 ↑154

[GP87] Cornelius Greither and Bodo Pareigis, *Hopf Galois theory for separable field extensions*, J. Algebra **106** (1987), no. 1, 239–258, DOI 10.1016/0021-8693(87)90029-9. MR878476 ↑4, 13, 14, 26, 83, 86, 89, 100, 109, 114, 117, 126, 140, 166

[GR21] D. Gil Muñoz and A. Rio, *On induced Hopf Galois structures and its local Hopf Galois modules*, Publ. Mat. (Barcelona) (2021), (to appear). ↑178

[Gre92] C. Greither, *Extensions of finite group schemes, and Hopf Galois theory over a complete discrete valuation ring*, Math. Z. **210** (1992), no. 1, 37–67, DOI 10.1007/BF02571782. MR1161169 ↑6, 172, 195, 210, 221, 228

[Gro79] Benedict H. Gross, *Ramification in p-adic Lie extensions*, Journées de Géométrie Algébrique de Rennes (Rennes, 1978), Astérisque, vol. 65, Soc. Math. France, Paris, 1979, pp. 81–102. MR563473 ↑242

[GRRS99] Cornelius Greither, Daniel R. Replogle, Karl Rubin, and Anupam Srivastav, *Swan modules and Hilbert-Speiser number fields*, J. Number Theory **79** (1999), no. 1, 164–173, DOI 10.1006/jnth.1999.2425. MR1718724 ↑183

[GV17] L. Guarnieri and L. Vendramin, *Skew braces and the Yang-Baxter equation*, Math. Comp. **86** (2017), no. 307, 2519–2534, DOI 10.1090/mcom/3161. MR3647970 ↑4, 9, 14, 23, 25

[Hal59] Marshall Hall Jr., *The theory of groups*, The Macmillan Co., New York, N.Y., 1959. MR0103215 ↑38, 86

[Hee71] Nickolas Heerema, *A Galois theory for inseparable field extensions*, Trans. Amer. Math. Soc. **154** (1971), 193–200, DOI 10.2307/1995437. MR269632 ↑156

[Hei96] Volker Heiermann, *De nouveaux invariants numériques pour les extensions totalement ramifiées de corps locaux* (French, with French summary), J. Number Theory **59** (1996), no. 1, 159–202, DOI 10.1006/jnth.1996.0092. MR1399703 ↑241, 248

[Hel90] Charles Helou, *Non-Galois ramification theory of local fields*, Algebra Berichte [Algebra Reports], vol. 64, Verlag Reinhard Fischer, Munich, 1990. MR1076620 ↑237

[Hel91] Charles Helou, *On the ramification breaks*, Comm. Algebra **19** (1991), no. 8, 2267–2279, DOI 10.1080/00927879108824258. MR1123123 ↑237

[Her61] I. N. Herstein, *Theory of rings*, Math Lecture Notes, Univ. of Chicago, 1961. ↑96

[Hil65] David Hilbert, *Gesammelte Abhandlungen. Dritter Band. Analysis, Grundlagen der Mathematik, Physik, Verschiedenes, nebst einer Lebensgeschichte* (German), Chelsea Publishing Co., New York, 1965. MR0188048 ↑183

[HP86] Rudolf Haggenmüller and Bodo Pareigis, *Hopf algebra forms of the multiplicative group and other groups*, Manuscripta Math. **55** (1986), no. 2, 121–136, DOI 10.1007/BF01168681. MR833240 ↑124

[HS71] Peter John Hilton and Urs Stammbach, *A course in homological algebra*, Springer-Verlag, New York-Berlin, 1971. Graduate Texts in Mathematics, Vol. 4. MR0346025 ↑211

[Hyo87] Osamu Hyodo, *Wild ramification in the imperfect residue field case*, Galois representations and arithmetic algebraic geometry (Kyoto, 1985/Tokyo, 1986), Adv. Stud. Pure Math., vol. 12, North-Holland, Amsterdam, 1987, pp. 287–314, DOI 10.2969/aspm/01210287. MR948250 ↑241

[Itô55] Noboru Itô, *Über das Produkt von zwei abelschen Gruppen* (German), Math. Z. **62** (1955), 400–401, DOI 10.1007/BF01180647. MR71426 ↑36

[Jac37] Nathan Jacobson, *Abstract derivation and Lie algebras*, Trans. Amer. Math. Soc. **42** (1937), no. 2, 206–224, DOI 10.2307/1989656. MR1501922 ↑152

[Jac44] N. Jacobson, *Galois theory of purely inseparable fields of exponent one*, Amer. J. Math. **66** (1944), 645–648, DOI 10.2307/2371772. MR11079 ↑152, 153

[Jac62] Nathan Jacobson, *Lie algebras*, Interscience Tracts in Pure and Applied Mathematics, No. 10, Interscience Publishers (a division of John Wiley & Sons), New York-London, 1962. MR0143793 ↑152

[Jac75] Nathan Jacobson, *Lectures in abstract algebra. III*, Springer-Verlag, New York-Heidelberg, 1975. Theory of fields and Galois theory; Second corrected printing; Graduate Texts in Mathematics, No. 32. MR0392906 ↑160

[Joh15] Henri Johnston, *Explicit integral Galois module structure of weakly ramified extensions of local fields*, Proc. Amer. Math. Soc. **143** (2015), no. 12, 5059–5071, DOI 10.1090/proc/12634. MR3411126 ↑175

[JY88] C. U. Jensen and N. Yui, *Quaternion extensions*, Algebraic geometry and commutative algebra, 1988, pp. 155–182. ↑185

[Kea20] Kevin Keating, *Galois scaffolds and semistable extensions*, J. Number Theory **207** (2020), 110–121, DOI 10.1016/j.jnt.2019.07.002. MR4017940 ↑9, 291

[Keg61] Otto H. Kegel, *Produkte nilpotenter Gruppen* (German), Arch. Math. (Basel) **12** (1961), 90–93, DOI 10.1007/BF01650529. MR133365 ↑36

[Ker90] Ina Kersten, *Brauergruppen von Körpern* (German), Aspects of Mathematics, D6, Friedr. Vieweg & Sohn, Braunschweig, 1990. MR1137014 ↑132

[KH75] H. F. Kreimer and N. Heerema, *Modularity vs. separability for field extensions*, Canadian J. Math. **27** (1975), no. 5, 1176–1182. ↑154

[KKTU19a] Alan Koch, Timothy Kohl, Paul J. Truman, and Robert Underwood, *Normality and short exact sequences of Hopf-Galois structures*, Comm. Algebra **47** (2019), no. 5, 2086–2101, DOI 10.1080/00927872.2018.1529237. MR3977722 ↑103, 104, 111

[KKTU19b] Alan Koch, Timothy Kohl, Paul J. Truman, and Robert Underwood, *Isomorphism problems for Hopf-Galois structures on separable field extensions*, J. Pure Appl. Algebra **223** (2019), no. 5, 2230–2245, DOI 10.1016/j.jpaa.2018.07.014. MR3906546 ↑118, 127, 138, 139, 140, 160

[KKTU19c] A. Koch, T. Kohl, P. J. Truman, and R. Underwood, *The structure of Hopf algebras acting on dihedral extensions*, Advances in algebra, 2019, pp. 201–218. ↑118, 140, 141

[KO74] Max-Albert Knus and Manuel Ojanguren, *Théorie de la descente et algèbres d'Azumaya* (French), Lecture Notes in Mathematics, Vol. 389, Springer-Verlag, Berlin-New York, 1974. MR0417149 ↑117, 119, 120, 122, 123

[Koc14] Alan Koch, *Hopf Galois structures on primitive purely inseparable extensions*, New York J. Math. **20** (2014), 779–797. MR3262032 ↑160, 161

[Koc17] Alan Koch, *Primitively generated Hopf orders in characteristic p*, Comm. Algebra **45** (2017), no. 6, 2673–2689, DOI 10.1080/00927872.2016.1233235. MR3594547 ↑225, 226, 227

[Koc21a] A. Koch, *Abelian maps, bi-skew braces, and opposite pairs of Hopf-Galois structures*, Proc. Amer. Math. Soc. Series B (2021), (to appear). arXiv:2007.08967. ↑56, 57

[Koc21b] A. Koch, *Abelian maps, brace blocks, and solutions to the Yang-Baxter equation*, 2021. arXiv:2102.06104. ↑57

[Koh07] Timothy Kohl, *Groups of order 4p, twisted wreath products and Hopf-Galois theory*, J. Algebra **314** (2007), no. 1, 42–74, DOI 10.1016/j.jalgebra.2007.04.001. MR2331752 ↑70, 72, 74, 75, 109, 113

[Koh13] Timothy Kohl, *Regular permutation groups of order mp and Hopf Galois structures*, Algebra Number Theory **7** (2013), no. 9, 2203–2240, DOI 10.2140/ant.2013.7.2203. MR3152012 ↑35, 39, 53, 70, 72, 75, 76

[Koh15] Timothy Kohl, *Multiple holomorphs of dihedral and quaternionic groups*, Comm. Algebra **43** (2015), no. 10, 4290–4304, DOI 10.1080/00927872.2014.943842. MR3366576 ↑82

[Koh16] Timothy Kohl, *Hopf-Galois structures arising from groups with unique subgroup of order p*, Algebra Number Theory **10** (2016), no. 1, 37–59, DOI 10.2140/ant.2016.10.37. MR3463035 ↑70, 71, 72, 75, 77, 78

[Koh19] Timothy Kohl, *Characteristic subgroup lattices and Hopf-Galois structures*, Internat. J. Algebra Comput. **29** (2019), no. 2, 391–405, DOI 10.1142/S0218196719500073. MR3934792 ↑35, 36, 84, 89, 90

[Koh20] Timothy Kohl, *Enumerating dihedral Hopf-Galois structures acting on dihedral extensions*, J. Algebra **542** (2020), 93–115, DOI 10.1016/j.jalgebra.2019.08.040. MR4018326 ↑82

[Koh98] Timothy Kohl, *Classification of the Hopf Galois structures on prime power radical extensions*, J. Algebra **207** (1998), no. 2, 525–546, DOI 10.1006/jabr.1998.7479. MR1644203 ↑15, 27, 28, 33, 59, 139

[KP69] Robert L. Kruse and David T. Price, *Nilpotent rings*, Gordon and Breach Science Publishers, New York-London-Paris, 1969. MR0266956 ↑66

[Kra37] M. Krasner, *Sur la primitivité des corps $\mathfrak{P}$-adiques*, Mathematica (Cluj) **13** (1937), 72–191. ↑237

[Kru70] R. L. Kruse, *On the circle group of a nilpotent ring*, Amer. Math. Monthly **77** (1970), 168–170, DOI 10.2307/2317333. MR257130 ↑38, 39

[KT20] Alan Koch and Paul J. Truman, *Opposite skew left braces and applications*, J. Algebra **546** (2020), 218–235, DOI 10.1016/j.jalgebra.2019.10.033. MR4033084 ↑91, 109, 110

[Lam05] T. Y. Lam, *Introduction to quadratic forms over fields*, Springer Graduate Studies in Mathematics, vol. 67, Springer, 2005. ↑130

[Lan65] Serge Lang, *Algebra*, Addison-Wesley Publishing Co., Inc., Reading, Mass., 1965. MR0197234 ↑191, 219

[Lar67] Richard G. Larson, *Group rings over Dedekind domains*, J. Algebra **5** (1967), 358–361, DOI 10.1016/0021-8693(67)90045-2. MR209368 ↑193

[Lar76] Richard Gustavus Larson, *Hopf algebra orders determined by group valuations*, J. Algebra **38** (1976), no. 2, 414–452, DOI 10.1016/0021-8693(76)90232-5. MR404413 ↑197, 199, 202

[Leo59] Heinrich-Wolfgang Leopoldt, *Über die Hauptordnung der ganzen Elemente eines abelschen Zahlkörpers* (German), J. Reine Angew. Math. **201** (1959), 119–149, DOI 10.1515/crll.1959.201.119. MR108479 ↑3, 165

[Li03] Cai Heng Li, *The finite primitive permutation groups containing an abelian regular subgroup*, Proc. London Math. Soc. (3) **87** (2003), no. 3, 725–747, DOI 10.1112/S0024611503014266. MR2005881 ↑37

[Lub13] Jonathan Lubin, *Elementary analytic methods in higher ramification theory*, J. Number Theory **133** (2013), no. 3, 983–999, DOI 10.1016/j.jnt.2012.02.017. MR2997782 ↑237, 244

[Mar71] Jacques Martinet, *Modules sur l'algèbre du groupe quaternionien* (French), Ann. Sci. École Norm. Sup. (4) **4** (1971), 399–408. MR291208 ↑183, 185

[MBD61] G. A. Miller, H. F. Blichfeldt, and L. E. Dickson, *Theory and applications of finite groups*, Dover Publications, Inc., New York, 1961. MR0123600 ↑73

[Miy07] Yoshimasa Miyata, *On Galois structure of the integers in elementary abelian extensions of local number fields*, J. Number Theory **125** (2007), no. 2, 442–458, DOI 10.1016/j.jnt.2006.12.005. MR2332598 ↑174

[Miy98] Y. Miyata, *On the module structure of rings of integers in $\mathfrak{p}$-adic number fields over associated orders*, Math. Proc. Cambridge Philos. Soc. **123** (1998), no. 2, 199–212, DOI 10.1017/S0305004197002016. MR1490195 ↑173

[MM84] M. Ram Murty and V. Kumar Murty, *On groups of squarefree order*, Math. Ann. **267** (1984), no. 3, 299–309, DOI 10.1007/BF01456092. MR738255 ↑35, 52

[Mon93] Susan Montgomery, *Hopf algebras and their actions on rings*, CBMS Regional Conference Series in Mathematics, vol. 82, Published for the Conference Board of the Mathematical Sciences, Washington, DC; by the American Mathematical Society, Providence, RI, 1993, DOI 10.1090/cbms/082. MR1243637 ↑153

[MRT13] A. Mézard, M. Romagny, and D. Tossici, *Models of group schemes of roots of unity* (English, with English and French summaries), Ann. Inst. Fourier (Grenoble) **63** (2013), no. 3, 1055–1135, DOI 10.5802/aif.2784. MR3137480 ↑218

[Nas19] Timur Nasybullov, *Connections between properties of the additive and the multiplicative groups of a two-sided skew brace*, J. Algebra **540** (2019), 156–167, DOI 10.1016/j.jalgebra.2019.05.005. MR4003478 ↑37

[Nej18] K. Nejabati Zenouz, *On Hopf-Galois structures and skew braces of order p^3*, Ph.D. Thesis, 2018. ↑60, 67, 111

[NZ19] Kayvan Nejabati Zenouz, *Skew braces and Hopf-Galois structures of Heisenberg type*, J. Algebra **524** (2019), 187–225, DOI 10.1016/j.jalgebra.2019.01.012. MR3905210 ↑25, 55, 67, 136, 138

[Noe32] Emmy Noether, *Normalbasis bei Körpern ohne höhere Verzweigung* (German), J. Reine Angew. Math. **167** (1932), 147–152, DOI 10.1515/crll.1932.167.147. MR1581331 ↑3, 165

[NSW08] Jürgen Neukirch, Alexander Schmidt, and Kay Wingberg, *Cohomology of number fields*, 2nd ed., Grundlehren der Mathematischen Wissenschaften [Fundamental Principles of Mathematical Sciences], vol. 323, Springer-Verlag, Berlin, 2008, DOI 10.1007/978-3-540-37889-1. MR2392026 ↑132

[Osi15] D. V. Osipov, *The discrete Heisenberg group and its automorphism group* (Russian), Mat. Zametki **98** (2015), no. 1, 152–155, DOI 10.4213/mzm10694; English transl., Math. Notes **98** (2015), no. 1-2, 185–188. MR3399166 ↑136

[Par90] Bodo Pareigis, *Forms of Hopf algebras and Galois theory*, Topics in algebra, Part 1 (Warsaw, 1988), Banach Center Publ., vol. 26, PWN, Warsaw, 1990, pp. 75–93. MR1171227 ↑142

[Poo08a] Bjorn Poonen, *Isomorphism types of commutative algebras of finite rank over an algebraically closed field*, Computational arithmetic geometry, Contemp. Math., vol. 463, Amer. Math. Soc., Providence, RI, 2008, pp. 111–120, DOI 10.1090/conm/463/09050. MR2459993 ↑68

[Poo08b] Bjorn Poonen, *The moduli space of commutative algebras of finite rank*, J. Eur. Math. Soc. (JEMS) **10** (2008), no. 3, 817–836, DOI 10.4171/JEMS/131. MR2421162 ↑66, 98

[Ras71] Richard Rasala, *Inseparable splitting theory*, Trans. Amer. Math. Soc. **162** (1971), 411–448, DOI 10.2307/1995762. MR284421 ↑154

[Ray74] Michel Raynaud, *Schémas en groupes de type* $(p,\dots,p)$ (French), Bull. Soc. Math. France **102** (1974), 241–280. MR419467 ↑216

[Ree55] D. Rees, *Valuations associated with a local ring. I*, Proc. London Math. Soc. (3) **5** (1955), 107–128, DOI 10.1112/plms/s3-5.1.107. MR67095 ↑270

[Rei75] I. Reiner, *Maximal orders*, Academic Press [A subsidiary of Harcourt Brace Jovanovich, Publishers], London-New York, 1975. London Mathematical Society Monographs, No. 5. MR0393100 ↑181, 182

[Rem32] Robert Remak, *Über Unterguppen direkter Produkte von drei Faktoren* (German), J. Reine Angew. Math. **166** (1932), 65–100, DOI 10.1515/crll.1932.166.65. MR1581299 ↑92

[Rob93] Derek J. S. Robinson, *A course in the theory of groups*, Graduate Texts in Mathematics, vol. 80, Springer-Verlag, New York, 1993. MR1261639 ↑31

[Rot17] Joseph J. Rotman, *Advanced modern algebra. Part 2*, 3rd ed., Graduate Studies in Mathematics, vol. 180, American Mathematical Society, Providence, RI, 2017. With a foreword by Bruce Reznick, DOI 10.1090/gsm/180. MR3677125 ↑211

[Rum07a] Wolfgang Rump, *Braces, radical rings, and the quantum Yang-Baxter equation*, J. Algebra **307** (2007), no. 1, 153–170, DOI 10.1016/j.jalgebra.2006.03.040. MR2278047 ↑4, 9, 14, 26

[Rum07b] Wolfgang Rump, *Classification of cyclic braces*, J. Pure Appl. Algebra **209** (2007), no. 3, 671–685, DOI 10.1016/j.jpaa.2006.07.001. MR2298848 ↑28, 30, 32, 33

[Rum19] Wolfgang Rump, *Classification of cyclic braces, II*, Trans. Amer. Math. Soc. **372** (2019), no. 1, 305–328, DOI 10.1090/tran/7569. MR3968770 ↑32

[Sch77] Hans-Jürgen Schneider, *Cartan matrix of liftable finite group schemes*, Comm. Algebra **5** (1977), no. 8, 795–819, DOI 10.1080/00927877708822195. MR439857 ↑170, 181

[Ser62] Jean-Pierre Serre, *Corps locaux* (French), Publications de l'Institut de Mathématique de l'Université de Nancago, VIII, Actualités Sci. Indust., No. 1296. Hermann, Paris, 1962. MR0150130 ↑168, 237, 241

[ST68] D. A. Suprunenko and R. I. Tyškevič, *Commutative matrices*, Academic Press, New York, 1968. ↑68

[SV18] Agata Smoktunowicz and Leandro Vendramin, *On skew braces (with an appendix by N. Byott and L. Vendramin)*, J. Comb. Algebra **2** (2018), no. 1, 47–86, DOI 10.4171/JCA/2-1-3. MR3763907 ↑4, 9, 14, 25, 31, 51, 66, 108, 185

[Swa60] Richard G. Swan, *Induced representations and projective modules*, Ann. of Math. (2) **71** (1960), 552–578, DOI 10.2307/1969944. MR138688 ↑170

[Swe68] Moss Eisenberg Sweedler, *Structure of inseparable extensions*, Ann. of Math. (2) **87** (1968), 401–410, DOI 10.2307/1970711. MR223343 ↑154, 155, 156

[Tay20] S. Taylor, *Hopf-Galois module structure of a class of tame quaternionic fields*, Ph.D. Thesis, 2020. ↑108, 170, 176, 185

[Tay81] M. J. Taylor, *On Fröhlich's conjecture for rings of integers of tame extensions*, Invent. Math. **63** (1981), no. 1, 41–79, DOI 10.1007/BF01389193. MR608528 ↑183

[Tho08] Lara Thomas, *A valuation criterion for normal basis generators in equal positive characteristic*, J. Algebra **320** (2008), no. 10, 3811–3820, DOI 10.1016/j.jalgebra.2008.05.024. MR2457723 ↑168

[TO70] John Tate and Frans Oort, *Group schemes of prime order*, Ann. Sci. École Norm. Sup. (4) **3** (1970), 1–21. MR265368 ↑6, 171, 205, 225, 227

[Tos10] Dajano Tossici, *Models of $\mu_{p^2,K}$ over a discrete valuation ring*, J. Algebra **323** (2010), no. 7, 1908–1957, DOI 10.1016/j.jalgebra.2010.01.012. With an appendix by Xavier Caruso. MR2594655 ↑218

[TQ20] Cindy Tsang and Chao Qin, *On the solvability of regular subgroups in the holomorph of a finite solvable group*, Internat. J. Algebra Comput. **30** (2020), no. 2, 253–265, DOI 10.1142/S0218196719500735. MR4077413 ↑36, 37

[Tru11] Paul J. Truman, *Towards a generalisation of Noether's theorem to nonclassical Hopf-Galois structures*, New York J. Math. **17** (2011), 799–810. MR2862153 ↑181, 183, 184

[Tru12] Paul J. Truman, *Hopf-Galois module structure of tame biquadratic extensions* (English, with English and French summaries), J. Théor. Nombres Bordeaux **24** (2012), no. 1, 173–199. MR2914905 ↑170, 184

[Tru13] Paul J. Truman, *Integral Hopf-Galois structures for tame extensions*, New York J. Math. **19** (2013), 647–655. MR3119101 ↑180

[Tru16a] Paul J. Truman, *Canonical nonclassical Hopf-Galois module structure of non-abelian Galois extensions*, Comm. Algebra **44** (2016), no. 3, 1119–1130, DOI 10.1080/00927872.2014.999930. MR3463133 ↑109, 166, 174, 175

[Tru16b] Paul J. Truman, *Hopf-Galois module structure of tame $C_p \times C_p$ extensions* (English, with English and French summaries), J. Théor. Nombres Bordeaux **28** (2016), no. 2, 557–582. MR3509724 ↑170, 184

[Tru18a] Paul J. Truman, *Commutative Hopf-Galois module structure of tame extensions*, J. Algebra **503** (2018), 389–408, DOI 10.1016/j.jalgebra.2018.01.047. MR3780002 ↑115, 179, 180, 183

[Tru18b] Paul J. Truman, *Commuting Hopf-Galois structures on a separable extension*, Comm. Algebra **46** (2018), no. 4, 1420–1427, DOI 10.1080/00927872.2017.1346107. MR3780516 ↑109, 166, 174, 181

[Tru20] Paul J. Truman, *Hopf-Galois module structure of tamely ramified radical extensions of prime degree*, J. Pure Appl. Algebra **224** (2020), no. 5, 106231, 13, DOI 10.1016/j.jpaa.2019.106231. MR4046237 ↑186, 187

[Tsa19a] Cindy Tsang, *Hopf-Galois structures of isomorphic-type on a non-abelian characteristically simple extension*, Proc. Amer. Math. Soc. **147** (2019), no. 12, 5093–5103, DOI 10.1090/proc/14627. MR4021072 ↑34, 45

[Tsa19b] Cindy Tsang, *Hopf-Galois structures on a Galois S_n-extension*, J. Algebra **531** (2019), 349–360, DOI 10.1016/j.jalgebra.2019.05.006. MR3953015 ↑35

[Tsa19c] Cindy Tsang, *Non-existence of Hopf-Galois structures and bijective crossed homomorphisms*, J. Pure Appl. Algebra **223** (2019), no. 7, 2804–2821, DOI 10.1016/j.jpaa.2018.09.016. MR3912948 ↑34, 45

[Tsa20] Cindy Tsang, *Hopf-Galois structures on finite extensions with almost simple Galois group*, J. Number Theory **214** (2020), 286–311, DOI 10.1016/j.jnt.2020.04.003. MR4105712 ↑45

[Tsa21] Cindy Tsang, *Hopf-Galois structures on finite extensions with quasisimple Galois group*, Bull. Lond. Math. Soc. **53** (2021), no. 1, 148–160, DOI 10.1112/blms.12407. MR4224519 ↑45

[TT19] Stuart Taylor and Paul J. Truman, *The structure of Hopf algebras giving Hopf-Galois structures on quaternionic extensions*, New York J. Math. **25** (2019), 219–237. MR3933762 ↑118, 130, 131, 133, 185

[TT20] S. Taylor and P. J. Truman, *On associated orders in separable Hopf-Galois extensions*, 2020. ↑176, 178

[UC06] Robert G. Underwood and Lindsay N. Childs, *Duality for Hopf orders*, Trans. Amer. Math. Soc. **358** (2006), no. 3, 1117–1163, DOI 10.1090/S0002-9947-05-03728-1. MR2187648 ↑212, 214, 215, 221, 222, 223, 224

[Und08] Robert G. Underwood, *Realizable Hopf orders in KC_{p^3}*, J. Algebra **319** (2008), no. 11, 4426–4455, DOI 10.1016/j.jalgebra.2008.02.029. MR2416729 ↑7, 224

[Und11] R. Underwood, *An introduction to Hopf algebras*, Springer, New York, 2011. ↑190, 191, 192, 193, 195, 197, 204, 209, 210, 212, 219, 220

[Und15] Robert G. Underwood, *Fundamentals of Hopf algebras*, Universitext, Springer, Cham, 2015, DOI 10.1007/978-3-319-18991-8. MR3379140 ↑86, 189, 191, 192, 193, 223, 232

[Und16] R. Underwood, *The structure of Hopf algebras acting on Galois extensions*, 2016. talk at the Hopf Galois Theory conference Omaha. ↑127

[Und94] Robert G. Underwood, *R-Hopf algebra orders in KC_{p^2}*, J. Algebra **169** (1994), no. 2, 418–440, DOI 10.1006/jabr.1994.1293. MR1297158 ↑6, 172, 202, 212

[Und96] Robert Underwood, *The valuative condition and R-Hopf algebra orders in KC_{p^3}*, Amer. J. Math. **118** (1996), no. 4, 701–743. MR1400057 ↑202, 214, 215

[Und97] Robert Underwood, *The group of Galois extensions over orders in KC_{p^2}*, Trans. Amer. Math. Soc. **349** (1997), no. 4, 1503–1514, DOI 10.1090/S0002-9947-97-01914-4. MR1407713 ↑218

[Ven19] Leandro Vendramin, *Problems on skew left braces*, Adv. Group Theory Appl. **7** (2019), 15–37, DOI 10.32037/agta-2019-003. MR3974481 ↑9, 30, 31, 36

[VL15] Michael Vaughan-Lee, *Groups of order p^8 and exponent p*, Int. J. Group Theory **4** (2015), no. 4, 25–42. MR3416635 ↑30

[Wat79] William C. Waterhouse, *Introduction to affine group schemes*, Graduate Texts in Mathematics, vol. 66, Springer-Verlag, New York-Berlin, 1979. MR547117 ↑145, 146, 151, 197, 207

[Wil09] Robert A. Wilson, *The finite simple groups*, Graduate Texts in Mathematics, vol. 251, Springer-Verlag London, Ltd., London, 2009, DOI 10.1007/978-1-84800-988-2. MR2562037 ↑65

[Win72] David L. Winter, *The automorphism group of an extraspecial p-group*, Rocky Mountain J. Math. **2** (1972), no. 2, 159–168, DOI 10.1216/RMJ-1972-2-2-159. MR297859 ↑136

[Wit37] Ernst Witt, *Zyklische Körper und Algebren der Charakteristik p vom Grad p^n. Struktur diskret bewerteter perfekter Körper mit vollkommenem Restklassenkörper der Charakteristik p* (German), J. Reine Angew. Math. **176** (1937), 126–140, DOI 10.1515/crll.1937.176.126. MR1581526 ↑280

Index

SELECTED PUBLISHED TITLES IN THIS SERIES

260 **Lindsay N. Childs, Cornelius Greither, Kevin P. Keating, Alan Koch, Timothy Kohl, Paul J. Truman, and Robert G. Underwood,** Hopf Algebras and Galois Module Theory, 2021

258 **Pramod N. Achar,** Perverse Sheaves and Applications to Representation Theory, 2021

257 **Juha Kinnunen, Juha Lehrbäck, and Antti Vähäkangas,** Maximal Function Methods for Sobolev Spaces, 2021

256 **Michio Jimbo, Tetsuji Miwa, and Fedor Smirnov,** Local Operators in Integrable Models I, 2021

255 **Alexandre Boritchev and Sergei Kuksin,** One-Dimensional Turbulence and the Stochastic Burgers Equation, 2021

254 **Karim Belabas and Henri Cohen,** Numerical Algorithms for Number Theory, 2021

253 **Robert R. Bruner and John Rognes,** The Adams Spectral Sequence for Topological Modular Forms, 2021

252 **Julie Déserti,** The Cremona Group and Its Subgroups, 2021

251 **David Hoff,** Linear and Quasilinear Parabolic Systems, 2020

250 **Bachir Bekka and Pierre de la Harpe,** Unitary Representations of Groups, Duals, and Characters, 2020

249 **Nikolai M. Adrianov, Fedor Pakovich, and Alexander K. Zvonkin,** Davenport–Zannier Polynomials and Dessins d'Enfants, 2020

248 **Paul B. Larson and Jindrich Zapletal,** Geometric Set Theory, 2020

247 **István Heckenberger and Hans-Jürgen Schneider,** Hopf Algebras and Root Systems, 2020

246 **Matheus C. Bortolan, Alexandre N. Carvalho, and José A. Langa,** Attractors Under Autonomous and Non-autonomous Perturbations, 2020

245 **Aiping Wang and Anton Zettl,** Ordinary Differential Operators, 2019

244 **Nabile Boussaïd and Andrew Comech,** Nonlinear Dirac Equation, 2019

243 **José M. Isidro,** Jordan Triple Systems in Complex and Functional Analysis, 2019

242 **Bhargav Bhatt, Ana Caraiani, Kiran S. Kedlaya, Peter Scholze, and Jared Weinstein,** Perfectoid Spaces, 2019

241 **Dana P. Williams,** A Tool Kit for Groupoid C^*-Algebras, 2019

240 **Antonio Fernández López,** Jordan Structures in Lie Algebras, 2019

239 **Nicola Arcozzi, Richard Rochberg, Eric T. Sawyer, and Brett D. Wick,** The Dirichlet Space and Related Function Spaces, 2019

238 **Michael Tsfasman, Serge Vlăduţ, and Dmitry Nogin,** Algebraic Geometry Codes: Advanced Chapters, 2019

237 **Dusa McDuff, Mohammad Tehrani, Kenji Fukaya, and Dominic Joyce,** Virtual Fundamental Cycles in Symplectic Topology, 2019

236 **Bernard Host and Bryna Kra,** Nilpotent Structures in Ergodic Theory, 2018

235 **Habib Ammari, Brian Fitzpatrick, Hyeonbae Kang, Matias Ruiz, Sanghyeon Yu, and Hai Zhang,** Mathematical and Computational Methods in Photonics and Phononics, 2018

234 **Vladimir I. Bogachev,** Weak Convergence of Measures, 2018

233 **N. V. Krylov,** Sobolev and Viscosity Solutions for Fully Nonlinear Elliptic and Parabolic Equations, 2018

232 **Dmitry Khavinson and Erik Lundberg,** Linear Holomorphic Partial Differential Equations and Classical Potential Theory, 2018

231 **Eberhard Kaniuth and Anthony To-Ming Lau,** Fourier and Fourier-Stieltjes Algebras on Locally Compact Groups, 2018

For a complete list of titles in this series, visit the
AMS Bookstore at **www.ams.org/bookstore/survseries/**.